Advanced PID Control

Karl J. Åström Tore Hägglund

Department of Automatic Control
Lund Institute of Technology
Lund University

International Society of Automation
67 Alexander Drive
P.O. Box 12277
Research Triangle Park, NC 27709

Printed in the United States of America.
10 9 8 7 6

ISBN-10: 1-55617-942-1 ISBN-13: 978-1-55617-942-6

Notice

The information presented in this publication is for the general education of the reader. Because neither the author nor the publisher has any control over the use of the information by the reader, both the author and the publisher disclaim any and all liability of any kind arising out of such use. The reader is expected to exercise sound professional judgment in using any of the information presented in a particular application.

Additionally, neither the author nor the publisher have investigated or considered the effect of any patents on the ability of the reader to use any of the information in a particular application. The reader is responsible for reviewing any possible patents that may affect any particular use of the information presented.

Any references to commercial products in the work are cited as examples only. Neither the author nor the publisher endorses any referenced commercial product. Any trademarks or tradenames referenced belong to the respective owner of the mark or name. Neither the author nor the publisher makes any representation regarding the availability of any referenced commercial product at any time. The manufacturer's instructions on use of any commercial product must be followed at all times, even if in conflict with the information in this publication.

Library of Congress Cataloging-in-Publication Data

Åström, Karl J. (Karl Johan), 1934-

Advanced PID control / Karl Johan Astrom and Tore Hagglund.

p. cm.

Includes bibliographical references and index.

ISBN 1-55617-942-1 (pbk.)

1. PID controllers. I. Hägglund, Tore. II. Title.

TJ223.P55A85 2006

629.8'3--dc22

2005014664

Preface

The PID controller is the most common solution to practical control problems. Although controllers with proportional and integral action have been used from the time when windmills and steam engines were the dominant technologies, the current form of the PID controller emerged with the pneumatic controllers in the 1930s. The controllers have been implemented in many different ways using mechanical, pneumatic, electronic, and computer technology. The development accelerated when the microprocessor implementations appeared in the 1980s. One reason was that the computer implementations made it possible to add features like auto-tuning and diagnostics, which are very beneficial for users. From an engineering perspective, it is particularly interesting to analyze what happened at the technology shifts, when some important features were rediscovered and others were added.

This book has grown out of more than 25 years of development of auto-tuners for PID controllers in close collaboration with industry. Through this work, we have been exposed to a large number of real industrial control problems. We have benefited much from participating in development, commissioning, and troubleshooting of industrial controllers. The practical work has also inspired research.

This book is the last part of a trilogy. The first book, *Automatic Tuning of PID Controllers, 1988*, which had 6 chapters, gave a short description of our early experiences with development of relay auto-tuners. The second book, *PID Controllers: Theory, Design, and Tuning, 1995*, which has 7 chapters, grew out of the need for a broader coverage of many aspects of PID control. In particular, it reviews many design methods for PID controllers that we investigated in connection with our work on auto-tuners.

The knowledge about PID control in 1995 still was not satisfactory for design of auto-tuners. One drawback was that the user had to provide the controller with design choices. It is particularly difficult for a user to assess if dynamics is dead-time or lag dominated. This question stimulated further research. Because of the drastic increase in computing power, it was also possible to use design algorithms that require more computations.

Tuning and design of PID controllers have traditionally been based on special techniques. Robust control was a major development of control theory that matured in the late 1990s, resulting in powerful design methods based on robust loop shaping. This stimulated us to initiate a research program to adapt these methods to PID control. At the same time, it seemed natural to bring PID control closer to the mainstream ideas in control. When working with industrial auto-tuners, we also saw a great need to include diagnostics in the controller, because it is no use to tune a controller if the process has severe malfunctions. The present book, *Advanced PID Control*, is the result of this effort.

With a total of 13 chapters, this new book substantially expands on some of the topics covered in the previous versions and provides several new chapters that deal with controller design, feedforward design, replacement of the Ziegler-Nichols tuning rules, predictive control, loop and performance assessment, and interaction. At this point in our book trilogy, we assume that the reader is highly familiar with control theory.

Our research has given a deeper understanding of the trade-offs between load disturbance attenuation, injection of measurement noise, and set-point response. We have also been able to answer questions like: Should a controller be tuned for response to load disturbances or set points? What information is required to design a PID controller? When can derivative action give significant improvements? When are more complicated controllers justified? When is it justified to develop more accurate process models? With the knowledge developed, it is now possible to design auto-tuners that can make these assessments autonomously. In addition, we have developed new simple methods for designing PID controllers.

As an example of the insight gained we can mention that control theory tells that it is not necessary to make a compromise between tuning for load disturbance response and set-point response. Both requirements can be satisfied by using a controller with two degrees of freedom, which combines feedback and feedforward. The feedback gains should be chosen to satisfy requirements on disturbance attenuation and robustness. The desired response to set-point changes can then be obtained by proper use of feedforward. Set-point weighting is a simple form of feedforward for PID control. In some cases, it is justified to use more elaborate feedforward. For this reason, we have included a chapter on controller design and another chapter on feedforward in the new book.

The robustness analysis also shows the advantage of having low controller gain at high frequency, high frequency roll-off. This can be accomplished by filtering the process output by a second order filter. Based on the insight obtained, we recommend extended use of set-point weighting or more advanced feedforward. We also recommend that the process output is filtered using a second order filter.

We would like to thank many people who have given knowledge, insight, and inspiration. Our interest in PID control was inspired by Axel Westrenius and Mike Somerville of Eurotherm in the early 1980s. We have learned much from working with students; particular thanks are due to Lars Göran Elfgren (Eurotherm), Göran Grönhammar (LTH), Ari Ingimundarson (UPC), Oskar Nordin (Volvo), Helene Panagopoulos (Volvo), Per Persson (Volvo), Mikael Petersson (ABB), Ola Slättke (ABB), and Anders Wallén (Ericsson Mobile Platforms), who continue to give us valuable insight even if they are now pursuing careers in industry.

We are very grateful to Sune Larsson and Lars Bååth, formerly of NAF

Controls, with whom we developed the first industrial relay auto-tuner. The company NAF Controls was merged several times and is now part of ABB, where we have enjoyed interactions with Göran Arinder, Alf Isaksson, Per Erik Modén, Lars Pernebo, and Thomas Vonheim. We have shared the joy and challenges in moving techniques for auto-tuning and diagnostics into a wide range of industrial products. Many stimulating discussions with our colleagues Anton Cervin (LTH), Sebastian Dormido (UNED), Guy Dumont (UBC), Chang Chieh Hang (NUS), Karl Henrik Johansson (KTH), Birgitta Kristiansson (CTH), Bengt Lennartsson (CTH), Manfred Morari (ETH), Dale Seborg (UCSB), Sigurd Skogestad (NTNU), Björn Wittenmark (LTH), and Karl-Erik Årzén (LTH) from academia are also highly appreciated.

Our friends in industry Bill Bialkowski, Terry Blevins, Greg McMillan, and Willy Wojsznis from Emerson, Edgar Bristol, Sigifredo Niño, and Greg Shinskey from Foxboro, Börje Eriksson (M-real), Krister Forsman (Perstorp), Ken Goff (Leeds and Northrup), Niklas Karlsson (Evolution Robotics), Joseph Lu (Honeywell), Tor Steinar Schei (Cybernetica), Stefan Rönnbäck (Optimation), have generously shared their knowledge and insight with us. We are particularly grateful to Peter Hansen, formerly of Foxboro, who read the complete manuscript and gave us very good feedback.

We are very grateful to Leif Andersson who made the layout of the text and gave much assistance with TeX, Agneta Tuszyński who translated much of the text to LaTeX, and Eva Dagnegård who drew several of the figures.

Finally, we would like to thank the Swedish Research Council (VR), the Swedish Agency for Innovation Systems (VINNOVA), and the Swedish Foundation for Strategic Research (SSF) who have supported our research for many years.

KARL JOHAN ÅSTRÖM
TORE HÄGGLUND

Department of Automatic Control
Lund Institute of Technology
Box 118, SE-221 00 Lund, Sweden

karl_johan.astrom@control.lth.se
tore.hagglund@control.lth.se

Contents

1

Introduction

1.1 Introduction

The idea of feedback is deceptively simple and, yet, extremely powerful. Feedback can reduce the effects of disturbances, it can make a system insensitive to process variations and it can make a system follow commands faithfully. Feedback has also had a profound influence on technology. Application of the feedback principle has resulted in major breakthroughs in control, communication, and instrumentation. Many patents have been granted on the idea.

The PID controller is a simple implementation of feedback. It has the ability to eliminate steady-state offsets through integral action, and it can anticipate the future through derivative action. PID controllers, or even PI controllers, are sufficient for many control problems, particularly when process dynamics are benign and the performance requirements are modest. PID controllers are found in large numbers in all industries. The controllers come in many different forms. There are stand-alone systems in boxes for one or a few loops. The PID controller is a key part of systems for motor control. The PID controller is an important ingredient of distributed systems for process control. The controllers are also embedded in many special-purpose control systems. They are found in systems as diverse as CD and DVD players, cruise control for cars, and atomic force microscopes. In process control, more than 95 percent of the control loops are of PID type; most loops are actually PI control. Many useful features of PID control have not been widely disseminated because they have been considered trade secrets. Typical examples are techniques for mode switches and anti-windup.

PID control is often combined with logic, sequential functions, selectors, and simple function blocks to build the complicated automation systems used for energy production, transportation, and manufacturing. Many sophisticated control strategies, such as model predictive control, are also organized hierarchically. PID control is used at the lowest level; the multivariable controller gives the set points to the controllers at the lower level. The PID controller can thus be said to be the "bread and butter" of control engineering. It is an important component in every control engineer's toolbox.

PID controllers have survived many changes in technology, ranging from pneumatics to microprocessors via electronic tubes, transistors, and integrated circuits. The microprocessor has had a dramatic influence on the PID controller. Practically all PID controllers made today are based on microprocessors. This has created opportunities to provide additional features like automatic tuning, gain scheduling, continuous adaptation, and diagnostics. Most new PID controllers that are produced today have some capability for automatic tuning. Tuning and adaptation can be done in many different ways. The simple controller has in fact become a test bench for many new ideas in control. There has also been a renaissance of analog implementation in micro-mechanical systems because analog implementation requires less silicon surface than digital implementations. The PID controller is also implemented using field programmable gate arrays for applications where very fast control is required.

A large number of instrument and process engineers are familiar with PID control. There is a well-established practice of installing, tuning, and using the controllers. In spite of this there are substantial potentials for improving PID control. Evidence for this can be found in the control rooms of any industry. Many controllers are put in manual mode, and among those controllers that are in automatic mode, derivative action is frequently switched off for the simple reason that it is difficult to tune properly. The key reasons for poor performance are equipment problems in valves and sensors, process constraints and bad tuning practice. The valve problems include wrong sizing, hysteresis, and stiction. The measurement problems include poor or no anti-aliasing filters; excessive filtering in "smart" sensors, excessive noise, and improper calibration. Substantial improvements can be made. The incentive for improvement is emphasized by demands for improved quality, which is manifested by standards such as ISO 9000. Knowledge and understanding are the key elements for improving performance of the control loop. Specific process knowledge is required as well as knowledge about PID control.

Based on our experience, we believe that a new era of PID control is emerging. This book will take stock of the development, assess its potential, and try to speed up the development by sharing our experiences in this exciting and useful field of automatic control. The goal of the book is to provide the technical background for understanding PID control.

1.2 Feedback

A simple feedback system is illustrated by the block diagram in Figure 1.1. The system has two major components, the process and the controller, represented as boxes with arrows denoting the causal relation between inputs and outputs. The process has one input, the manipulated variable (MV), also called the control variable. It is denoted by u. The control variable influences the process via an actuator, which typically is a valve or a motor. The process output is called process variable (PV) and is denoted by y. This variable is measured by a sensor. In Figure 1.1 the actuator and the sensor are considered part of the block labeled "Process". The desired value of the process variable is called the set point (SP) or the reference value. It is denoted by y_{sp}. The control error e is the difference between the set point and the process variable, i.e., $e = y_{sp} - y$.

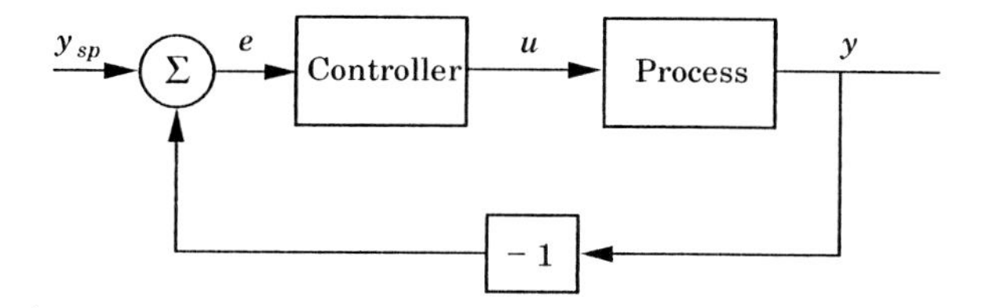

Figure 1.1 Block diagram of a process with a feedback controller.

Assume for simplicity that the process is such that the process variable increases when the manipulated variable is increased. The principle of feedback can then be expressed as follows:

> Increase the manipulated variable when the error is positive, and decrease it when the error is negative.

This type of feedback is called *negative feedback* because the manipulated variable moves in opposite direction to the process variable since $e = y_{sp} - y$.

The PID controller is by far the most common form of feedback. This type of controller has been developed over a long period of time, and it has survived many changes in technology, from mechanical and pneumatic to electronic and computer based. Some insight into this is useful in order to understand its basic properties as is discussed in Section 1.4.

Some properties of feedback can be understood intuitively from Figure 1.1. If the feedback works well the error will be small, and ideally it will be zero. When the error is small the process variable is also close to the set point irrespective of the properties of the process. To realize feedback it is necessary to have appropriate sensors and actuators and a mechanism that performs the control actions.

Feedback has some interesting and useful properties.

- Feedback can reduce effects of disturbances
- Feedback can make a system insensitive to process variations
- Feedback can create well-defined relations between variables in a system

1.3 Simple Forms of Feedback

Many of the nice properties of feedback can be accomplished with simple controllers. In this section we will discuss some simple forms of feedback, namely, on-off control, proportional control, integral control, and PID control.

On-Off Control

The feedback can be arranged in many different ways. A simple feedback mechanism can be described as

$$u = \begin{cases} u_{\max}, & \text{if } e > 0 \\ u_{\min}, & \text{if } e < 0, \end{cases} \tag{1.1}$$

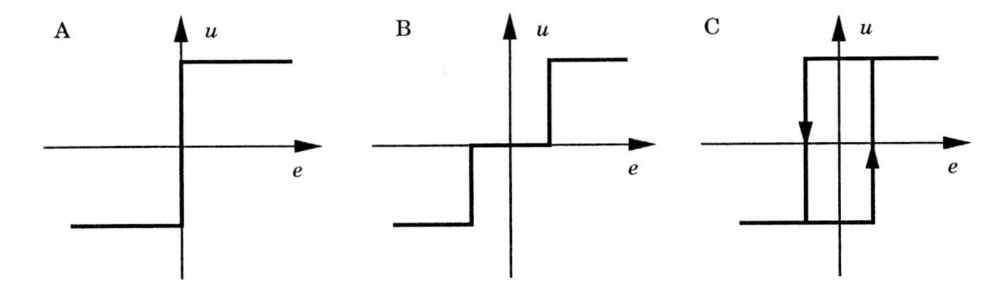

Figure 1.2 Controller characteristics for ideal on-off control (A), and modifications with dead zone (B) and hysteresis (C).

where $e = y_{sp} - y$ is the control error. This control law implies that maximum corrective action is always used. This type of feedback is called *on-off control*. It is simple and there are no parameters to choose. On-off control often succeeds in keeping the process variable close to the set point, but it will typically result in a system where the variables oscillate. Notice that in Equation 1.1 the control variable is not defined when the error is zero. It is common to have some modifications either by introducing hysteresis or a dead zone (see Figure 1.2).

Proportional Control

The reason why on-off control often gives rise to oscillations is that the system overreacts since a small change in the error will make the manipulated variable change over the full range. This effect is avoided in proportional control where the characteristic of the controller is proportional to the control error for small errors. This can be achieved by making the control signal proportional to the error

$$u = K(y_{sp} - y) = Ke, \tag{1.2}$$

where K is the controller gain.

Integral Control

Proportional control has the drawback that the process variable often deviates from the set point. This can be avoided by making the control action proportional to the integral of the error

$$u(t) = k_i \int_0^t e(\tau)d\tau, \tag{1.3}$$

where k_i is the integral gain. This strategy is called integral control. Integral control has an amazing property. Assume that there is a steady state with constant error e_0 and constant control signal u_0. It follows from the above equation that

$$u_0 = k_i e_0 t.$$

Since u_0 is a constant it follows that e_0 must be zero. We thus find that if there is a steady state and a controller has integral action the steady-state error is

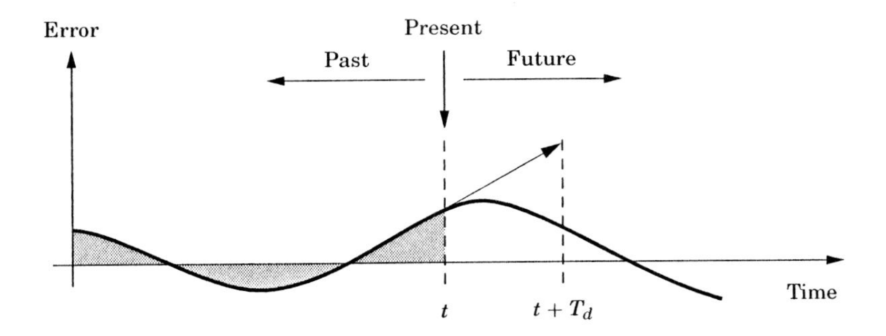

Figure 1.3 A PID controller takes control action based on past, present, and future control errors.

always zero. It follows that this is also true for the PI controller

$$u(t) = Ke(t) + k_i \int_0^t e(\tau)d\tau. \tag{1.4}$$

This is one of the reasons why PI controllers are so common.

PID Control

An additional refinement is to provide the controller with an anticipative ability by using a prediction of the output based on linear extrapolation. See Figure 1.3. This can be expressed mathematically as

$$u(t) = K\left(e(t) + \frac{1}{T_i}\int_0^t e(\tau)d\tau + T_d\frac{de(t)}{dt}\right). \tag{1.5}$$

The control action is thus a sum of three terms representing the past by the integral of the error (the I-term), the present (the P-term) and the future by a linear extrapolation of the error (the D-term). The term $e + T_d\frac{de}{dt}$ is a linear prediction of the error T_d time units in the future. The parameters of the controller are called: proportional gain K, integral time T_i, and derivative time T_d.

It has been found empirically that the PID controller is capable of solving a wide range of control problems. There are more complicated controllers that differ from the PID controller by using more sophisticated methods for prediction.

1.4 How the PID Controller Developed

The PID controller has developed over a period of time that stretches over at least 250 years. It is useful to have some perspective of this development in order to understand many of the issues. The technology used to implement the controllers has naturally changed significantly over the years. The first controllers were mechanical devices (centrifugal governors) used to control windmills and steam engines. Sensing of angular velocity was combined with actuation of valves. A great deal of cleverness was involved in devising integral action.

A significant change occurred in connection with the development of industrial process control. The functions of sensing, control, and actuation were then separated and special devices that performed the control actions represented by Equation 1.5 were built. An interesting feature was that signal transmission and computing were done pneumatically. A major advance occurred when the tubes used to transmit the pressure and the pressure levels were standardized to 3–15 PSI. This made it possible to combine sensors, controllers, and actuators from different suppliers. It also made it possible to concentrate controllers in separate control rooms that were located far away from sensors and actuators. Much cleverness was again used to obtain the controllers. The use of feedback in the controllers themselves was a major improvement. In this way it was possible to obtain linear action out of components that had strongly nonlinear characteristics.

Starting in 1950s, electronic versions of the PID controller became available. The control actions represented by Equation 1.5 were then obtained by a simple analog computer based on operational amplifiers. The signal transmission was also standardized as current signals in the range 4–20 mA. To represent zero by a nonzero current was useful for diagnostics.

Yet another advance occurred when digital computers were used to implement controllers. Strongly centralized systems were first used when computer control emerged, because digital computing was only cost effective in large systems. With the advent of microprocessors in the 1970s even simple controllers were implemented using computers. When a digital computer is used it is also possible to add many functions such as automatic tuning, adaptation, and diagnostics. This is an area of very active development.

Today we are experiencing other shifts in technology. Analog implementations are reappearing in micro-mechanical electrical systems (MEMS), and digital controllers are also implemented using field programmable gate arrays (FPGA), which admit very short sampling periods. The FPGAs differ significantly from digital computers because they are highly parallel.

Today we find PID controllers in many forms. There are dedicated controllers that can control one or a few loops. PID functions are found in Programmable Logic Controllers that were originally designed to replace relays. There are systems that contain many PID controllers implemented in computers ranging from small systems for a few dozen loops to large distributed systems for process control. PID controllers are commonly used in dedicated systems for motion control. There are also a whole range of special controllers such as autopilots and control systems for CD and DVD players and optical memories that are based on PID control.

1.5 Technology Changes and Knowledge Transfer

The PID controller is an interesting case study for management of technology, because it has a long history and it has experienced many technology changes. Since we have had personal experiences of several technology shifts, we will present some personal reflections where we discuss creation and destruction of knowledge and the role of key people and documentation.

Technology transfers are often abrupt and unplanned. The reason why a company decides to change technology may be drastic drops in hardware costs or pressure from competitors and customers. A switch in technology often means that R&D staff has to be replaced by new people that are familiar with the new technology, but often not with the old one. This means that there is a high risk that information is lost during the transition. Since the technology transfers often have to be done fast, there is also a high risk that the potential of the new technology is not utilized.

Early temperature controllers were of the on-off type. The on-off controllers are simple and cheap, but there are unavoidable oscillations. The amplitude of the oscillations can be kept at reasonable levels, since the dynamics of many thermal systems is lag dominated. When electronics became cost effective, there was a transition from on-off to continuous PID control. The development of the analog PID controller is well documented in publically available material from Eurotherm, which was started by faculty from the University of Manchester. The controllers were developed based on solid knowledge of modeling and control. Theory helps, because many applications of temperature admit high gains and derivative action can be very beneficial. Tuning rules were also provided and protection for windup was developed under the name of *integrator desaturation* and *crossed time-constants*. The result of the development was a drastic improvement of the performance of temperature controllers. It is interesting to observe that it took a long time before the interesting and important phenomena of integrator windup received any attention in academia.

When computer based process control emerged in the early 1960s, the focus was initially on higher-level control functions. Analog PID controllers were used at the base level and the computer supplied set point to the controllers. As the systems developed, the attention focused again on PID control where many PID loops were implemented in one computer, so called Direct Digital Control. The technical development focused on discretization of the PID algorithm, one reason being that computing resources was a bottleneck. Little attention was given to integrator windup, and some attention was given to filtering of sensor signals.

The emergence of the microprocessor made digital computing cheaply available in small quanta, a development which had a major impact on the PID controller. It resulted in small single loop controllers, controllers for a few loops, and large distributed systems. The development was slow for two reasons. Many new persons without previous experience of analog control entered the arena, and many old-timers were unwilling to learn the new technology. Important aspects such as integrator windup and filtering were not documented in a way that was easily accessible. Therefore, it took some time before the appropriate knowledge and experience was recaptured. There was also a tendency to simply implement old ideas in new technology without considering the opportunities offered by the new technology. Gradually the potentials of the digital computer were exploited by incorporating features like, auto-tuning, adaptation, and diagnostics into the systems.

When distributed control system (DCS) systems replaced the analog systems, the distributed architecture was retained. Analog controllers and function modules were represented as blocks in the DCS programs. This was probably a good idea, but the opportunities given by the fact that all signals were available in one computer was not utilized. It took over a decade before DCS systems that handle anti-windup above the loop level were presented.

A couple of conclusions that can be drawn are that documentation, and open-minded persons who can bridge the gap between different technologies, are important. When new technologies are available it is also useful to stop and think to find out how the new technology can be exploited rather than to quickly implement old ideas in the new technology. It is also important to filter out the essence of the old systems so that good features are not lost. Finally, it is important to document ideas, write books, and ensure that information is not only transferred from human to human, but widespread.

1.6 Outline of the Contents of the Book

The reader is advised to look at the table of contents to see the overall structure of the book. Process dynamics is a key for understanding any control problem. Chapter 2 presents concepts that are useful for describing the behavior of processes. Static models are mentioned briefly, but the main focus of the chapter is on process dynamics. Representations in terms of time- and frequency responses are given. These dual views are very useful to gain a good understanding of dynamics. The notions of step responses and transfer functions are used throughout the book. A number of typical models that are used for PID control are discussed in detail. Models for disturbances are also treated as well as techniques for experimental determination of the models.

An in-depth presentation of the PID controller is given in Chapter 3. This includes principles as well as many implementation details, such filtering to provide high-frequency roll-off, anti-windup, improvement of set-point response, etc. The PID controller can be structured in different ways. Commonly used forms are the series and the parallel forms. The differences between these and the controller parameters used in the different structures are treated in detail. The limitations of PID control are also described. Typical cases where more complex controllers are worthwhile are systems that have long dead time and oscillatory systems. Extensions of PID control to deal with such systems are discussed briefly.

Chapter 4 treats controller design in general. There is a rich variety of control problems with very diverse goals. The chapter gives an overview of ideas and concepts that are relevant for PID control. It is attempted to bring design of PID controllers into the mainstream of control design. Topics such as fundamental limitations, stability, robustness, and specifications are treated.

Feedforward control, a simple and powerful technique that complements feedback, is treated in Chapter 5. A systematic design of feedforward control to improve set-point responses is given as well as a discussion of design of model-following systems. The special case of set-point weighting is discussed in detail, and methods for determining the set-point weights are provided. The chapter also shows how feedforward can be used to to reduce the effect of disturbances that can be measured.

Chapter 6 describes methods for the design of PID controllers. Many different methods for tuning PID controllers that have been developed over the years are presented. Their properties are discussed thoroughly. It has been attempted to strike a balance by providing both an historical perspective and to present powerful methods.

A reasonable design method should consider load disturbances, model uncertainty, measurement noise, and set-point response. A drawback of many of the traditional tuning rules for PID control is that such rules do not consider all these aspects in a balanced way. New tuning techniques that do consider all these criteria are presented in Chapter 7.

Chapter 8 treats model predictive controllers. The Smith predictor, which is a special case, is first presented and analysed, and modifications to treat integrating processes are provided. Then other types of model predictive controllers are presented, such as the MPC controller, the Dahlin-Higham controller, dynamic matrix control, and minimum variance control.

In Chapter 9 we discuss some techniques for adaptation and automatic tuning of PID controllers. This includes methods based on parametric models and non-parametric techniques. Supervision of adaptive controllers and iterative feedback tuning are also discussed. A number of commercial controllers are described to illustrate the different techniques.

Chapter 10 treats methods for commissioning, supervision and diagnosis of control loops. Loop assessment procedures are used to investigate properties of the control loop, e.g. signal levels, noise levels, nonlinearities, and equipment conditions. Performance assessment procedures are used to supervise the control loops during operation, and ensure that they meet the specifications.

The PID controller is typically used as a single-loop controller. In practice, there are often interactions between the loops. Some key issues about interacting loops that are of particular relevance for PID control are discussed in Chapter 11. In particular it is shown that controller parameters in one loop may have significant input on dynamics of other loops. Bristol's relative gain array, which is a simple way to characterize the interactions, is also introduced. The problem of pairing inputs and outputs is discussed, and a design method based on decoupling, which is a natural extension of the tuning methods for single input single output systems, is presented.

In Chapter 12 it is shown how complex control problems can be solved by combining simple controllers in different ways. The control paradigms of repetitive control, cascade control, mid-range and split-range control, ratio control, and control with selectors are discussed. Use of currently popular techniques such as neural networks and fuzzy control are also covered briefly.

Chapter 13 presents implementation issues related to PID control. A short overview of the early analog pneumatic and electronic implementations are first given. A detailed presentation of computer implementation aspects such as sampling, pre-filtering, and discretization of the PID algorithm is then given. Operational aspects such as bumpless transfers are presented, and the chapter ends with a discussion about the different controller outputs that have to be used depending on which actuating device is used.

1.7 Summary

In this section we have given a brief description of the concept of feedback. The application of feedback has had very useful and sometimes revolutionary impact. Some of the useful properties of feedback, its ability to reduce disturbances, insensitivity to process variations, linearity between set point and process variable, have been discussed. We have also briefly described some simple forms of feedback such as on-off control and PID. The development of PID controllers has been discussed briefly, and the contents of the book have finally been outlined.

1.8 Notes and References

PID controllers were used extensively in the early development of control from the 1870s through 1920. The modern form of the PID controller emerged in the development of process control in the 1930s and 1940s, as discussed in [Bennett, 1979] and [Bennett, 1993]. The PID controller is still the standard tool for solving industrial control problems. In a detailed study of the state of the art in industrial process control by the Electric Measuring Instrument Manufacturer in Japan from 1989 it is found that more than than 90 percent of the control loops were of the PID type; see [Yamamoto and Hashimoto, 1991]. The paper [Desbourough and Miller, 2002] surveyed the U.S. industry. It is found that there are more than 8 million facilities in the petrochemical, pulp and paper, power, and metals industries. Each facility has between 500 and 5000 regulatory control loops, 97 percent of them are of the PID type. The PID controllers are manufactured in large quantities in other industries too. Optical memories for CD and DVD contain three PID loops for control of rotation speed, focus, and track following. About 140 million units were manufactured in 2002; see [Akkermans and Stan, 2002]. In addition, there is a large number of PID controllers for motor drives and positioning systems. It is therefore safe to say that the PID controller is one of the most common tools for control.

PID control is discussed in most textbooks on process control such as [Luyben, 1990; Shinskey, 1994; Marlin, 2000; Bequette, 2003; Seborg *et al.*, 2004], and there are also books that focus on PID control [McMillan, 1983; Corripio, 1990; Suda *et al.*, 1992; Wang and Cluett, 2000; Quevedo and Escobet, 2000; Wang *et al.*, 2000; O'Dwyer, 2003; Michael and Moradi, 2005].

The theory of PID controllers was for a long time based on special techniques. Lately there have been efforts to bring PID control into the the mainstream of control theory. A notable effort was made in the year 2000 when the International Federation of Automatic Control (IFAC) arranged a workshop on the past, present, and future of PID control; see [Quevedo and Escobet, 2000]. A collection of papers from this workshop was also published as a special issue of Control Engineering Practice. The papers [Bennett, 2000] and [Åström and Hägglund, 2001] give a perspective on the development of PID control.

Because of the large number of PID controllers and their widespread use there are still significant benefits in improving the practice of PID control. Such an improvement requires attention to the complete control loop and not just the controller itself as is demonstrated in the paper [Bialkowski, 1994] which describes audits of paper mills in Canada. A typical mill has more than 2000 control loops, 97 percent of the loops are based on PI control. It was found that only 20 percent of the control loops worked well and decreased process variability. The reasons why performance is poor are bad tuning (30 percent) and valve problems (30 percent). The remaining 20 percent of the controllers functioned poorly for a variety of reasons such as: sensor problems, bad choice of sampling rates, and poor or non-existing anti-aliasing filters. Similar observations are given in [Ender, 1993], where it is claimed that 30 percent of installed process controllers operate in manual mode, that 20 percent of the loops use default parameters set by the controller manufacturer (so-called "factory tuning"), and that 30 percent of the loops function poorly because of equipment problems in valves and sensors.

2

Process Models

2.1 Introduction

Mathematical models are commonly used to describe the behavior of processes. Models give a unified way to treat systems of widely different types, and they make it possible to introduce a number of useful concepts. The models are also essential for simulation and control design. In this chapter we will review some of the models that are commonly used for PID control. The models try to capture some aspects of the process that are relevant for control. Many different types of models are used.

The steady-state behavior of a process can be captured by a function that tells the steady-state value of the process variable for given values of the manipulated variable. Such models are discussed in Section 2.2.

To control a system it is necessary to describe process dynamics. For the purpose of control it is often sufficient to describe small deviations from an equilibrium. In this case the behavior can be modeled as a linear dynamical system. This is a very rich field with many useful concepts and tools, which form the core of control theory. Different ways to describe process dynamics are discussed in Section 2.3. The ideas of transient response and frequency response are introduced as well as the important concepts of step response, impulse responses, and transfer functions.

Special techniques for modeling process dynamics have traditionally been used in PID control. The idea is to characterize process dynamics by a few features. This is discussed in Section 2.4, where features such as average residence time, apparent time delay, apparent time constant, normalized delay, ultimate gain, ultimate frequency, and gain ratio are introduced.

In Section 2.5 we introduce some particular models that are widely used for PID control. These models are introduced in terms of their transfer functions. The important concepts of normalization are also introduced in that section, as well as nonlinearities. The examples introduced in Section 2.5 will be used extensively in the book.

Disturbances are an important aspect of a control problem. In Section 2.6 we describe some models that are used to describe disturbances. Section 2.7 describes simple methods for obtaining the models, and Section 2.8 describes

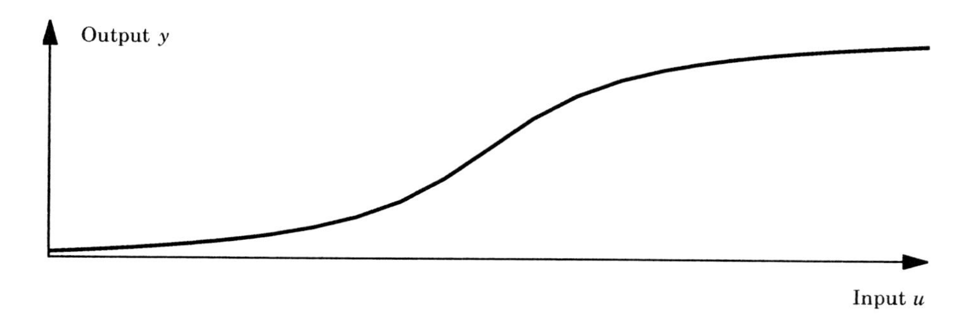

Figure 2.1 Static process characteristic, which shows process output y as a function of process input u under steady-state conditions.

some techniques used to simplify a complicated model. The chapter is summarized in Section 2.9, and references are given in Section 2.10.

2.2 Static Models

It is natural to start by describing the stationary behavior of the process. This can be done by a curve that shows the steady-state value of the process variable y (the output) for different values of the manipulated variable u (the input); see Figure 2.1. This curve is called a static model or a static process characteristic. All process investigations should start with a determination of the static process model. It can be used to determine the range of control signals required to change the process output over the desired range, to size actuators, and to select sensor resolution. The slope of the curve in Figure 2.1 tells how much the process variable changes for small changes in the manipulated variable. This slope is called the static gain of the process. Large variations in the gain indicate that the control problem may be difficult.

The static model can be obtained experimentally in several ways. A natural way is to keep the input at a constant value and measure the steady-state output. This gives one point on the process characteristics. The experiment is repeated to cover the full range of inputs.

An alternative procedure is to make a closed-loop experiment where the output of the system is kept constant by feedback and the steady-state value of the input is measured.

The experiments required to determine the static process model often give a good intuitive feel for how easy it is to control the process, and if there are many disturbances. Data for steady-state models can also be obtained from on-line measurements.

Sometimes process operations do not permit the experiments to be done as described above. Small perturbations are normally permitted, but it may not be possible to move the process over the full operating range. In such a case the experiment must be done over a long period of time. It is possible to provide a control system with facilities to automatically determine the static process model during operation; see Chapter 10.

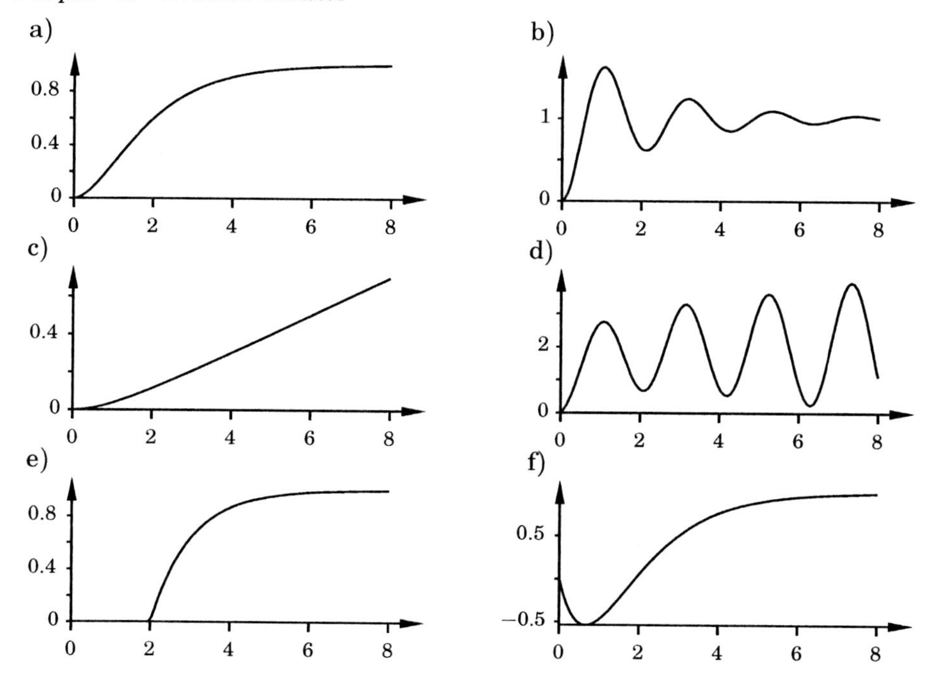

Figure 2.2 Open-loop step responses.

2.3 Dynamic Models

A static process model like the one discussed in the previous section tells the steady-state relation between the input and the output signal. A dynamic model should give the relation between the input and the output signal during transients. It is naturally much more difficult to capture dynamic behavior. This is, however, essential when dealing with control problems.

Qualitative Characterization of Process Dynamics

Before attempting to model a system it is often useful to give a crude characterization of its dynamical behavior. To describe the dynamical behavior we will simply show the response of the system to a step change in the manipulated variable. This is called the step response of the system or the process reaction curve.

One distinction is between stable and unstable systems. The step response of a stable system goes to a constant value. An unstable system will not reach a steady state after a step change. Systems with integrating action are a typical example of an unstable system. In early process control literature, stable systems were called self-regulating systems.

Many properties of a system can be obtained directly from the step response. Figure 2.2 shows step responses that are typically encountered in process control.

In Figure 2.2a, the process output is monotonically changed to a new stationary value. This is the most common type of step response encountered in process control. In Figure 2.2b, the process output oscillates around its final stationary value. This type of process is uncommon in process control. One case where it occurs is in concentration control of recirculation fluids. In mechanical designs, however, oscillating processes are common where elastic materials are used, e.g., weak axles in servos, spring constructions, etc. The systems in Figures 2.2a and 2.2b are stable, whereas the system shown in Figures 2.2c and 2.2d are unstable. The system in Figure 2.2c is an integrating process. Examples of integrating processes are level control, pressure control in a closed vessel, concentration control in batches, and temperature control in well isolated chambers. The common factor in all these processes is that some kind of storage occurs in them. In level, pressure, and concentration control, storage of mass occurs, while in the case of temperature control there is a storage of energy. The system in Figure 2.2e has a long time delay. The time delay occurs when there are transportation delays in the process. The system in Figure 2.2f is a non-minimum phase system. Notice that the output initially moves in the wrong direction. The water level in boilers often reacts like this after a step change in feed water flow.

Linear Time-Invariant Systems

There is a restricted class of models, called linear time-invariant systems, that can often be used. Such models describe the behavior of systems for small deviations from an equilibrium. Time-invariant means that the behavior of the system does not change with time. Linearity means that the superposition principle holds. This means that if the input u_1 gives the output y_1 and the input u_2 gives the output y_2 it then follows that the input $au_1 + bu_2$ gives the output $ay_1 + by_2$.

A nice property of linear time-invariant systems is that their response to an arbitrary input can be completely characterized in terms of the response to a simple signal. Many different signals can be used to characterize a system. Broadly speaking, we can differentiate between transient and frequency responses.

In a control system we typically focus on only two signals, the control signal and the measured variable. Process dynamics deals with the relation between those signals. This means that it includes dynamics in actuators, process, and sensors. The dynamics are often dominated by process dynamics. In some cases it is, however, the sensors and actuators that give the major contribution to the dynamics. For example, it is very common that there are long filter-time constants in temperature sensors. There may also be measurement noise and other imperfections. There may also be significant dynamics in the actuators. To do a good job of control, it is necessary to be aware of the physical origin of the process dynamics to judge if a good response in the measured variable actually corresponds to a good response in the physical process variable. Even if the attention is focused on the measured variable it is useful to always keep in mind that the process variable is the signal that really matters.

Physical Modeling—Differential Equations

A traditional way to obtain a process model is to use basic physical laws such as mass, momentum and energy balances. Such descriptions typically lead to

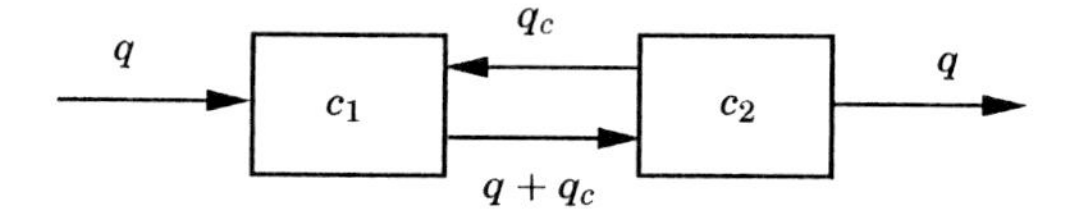

Figure 2.3 Schematic diagram of a system consisting of two tanks.

a mathematical model in terms of a differential equation. We illustrate this with two examples.

EXAMPLE 2.1—STIRRED TANK
Consider an ideal stirred tank reactor. Let the reactor volume be V and the volume flow rate through the reactor be q. The manipulated variable is the concentration u of the inflow, and the process variable y is the concentration in the reactor. A mass balance for the reactor gives

$$V\frac{dy}{dt} = q(u - y).$$

The parameter $T = V/q$, which has dimension time, is the average residence time of particles that enter the reactor. It is also called the time constant of the system. □

The system in Example 2.1 is of first order because only one variable is required to account for the storage in the tank. This is possible because the tank is well stirred so the concentration is constant throughout the tank. In more complicated cases many variables are required to account for the storage of mass, energy, and momentum. This is illustrated in the next example.

EXAMPLE 2.2—COUPLED TANKS
Consider the system shown in Figure 2.3, which is composed of two well-stirred tanks. Assume that each tank has volume V, that the inflow and the outflow are q, and that the reflux flow is q_c. Furthermore, let the input be the concentration of the inflow $u = c_{in}$, and let the output be the concentration of the outflow, $y = c_{out}$. When the tanks are well stirred the mass balance can be characterized by the concentrations in the tanks. The mass balances for the tanks become

$$\begin{aligned}
V\frac{dc_1}{dt} &= -(q + q_c)c_1 + q_c c_2 + qu \\
V\frac{dc_2}{dt} &= (q + q_c)c_1 - (q + q_c)c_2 \\
y &= c_2.
\end{aligned}$$

□

The model in the example consists of two differential equations of first order. There are two differential equations because the system is completely described by mass balances and the storage of mass can be captured by two variables. Similar descriptions are obtained for more complicated systems, but the number of equations increases with the complexity of the system. The differential equation may also be nonlinear if there are nonlinear transport phenomena.

The model in Example 2.2 consists of a system of first-order differential equations. If we are only interested in the relations between the input u and the output y a linear model can also be described by a differential equation of higher order, i.e.,

$$\frac{d^n y}{dt^n} + a_1 \frac{d^{n-1} y}{dt^{n-1}} + \ldots + a_n y = b_1 \frac{d^{n-1} u}{dt^{n-1}} + \ldots + b_n u. \tag{2.1}$$

The number n is equal to the number of variables required to account for the storage. This is one of the standard models used in automatic control.

The differential equation (2.1) is characterized by two polynomials

$$\begin{aligned} a(s) &= s^n + a_1 s^{n-1} + \cdots + a_n \\ b(s) &= b_1 s^{n-1} + \cdots + b_n, \end{aligned} \tag{2.2}$$

where the polynomial $a(s)$ is called the characteristic polynomial. The zeros of the polynomial $a(s)$ are called the *poles* of the system, and the zeros of the polynomial $b(s)$ are called the *zeros* of the system.

The differential equation (2.1) has a solution of the form

$$y(t) = \sum kC_k(t)e^{\alpha_k t} + \int_0^t g(t - \tau)u(\tau)d\tau, \tag{2.3}$$

where α_k are the poles of the system and $C_k(t)$ are polynomials (constants if the poles are distinct). The first term of the above equation depends on the initial conditions and the second on the input. The function g has the same form as the first term of the right-hand side of (2.3). The poles thus give useful qualitative insight into the properties of the system.

In more complicated situations it may be more difficult to account for the storage of mass momentum and energy. We illustrate with a simple example.

EXAMPLE 2.3—TIME DELAY

Consider a system where mass is transported on a conveyor belt. Let the input $u(t)$ be the mass flow rate onto the belt, and let the output $y(t)$ be the mass flow out of the belt. The input-output relation for the system is then

$$y(t) = u(t - L), \tag{2.4}$$

where L is the time it takes for a particle to pass the belt. To account for the storage of mass on the belt it is necessary to specify the mass distribution on the belt. The output is thus simply the delayed input. This system is therefore called a time delay or a transport delay. A time delay is also called a dead time. The model (2.4) also describes the concentration in a pipe with no mixing. □

Other physical systems such as heat conduction and diffusion give rise to models in terms of partial differential equations; examples of such models are given in Section 2.5.

An attractive feature of physical models is that the parameters of the equation can be related to physical quantities such as volumes, flows, and material constants. Complicated models can also be constructed by dividing a system into subsystems, deriving simple models for each subsystem and combining the simple models.

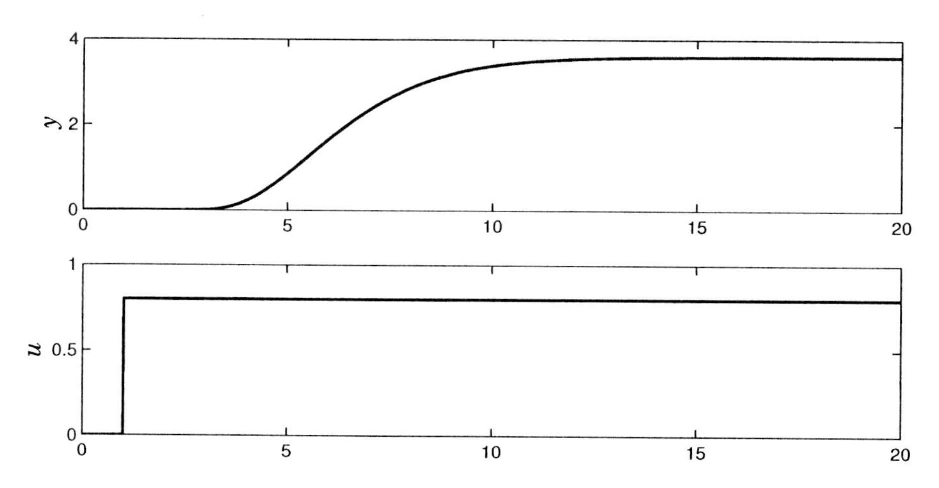

Figure 2.4 The lower curve shows an input signal in the form of a step, and the upper curve shows the response of the system to the step.

State Models

The notion of *state* is an important concept in system dynamics. The state is a collection of variables that summarize the past behavior of the system and admits a prediction of the future under the assumption that future inputs are known. For the system in Example 2.2, which consists of two tanks, the state is simply the concentrations c_1 and c_2 in the tanks. In general, the state is the variables required to describe storage of mass, momentum, and energy of a system. Sometimes it is necessary to use infinitely many variables to describe storage in a system. For the system in Example 2.3 the state at time t is the past inputs over an interval of length L, i.e., $\{u(\tau), t - L \leq \tau < t\}$.

Transient Responses

An alternative to describing models by differential equations is to focus directly on the input-output behavior. Dynamics can in principle be described by a large table of input signals and corresponding output signals. This approach, which is called transient response, is perhaps the most intuitive way to characterize process dynamics. A very nice property of linear time-invariant systems is that the table can be described by one pair of signals. The particular input signal is often chosen so that it is easy to generate experimentally. Typical examples are steps, pulses, and impulses. Recall that typical step responses were shown in Figure 2.2.

Because of the superposition principle the amplitude of the signals can be normalized. For simplicity it is common practice to normalize by dividing the output with the magnitude of the input step. It is also common practice to translate the curve so that the step starts at time $t = 0$. It is then sufficient to show the output only. This practice will be followed in this book. For example, in Figure 2.4 the output should be divided by 0.8 and translated one unit to the left. In early process control literature the step response was also called the reaction curve.

The output generated by an arbitrary input can be computed from the step response. Let $h(t)$ be the response to a unit step. The output $y(t)$ to an arbitrary input signal $u(t)$ is then given by

$$y(t) = \int_0^t u(\tau)\,\frac{dh(t-\tau)}{dt}\,d\tau = \int_0^t u(\tau)g(t-\tau)d\tau, \tag{2.5}$$

where we have introduced $g(t)$ as the derivative of the step response $h(t)$. The function $g(t)$ is called the impulse response of the system because it can be interpreted as the response of the system to a very short impulse with unit area.

The Transfer Function

The formula (2.5) can be simplified significantly by introducing Laplace transforms. The Laplace transform $F(s)$ of a time function $f(t)$ is defined as

$$F(s) = \int_0^\infty e^{-st} f(t)dt. \tag{2.6}$$

Assuming that the system is initially at rest, i.e., $y(t) = 0$ and $u(t) = 0$ for $t \leq 0$, and using Laplace transforms, Equation 2.5 can be written as

$$Y(s) = G(s)U(s), \tag{2.7}$$

where $U(s)$, $Y(s)$, and $G(s)$ are the Laplace transforms of $u(t)$, $y(t)$, and $g(t)$, respectively. The function $G(s)$ is called the transfer function of the system. The transfer function $G(s)$ is also the Laplace transform of the impulse response $g(t)$.

The formula given by (2.7) has a strong intuitive interpretation. The Laplace transform of the output is simply the Laplace transform of the input multiplied by the transfer function of the system. This is one of the main reasons for using Laplace transforms when analyzing linear systems. Analysis of linear systems is reduced to pure algebra. A nice feature is that processes, controllers, and signals are described in the same way.

Equation 2.7 can also be used to define the transfer function as the ratio of the Laplace transforms of the input and the output of a system. As illustrations we will give the transfer function for some systems.

EXAMPLE 2.4—STIRRED TANK
The stirred tank in Example 2.1 has the transfer function

$$G(s) = \frac{1}{sV/q+1} = \frac{1}{sT+1}, \tag{2.8}$$

where the quantity $T = V/q$, which has dimension time, is called the time constant of the system. □

EXAMPLE 2.5—TIME DELAY
Consider the system describing a transport delay in Example 2.3. Assuming that $u(t) = 0$ for $-L \leq t \leq 0$ we find

$$Y(s) = \int_0^\infty e^{-st} y(t)dt = \int_0^\infty e^{-st} u(t-L)dt = e^{-sL} U(s).$$

The transfer function of a transport delay is thus

$$G(s) = e^{-sL}. \tag{2.9}$$

□

Equation 2.7 implies that it is easy to obtain the transfer function of interconnected system. This is illustrated by the following example.

EXAMPLE 2.6—FIRST-ORDER SYSTEM WITH TIME DELAY (FOTD)
Consider a system that is a stirred tank that is fed by a pipe with no mixing. Multiplying the transfer function of the tank in Example 2.4 with the transfer function of a time delay in Example 2.5 we find that the system has the transfer function

$$G(s) = \frac{1}{1+sT} e^{-sL}. \tag{2.10}$$

This model is very common in process control. It is called a first-order system with a time delay or a FOTD system for short. □

Another nice property of Laplace transforms is that the transform of a derivative is given by the formula

$$\int_0^\infty e^{-st} f'(t)dt = s \int_0^\infty e^{-st} f(t)dt - f(0) = sF(s) - f(0).$$

If the initial value of the time function is zero it follows that differentiation of a time function corresponds to multiplication of the Laplace transform with s. Similarly, it can be shown that integration of a signal corresponds to dividing the Laplace transform with s. This gives a very simple rule for manipulating differential equations where initial values are zero. Simply replace functions with their corresponding Laplace transforms and derivatives by s. The relation between signals is then obtained by simple algebra.

EXAMPLE 2.7—GENERAL DIFFERENTIAL EQUATION
Consider the system described by the differential equation (2.1). Assuming that the system is initially at rest and taking Laplace transforms of (2.1) we get

$$(s^n + a_1 s^{n-1} + \ldots + a_n)Y(s) = (b_1 s^{n-1} + b_2 s^{n-2} + \ldots + b_n)U(s),$$

where $Y(s)$ is the Laplace transform of the output, and $U(s)$ the Laplace transform of the input. The transfer function of the system is the ratio of the Laplace transforms of output and input, i.e.,

$$G(s) = \frac{Y(s)}{U(s)} = \frac{b_1 s^{n-1} + b_2 s^{n-2} + \ldots + b_n}{s^n + a_1 s^{n-1} + \ldots + a_n}. \tag{2.11}$$

□

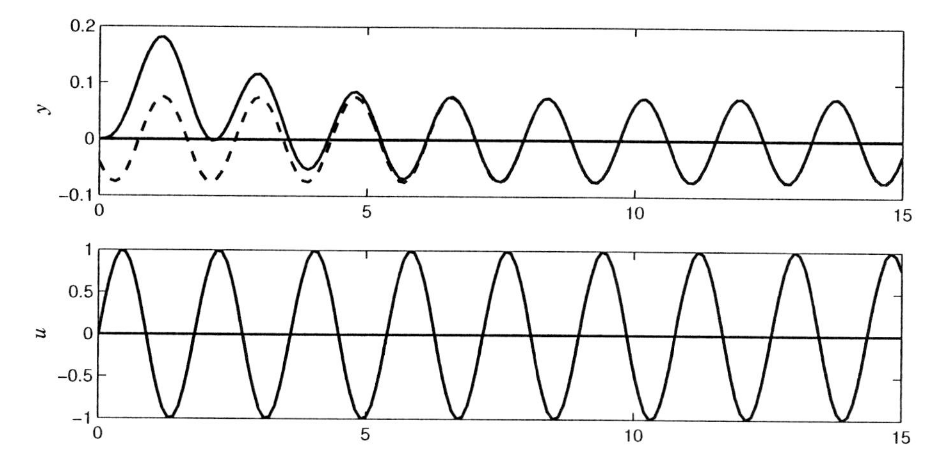

Figure 2.5 Illustration of frequency response. The input signal u is a sinusoid, and the output signal y becomes sinusoidal after a transient. The dashed line shows the steady-state response to the sinusoidal input.

EXAMPLE 2.8—PID CONTROLLER
The PID controller given by Equation 1.5 is a dynamical system with the transfer function

$$C(s) = \frac{U(s)}{E(s)} = K\left(1 + \frac{1}{sT_i} + sT_d\right). \tag{2.12}$$

□

The last two examples illustrate that transfer functions can be obtained from differential equations by inspection. The rule is simply to replace derivatives by s, integrals by $1/s$, and time functions by their transforms. The transfer functions are then obtained as the ratio between signals.

Frequency Response

Another way to characterize the dynamics of a linear time-invariant system is to investigate the response to sinusoidal input signals, an idea that goes back to the French mathematician Fourier. Frequency response is less intuitive than transient response, but it gives other insights.

Consider a stable linear system. If the input signal to the system is a sinusoid, then the output signal will also be a sinusoid after a transient (see Figure 2.5). The output will have the same frequency as the input signal. Only the phase and the amplitude are different. If the input signal is $u(t) = u_0 \sin \omega t$ the steady-state output is

$$y(t) = a(\omega)u_0 \sin\left(\omega t + \varphi(\omega)\right).$$

The steady-state relations between the output and a sinusoidal input with frequency ω can be described by two numbers: the amplitude ratio and the phase. The amplitude ratio is the output amplitude divided by the input amplitude,

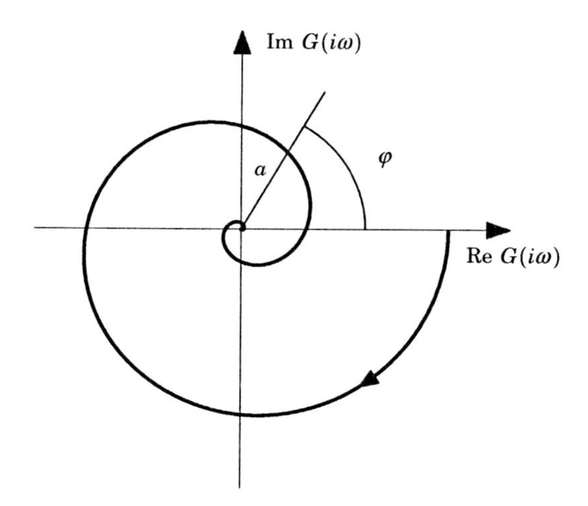

Figure 2.6 The Nyquist curve of a system is the locus of the complex number $G(i\omega)$ as ω goes from 0 to ∞.

and the phase is the phase shift of the output in relation to the input. The functions $a(\omega)$ and $\varphi(\omega)$ give amplitude ratio and phase for all frequencies. The functions $a(\omega)$ and $\varphi(\omega)$ are related to the transfer function in the following way.

$$G(i\omega) = a(\omega)e^{i\varphi(\omega)}. \tag{2.13}$$

The values of the transfer function for imaginary arguments thus describe the steady-state transmission of sinusoidal signals, and $G(i\omega)$ is called the frequency response function of the system.

The Nyquist Plot

There are very useful graphical illustrations of the frequency response. The complex number $G(i\omega)$ can be represented by a vector with length $a(\omega)$ that forms angle $\varphi(\omega)$ with the real axis (see Figure 2.6). When the frequency goes from 0 to ∞, the vector describes a curve in the plane, which is called the frequency curve or the Nyquist curve.

The Nyquist curve gives a complete description of the system. It can be determined experimentally by sending sinusoids of different frequencies through the system. This may, however, be time consuming. It can also be determined from other signals.

The Bode Plot

The Bode plot is another graphical representation of the transfer function. The Bode plot of a transfer function consists of two curves, the gain curve and the phase curve; see Figure 2.7. The amplitude or gain curve shows the amplitude ratio $a(\omega) = |G(i\omega)|$ as a function of the frequency ω. The phase curve shows the phase $\varphi(\omega) = \arg G(i\omega)$ as a function of the frequency ω. The frequency is given in logarithmic scales on both curves, either in rad/s or Hz. The gain is also given in logarithmic scales. The angle is given in linear scales. The Bode

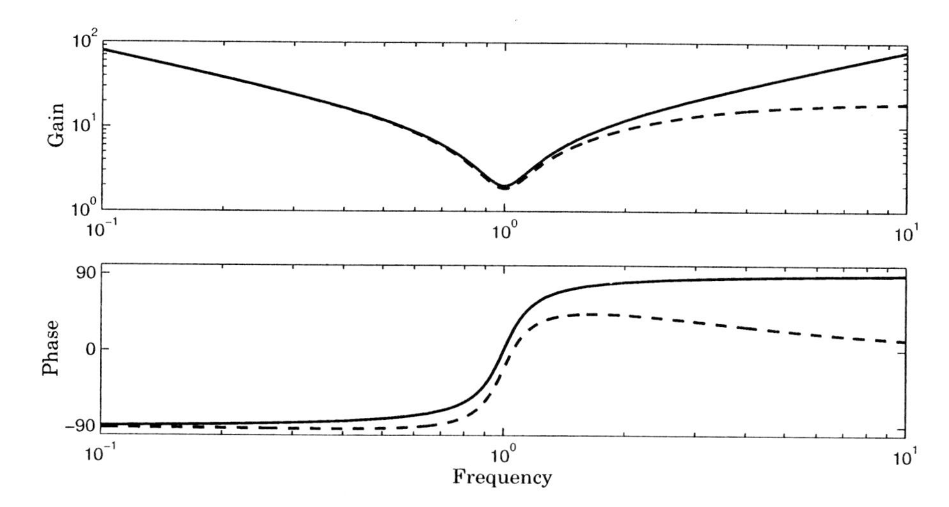

Figure 2.7 Bode plot of an ideal PID controller (solid lines) and a controller with a filter (dashed). The upper curve shows the gain curve $|G(i\omega)|$, and the lower diagram shows the phase curve $\arg G(i\omega)$. The controller has high gain for low frequencies, and the phase is $-90°$. The ideal controller also has high gain at high frequencies and the phase is $90°$. The controller with a filter has constant gain for high frequencies.

plot gives a good overview of the properties of a system over a wide frequency range. Because of the scales the gain curve also has linear asymptotes.

2.4 Feature-Based Models

Sometimes it is desirable to have a crude characterization of a process based on only a few features. The features should be chosen so that they are meaningful with good physical interpretation. They should also be easy to determine experimentally. This way of describing dynamics has a long tradition in process control. It is useful to start with a crude classification of step responses as illustrated in Figure 2.2.

Process Gain

For stable processes the steady-state behavior can be described by one parameter, the process gain K_p. For processes with integration a constant input gives in steady state an output that changes with a constant rate. This behavior can be captured by the rate constant K_v.

Average Residence Time

It is also useful to find a few parameters to characterize process dynamics. The time behavior of stable system with positive impulse response can be characterized with the parameter

$$T_{ar} = \frac{\int_0^\infty t g(t) dt}{\int_0^\infty g(t) dt}, \tag{2.14}$$

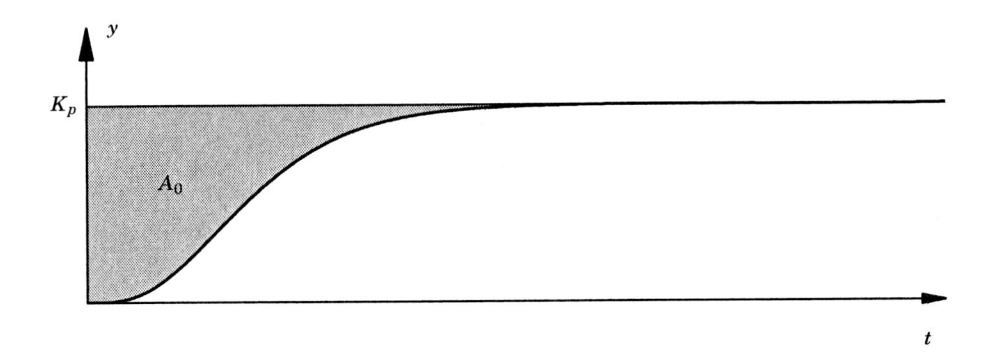

Figure 2.8 Illustrates the area method for determining the average residence time.

which is called the average residence time. The average residence time is a rough measure of how long it takes for the input to have a significant influence on the output. Notice that the function $g(t)/\int g(t)dt$ can be interpreted as a probability density if $g(t) \geq 0$.

The average residence time can be calculated from the step response in the following way

$$T_{ar} = \frac{\int_0^\infty (h(\infty) - h(t))dt}{K_p} = \frac{A_0}{K_p}, \tag{2.15}$$

where $h(t)$ is the step response and $K_p = G(0)$ is the static process gain. Notice that $K_p = h(\infty)$ and that A_0 is the shaded area in Figure 2.8.

Average Residence Time and Transfer Functions

The average residence time can be computed very conveniently from the transfer function. Since the transfer function is the Laplace transform of the impulse response we have

$$G(s) = \int_0^\infty e^{-st} g(t)dt.$$

Differentiation of this expression with respect to s gives

$$G'(s) = -\int_0^\infty e^{-st} t g(t)dt.$$

Setting $s = 0$ in these expressions it then follows from the definition of the average residence time (2.14) that

$$T_{ar} = -\frac{G'(0)}{G(0)}. \tag{2.16}$$

This formula will now be illustrated by a few examples.

EXAMPLE 2.9—AVERAGE RESIDENCE TIME FOR STIRRED TANK
The transfer function for the stirred tank in Example 2.4 is

$$G(s) = \frac{1}{1+sT}.$$

We have

$$G'(s) = -\frac{T}{(1+sT)^2},$$

and it follows from (2.16) that the average residence time is

$$T_{ar} = T = \frac{V}{q}.$$

The average residence time is thus the ratio of the volume and the flow through the tank. □

EXAMPLE 2.10—AVERAGE RESIDENCE TIME FOR TIME DELAY
The transfer function for the time delay in Example 2.5 is

$$G(s) = e^{-sL}.$$

We have

$$G'(s) = -Le^{-sL},$$

and it follows from (2.16) that the average residence time is

$$T_{ar} = L.$$

The average residence time is thus equal to the time delay. □

EXAMPLE 2.11—AVERAGE RESIDENCE TIME FOR CASCADED SYSTEMS
A system that is the cascade combination of two stable linear systems with transfer functions $G_1(s)$ and $G_2(s)$ has the transfer function

$$G(s) = G_1(s)G_2(s).$$

Differentiation gives

$$G'(s) = G_1'(s)G_2(s) + G_1(s)G_2'(s).$$

It follows from (2.16) that the average residence time is

$$T_{ar} = -\frac{G_1'(0)G_2(0) + G_1(0)G_2'(0)}{G_1(0)G_2(0)} = -\frac{G_1'(0)}{G_1(0)} - \frac{G_2'(0)}{G_2(0)}.$$

The average residence time is the sum of the residence times of each system. □

It follows from this example that the average residence time for the FOTD model in Example 2.6 is $T_{ar} = L + T$.

A system with the transfer function

$$G(s) = \frac{K_p(1+sT_1)(1+sT_2)}{(1+sT_3)(1+sT_4)(1+sT_5)}e^{-sL}.$$

has the average residence time $T_{ar} = T_3 + T_4 + T_5 + L - T_1 - T_2$.

Models with Two Parameters

A very simple way to characterize the dynamics of a stable process is to use the gain K_p and the average residence time T_{ar}. This gives the following models

$$G(s) = \frac{K_p}{1 + sT_{ar}}$$
$$G(s) = K_p\, e^{-sT_{ar}}, \tag{2.17}$$

where dynamics is either represented by a lag or a time delay.

Apparent Time Delay and Apparent Time Constant

Systems with essentially monotone step responses are very common in process control. Such systems can be modeled as first-order systems with time delay with the transfer function

$$G(s) = \frac{K_p}{1 + sT}\, e^{-sL}. \tag{2.18}$$

To emphasize that the parameters L and T are approximate they are referred to as the *apparent time delay* and the *apparent time constant*, or the *apparent lag*, respectively. The average residence time is $T_{ar} = L + T$. The parameter

$$\tau = \frac{L}{T_{ar}} = \frac{L}{L + T}, \tag{2.19}$$

which has the property $0 \leq \tau \leq 1$, is called the *normalized time delay* or the *normalized dead time*. This parameter can be used to characterize the difficulty of controlling a process. It is sometimes also called the *controllability ratio*. Roughly speaking, processes with small τ are easy to control, and the difficulty in controlling the system increases as τ increases. Systems with $\tau = 1$ correspond to processes with pure time delay, which are difficult to control well.

Ultimate Gain and Ultimate Period

So far we have used features that are based on the transient response. It is also possible to use features of the frequency response. Models can be characterized in terms of their phase lags and the frequency, where the systems have a given phase lag. For this purpose, we introduce ω_φ to denote the frequency where the phase lag is φ degrees, and we introduce $K_\varphi = |G(i\omega_\varphi)|$ to denote the process gain at ω_φ. The frequencies ω_{90} and ω_{180} and the corresponding process gains K_{90} and K_{180} are of particular interest for PID control. These frequencies correspond to the intersections of the Nyquist curve with the negative imaginary and real axes; see Figure 2.9. They also have nice physical interpretations. Consider a process with pure proportional control. If the controller gain is increased the process will start to oscillate, and it will reach the stability limit when the controller gain is $K_u = 1/K_{180}$. The oscillation will have the frequency ω_{180}. This frequency is called the ultimate frequency. The parameter K_u is called the ultimate gain or the critical gain. The parameters K_{90} and ω_{90} have similar interpretations for a process with pure integral control.

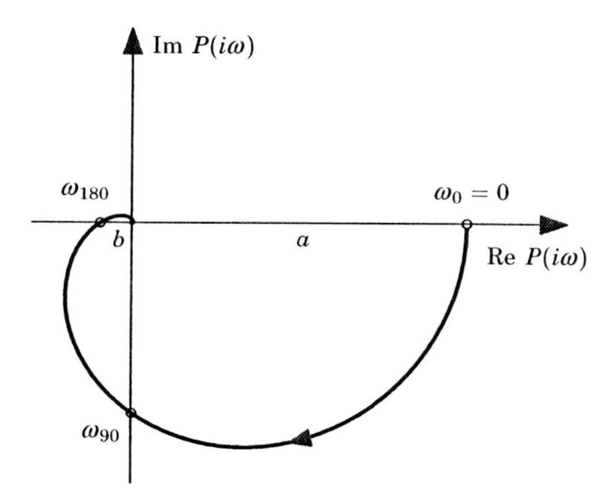

Figure 2.9 Nyquist curve with the points ω_0, ω_{90} and ω_{180}. The gain ratio κ is the ratio of the distances a and b.

The Gain Ratio

The *gain ratio* is an additional parameter that gives useful information about the system. This parameter is defined as

$$\kappa = \frac{K_{180}}{K_p} = \frac{|G(i\omega_{180})|}{G(0)}. \tag{2.20}$$

It is an indicator of how difficult it is to control the process. Processes with a small κ are easy to control. The difficulty increases with increasing κ. The parameter is also the ratio between the distances a and b in the Nyquist plot; see Figure 2.9.

Parameter κ is also related to the normalized time delay τ. For the FOTD model given by Equation 2.18 the parameters τ and κ are related in the following way:

$$\tau = \frac{\pi - \arctan\sqrt{1/\kappa^2 - 1}}{\pi - \arctan\sqrt{1/\kappa^2 - 1} + \sqrt{1/\kappa^2 - 1}}. \tag{2.21}$$

This relation is close to linear as is shown in Figure 2.10. This relation holds approximately for many other systems. As a crude approximation we can thus equate κ and τ. For small values a better approximation is given by $\kappa = 1.6\tau$. For the FOTD model it is also possible to find the parameters L and T from κ and ω_{180} using the following equations

$$\begin{aligned}
T &= \frac{1}{\omega_{180}}\sqrt{\kappa^{-2} - 1} \\
L &= \frac{1}{\omega_{180}}(\pi - \arctan\kappa^{-2} - 1) \\
K &= \frac{|G(i\omega_{180})|}{\kappa}.
\end{aligned} \tag{2.22}$$

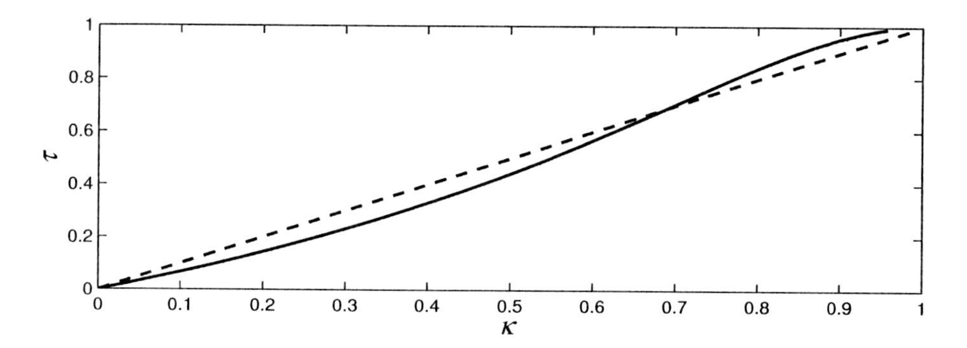

Figure 2.10 The normalized time delay τ as a function of gain ratio κ for the system (2.18). The dashed line shows the straight line approximation $\kappa = \tau$.

2.5 Typical Process Models

Much of the dynamic behavior encountered in control is relatively simple. Processes are designed in such a way that they should be easy to control. If PID control is used it is thus natural that simple process models are used. In this section we will discuss some of the models that are commonly used in connection with PID control. Most of these models are characterized by a few parameters only.

The FOTD Model

A process model that is commonly used in process control has the transfer function (2.18). It is simple and it describes the dynamics of many industrial processes approximately. A comparison with Examples 2.3 and 2.4 shows that it can represent the dynamics of a stirred tank with a pipe without mixing. The model is characterized by three parameters: the (static) gain K_p, the time constant T, and the time delay L. The time constant T is also called the lag. The step response of the model (2.18) is

$$h(t) = K_p \left(1 - e^{-(t-L)/T}\right).$$

Since the average residence time is $T_{ar} = L + T$, the value of the step response at this time becomes

$$h(T_{ar}) = K_p \left(1 - e^{-1}\right) \approx 0.63 K_p.$$

The average residence time can thus be determined as the time when the step response has reached 63 percent of its steady-state value.

Two parameters of the model (2.18) correspond to scaling of the axes and can be reduced by normalization. They can be chosen as the gain and the average residence time. This means that if the output is scaled by the gain $K_p = G(0)$ and time by the average residence time T_{ar} the response is completely characterized by one parameter, the normalized dead time τ. The system is a pure time delay for $\tau = 1$ and a first-order system or a pure lag for $\tau = 0$.

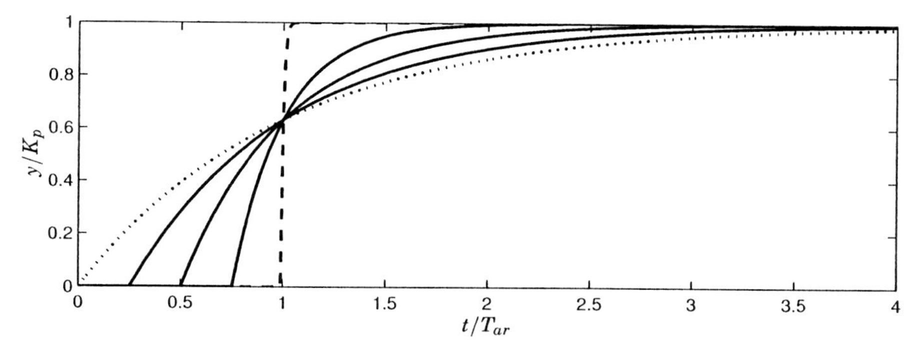

Figure 2.11 Normalized step responses of the FOTD model (2.18) for different values of the normalized time delay. The normalized time delay is $\tau = 0$ (dotted), 0.25, 0.5, 0.75 and 0.99 (dashed).

Figure 2.11 shows the normalized step responses for different values of τ. Notice that all curves intersect at one point $t = T_{ar}$ because of the normalization.

Noninteracting Tanks or Multiple Lags

The transfer function (2.8) represents the dynamics of a simple tank. The upper part of Figure 2.12 shows a system that is a cascade combination of n tanks. This system has the transfer function

$$G_n(s) = \frac{K_p}{(1+sT)^n}. \tag{2.23}$$

where n is the number of tanks. Since a first-order system is also called a lag the system is also called a multiple lag system. Notice that this formula holds only if the outflow of each tank only depends on its level. This means that there is no interaction between the tanks.

The average residence time is

$$T_{ar} = -\frac{G'(0)}{G(0)} = nT.$$

The model (2.23) has the impulse response

$$g(t) = \frac{K_p}{(n-1)!}\frac{t^{n-1}}{T^n}e^{-t/T}, \tag{2.24}$$

which has its maximum

$$\max g(t) = \frac{K_p(n-1)^{n-2}}{T(n-2)!}e^{-n+1},$$

for $t = (n-1)T$. The unit step response is

$$h(t) = K_p\left(1 - \left(1 + \frac{t}{T} + \frac{t^2}{2T^2} + \ldots + \frac{t^{n-1}}{(n-1)!T^{n-1}}\right)\right)e^{-t/T}. \tag{2.25}$$

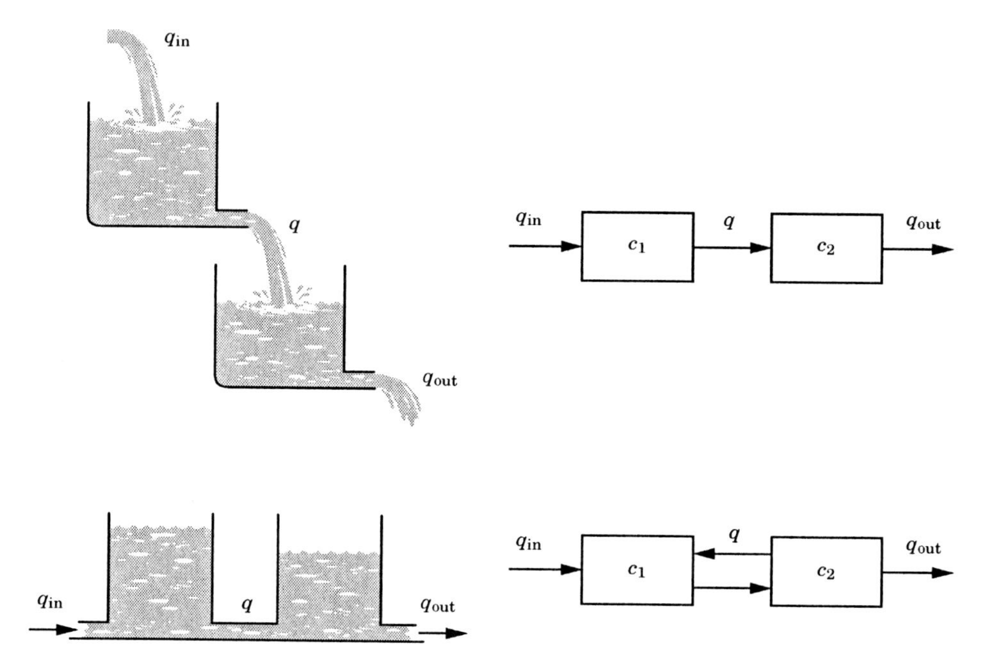

Figure 2.12 Cascaded tanks and corresponding block diagram representations. The upper tanks are noninteracting, and the lower are interacting.

The step response is characterized by three parameters, K_p, n, and T. The number of parameters can be reduced by normalization. Parameters K_p and T only influence the scaling of the axes. The shape of the step response is thus uniquely given by the parameter n. Normalized step responses for different values of n are shown in Figure 2.13. The step responses are close but not equal at $t = T_{ar}$. As n goes to infinity we have

$$\lim_{n\to\infty} G_n(s) = K_p e^{-t/T_{ar}}.$$

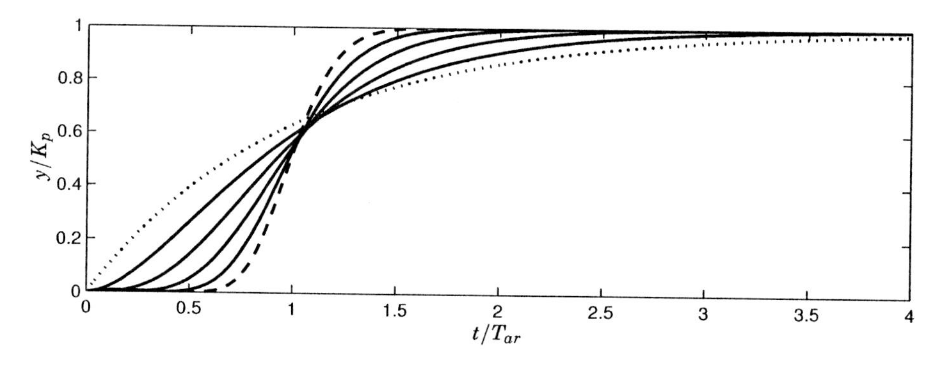

Figure 2.13 Normalized step responses for the processes $G_n(s) = 1/(1 + sT)^n$ for $n = 1$ (dotted), 2, 4, 8, 16, and 32 (dashed).

Table 2.1 Apparent time constant T_e, apparent time delay L_e, average residence time T_{ar}, and normalized time delay τ for the process (2.23).

n	2	3	4	5	6	7	8	16	32
T_e	1.86	2.44	2.91	3.32	3.68	4.01	4.31	6.23	8.90
L_e	0.28	0.81	1.43	2.10	2.81	3.55	4.31	10.78	24.67
T_{ar}	2.14	3.25	4.34	5.42	6.49	7.56	8.62	17.02	33.57
τ	0.13	0.25	0.33	0.39	0.43	0.47	0.50	0.63	0.73

For large n the system thus approaches a pure time delay. Figure 2.13 shows, however, that very large values of n are required to get a good approximation of the step response of a time delay.

The transfer function $G_n(s)$ can be approximated by an FOTD system. The apparent time constants and time delays for the approximation are given in Table 2.1.

Multiple Interacting Tanks—Distributed Lags

The dynamics of cascaded tanks are very different if the tanks are interacting. In the system shown in the lower part of Figure 2.12 the outflow of a tank depends on the levels of the neighboring tanks. Let x_k be the level of tank k. The control variable u is the inflow to the first tank, and let the output be the outflow of tank n. Assume that the tanks have unit cross-section, and assume that the flow from tank k to tank $k+1$ is $x_k - x_{k-1}$. The mass balances for the tanks are

$$
\begin{aligned}
\frac{dx_1}{dt} &= -x_1 + x_2 + u \\
&\vdots \\
\frac{dx_k}{dt} &= x_{k-1} - 2x_k + x_{k+1} \\
&\vdots \\
\frac{dx_n}{dt} &= x_{n-1} - 2x_n.
\end{aligned}
\tag{2.26}
$$

This is a state model with n states. The state variables represent the levels in the different tanks. The system is also called a distributed lag. With a unit step input the equilibrium values of the states are $x_k = n - k + 1$. The characteristic polynomials of systems having different order are

$$
\begin{aligned}
d_1 &= s + 1 \\
d_2 &= s^2 + 3s + 1 \\
d_n &= (s+2)d_{n-1} - d_{n-2}.
\end{aligned}
$$

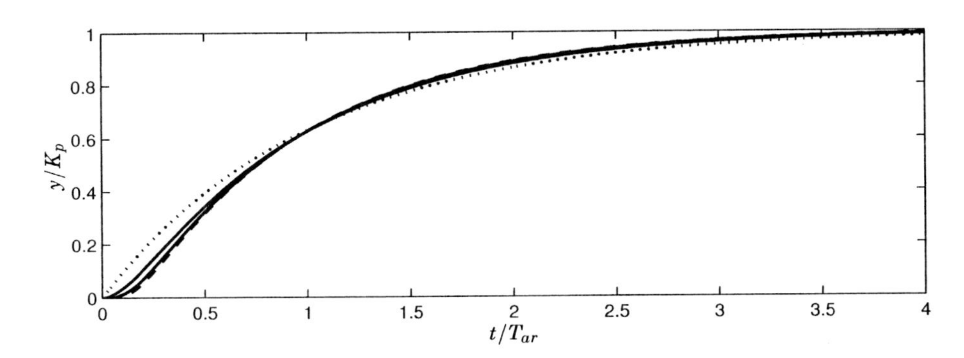

Figure 2.14 Normalized step responses for interacting tanks, (2.26), for $n = 1$ (dotted), 2, 4, and 8 (dashed).

The transfer functions for a few values of n are given by

$$\begin{aligned}
G_1(s) &= \frac{1}{s+1}\\
G_2(s) &= \frac{1}{s^2+3s+1}\\
G_4(s) &= \frac{1}{s^4+7s^3+15s^2+10s+1}\\
G_8(s) &= \frac{1}{s^8+15s^7+91s^6+286s^5+495s^4+462s^3+210s^2+36s+1}.
\end{aligned}$$

The average residence time is the ratio of the total steady-state volume to the flow, hence

$$T_{ar} = \frac{n(n+1)}{2}.$$

This is also the coefficient of the s-term in the denominator of the transfer function. As the number of tanks increases we have asymptotically for large n

$$G_n(s) \approx \frac{1}{\cosh\sqrt{2T_{ar}s}}.$$

These transfer functions are very different from the transfer function (2.23) of noninteracting tanks.

Figure 2.14 shows normalized step responses for interacting tanks. Notice that the responses are very similar for larger values of n. A comparison with Figure 2.13 shows that there is a significant difference between interacting and noninteracting tanks.

Another Version of Interacting Tanks

The model (2.26) is not the only way to interconnect tanks. Another configuration is shown in Figure 2.15. For simplicity we have shown a system with

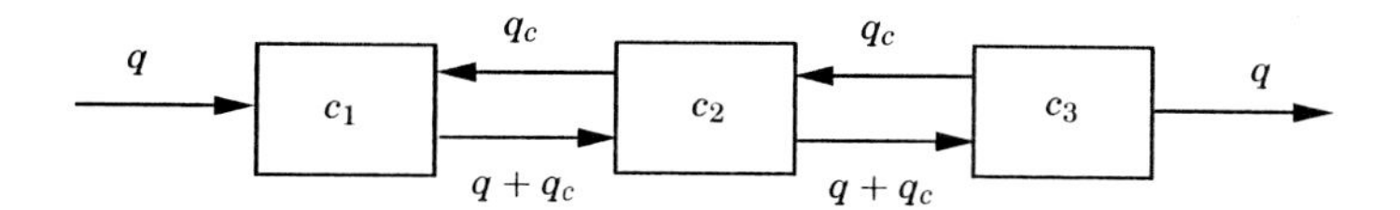

Figure 2.15 Schematic diagram of three cascaded tanks with recirculation.

three tanks. The system consists of identical stirred tanks with a forward flow q and a recirculation flow q_c. Let V be the tank volume, u the concentration of the inflow, c_k the concentration in the kth tank, and $y = c_n$ the concentration in the outflow. The mass balances for a system with n tanks are

$$
\begin{aligned}
V\frac{dc_1}{dt} &= -(q+q_c)c_1 + q_c c_2 + qu \\
&\vdots \\
V\frac{dc_k}{dt} &= (q+q_c)c_{k-1} - (q+2q_c)c_k + q_c c_{k+1} \\
&\vdots \\
V\frac{dc_n}{dt} &= (q+q_c)c_{n-1} - (q+q_c)c_n \\
y &= c_n.
\end{aligned}
$$

This is also a state model where the states are the concentrations in the different tanks. The transfer functions for a few values of n are

$$
\begin{aligned}
G_1(s) &= \frac{q}{Vs+q} \\
G_2(s) &= \frac{q(q+q_c)}{Vs^2 + 2V(q+q_c)s + q(q+q_c)} \\
G_3(s) &= \frac{2q_c(q+q_c)^2}{(Vs+q+q_c)(V^2s^2 + (2q+3q_c)s + 2q_c(q+q_c))} \\
&= \frac{2q_c(q+q_c)^2}{V^3s^3 + (3q+4q_c)V^2s^2 + (q+q_c)(q+3q_c)Vs + 2q_c(q+q_c)^2}.
\end{aligned}
$$

The static gain is $K_p = 1$ and the average residence time is

$$
T_{ar} = \frac{nV}{q}.
$$

The recirculation flow has a major impact on the dynamics. For $q_c = 0$ there is no interaction, and the system is equivalent to the model given by (2.23). As $q_c/q \to \infty$ the model is equivalent to the model given by (2.26). The model with recirculation thus makes it possible to interpolate between the models with noninteracting and distributed lags.

Figure 2.16 shows the step response of a system of n:th order for different values of the recirculation flow.

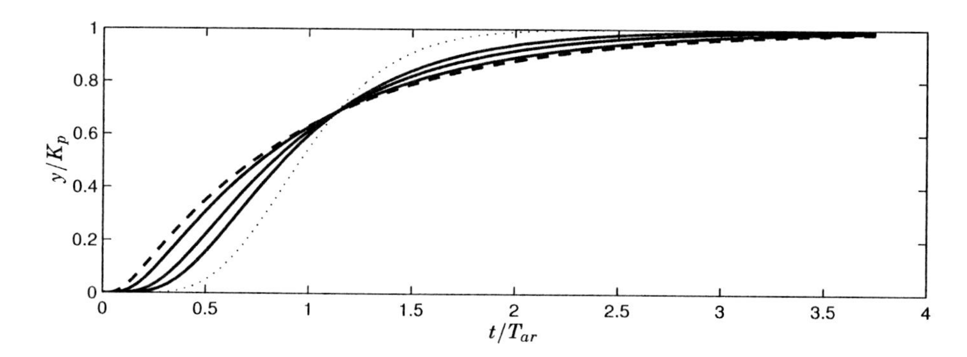

Figure 2.16 Normalized step responses for eight tanks with recirculation. The recirculation ratio is $q_c/q = 0$ (dotted), 1, 2, 5 and 10 (dashed).

Oscillatory Systems

The model (2.18) cannot describe systems with oscillatory responses. A simple model for such systems is given by the transfer function

$$G(s) = \frac{K_p}{1 + 2\zeta sT + (sT)^2}. \tag{2.27}$$

This model has three parameters: static gain K_p, time constant T, and relative damping ζ. The parameter $1/T$ is also called undamped natural frequency. The step responses can be normalized by the gain and the time constant. Its shape is then determined by one parameter only. The step responses are shown in Figure 2.17. For $\zeta < 1$ the step response has its maximum

$$M = K_p e^{-\frac{\pi\zeta}{\sqrt{1-\zeta^2}}},$$

which occurs at

$$t_{max} = \frac{2\pi T}{\sqrt{1-\zeta^2}}.$$

The position of the maximum increases with increasing ζ, and it becomes infinite for $\zeta = 1$ when the overshoot disappears. The transfer function is then

$$G(s) = \frac{K_p}{(1 + sT)^2},$$

and the step response is

$$h(t) = K_p\Big(1 - e^{-t/T} - \frac{t}{T}e^{-t/T}\Big).$$

The Bode plots of the systems are shown in Figure 2.18.

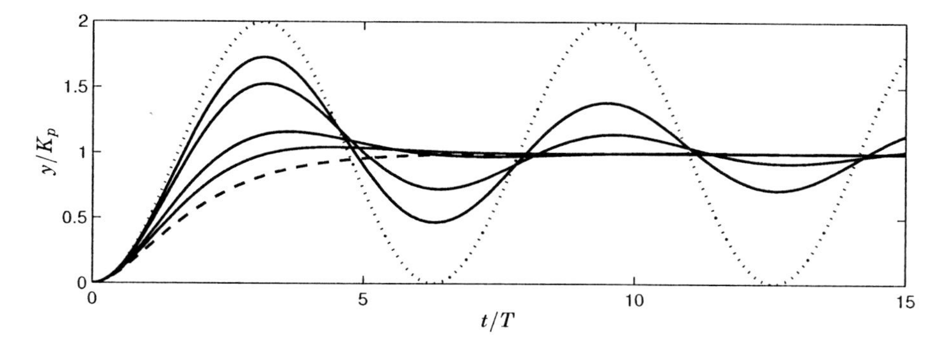

Figure 2.17 Normalized step responses of oscillatory systems (2.27) with $\zeta = 0$ (dotted), 0.1, 0.2, 0.5, 0.7, and 1.0 (dashed).

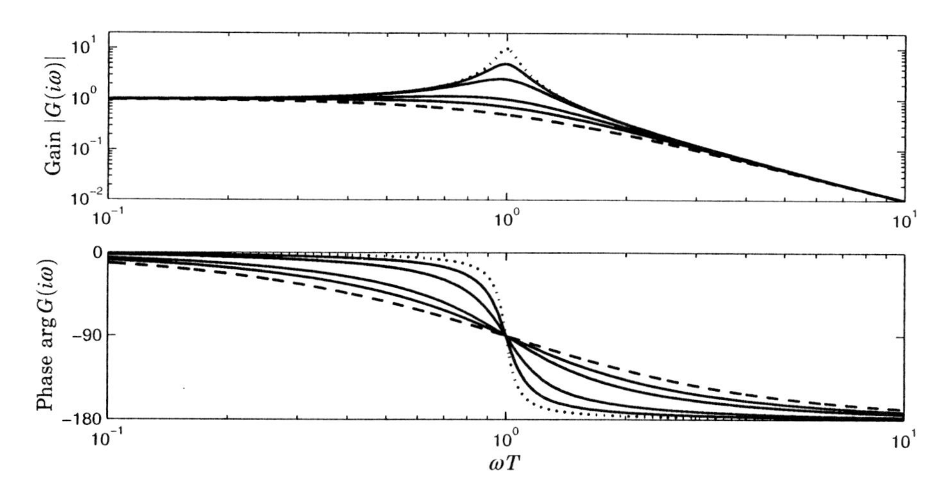

Figure 2.18 Bode plots of oscillatory systems (2.27) with $\zeta = 0.05$ (dotted), 0.1, 0.2, 0.5, 0.7, and 1.0 (dashed).

Processes with Integration

Integrating processes will not reach steady state during open-loop conditions. In practice, the same is true for processes with very long time constants. Asymptotically, the output will change at constant rate after a step change in the control signal. In the early process control literature these systems were said to be without self-regulation because the process variable did not reach a steady state after a disturbance. Many methods for PID tuning also treat such systems separately. Models for such systems are obtained simply by dividing the transfer function of a process with self-regulation by s.

A combination of an integrator and a time delay is a common model. The transfer function is

$$G(s) = \frac{K_v}{s} e^{-sL}. \tag{2.28}$$

This model is characterized by two parameters, a gain and a time delay. The

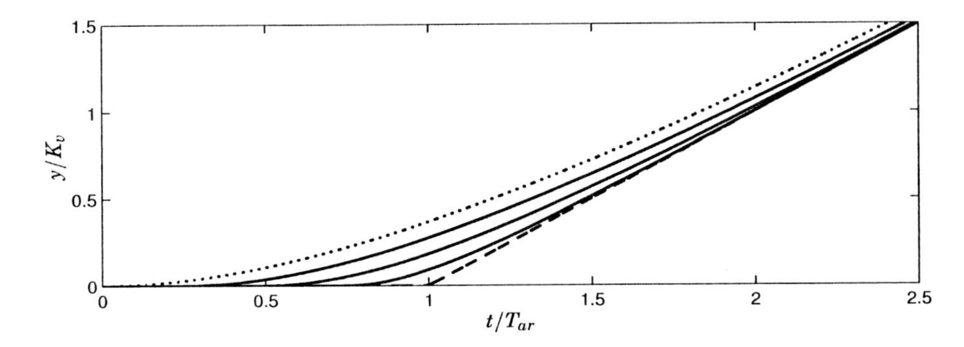

Figure 2.19 Normalized step responses for the FOTDI model (2.30) for $\tau = 0$ (dotted), 0.25, 0.5, 0.75, and 0.99 (dashed).

integrating gain is denoted by a special symbol K_v, which tells how fast the output increases in steady state after a unit input step change. The parameter K_v has dimension frequency.

A combination of a lag and an integrator is a model that is commonly used to describe simple drive systems. This model has the transfer function

$$G(s) = \frac{K_v}{s(1+sT)}. \tag{2.29}$$

The transfer function $K_v/(1+sT)$ represents the transfer function from the voltage of the drive system to the rate of rotation, and the integrator represents the relation between angular rate and angle.

A slightly more complicated model is obtained by adding integration to the standard model (2.18).

$$G(s) = \frac{K_v}{s(1+sT)} e^{-sL}. \tag{2.30}$$

We call this the FOTD model with integration or FOTDI for short. This process can be normalized in the same way as the model (2.18) by introducing the normalized time delay given by (2.19). The normalized step responses of the FOTDI model are shown in Figure 2.19.

Systems with Inverse Responses

The systems discussed so far do not have any zeros. Systems that are represented as a parallel connection of several systems can have transfer functions of the type

$$G(s) = \frac{1+sT}{s^2+1.4s+1}. \tag{2.31}$$

This system has a zero at $s = -1/T$, which may have a significant influence on the response of the system. Figure 2.20 shows the step response of this system for $T = -2, -1, 0, 1,$ and 2. Notice that the overshoot of the step response increases with increasing positive values of T. Also notice that the output signal initially moves in the wrong direction when T is negative. Such

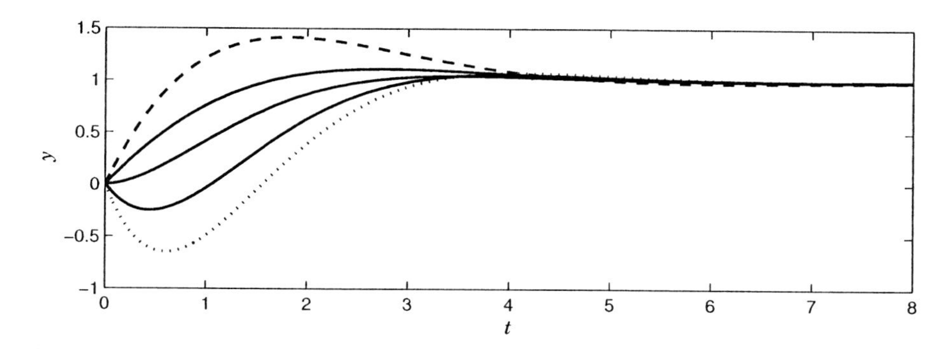

Figure 2.20 Step responses for the model (2.31) for $T = -2$ (dotted), -1, 0, 1 and 2 (dashed).

systems are said to have inverse responses. Systems with inverse responses are difficult to control. Examples of such systems are level dynamics in steam generators, dynamics of hydroelectric power stations, dynamics of backing cars, etc.

Heat Conduction

Temperature control is a very common application of PID control. Some models that are directly based on physics will now be discussed. Consider an infinitely long rod with thermal diffusivity λ. Assume that there is no radial heat transfer and that the input is the temperature at the left end of the rod. The transfer function to a point at the distance a from the left end point is

$$G(s) = e^{-\sqrt{sT}}, \tag{2.32}$$

where $T = a^2/\lambda$. The impulse response of the system is given by

$$h(t) = \frac{\sqrt{T}}{2\sqrt{\pi}t^{3/2}}\, e^{-\frac{T}{4t}}. \tag{2.33}$$

This impulse response has the property that all its derivatives are zero for $t = 0$, which means that the initial response of the system is very slow. The impulse response has a maximum at $t = T/6$. For large values of t the impulse response decays very slowly as $t^{-1.5}$. The step response of the system is

$$y(t) = 1 - \operatorname{erf}\sqrt{\frac{T}{4t}} = 1 - \frac{2}{\sqrt{\pi}}\int_{-\infty}^{\sqrt{T/4t}} e^{-x^2}dx. \tag{2.34}$$

The step and impulse responses are shown in Figure 2.21. Notice that the temperature starts to rise very slowly initially. After a rapid rise it also approaches the steady state very slowly.

We will now instead consider the situation when the right-hand side is isolated. The transfer function then becomes

$$G(s) = \frac{1}{\cosh\sqrt{sT}}. \tag{2.35}$$

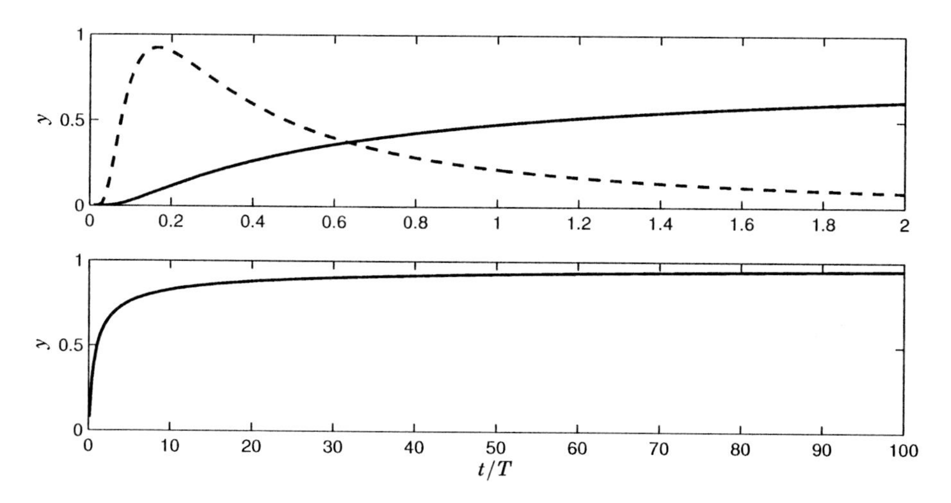

Figure 2.21 Step response (solid) and impulse response (dashed) for a system with the transfer function $e^{-\sqrt{sT}}$. The upper curves show the step responses and the impulse response. The lower curve shows the step response in a different time scale.

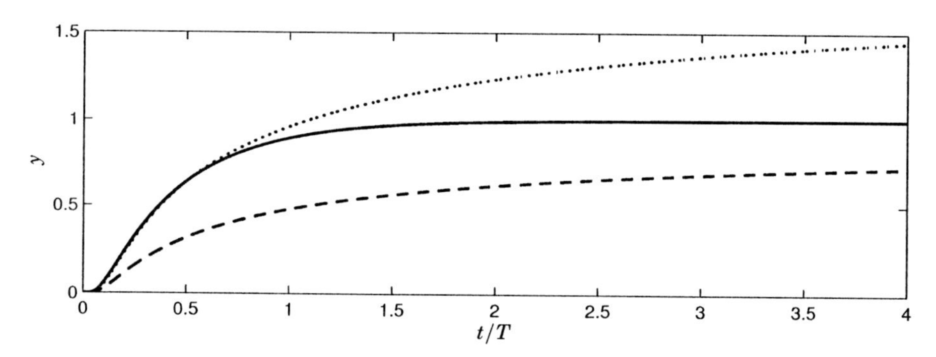

Figure 2.22 Step responses for the transfer function $1/\cosh\sqrt{sT}$ (solid), $e^{-\sqrt{sT}}$ (dashed) and $2e^{-\sqrt{sT}}$ (dotted).

This is a system with infinitely many lags with time constants $4T/\pi^2$, $T/9\pi^2$, $4T/25\pi^2$, $4T/49\pi^2$,.... This transfer function is also called a distributed lag.

The step response of this transfer function is shown in Figure 2.22. Notice that the response approaches the steady-state value faster than the system (2.32). The step response of the systems are thus quite different. The isolation of the right end of the rod makes it much easier to transfer heat into the system. A simple calculation shows that the average residence time for the system is

$$T_{ar} = -G'(0) = \frac{T}{2}. \tag{2.36}$$

The system (2.32) with the transfer function $e^{-\sqrt{sT}}$ has infinite residence time which reflects the fact that the impulse response decays very slowly; compare with Figure 2.21.

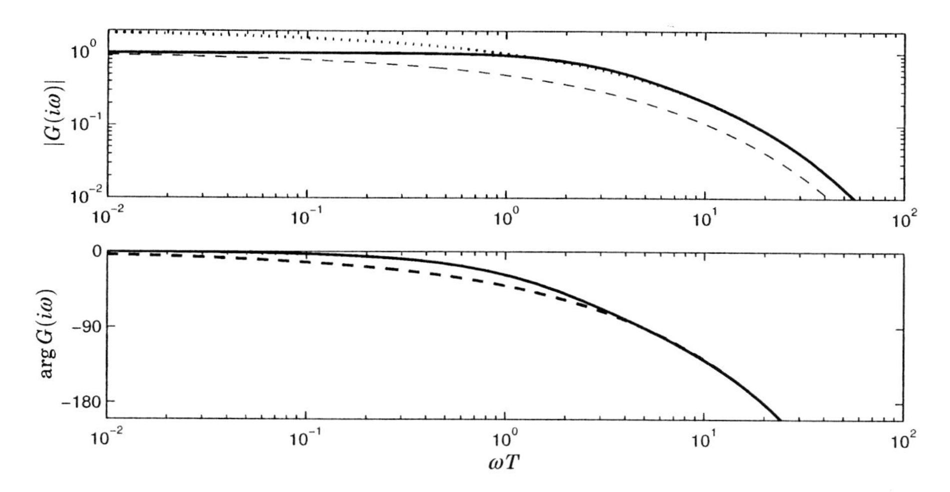

Figure 2.23 Bode plots for the transfer functions $1/\cosh\sqrt{sT}$ (solid lines), $e^{-\sqrt{sT}}$ (dashed). For comparison we also show the gain curve for the transfer function $2e^{-\sqrt{s}}$ (dotted).

We will now investigate the frequency responses of the systems (2.32) and (2.35). It can be shown that both transfer functions have a phase lag of 180° at the same frequency.

$$\omega_{180} = \frac{2\pi^2}{T}. \tag{2.37}$$

The magnitudes of the transfer functions at ω_{180} are given by

$$|e^{-\sqrt{i\omega_{180}T}}| = e^{-\pi} \approx 0.04321$$
$$\frac{1}{|\cosh\sqrt{i\omega_{180}T}|} = \frac{2e^{-\pi}}{1+e^{-2\pi}} \approx 0.08627.$$

At the frequency where the phase lag is 180° the gain of the system (2.35) is thus very close to twice as high as the gain for the system (2.32). The Bode plots of the system are shown in Figure 2.23. Notice that for frequencies above 2 rad/s there are very small differences between the transfer functions $2e^{-\sqrt{s}}$ and $1/\cosh\sqrt{sT}$ even if the step responses differ significantly. This observation is very important for the design of control systems. Figure 2.22 also shows that the step responses for the transfer functions $2e^{-\sqrt{s}}$ and $1/\cosh\sqrt{sT}$ are very close.

A Heat Exchanger

The transfer function from input temperature to output temperature of an ideal heat exchanger is

$$G(s) = \frac{1}{sT}(1 - e^{-sT}). \tag{2.38}$$

The step and impulse responses of this system are shown in Figure 2.24. Notice that the step response settles to the final value at time $t = T$ and that the

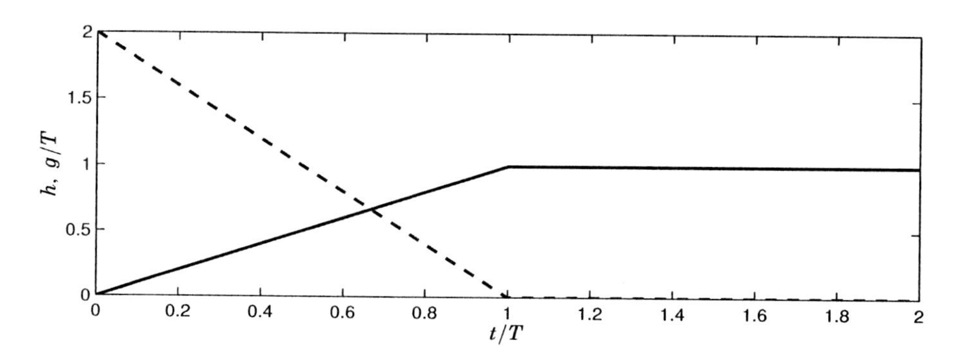

Figure 2.24 Normalized step (solid) and impulse (dashed) responses for the transfer function (2.38) of an ideal heat exchanger.

impulse response is zero after that time. This reflects the fact that once liquid has passed through the heat exchanger its temperature is no longer influenced. The average residence time of the system is

$$T_{ar} = \frac{T}{2}.$$

The frequency response of the system is

$$G(i\omega) = \frac{1}{i\omega T}\left(1 - e^{-i\omega T}\right).$$

The transfer function is zero for $\omega T = 2n\pi$. This is clearly seen in the Nyquist curve of the transfer function in Figure 2.25. An interesting property of this transfer function is that

$$\arg G(i\omega) = -\frac{\omega T}{2}, \quad \text{for } \omega T < 2\pi.$$

A Continuous Stirred Tank Reactor

Consider a continuous-time stirred tank reactor where the reaction $\mathcal{A} \rightarrow \mathcal{B}$ takes place. The reaction is exothermic, and reaction heat is removed by a coolant. The system is modeled by mass and energy balances. The mass balance is

$$\frac{dc}{dt} = \frac{q}{V}(c_f - c) - k(T)c, \tag{2.39}$$

where c [kmol/m^3] is the concentration of species $\mathcal{A}$, c_f the concentration of $\mathcal{A}$ in the feed, q [m^3/s] the volume flow rate, V [m^3] the reactor volume, and $k(T)$ [s^{-1}] the reaction rate which is a function of temperature

$$k(T) = k_0 e^{-E/RT}. \tag{2.40}$$

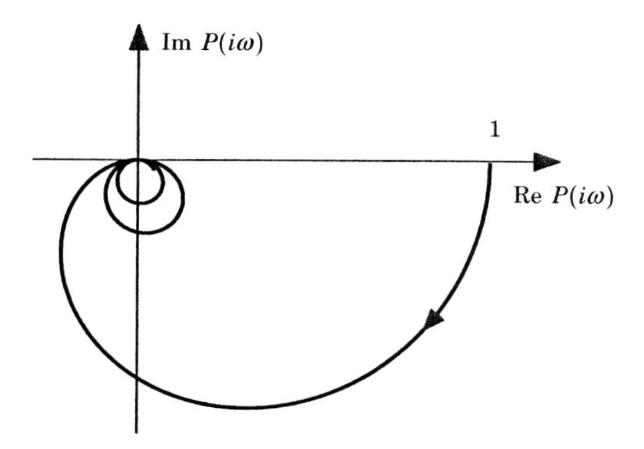

Figure 2.25 Nyquist plot of the transfer function $G(s)$ (2.38) of an ideal heat exchanger.

Table 2.2 Parameters of the exothermic continuous-time stirred tank reactor.

q	$0.002\ m^3/s$	T_f, T_c	300 K
V	$0.1\ m^3$	E/R	8750 K
ρ	$1000\ kg/m^3$	c_f	$2\ \text{kmol}/m^3$
C_p	4 kJ/kgK	UA	50 W/K
k_0	$3\times10^8\ s^{-1}$	ΔH	-5×10^5 kJ/kmol

The first term on the right-hand side represents the mass flow rate and the second term represents the rate of removal of $\mathcal{A}$ through the reaction.

The energy balance can be written as

$$\frac{dT}{dt} = \frac{q}{V}(T_f - T) + k(T)\frac{-\Delta H}{\rho C_p}c + \frac{UA}{\rho V C_p}(T_c - T), \tag{2.41}$$

where ΔH $[kJ/\text{kmol}]$ is the reaction heat, ρ $[kg/m^3]$ the density of the species $\mathcal{A}$, C_p $[kJ/kgK]$ specific heat, U $[J/min/K/m^2]$ the heat transfer coefficient, A $[m^2]$ the area, T_c $[K]$ the cooleant temperature, and T_f $[K]$ the feed temperature. The first term on the right-hand side represents the energy flow rate of the system, the second term represents the power generated by the reaction, and the last term represents the energy removal rate through cooling. Typical parameters are given in Table 2.2.

We will first analyse the steady-state solutions. In steady state it follows from (2.40) that

$$c = \frac{1}{1 + Vk(T)/q}c_f.$$

The power generated by the reaction is

$$P_g = \frac{k(T)}{1 + Vk(T)/q}(-\Delta H)c, \tag{2.42}$$

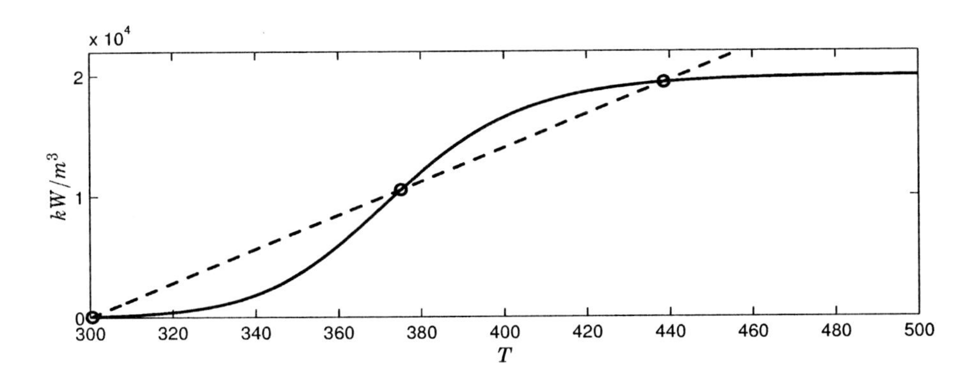

Figure 2.26 Steady-state heat generation rate (solid) and heat removal rate (dashed) as function of temperature. The equilibria are marked with ○.

and the rate of removal of energy is

$$P_r = \frac{q\rho C_p}{V}(T - T_f) + UA(T - T_c). \tag{2.43}$$

Equating P_g and P_r gives an equation in one variable to determine the reaction temperature T. A graphical solution gives insight as illustrated in Figure 2.26, which shows P_g and P_r as functions of temperature. There are three equilibria where the curves intersect at $T = 300.5$, 375.1, and 438.6. The equilibrium at $T = 375.1$ is unstable because the rate of heat generation is larger than the rate heat removal if temperature is increased. The other equilibria are stable. Approximating the dynamics in the neighborhood of the unstable equilibrium gives the following linear model of the system

$$\begin{aligned} \frac{dx_1}{dt} &= -0.0422x_1 + 0.0013x_2 \\ \frac{dx_2}{dt} &= 2.7746x_1 - 0.0064x_2 + 0.15u, \end{aligned} \tag{2.44}$$

where $x_1 = c - c_0$, $x_2 = T - T_0$, and $u = T_c - T_{c_0}$ and c_0, T_0, and T_{c_0} are the equilibrium values. The transfer function is

$$P(s) = -\frac{0.15s + 0.0063}{s^2 + 0.048631s - 0.003359} = -0.15\frac{s + 0.04220}{(s + 0.08717)(s - 0.03854)}$$

The system has the pole $s = 0.03854$ in the right half plane.

Nonlinear Black Models

The static model discussed in Section 2.2 could be nonlinear. The dynamic models discussed so far have, however, been linear. Since nonlinearities are common in practice it is highly desirable to have nonlinear models. Valves, actuators, and sensors may be nonlinear; the process dynamics itself can also be

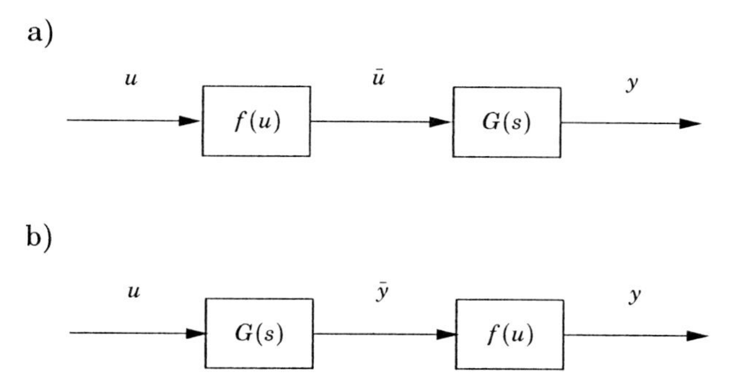

Figure 2.27 A Hammerstein model a), and a Wiener model b).

nonlinear. General models for nonlinear dynamics are complicated, and there are no good methods for designing PID controllers for such systems.

Fortunately, there are special classes of models that are well suited for PID control. A system may be represented as a combination of a static nonlinearity and a linear dynamical system. Such models are quite simple, and they are nicely adapted to PID control, but there are nonlinear systems that cannot be modeled well using this approach.

The nonlinearity can be before the linear part as shown in Figure 2.27a. This model is called a Hammerstein model. It is a good model for a system with a nonlinear actuator, for example, a nonlinear valve.

The nonlinearity can also be placed after the linear dynamical system. This gives a Wiener model, which is illustrated in the block diagram in Figure 2.27b. The Wiener model is a good representation for a system with a nonlinear sensor, for example, a pH electrode.

If the process is nonlinear the dynamics are varying with the operating conditions. Ideally, the controller should be tuned with respect to these variations. A conservative approach is to tune the controller for the worst case and accept degraded performance at other operating conditions. Another approach is to find a measurable variable that is well correlated with the process nonlinearity. Such a variable is called a scheduling variable. The controller is then tuned for a few values of the scheduling variable. Controller parameters for intermediate values may be obtained by interpolation. This approach to generating a nonlinear controller is called gain scheduling. It will be discussed in more detail in Section 9.3.

It is easy to compensate for the nonlinearity for a system that is described by a Wiener or a Hammerstein model by using a nonlinear controller composed of a PID controller and a static nonlinearity. The linear PID controller is designed as if the system was linear. When the process has a nonlinearity at the input we simply pass the control signal through the inverse of the nonlinearity. If the nonlinearity is at the output, as for the Wiener model, we simply pass the sensor signal through an inverse of the nonlinearity before feeding the measured signal to the controller. Many PID controllers have a facility to introduce a nonlinearity characterized as a piecewise linear function.

2.6 Models for Disturbances

So far, we have only discussed models of process dynamics. Disturbances are another important aspect of the control problem. In fact, without disturbances and process uncertainty there would be no need for feedback. There is a special branch of control, stochastic control theory, that deals explicitly with disturbances. This has had little impact on the tuning and design of PID controllers. For PID control, disturbances have mostly been considered indirectly, e.g., by introducing integral action. As our ambitions increase and we strive for control systems with improved performances it will be useful to consider disturbances explicitly. In this section, therefore, we will present some models that can be used for this purpose. Models for disturbances are useful for simulation, diagnostics, and performance evaluation.

The Nature of Disturbances

We distinguish between three types of disturbances, namely, set-point changes, load disturbances, and measurement noise. In process control, most control loops have set points that are constant over long periods of time with occasional changes. An appropriate model is therefore a piecewise constant signal. Set-point changes are typically known beforehand. Good response to set-point changes is the major issue in drive systems.

Load disturbances are disturbances that enter the control loop somewhere in the process and drive the system away from its desired operating point. Load disturbances typically have low frequency. Efficient reduction of load disturbances is a key issue in process control systems.

Measurement noise represents disturbances that distort the information about the process variables obtained from the sensors. Measurement noise is often a high-frequency disturbance. It is often attempted to filter the measured signals to reduce the measurement noise. Filtering does, however, add dynamics to the system.

The Character of Disturbances

One way to get a first estimate of the disturbances is to log the measured variable. The measured signal has contributions both from load disturbances and measurement noise. If there are large variations it is often useful to investigate the sensor to reduce some of the measurement noise. Filtering may also be useful. Filtering should be done in such a way that it does not impair control.

The process variations may have very different character. Some examples are given in Figure 2.28. The disturbances can be classified as pulses (a), steps (b), ramps (c), and periodic (d). It is useful to compute statistics such as mean values, variances, and maximum deviation. It is also useful to plot a histogram of the amplitude distribution of the disturbances.

Simple Models

It is useful to have simple models for disturbances for simulation and evaluation of control strategies. Models that are typically used are shown in Figure 2.28. The impulse is a mathematical idealization of a pulse whose duration

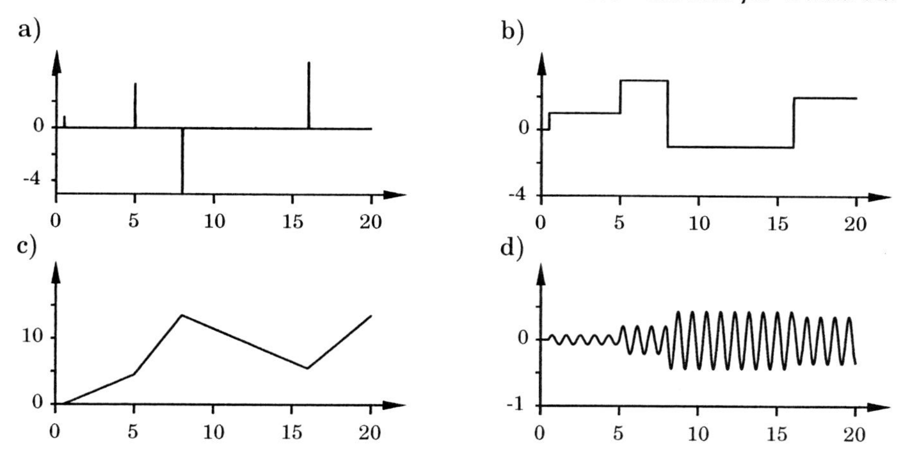

Figure 2.28 Different types of disturbances: a) impulses, b) steps, c) ramps, and d) sinusoids.

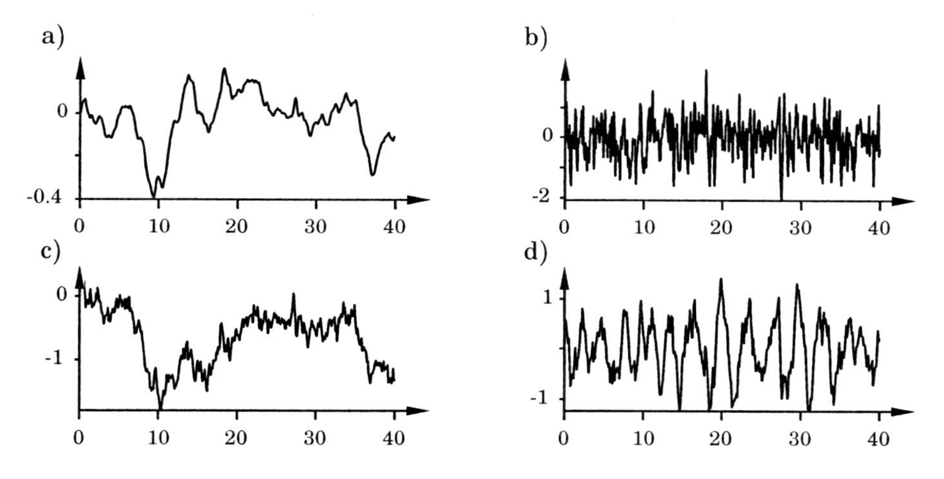

Figure 2.29 Examples of stochastic disturbances.

is short in comparison with the time scale. The signals are essentially deterministic. The only uncertain elements in the impulse, step, and ramp are the times when they start and the signal amplitude. The uncertain elements of the sinusoid are frequency, amplitude, and phase.

Random Fluctuations

Disturbances may also be more irregular as is shown in Figure 2.29. There are well developed concepts and techniques for dealing with random fluctuations that are described as stochastic processes. There are both time domain and frequency domain characterizations. In the frequency domain the random disturbances are characterized by the spectral density function $\phi(\omega)$. The variance

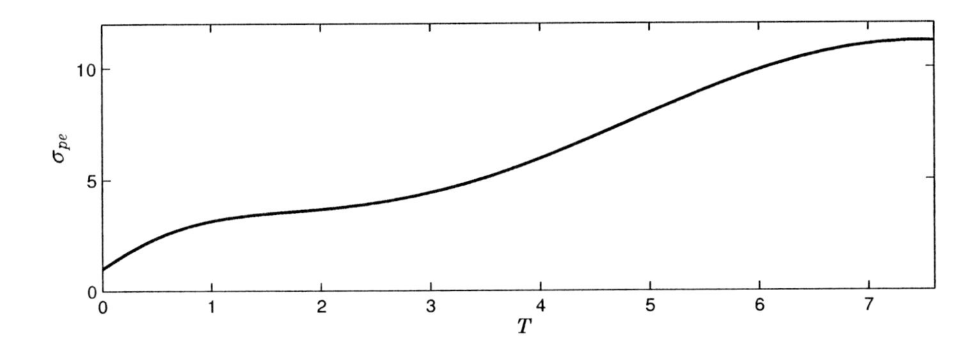

Figure 2.30 Prediction error σ_{pe} as a function of prediction time T_p.

of the signal is given by

$$\sigma^2 = \int_{-\infty}^{\infty} \phi(\omega) d\omega.$$

The spectral density tells how the variation of the signal is distributed on different frequencies. The value

$$2\phi(\omega)\Delta\omega$$

is the average energy in a narrow band of width $\Delta\omega$ centered around ω. A signal where $\phi(\omega)$ is constant is called white noise. Such a signal has its energy equally distributed on all frequencies.

There are efficient techniques to compute the spectral density of a given function. If the spectral density is known it is possible to evaluate how the variations in the process variable are influenced by different control strategies.

Prediction of Disturbances

When controlling important quality variables in a process it is often of interest to assess the improvements that can be achieved and to determine if a particular control strategy gives a performance that is close to the achievable limits. This can be done as follows. The process variable $y(t)$ is logged during normal operation with or without control. By analyzing the fluctuations it is possible to determine how accurately the process variable can be predicted T_p time units into the future based on present and past values of y. Let $\hat{y}(t + T_p|t)$ be the best prediction of $y(t + T_p)$ based on $y(\tau)$ for all $\tau < t$. By plotting the variance of the prediction error $y(t + T_p) - \hat{y}(t + T_p|t)$ as a function of the prediction time we obtain the curve shown in Figure 2.30. For large prediction times the prediction error is equal to the variance of the process variable, approximately $\sigma_{pe} = 12$ in the figure. The best control error that can be achieved is the prediction error at a prediction time T_p corresponding to the time delay of the process and the sampling time of the controller. This can be achieved with a so called minimum variance controller. See Section 8.6. The figure indicates that variances less that 5 can be obtained if T_p is less than 3.4. Further reductions are possible for smaller T_p, but variances less than 1 cannot be achieved even if T_p is very short. By comparing this with the actual variance we get an assessment of the achievable performance. This is discussed in more detail in Chapter 10. There is efficient software for computing the prediction error and its variance from process data.

2.7 How to Obtain the Models

In previous sections we have briefly mentioned how the models can be obtained. In this section we will give a more detailed discussion of methods for determining the models. There are two broad types of methods that can be used. One is physical modeling, and the other is modeling from data.

Physical modeling uses first principles to derive the equations that describe the system. The physical laws express conservation of mass, momentum, and energy. They are combined with constitutive equations that describe material properties. When deriving physical models a system is typically split into subsystems. Equations are derived for each subsystem, and the results are combined to obtain a model for the complete system. Simple examples were given in Section 2.3. Physical modeling is often very time consuming. There are often difficult decisions on suitable approximations. The models obtained can, however, be very useful since they have a sound physical basis. They also give considerable insight into the dependence of the model on the physical parameters. A simple way to start is to model dynamics as first-order systems where the time constants are the ratio of storage and flow.

Modeling from data is an experimental procedure. Data is generated by perturbing the input signal (the manipulated variable) and recording the system output. The experiment can also be performed under closed-loop conditions, for example, by perturbing the set point of a controller or the controller output. It is then attempted to find a model that fits the data well. There are several important issues to consider; selection of input signals, selection of a suitable model structure, parameter adjustments, and model validation. Ideally, the experimental conditions should be chosen to be as similar as possible to the intended use of the model. The parameter adjustment can be made manually for crude models or by using optimization techniques.

Static Models

Static models are easy to obtain by observing the relation between the input and the output in steady state. For stable, well-damped processes the relation can be obtained by setting the input to a constant value and observing the steady-state output. The procedure is then repeated for different values of the input until the full range is covered. For systems with integration it is convenient to use a controller to keep the output at a constant value. The set point of the controller is then changed so that the full signal range is covered. Effects of disturbances can be reduced by taking averages.

The Bump Test

The bump test is a simple procedure that is commonly used in process control. It is based on an experimental determination of the step response. To perform

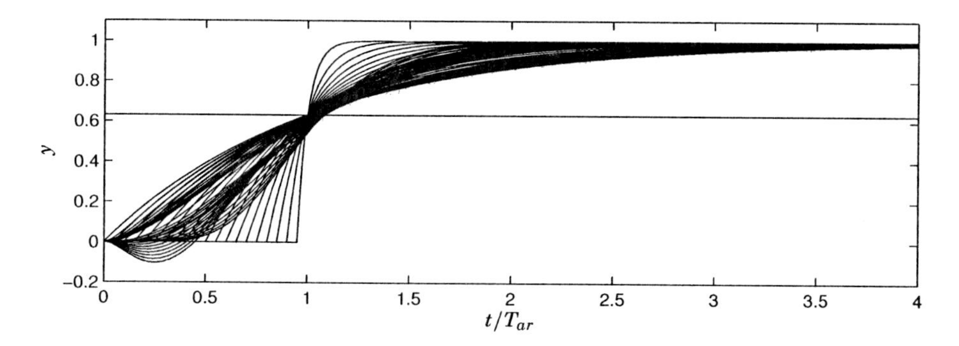

Figure 2.31 Step responses for a large batch of stable systems. The responses have been normalized to give the same average residence time.

the experiment the system is first brought to steady state. The manipulated variable is changed rapidly to a new constant value and the output is recorded. The measured data is scaled to correspond to a unit step. The change in the manipulated variable should be large in order to get a good signal-to-noise ratio but it should not be so large that the process behavior is not linear. The allowable magnitude is also limited by process operation. It is also useful to record the fluctuations in the measurement signal when the control signal is constant. This gives data about the process noise. It is good practice to repeat the experiment for different amplitudes of the input signal and at different operating conditions. This gives an indication of the signal ranges when the model is linear. It also indicates if the process changes with the operating conditions.

By inspection of the step response it is possible to make a crude classification of the dynamics of the system into the categories shown in Figure 2.2. A model with a few parameters is then fitted to the data.

The Average Residence Time

The average residence time is a simple way to characterize the response time of systems with essentially monotone step responses. Figure 2.31 step responses for a large batch of systems that are normalized to give the same average residence time. (The transfer functions for the systems are given in Section 7.1.) The figure shows that all step responses are close for $t = T_{ar}$. For all processes in the test batch we have $0.99 < T_{63}/T_{ar} < 1.08$. The average residence time can thus be estimated as the time T_{63} where the step response has reached 63 percent of its final value.

The FOTD model

The parameters of the FOTD model given by Equation 2.18 can be determined from a bump test as illustrated in Figure 2.32. The static gain K_p is simply determined from the steady-state values of the signals before and after the step change. The apparent time delay L is given by the point where the steepest tangent intersects the steady-state level before the step change. The average residence time $T_{ar} = T + L$ is determined as the time T_{63} where the step

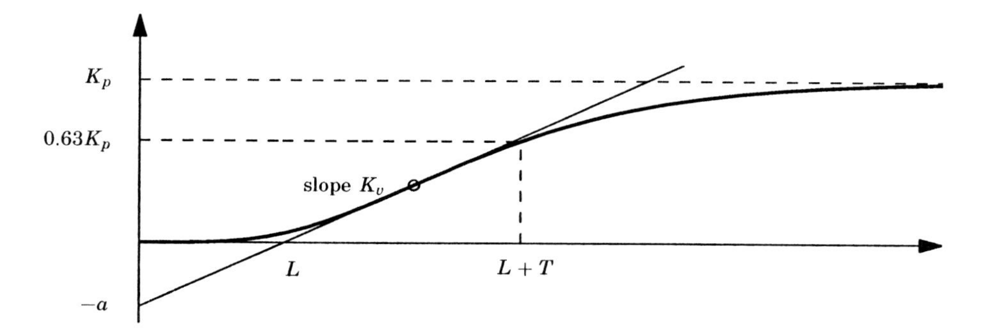

Figure 2.32 Unit step response of a process and a procedure used to determine the process parameters K_p, L, T, and K_v of an FOTD model. The point of largest slope is denoted by ∘.

response has reached 63 percent of its final steady-state value. This gives the correct results for the FOTD-model (see Figure 2.11) and approximate results for many other models (see Figures 2.13, 2.14, 2.16, and 2.31). The velocity gain K_v is the slope of the steepest tangent.

Similar methods can be used when the input signal is a pulse instead of a step. Pulses may be used when it is not permitted to use a step. This is common in medical and biological applications and is less common in process control. Ramp response analysis is common when analyzing servo drives and hydraulic systems.

The Integral and Time Delay Approximation

The model parameters of the model (2.28), which has the transfer function

$$G(s) = \frac{K_v}{s} e^{-sL} = \frac{a}{sL} e^{-sL}, \tag{2.45}$$

can also be determined from a bump test as indicated in Figure 2.32. The velocity constant K_v is the steepest slope of the step response, and the intersections of this tangent with the vertical and horizontal axes give a and L, respectively. The model given by Equation 2.45 is the basis for the Ziegler-Nichols tuning procedure discussed in Chapter 6.

The Doublet-Pulse Method

A variation of the bump test is to excite the process by a doubled pulse as is illustrated in Figure 2.33. The pulse amplitude a is chosen so that the response is well above the noise level, and the pulse width T_p is chosen a little longer than the time delay of the process. The maximum $y_{\max}$ and the minimum $y_{\min}$ and the times $t_{\max}$ and $t_{\min}$ when they occur are determined. Simple calculations show that for an FOTD system with the transfer function (2.10) we

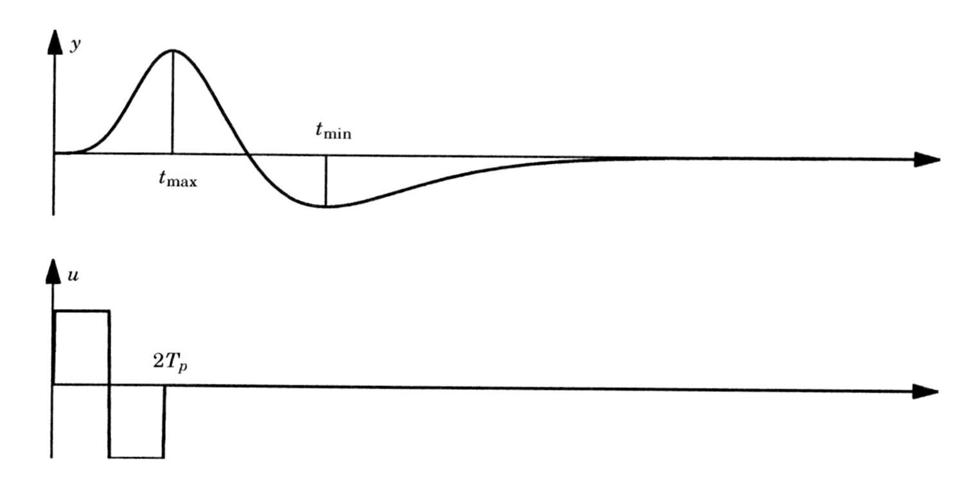

Figure 2.33 Determination of the parameters of an FOTD model by exciting the process by a doublet pulse.

have

$$
\begin{aligned}
y_{\max} &= aK_p\left(1 - e^{-T_p/T}\right)\\
y_{\min} &= -aK_p\left(1 - e^{-T_p/T}\right)^2\\
t_{\max} &= L + T_p\\
t_{\min} &= L + 2T_p.
\end{aligned}
$$

It follows from these equations that

$$
\begin{aligned}
\frac{y_{\min}}{y_{\max}} &= -1 + e^{-T_p/T}\\
\frac{y_{\max}^2}{y_{\min}} &= -aK_p,
\end{aligned}
$$

and we get the following simple equations for the parameters of the model

$$
\begin{aligned}
K &= -\frac{y_{\max}^2}{a\,y_{\min}}\\
T &= \frac{T_p}{\log(1 + y_{\max}/y_{\min})}\\
L &= t_{\max} - T_p\\
L &= t_{\min} - 2T_p.
\end{aligned}
\tag{2.46}
$$

The fact that the time delay L can be estimated in two ways can be used to asses if a process can be modeled by an FOTD model.

The selection of the pulse time T_p can be determined automatically, for example, as the time when the output has changed a specified amount. The method can be applied to SOTD models, but the formulas are more complicated.

The main advantages of using a doublet pulse is that the process output returns to its original value after the perturbation, and the time required to

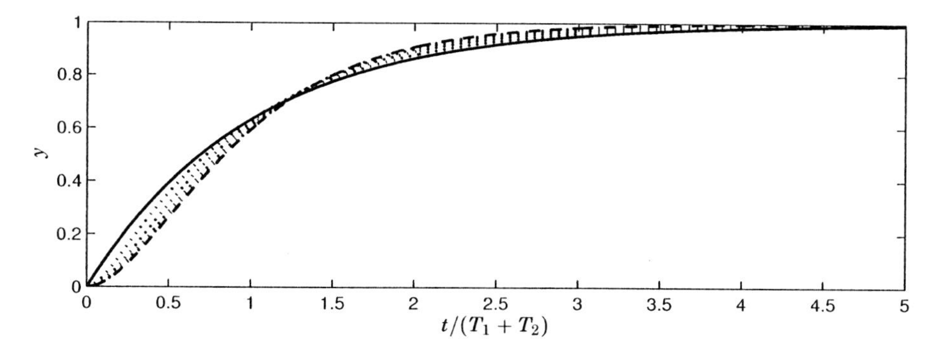

Figure 2.34 Normalized step responses for the system (2.48) for $T_2/T_1 = 0.1, \ldots, 1$.

determine the dynamics is short because it is not necessary to wait for steady state as for the bump test. The disadvantages of the method are that it is difficult to determine times when the extrema occurs accurately, and the estimate of the gain is poor because the excitation of the pulse is mainly in the high-frequency regime. Another disadvantage is that the method cannot be applied to oscillatory systems.

The SOTD Model

The model

$$G(s) = \frac{K_p}{(1 + sT_1)(1 + sT_2)} e^{-sL_1}, \tag{2.47}$$

which is a natural generalization of the FOTD model (2.18), is called the second-order model with time delay or SOTD model. Without loss of generality it can be assumed that $T_2 \leq T_1$. The step response of the system (2.47) is

$$y(t) = \begin{cases} K_p\left(1 - \dfrac{T_1}{T_1 - T_2} e^{-(t-L_1)/T_1} - \dfrac{T_2}{T_2 - T_1} e^{-(t-L_1)/T_2}\right) & \text{if } T_1 \neq T_2 \\ K_p\left(1 - e^{-(t-L_1)/T_1} - \dfrac{t}{T_1} e^{-(t-L_1)/T_1}\right) & \text{if } T_1 = T_2. \end{cases} \tag{2.48}$$

The normalized step responses for different ratios T_2/T_1 are shown in Figure 2.34. The responses have been normalized so that all systems have the same average residence time. All step responses are quite close, and they are almost identical for $t/(T_1 + T_2) \approx 1.3$. Since the separation of the curves is so small it is difficult to determine the parameters T_1 and T_2 robustly from the step response, particularly if there is a small amount of noise. Other inputs that excite the system better are necessary to determine the parameters reliably. The figure shows that it would be easier to determine the parameters based on an impulse response, which could be obtained by differentiating the step response.

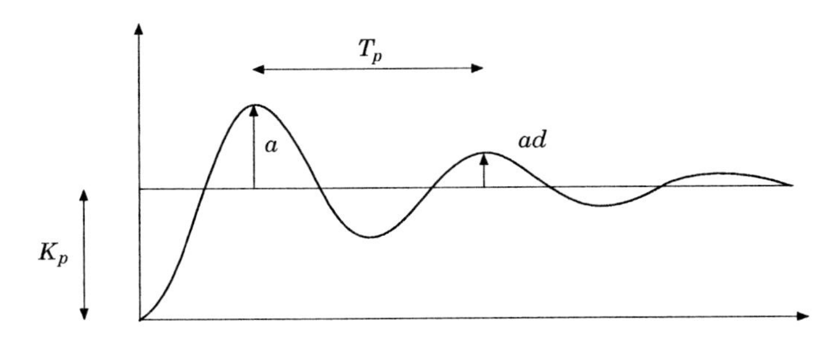

Figure 2.35 Graphical determination of mathematical models for systems with an oscillatory step response.

An Oscillatory System

The model (2.18) cannot model systems with oscillatory responses. A simple model for such systems is given by the transfer function (2.27), which has three parameters: the static gain K_p, the undamped natural frequency $1/T$, and the relative damping ζ. These parameters can be determined approximately from the step response as indicated in Figure 2.35. Parameters T and ζ are related to time period T_p and decay ratio d as follows.

$$d = e^{-2\zeta\pi/\sqrt{1-\zeta^2}} \qquad T_p = \frac{2\pi T}{\sqrt{1-\zeta^2}} \tag{2.49}$$

or

$$\zeta = \frac{1}{\sqrt{1+(2\pi/\log d)^2}} \qquad T = \frac{\sqrt{1-\zeta^2}}{2\pi} T_p. \tag{2.50}$$

The accuracy of the model is limited by the limited excitation obtained with a step or a pulse. Measurement errors and difficulty in obtaining steady state are other factors that limit the accuracy. Some improvements can be made by using optimization for fitting the parameters. Typically, it is difficult to determine more than three parameters from a step response unless the experimental conditions are exceptional.

Frequency Response

In frequency response analysis a sinusoidal signal is instead introduced, and the steady-state response is analyzed. An advantage with frequency response analysis is that very accurate measurements can be made by using correlation techniques. The long experimental times is a drawback.

It is also possible to introduce an arbitrary signal as a perturbation. The frequency response can be obtained as the ratio of the Fourier transforms of the output and the input signals. It is also possible to fit the parameters of a model with given structure to the data.

A nice feature of using signals other than steps is that it is possible to make a trade-off between signal amplitude and duration.

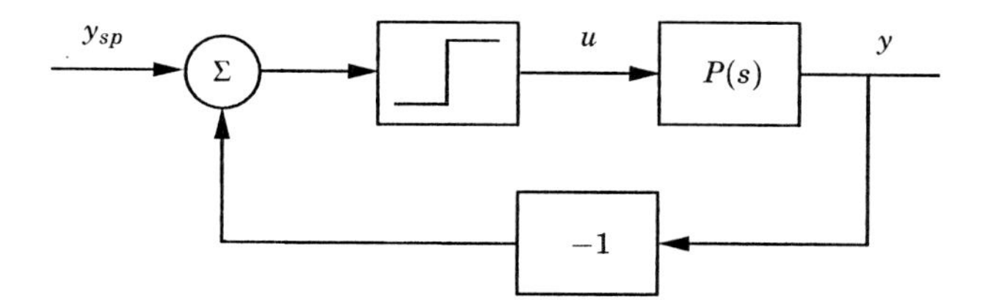

Figure 2.36 Block diagram of a process with relay feedback.

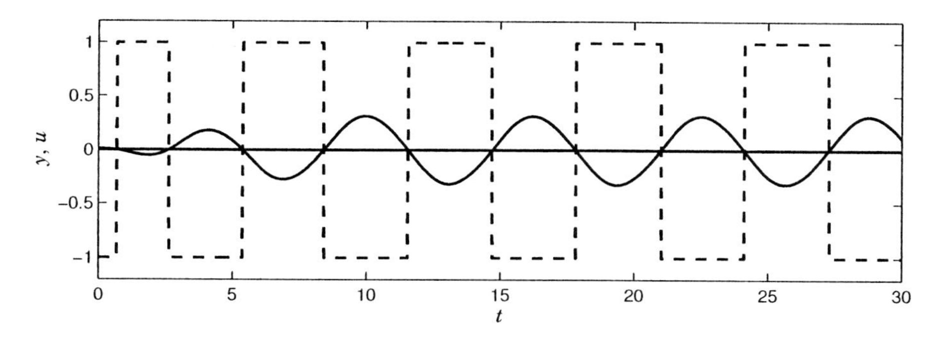

Figure 2.37 Relay output u (dashed) and process output y (solid) for a system under relay feedback.

Relay Feedback

There is a very special technique that is particularly suited to determine ω_{180} and K_{180}. This has been used very effectively for tuning PID controllers. The idea is the observation that it is possible to create an oscillation with the ultimate frequency automatically by using relay feedback.

To make the experiment the system is connected in a feedback loop with a relay function as shown in Figure 2.36. For many systems there will then be an oscillation (as shown in Figure 2.37) where the control signal is a square wave and the process output is close to a sinusoid. Notice that the process input and output have opposite phase.

To explain how the system works, assume that the relay output is expanded in a Fourier series and that the process attenuates higher harmonics effectively. It is then sufficient to consider the first harmonic component of the input only. The input and the output then have opposite phase, which means that the frequency of the oscillation is equal to ω_{180}. If d is the relay amplitude, the first harmonic of the square wave input has amplitude $4d/\pi$. Let a be the amplitude of the process output. The process gain at ω_{180} is then given by

$$K_{180} = \frac{\pi a}{4d}. \tag{2.51}$$

Notice that the relay experiment is easily automated. Since the amplitude of the oscillation is proportional to the relay output, it is easy to control it by adjusting the relay output. Also notice in Figure 2.37 that a stable oscillation is established very quickly. The amplitude and the period can be determined after about 20 s only, in spite of the fact that the system is started so far from the equilibrium that it takes about 8 s to reach the correct level. The average residence time of the system is 12 s, which means that it would take about 40 s for a step response to reach steady state.

The SOTD Model—Combined Step and Frequency Response

It was mentioned previously that the parameters of the SOTD model cannot be determined reliably from step response data. Good estimates can, however, be obtained by combining step and frequency response data. The idea is that the step response gives K_p and T_{63} and the frequency response method gives the ultimate frequency $\omega_u = \omega_{180}$ and the ultimate gain $K_u = 1/K_{180}$. This gives the equations

$$\begin{aligned} K_p^2 K_u^2 &= (1+\omega_u^2 T_1^2)(1+\omega_u^2 T_2^2) \\ \pi &= \arctan \omega_u T_1 + \arctan \omega_u T_2 + \omega_u L_1. \end{aligned} \tag{2.52}$$

Combined with the data K_p and T_{63} the parameters are then given by Equations 2.48 and 2.52 which gives four equations for the four unknown.

$$\begin{aligned} 0 &= \begin{cases} 0.37 - \dfrac{T_1}{T_1 - T_2} e^{-(T_{63}-L_1)/T_1} - \dfrac{T_2}{T_2 - T_1} e^{-(T_{63}-L_1)/T_2} & \text{if } T_1 \neq T_2, \\ 1 - e^{-(T_{63}-L_1)/T_1} - \dfrac{T_{63}}{T_1} e^{-(T_{63}-L_1)/T_1} - 0.63 & \text{if } T_1 = T_2 \end{cases} \\ 0 &= (1+\omega_u^2 T_1^2)(1+\omega_u^2 T_2^2) - K_p^2 K_u^2 \\ 0 &= \arctan \omega_u T_1 + \arctan \omega_u T_2 + \omega_u L_1 - \pi. \end{aligned} \tag{2.53}$$

These equations can be solved iteratively, but this is complicated since we have to take care of the special cases when the parameters T_1 and T_2 are equal or zero.

An alternative method is to iterate the ratio $a = T_2/T_1$ until the equations match. Parameter K_p is determined as the static gain of the step response. The equation (2.52) for the ultimate gain then becomes

$$(1+\omega_u^2 T_1^2)(1+a^2\omega_u^2 T_1^2) = K_p^2 K_u^2.$$

This equation has the solution

$$T_1 = \frac{1}{a\omega_u\sqrt{2}} \sqrt{\sqrt{4a^2 K_p^2 K_u^2 + (1-a^2)^2} - 1 - a^2}.$$

The parameters T_2 and L_1 are then given by

$$\begin{aligned} T_2 &= aT_1 \\ L_1 &= \frac{\pi - \arctan \omega_u T_1 - \arctan \omega_u T_2}{\omega_u}. \end{aligned}$$

The step response given by (2.48) can then be computed as a function of a, and the parameter a can be iterated to match the value of T_{63}.

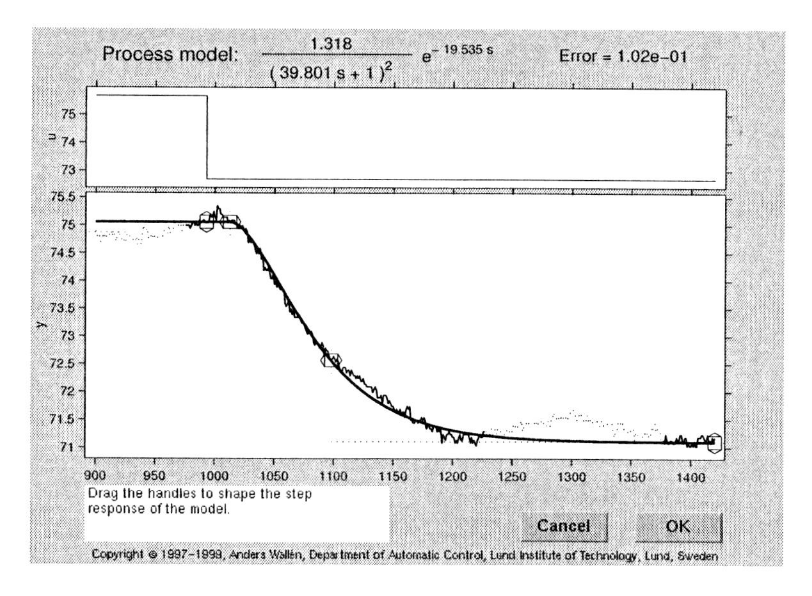

Figure 2.38 Computer screen from a tool for process modeling. From [Wallén, 2000]

Modeling Tools

There are several modeling tools that are very useful. They make it possible to enter process data in the form of sequences of input-output data from bump tests or other process experiments. Models of different structure can be selected, and their parameters can be fitted to the data using some optimization procedure. The tools also permit selection of parts of data sequences used in the analysis.

Figure 2.38 shows the computer screen for a particular system. A model structure can be chosen from a menu. When data has been entered a preliminary model can be fitted by manually dragging the handles shown in the figure. The handles represent the start of the step, the initial level, the final level, and the time when the response has reached 63 percent of its final value. The model parameters are displayed. Optimization can then be used to improve the fit.

The particular tool illustrated in the figure also allows use of a nonlinear model as illustrated in Figure 2.39. In this case a static model is first fitted to input-output data obtained from a static experiment. A dynamic model is then fitted as indicated in Figure 2.39. Both Wiener and Hammerstein models are tried to see which gives the best fit. The particular example is from a tank system where the outflow is a nonlinear function of level. In this case the Wiener model gives the best fit because the nonlinearity appears at the system output. Notice in Figure 2.39 that the input steps are of equal size, but the magnitude of the output response changes significantly. This data cannot be well matched by a linear model.

The interactive tools give a very good feel for the relations between the parameters and the response and the sensitivity of the parameters. It is also very effective to combine simple manual fits with numeric optimization. Most tools

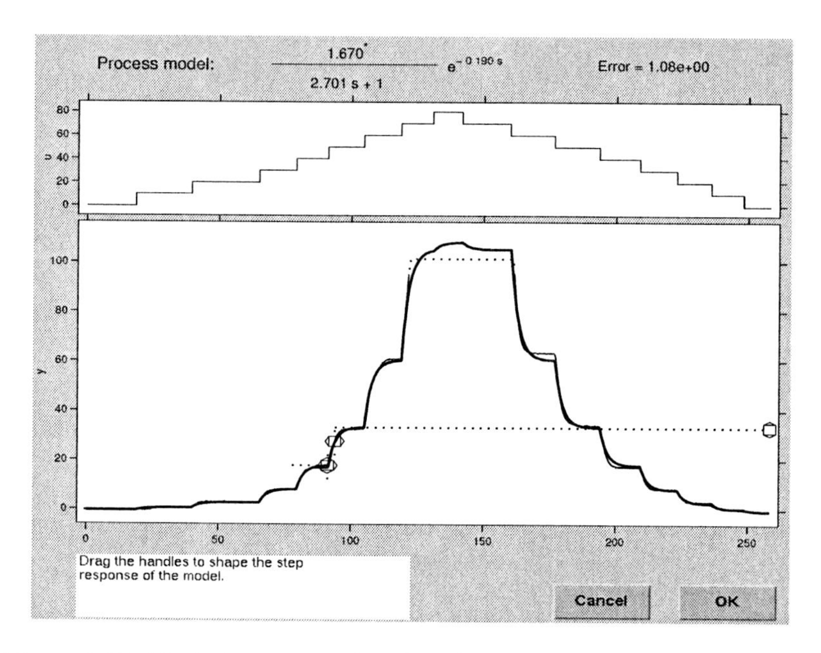

Figure 2.39 Illustrates computer-based nonlinear modeling. From [Wallén, 2000]

permit fitting a simple bump test, but there are also tools that permit general input signals. They also make it possible to determine noise characteristics and the prediction curve shown in Figure 2.30.

2.8 Model Reduction

Many methods for tuning PID controllers are based on simple models of process dynamics. To use such methods it is necessary to have methods for simplifying a complicated model. A typical case is when a model is obtained by combining models for subsystems. To find suitable approximations it is necessary to specify the purpose of the model. For tuning PID controllers this can be done by specifying the frequency range of interest. This can be done simply by specifying the highest frequency ω^* where the model is valid. For PI control the frequency ω^* is about ω_{145}, the frequency where the phase lag of the process is 145°. The reason for this is that a PI controller always has a phase lag. For a PID controller, which can provide phase lead, the frequency ω^* can be chosen as ω_{180}.

Model reduction starts with a model represented by the transfer function $G(s)$. The transfer function is first factored as

$$G(s) = G_l(s)\frac{1}{1+sT_s}G_h(s). \tag{2.54}$$

The low frequency factor $G_l(s)$ has all its poles and zeros and time delays at frequencies around ω^* or at lower frequencies. The high-frequency factor $G_h(s)$

has dynamics at frequencies higher than ω^*. The time constant T_s represents an intermediate pole. The factorization can always be done in such a way that the high frequency factor $G_h(s)$ has the property $G_h(0) = 1$.

For design of PID controllers the model (2.54) will be simplified to

$$
\begin{aligned}
G(s) &= \frac{K_p}{1+sT} e^{-sL} \\
G(s) &= \frac{K_p}{(1+sT_1)(1+sT_2)} e^{-sL}.
\end{aligned}
\tag{2.55}
$$

The reason for these choices is that there are methods for designing PID controllers for models of this type. These models are particularly suitable for typical process control problems where the dynamics have essentially monotone step responses.

The Low-Frequency Factor

The low-frequency factor will normally only contain one or two modes. If the system has multiple poles they can be approximated by the transfer function

$$
G_l(s) = \frac{K_p}{1+sT_e} e^{-sL_e}.
$$

where T_e and L_e are obtained from Table 2.1. In this way we obtain a low-frequency factor of first or second order, which is required for PID control. If the model is more complex it is necessary to reduce ω^* or to use a more complex controller.

Approximation of Fast Modes

There are several ways to approximate the fast modes. A simple way is to characterize the high-frequency part by its average residence time T_{arh}. This is illustrated by the following example.

EXAMPLE 2.12—APPROXIMATION OF FAST MODES
Consider a system where the high-frequency factor is

$$
G_h(s) = \frac{(1+sT_1)(1+sT_2)}{(1+sT_3)(1+sT_4)(1+sT_5)} e^{-sL}.
$$

This system has the average residence time

$$
T_{arh} = T_3 + T_4 + T_5 + L - T_1 - T_2.
$$

□

Compare this with Section 2.4, which shows how average residence times are computed. When using digital control half the sampling period should also be added to T_{arh}.

Skogestad's Half Rule

Having simplified the low- and high-frequency factors we have obtained a low-frequency factor of the form given by (2.55) or (2.58) and a characterization of the high-frequency factor by its average residence time T_{arh}. Skogestad has suggested that the intermediate time constant T_s in (2.54) is approximated by adding $T_s/2$ to the time delay of the model and $T_s/2$ to its time constant. The reduced model then becomes

$$
\begin{aligned}
G(s) &= \frac{K_p}{1+s(T+T_s/2)} e^{-s(L+T_{arh}+T_s/2)} \\
G(s) &= \frac{K_p}{(1+sT_1)(1+s(T_2+T_{arh}/2))} e^{-s(L+T_{arh}/2)}.
\end{aligned}
\tag{2.56}
$$

The model error is characterized by $T_{arh} + T_s/2$, which means that it must be required that $\omega^*(T_{arh} + T_s/2)$ is sufficiently small. A reasonable value is that it is less than 0.1 or 0.2, which means that the neglected dynamics has a phase lag of 6 to 12 degrees.

Approximating Slow Modes by Integrators

Modes that are much slower than ω^* can be approximated by integrators. For example, if $\omega^* T$ or $\omega^* T_1$ are larger than 5 to 10, the model (2.56) can be approximated by

$$
\begin{aligned}
G(s) &= \frac{K_p}{1+s(T+T_{arh}/2)} e^{-s(L+T_{arh}/2)} \approx \frac{K_p}{s(T+T_{arh}/2)} e^{-s(L+T_{arh}/2)} \\
G(s) &= \frac{K_p}{(1+sT_1)(1+s(T_2+T_{arh}/2))} e^{-s(L+T_{arh}/2)} \\
&\approx \frac{K_p}{sT_1(1+s(T_2+T_{arh}/2))} e^{-s(L+T_{arh}/2}.
\end{aligned}
\tag{2.57}
$$

Another Model Representation

For some design techniques it is desirable to have models of the form

$$
\begin{aligned}
G(s) &= \frac{b}{s+a} \\
G(s) &= \frac{b_1 s + b_2}{s^2 + a_1 s + a_2},
\end{aligned}
\tag{2.58}
$$

which do not have any time delays. These forms can also be used for oscillatory systems. The models given by (2.55) can be converted to the form (2.58) by using the approximation

$$
e^{-sT} \approx \frac{1 - sT/2}{1 + sT/2}.
\tag{2.59}
$$

Time delays and zeros in the right half plane are the features of a system that ultimately limits the achievable performance. These properties are preserved by the above approximation.

Examples

Model reduction will now be illustrated with a few examples.

EXAMPLE 2.13—MODEL REDUCTION
Consider a system described by the transfer function

$$G(s) = \frac{K_p}{(1+s)(1+0.1s)(1+0.01s)(1+0.001s)}. \tag{2.60}$$

We have $\omega_{90} = 3$ and $\omega_{180} = 31.6$, which gives the ranges of ω^*. Let us first consider model reduction for a design with $\omega^* = 3$. The low-frequency factor is

$$G_l(s) = \frac{K_p}{1+s},$$

the mid-frequency factor is $T_s = 0.1$, and the average residence time of the high-frequency part is $T_{arh} = 0.011$. Skogestad's half rule gives the model

$$\tilde{G}(s) = \frac{K_p}{1+1.05s} e^{-0.061s}.$$

Requiring that $\omega^*(T_{arh} + T_s/2) < 0.2$ we find that the model can be used for designs with $\omega^* < 3.3$.

For $\omega^* = 31.6$ the low-frequency factor becomes

$$G_l(s) = \frac{K_p}{(1+s)(1+0.1s)},$$

the mid-frequency time constant is $T_s = 0.01$, and the average residence time of the high-frequency part is then $T_{arh} = 0.001$. The half rule gives the model

$$\tilde{G}(s) = \frac{K_p}{(1+s)(1+0.105s)} e^{-0.006s}.$$

Requiring that $\omega^*(T_{arh} + T_s/2) < 0.2$ we find that the model can be used for designs with $\omega^* < 33$.

The approximations are illustrated in Figure 2.40. □

A Warning

The fact that the step responses of two systems are similar does not imply that the systems are similar under feedback control. This is illustrated by the following example.

EXAMPLE 2.14—SIMILAR OPEN LOOP – DIFFERENT CLOSED LOOP
Systems with the transfer functions

$$G_1(s) = \frac{100}{s+1}, \qquad G_2(s) = \frac{100}{(s+1)(1+0.025s)^2}$$

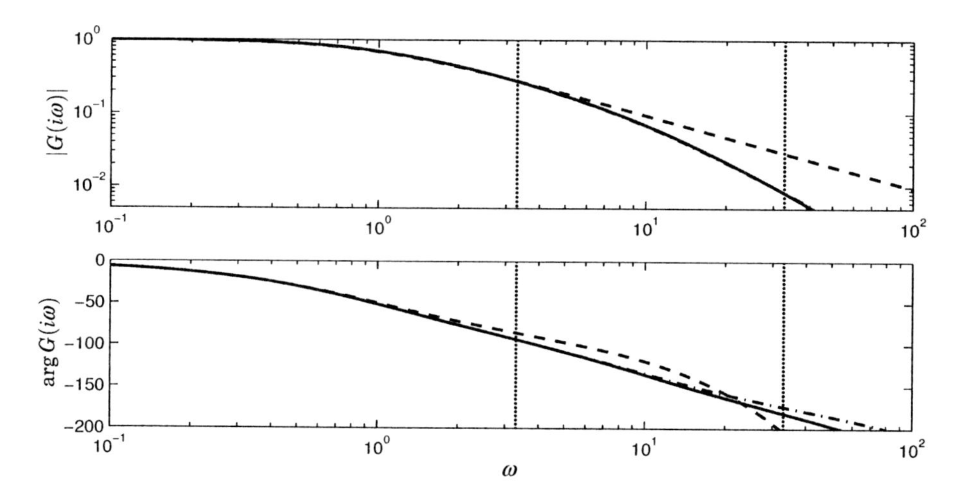

Figure 2.40 Bode plots of the original system (solid line) and the approximations for frequencies $\omega^* < 3.3$ (dashed) and $\omega^* < 33$ (dash-dotted).

have very similar open-loop responses as illustrated in Figure 2.41. The differences between the step responses are barely noticeable in the figure. The closed-loop systems obtained with unit feedback have the transfer functions

$$G_{1cl} = \frac{100}{s+101}, \qquad G_{2cl} = \frac{100}{(1+0.01192s)(1-0.001519s+0.0005193s^2)}.$$

The closed-loop systems are very different since the system P_{2cl} is unstable. □

It is also possible to have the opposite situation, namely, systems whose closed-loop behavior are very similar even if their open-loop behavior are very different.

EXAMPLE 2.15—DIFFERENT OPEN LOOP – SIMILAR CLOSED LOOP
The systems with the transfer functions

$$P_1(s) = \frac{100}{s+1}, \qquad P_2(s) = \frac{100}{s-1}$$

have very different open-loop properties because one system is unstable and the other is stable. The closed-loop systems obtained with unit feedback are, however,

$$P_{1cl}(s) = \frac{100}{s+101} \qquad P_{2cl}(s) = \frac{100}{s+99},$$

which are very close. □

The paradoxes in the examples can be resolved by considering the frequency ranges that are relevant for closed-loop control. In Example 2.14 the closed-loop system bandwidth of relevance is about 100 rad/s. This corresponds to time

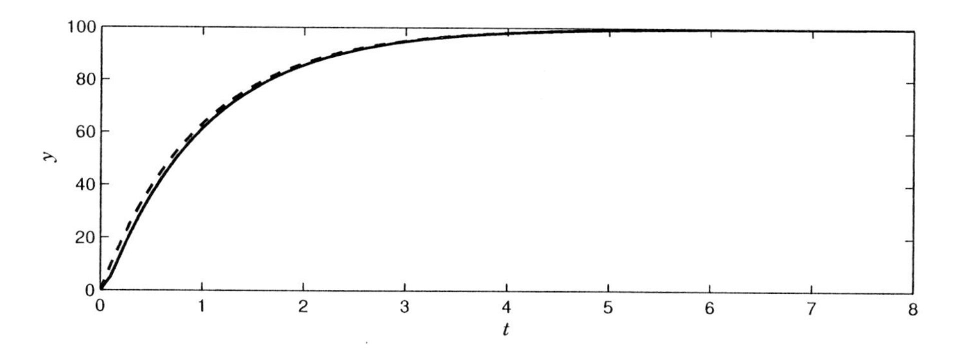

Figure 2.41 Step responses for systems with the transfer functions $G_1(s) = 100/(s+1)$ (dashed) and $G_2(s) = 100/((s+1)(1+0.025s)^2)$ (solid).

constants of about 0.01 s. A closer examination shows that the step responses in Figure 2.41 are indeed quite different at that time scale even if the general appearance of the step responses are very similar. In Example 2.15 the closed-loop bandwidth is also about 100 rad/s, which corresponds to a time scale of 0.01 s. At that time scale the open-loop systems are very similar even if one model is stable and the other unstable. It is a good rule to be aware of the relevant frequency ranges and to analyze the Bode plots. This is one of the main reasons for using frequency response.

2.9 Summary

Modeling is an important aspect of controller tuning. The models we need should describe how the process reacts to control signals. They should also describe the properties of the disturbances that enter the system. Most work on tuning of PID controllers has focused on the process dynamics, which is also reflected in the presentation in this chapter.

A number of methods for determining the dynamics of a process have been presented in this chapter. Some are very simple: they are based on a direct measurement of the step response and simple graphical constructions. Others are based on the frequency response. It has been shown that very useful information can be generated from relay feedback experiments. Such experiments are particularly useful because the process is brought into self-oscillation at the ultimate frequency, which is of considerable interest for design of controllers.

The simple methods are useful in field work when a controller has to be tuned and few tools are available. The methods are also useful to provide understanding as well as being references when more complicated methods are assessed. We have also presented more complicated methods that require significant computations.

Models of different complexity have been presented. Many models were characterized by a few parameters. Such models are useful for many purposes and are discussed in Chapter 6. When using such models it should be kept in mind that they are approximations.

When deriving the models we also introduced two dimension-free quantities, the normalized time delay τ and the gain ratio κ. These parameters make it possible to make a crude assessment of the difficulty of controlling the process. Processes with small values of κ and τ are easy to control. The difficulty increases as the values approach 1. Tuning rules based on τ and κ are provided in Chapter 7.

To summarize: When deriving a simple model to be used for PID controller tuning, it is important to ensure that the model describes the process well for the typical input signals obtained during the process operations. The amplitude and frequency distribution of the signal is of importance. Model accuracy may be poor if the process is nonlinear or time varying. Control quality can be improved by gain scheduling or adaptive control. It is also important to know what kind of disturbances are acting on the system and which limitation they impose.

2.10 Notes and References

Early efforts in modeling using differential equations were made independently by [Maxwell, 1868] and [Vyshnegradskii, 1876] in connection with analysis of engines with centrifugal governors. The idea of modeling a process by its reaction curve (step response) emerged in the 1930s. The reaction curve was approximated by an FOTD model (2.18) in [Callender *et al.*, 1936]. The reaction curve was also used in [Ziegler and Nichols, 1942]. Frequency response arguments were used in [Ivanoff, 1934] who investigated a temperature-control loop using the model given by (2.32). Frequency response was also used by [Ziegler and Nichols, 1942]. An early reference to the notion of block diagram is found in [Mason and Philbrick, 1940].

Process modeling is a key element in understanding and solving a control problem. Good presentations of modeling are found in standard textbooks on control, such as [Eckman, 1945; Buckley, 1964; Cannon, 1967; Smith, 1972; Luyben, 1990; Shearer and Kulakowski, 1990]. The books [Oquinnaike and Ray, 1994; Marlin, 2000; Bequette, 2003; Rawlings and Ekerdt, 2002; Seborg *et al.*, 2004] are of particular interest for process control. These books have much material on many different modeling techniques. Similar presentations are given in [Gille *et al.*, 1959; Harriott, 1964; Oppelt, 1964; Takahashi *et al.*, 1972; Deshpande and Ash, 1981; Shinskey, 1996; Stephanopoulos, 1984; Hägglund, 1991]. The books [Tucker and Wills, 1960] and [Lloyd and Anderson, 1971] are written by practitioners in control companies. There are also books that specialize in modeling for control system design; see [Wellstead, 1979; Nicholson, 1980; Nicholson, 1981; Close and Frederick, 1993].

By the mid-1950s frequency response was very well established as manifested by a symposium organized as part of the annual meeting of the American Society of Mechanical Engineering in 1953. The proceedings of the symposium were published in the book [Oldenburg, 1956]. A nice overview of step and frequency response methods is given in the paper [Rake, 1980]. Additional details are given in [Strejc, 1959; Anderssen and White, 1971; Anderssen and White, 1970]. The doublet method is discussed in [Shinskey, 1994], and the method of moments is described in [Gibilaro and Lees, 1969].

The relay method is introduced in [Åström and Hägglund, 1984b], and it is elaborated in [Hägglund and Åström, 1991; Schei, 1992; Hang and Åström, 2002]. The describing function method is well documented in [Atherton, 1975] and [Gelb and Velde, 1968]. A method to estimate what is today called an ARX model was developed in [Åström and Bohlin, 1965] and applied to modeling and control of paper machines in [Åström, 1967]. There are many books on parameter estimation: [Ljung and Söderström, 1983; Ljung, 1998; Söderström and Stoica, 1989; Bohlin, 1991; Johansson, 1993]. Many useful practical aspects on system identification are given in [Isermann, 1980].

Modeling has been greatly enhanced by simulation. The first simulation of a control system with PID control was made at the University of Manchester using a copy of the differential analyzer developed by Vannevar Bush; see [Callender *et al.*, 1936]. The differential analyser was also used in [Ziegler and Nichols, 1942] to develop tuning rules. Pneumatic simulators built from components of pneumatic controllers were used early by equipment manufacturers. The first electronic analog computer developed by Philbrick had a major impact, and the use of simulation increased drastically. The rapid development of digital computing has made it possible for every engineer to have simulation tools on his lap; see [Åström *et al.*, 1998]. Many of the simulation programs used today mimic the diagrams used to program early analog computers in the 1950s. There are major efforts underway to combine experiences of process modeling with advances in computing science to develop a new generation of languages and tools for process modeling; see [Elmqvist *et al.*, 1998] and [Tiller, 2001].

There are many methods for model reduction. Early work was reported in [Ziegler and Nichols, 1943]. A nice survey is found in [Glover, 1990]. One method that is geared to PID control is presented in [Fröhr and Orttenburger, 1982]. The half-rule was developed in [Skogestad, 2003] as a simple method that works well for the purpose of tuning PID controllers.

3

PID Control

3.1 Introduction

The PID controller is by far the most common control algorithm. Most feedback loops are controlled by this algorithm or minor variations of it. It is implemented in many different forms, as a stand-alone controller or as a part of a DDC (Direct Digital Control) package or a hierarchical distributed process control system. Many thousands of instrument and control engineers worldwide are using such controllers in their daily work. The PID algorithm can be approached from many different directions. It can be viewed as a device that can be operated with a few rules of thumb, but it can also be approached analytically.

This chapter gives an introduction to PID control. The basic algorithm and various representations are presented in detail. A description of the properties of the controller in a closed loop based on intuitive arguments is given. The phenomenon of reset windup, which occurs when a controller with integral action is connected to a process with a saturating actuator, is discussed, including several methods to avoid it. Filters to reduce noise influence and means to improve set-point responses are also provided.

Implementation aspects of the PID controller are presented in Chapter 13.

3.2 The PID Controller

The "textbook" version of the PID algorithm can be described as:

$$u(t) = K\left(e(t) + \frac{1}{T_i} \int_0^t e(\tau) d\tau + T_d \frac{de(t)}{dt} \right), \tag{3.1}$$

where u is the control signal and e is the control error ($e = y_{sp} - y$). The control signal is thus a sum of three terms: the P-term (which is proportional to the error), the I-term (which is proportional to the integral of the error), and the D-term (which is proportional to the derivative of the error). The controller

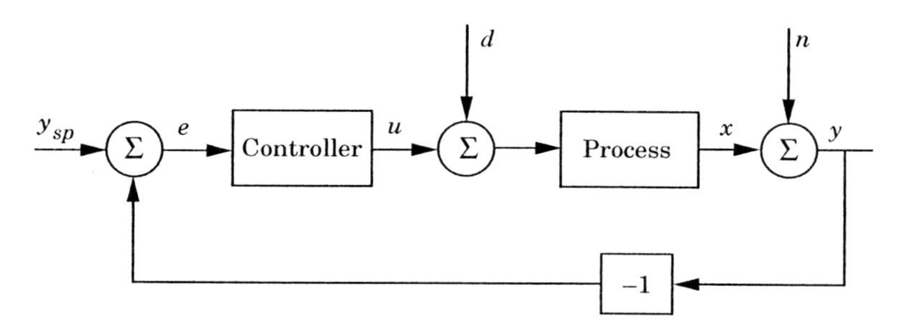

Figure 3.1 Block diagram of a simple feedback loop.

parameters are proportional gain K, integral time T_i, and derivative time T_d.

Proportional Action

In the case of pure proportional control, the control law of Equation 3.1 reduces to

$$u(t) = Ke(t) + u_b. \tag{3.2}$$

The control action is simply proportional to the control error. The variable u_b is a bias or a reset. When the control error e is zero, the control variable takes the value $u(t) = u_b$. Bias u_b is often fixed to $(u_{\max} + u_{\min})/2$, but can sometimes be adjusted manually so that the stationary control error is zero at a given set point.

Static Analysis Several properties of proportional control can be understood by the following argument, which is based on pure static considerations. Consider the simple feedback loop, shown in Figure 3.1, and composed of a process and a controller. Assume that the controller has proportional action and that the process is modeled by the static model

$$x = K_p(u + d), \tag{3.3}$$

where x is the process variable, u is the control variable, d is a load disturbance, and K_p is the static process gain. The following equations are obtained from the block diagram.

$$\begin{aligned} y &= x + n \\ x &= K_p(u + d) \\ u &= K(y_{sp} - y) + u_b. \end{aligned} \tag{3.4}$$

where n is measurement noise. Elimination of intermediate variables gives the following relation between process variable x, set point y_{sp}, load disturbance d, and measurement noise n:

$$x = \frac{KK_p}{1 + KK_p}(y_{sp} - n) + \frac{K_p}{1 + KK_p}(d + u_b). \tag{3.5}$$

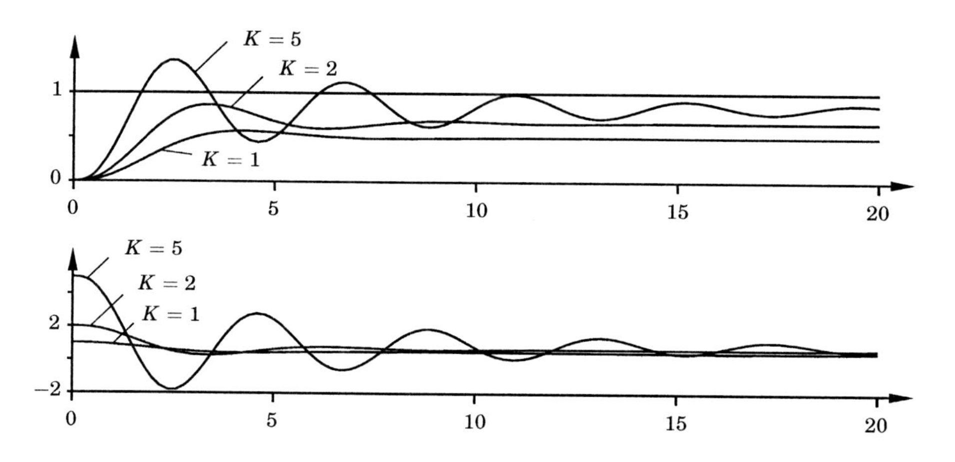

Figure 3.2 Simulation of a closed-loop system with proportional control. The process transfer function is $G(s) = (s+1)^{-3}$. The upper diagram shows set point $y_{sp} = 1$ and process output y for different values of controller gain K. The lower diagram shows control signal u for different controller gains.

Product KK_p is a dimensionless number called the *loop gain*. Several interesting properties of the closed-loop system can be read from Equation 3.5. First assume that n and u_b are zero. Then the loop gain should be high in order to ensure that process output x is close to set point y_{sp}. A high value of controller gain K also makes the system insensitive to load disturbance d. However, if n is nonzero, it follows from Equation 3.5 that measurement noise n influences the process output in the same way as set point y_{sp}. To avoid making the system sensitive to measurement noise, the loop gain should not be made too large. Further, the controller bias u_b influences the system in the same way as a load disturbance. It is, therefore, obvious that the design of the loop gain is a trade-off between different control objectives and that there is no simple answer to what loop gain is the best. This will depend on which control objective is the most important.

It also follows from Equation 3.5 that there will normally be a steady-state error with proportional control. This can be deduced intuitively from the observation following from Equation 3.4 that the control error is zero only when $u = u_b$ in stationarity. The error, therefore, can be made zero at a given operating condition by a proper choice of the controller bias u_b.

The static analysis given above is based on the assumption that the process can be described by a static model. This leaves out some important properties of the closed-loop system dynamics. The most important one is that the closed-loop system will normally be unstable for high-loop gains if the process dynamics are considered. In practice, the maximum loop gain is thus determined by the process dynamics. One way to describe process dynamics leads to descriptions like Equation 3.3 where the process gain is frequency-dependent. This was discussed in Chapter 2.

A typical example of proportional control is illustrated in Figure 3.2. The figure shows the behavior of the process output and the control signal after

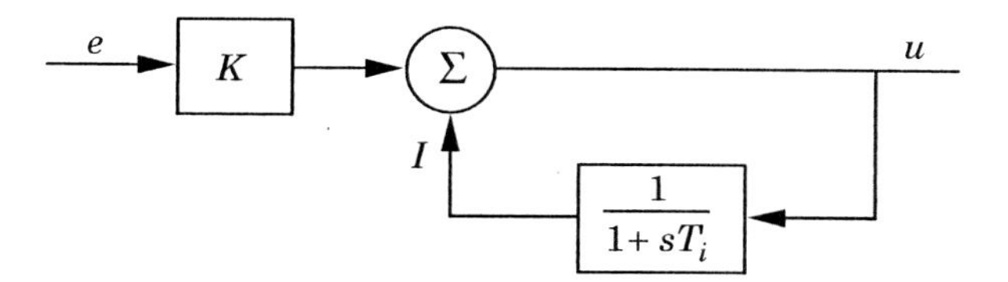

Figure 3.3 Implementation of integral action as positive feedback around a lag.

a step change in the set point. The steady-state error can be computed from Equation 3.5. The bias term u_b, the load d, and the noise n are all zero in the simulation. With a controller gain $K = 1$ and a static process gain $K_p = 1$, the error is therefore 50 percent. The figure shows that the steady-state error decreases with increasing controller gain as predicted by Equation 3.5. Notice also that the response becomes more oscillatory with increasing controller gain. This is due to the process dynamics.

Integral Action

The main function of the integral action is to make sure that the process output agrees with the set point in steady state. With proportional control, there is normally a control error in steady state. With integral action, a small positive error will always lead to an increasing control signal, and a negative error will give a decreasing control signal no matter how small the error is.

The following simple argument shows that the steady-state error will always be zero with integral action. Assume that the system is in steady state with a constant control signal (u_0) and a constant error (e_0). It follows from Equation 3.1 that the control signal is then given by

$$u_0 = K\left(e_0 + \frac{e_0}{T_i}t\right).$$

As long as $e_0 \neq 0$, this clearly contradicts the assumption that the control signal u_0 is constant. A controller with integral action will always give zero steady-state error.

Integral action can also be visualized as a device that automatically resets the bias term u_b of a proportional controller. This is illustrated in the block diagram in Figure 3.3, which shows a proportional controller with a reset that is adjusted automatically. The adjustment is made by feeding back a signal, which is a filtered value of the output, to the summing point of the controller. This was actually one of the early inventions of integral action, or "automatic reset," as it was also called. The implementation shown in Figure 3.3 is still used by many manufacturers.

Simple calculations show that the controller in Figure 3.3 gives the desired results. The following equation is obtained from the block diagram:

$$u = Ke + I = T_i\frac{dI}{dt} + I.$$

Hence,

$$T_i\frac{dI}{dt} = Ke,$$

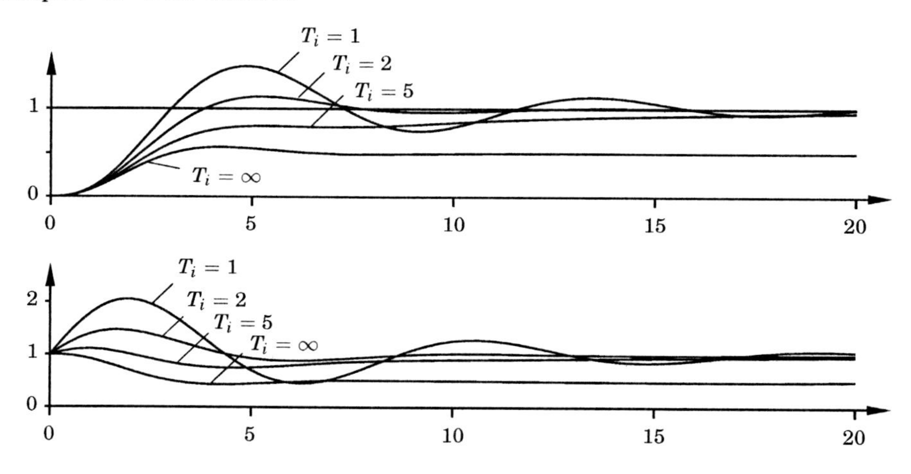

Figure 3.4 Simulation of a closed-loop system with proportional and integral control. The process transfer function is $G(s) = (s+1)^{-3}$, and the controller gain is $K = 1$. The upper diagram shows set point $y_{sp} = 1$ and process output y for different values of integral time T_i. The lower diagram shows control signal u for different integral times.

and we find that

$$u = K\left(e + \frac{1}{T_i}\int e(\tau)d\tau\right),$$

which is a PI controller.

The properties of integral action are illustrated in Figure 3.4, which shows a simulation of a system with PI control. The proportional gain is constant, $K = 1$, and the integral time is changed. The case $T_i = \infty$ corresponds to pure proportional control. This case is identical to the case $K = 1$ in Figure 3.2, where the steady-state error is 50 percent. The steady-state error is removed when T_i has finite values. For large values of the integration time, the response creeps slowly towards the set point. The approach is approximately exponential with time constant T_i/KK_p. The approach is faster for smaller values of T_i but is also more oscillatory.

Derivative Action

The purpose of the derivative action is to improve the closed-loop stability. The instability mechanism can be described intuitively as follows. Because of the process dynamics, it will take some time before a change in the control variable is noticeable in the process output. Thus, the control system will be late in correcting for an error. The action of a controller with proportional and derivative action may be interpreted as if the control is made proportional to the *predicted* process output, where the prediction is made by extrapolating the error by the tangent to the error curve (see Figure 3.5). The basic structure of a PD controller is

$$u(t) = K\left(e(t) + T_d\frac{de(t)}{dt}\right).$$

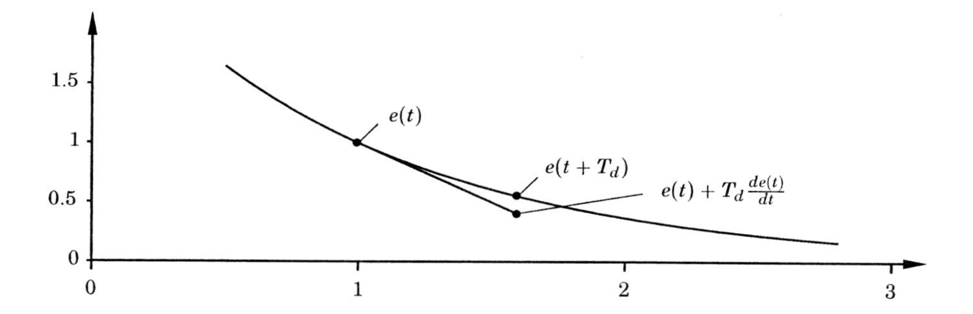

Figure 3.5 Interpretation of derivative action as predictive control, where the prediction is obtained by linear extrapolation.

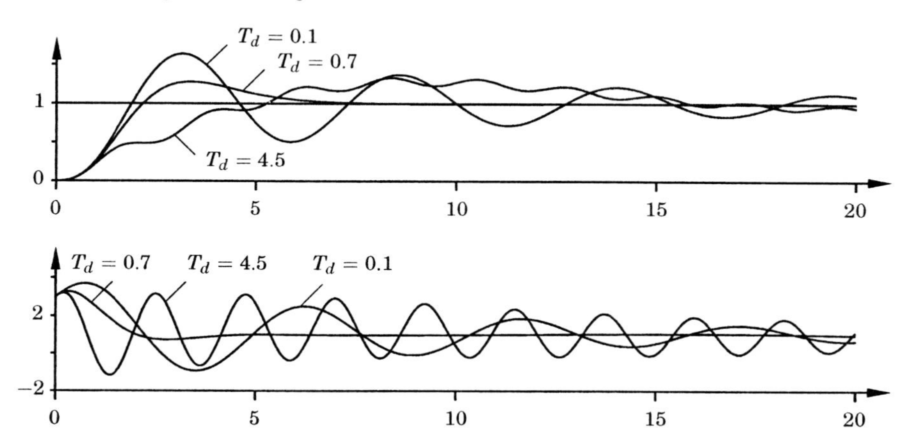

Figure 3.6 Simulation of a closed-loop system with proportional, integral, and derivative control. The process transfer function is $G(s) = (s+1)^{-3}$, the controller gain is $K = 3$, and the integral time is $T_i = 2$. The upper diagram shows set point $y_{sp} = 1$ and process output y for different values of derivative time T_d. The lower diagram shows control signal u for different derivative times.

A Taylor series expansion of $e(t + T_d)$ gives

$$e(t + T_d) \approx e(t) + T_d \frac{de(t)}{dt}.$$

The control signal is thus proportional to an estimate of the control error at time T_d ahead, where the estimate is obtained by linear extrapolation. The properties of derivative action are illustrated in Figure 3.6, which shows a simulation of a system with PID control.

Controller gain and integration time are kept constant, $K = 3$ and $T_i = 2$, and derivative time T_d is changed. For $T_d = 0$ we have pure PI control. The closed-loop system is oscillatory with the chosen parameters. Initially damping increases with increasing derivative time but decreases again when derivative time becomes too large.

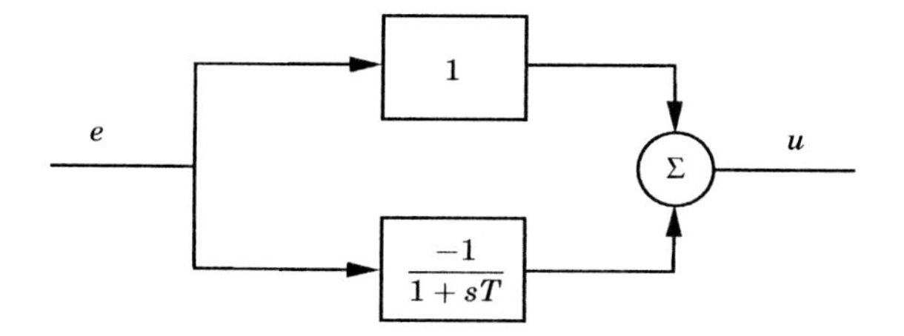

Figure 3.7 Classical implementation of derivative action.

Classical Implementation of Derivative Action

In Figure 3.3 it was shown that integral action originally was implemented by positive feedback around a first-order lag. Derivative action was also originally implemented using a first-order lag as is shown by the block diagram in Figure 3.7. The Laplace transform of the output is given by

$$U(s) = \left(1 - \frac{1}{1+sT}\right)E(s) = \frac{sT}{1+sT}E(s). \tag{3.6}$$

The system thus has the transfer function $G(s) = sT/(1+sT)$. Notice that filtering is obtained automatically with this implementation.

Alternative Representations

The PID algorithm given by Equation 3.1 can be represented by the transfer function

$$C(s) = K\left(1 + \frac{1}{sT_i} + sT_d\right). \tag{3.7}$$

A slightly different version is most common in commercial controllers. This controller is described by

$$C'(s) = K'\left(1 + \frac{1}{sT_i'}\right)(1 + sT_d'). \tag{3.8}$$

The two controller structures are presented in block diagram form in Figure 3.8. The controller given by Equation 3.7 is called non-interacting, and the one given by Equation 3.8 interacting. The reason for this nomenclature is that in the controller (3.7) the integral time T_i does not influence the derivative part, and the derivative time T_d does not influence the integral part. The parts are thus non-interacting. In the interacting controller, the derivative time T_d' does influence the integral part. Therefore, the parts are interacting.

The interacting controller (3.8) can always be represented as a non-interacting controller whose coefficients are given by

$$\begin{aligned} K &= K'\frac{T_i' + T_d'}{T_i'} \\ T_i &= T_i' + T_d' \\ T_d &= \frac{T_i'T_d'}{T_i' + T_d'}. \end{aligned} \tag{3.9}$$

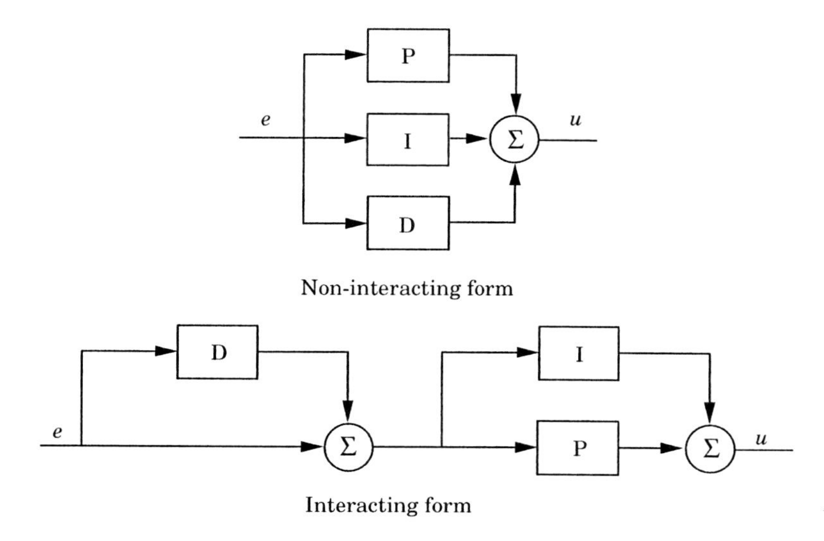

Figure 3.8 Interacting and non-interacting form of the PID algorithm.

An interacting controller of the form (3.8) that corresponds to a non-interacting controller can be found only if

$$T_i \geq 4T_d.$$

Then,

$$\begin{aligned} K' &= \frac{K}{2}\left(1+\sqrt{1-4T_d/T_i}\right) \\ T_i' &= \frac{T_i}{2}\left(1+\sqrt{1-4T_d/T_i}\right) \\ T_d' &= \frac{T_i}{2}\left(1-\sqrt{1-4T_d/T_i}\right). \end{aligned} \tag{3.10}$$

The non-interacting controller given by Equation 3.7 is more general, and we will use that in the future. It is, however, sometimes claimed that the interacting controller is easier to tune manually.

There is also a historical reason for preferring the interacting controller. Early pneumatic controllers were easier to build using the interacting form. See Chapter 13. When the controller manufacturers changed technology from pneumatic to analog electric and, finally, to digital technique, they kept the interactive form. Therefore, the interacting form is most common among single-loop controllers.

It is important to keep in mind that different controllers may have different structures. It means that if a controller in a certain control loop is replaced by another type of controller, the controller parameters may have to be changed. Note, however, that the interacting and the non-interacting forms are different only when both the I and the D parts of the controller are used. If we only use the controller as a P, PI, or PD controller, the two forms are equivalent. Yet another representation of the PID algorithm is given by

$$C''(s) = k + \frac{k_i}{s} + sk_d. \tag{3.11}$$

The parameters are related to the parameters of standard form through

$$k = K \qquad k_i = \frac{K}{T_i} \qquad k_d = KT_d. \tag{3.12}$$

The representation (3.11) is equivalent to the standard form, but the parameter values are quite different. This may cause great difficulties for anyone who is not aware of the differences, particularly if parameter $1/k_i$ is called integral time and k_d derivative time. The form given by Equation 3.11 is often useful in analytical calculations because the parameters appear linearly. The representation also has the advantage that it is possible to obtain pure proportional, integral, or derivative action by finite values of the parameters.

Summarizing, we have thus found that there are three different forms of the PID controller.

- The standard or non-interacting form given by Equation 3.7.
- The series or interacting form given by Equation 3.8.
- The parallel form given by Equation 3.11.

The standard form is sometimes called the ISA algorithm, or the ideal algorithm. The proportional, integral, and derivative actions are non-interacting in the time domain. This algorithm admits complex zeros, which is useful when controlling systems with oscillatory modes.

The series form is also called the classical form. This representation is obtained naturally when a controller is implemented as an analog device based on a pneumatic force balance system. The name classical reflects this. The series form has an attractive interpretation in the frequency domain because the zeros correspond to the inverse values of the derivative and integral times. All zeros of the controller are real. Pure integral or proportional action cannot be obtained with finite values of the controller parameters.

The parallel form is the most general form because pure proportional or integral action can be obtained with finite parameters. The controller can also have complex zeros. In this way it is the most flexible form. However, it is also the form where the parameters have little physical interpretation.

Summary

The PID controller has three terms. The proportional term P corresponds to proportional control. The integral term I gives a control action that is proportional to the time integral of the error. This ensures that the steady-state error becomes zero. The derivative term D is proportional to the time derivative of the control error. This term allows prediction of the future error. There are many variations of the basic PID algorithm that will substantially improve its performance and operability. They are discussed in the following sections.

3.3 Filtering the Derivative

A drawback with derivative action is that an ideal derivative has very high gain for high-frequency signals. This means that high-frequency measurement noise will generate large variations of the control signal. To see this we consider a measured output

$$y = \sin t + a \sin \omega t,$$

where the first term is the useful signal and the second term represents noise. The ratio of noise to signal is thus a. The derivative term of the controller is then

$$D = KT_d \frac{dy}{dt} = KT_d \big(\cos t + a\omega \cos \omega t\big). \tag{3.13}$$

The amplitude of the signal is KT_d, and the amplitude of the noise is $KT_d a\omega$. The ratio of noise and signal is $a\omega$. This can be arbitrarily large even if a is small if the frequency is sufficiently high. The effect of measurement noise can to some extent be eliminated by implementing the derivative term as

$$D = -\frac{sKT_d}{1 + sT_d/N} Y. \tag{3.14}$$

This can be interpreted as an ideal derivative that is filtered using a first-order system with the time constant T_d/N. For small s the transfer function is approximately sKT_d, and for large s it is equal to KN. The approximation acts as a derivative for low-frequency signal components, and the high-frequency gain is limited to KN. High-frequency measurement noise is amplified at most by a factor KN. Typical values of N are 2 to 20. Notice that the implementation of the derivative given in Figure 3.7 automatically gives a limitation of the high-frequency gain; see Equation 3.6.

The transfer function of a PID controller with a filtered derivative is

$$C(s) = K \left(1 + \frac{1}{sT_i} + \frac{sT_d}{1 + sT_d/N}\right). \tag{3.15}$$

The high-frequency gain of the controller is $K(1 + N)$. High frequency measurement noise can thus generate significant variations in the control signal. It is therefore advantageous to use heavier filtering.

Instead of filtering only the derivative it is possible to filter the measured signal and apply the filtered signal to an ideal PID controller. The equivalent controller transfer function is

$$C_{eq} = C(s)G_f(s) = K \left(1 + \frac{1}{sT_i} + sT_d\right) \frac{1}{1 + sT_f + (sT_f)^2/2}, \tag{3.16}$$

when a second-order filter with relative damping $\zeta = 1/\sqrt{2}$ is used. The filter-time constant T_f is typically chosen as T_i/N for PI control or as T_d/N for PID control, where N ranges from 2 to 20.

It follows from (3.16) that the controller gain goes to zero for high frequencies. This property, which is called *high frequency roll-off*, guarantees that high-frequency measurement noise will not generate large control signals. High-frequency roll-off also increases the robustness of the closed loop system.

3.4 Set-Point Weighting

The control system in Figure 3.1 is called a system with *error feedback* because the controller acts on the error, which is the difference between the set point and the output. A more flexible structure is obtained by treating the set point and the process output separately. A PID controller of this form is given by

$$u(t) = K\left(e_p + \frac{1}{T_i}\int_0^t e(s)ds + T_d\frac{de_d}{dt}\right), \tag{3.17}$$

where the error in the proportional part is

$$e_p = by_{sp} - y, \tag{3.18}$$

and the error in the derivative part is

$$e_d = cy_{sp} - y. \tag{3.19}$$

The error in the integral part must be the true control error

$$e = y_{sp} - y,$$

in order to avoid steady-state control errors. The controllers obtained for different values of b and c will respond to load disturbances and measurement noise in the same way. The response to set-point changes will depend on the values of b and c, which are called *set-point weights*.

The properties of a system where the controller has set-point weighting is illustrated in Figure 3.4, which shows the response of a PID controller to set-point changes, load disturbances, and measurement errors for different values of b. The figure shows clearly the effect of changing b. The overshoot for set-point changes is smallest for $b = 0$, which is the case where the reference is only introduced in the integral term, and increases with increasing b. Notice that a simulation like the one in Figure 3.4 is useful in order to give a quick assessment of the responses of a closed-loop system to set-point changes, load disturbances, and measurement errors.

The parameter c is normally chosen equal to zero to avoid large transients in the control signal due to sudden changes in the set point. An exception is when the controller is the secondary controller in a cascade coupling (see Section 12.4). In this case, the set-point changes smoothly because it is given by the primary controller output. Notice that if the integral action is implemented with positive feedback around a lag as in Figure 3.3, the parameter b is equal to one.

The controller with $b = 0$ and $c = 0$ is sometimes called an I-PD controller, and the controller with $b = 1$ and $c = 0$ is sometimes called a PI-D controller. We prefer to stick to the generic use of PID and give the parameters b and c, thereby making a small contribution towards reduction of three-letter abbreviations.

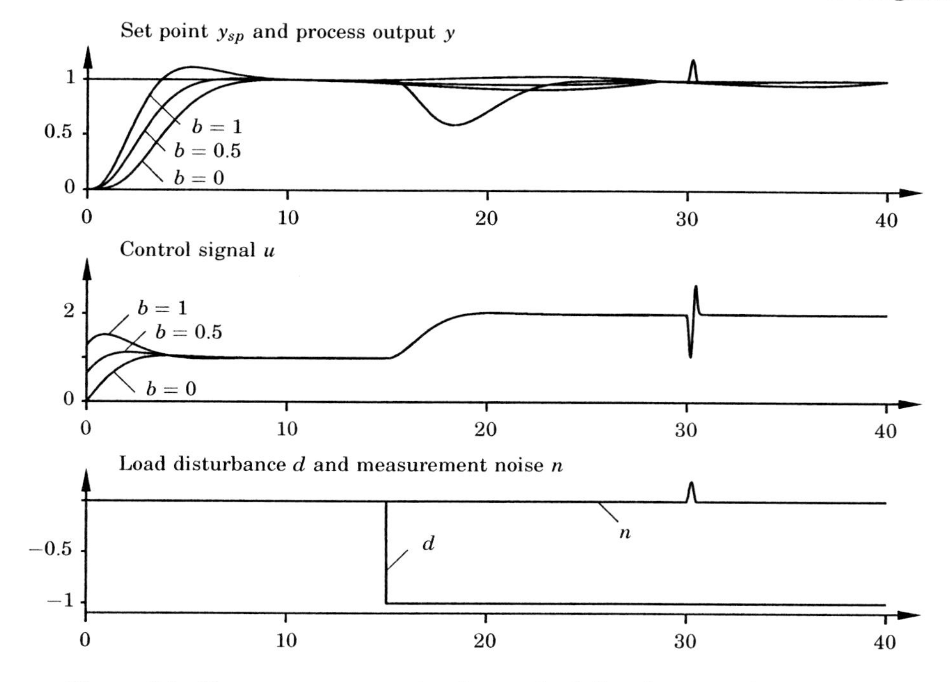

Figure 3.9 The response to set-point changes, load disturbances, and measurement errors for different values of set-point weighting b.

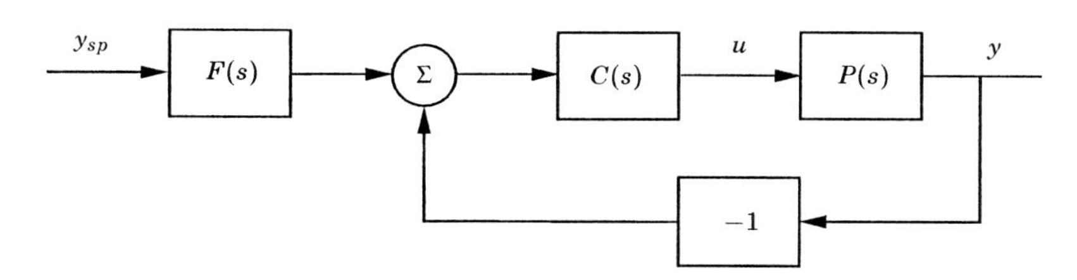

Figure 3.10 Block diagram of a simple feedback loop with a PID controller having a two-degree-of-freedom structure. The transfer function $P(s)$ is the process transfer function.

In the block diagram in Figure 3.1, the controller output is generated from the error $e = y_{sp} - y$. Notice that this diagram is no longer valid when the control law given by Equation 3.17 and the error definitions (3.18) and (3.19) are used. A block diagram for a system with PID control is instead given by Figure 3.10 where the transfer function $C(s)$ is given by (3.7) and

$$F(s) = \frac{b + \frac{1}{sT_i} + scT_d}{1 + \frac{1}{sT_i} + sT_d} = \frac{cT_iT_ds^2 + bsT_i + 1}{T_iT_ds^2 + sT_i + 1}. \tag{3.20}$$

System with Two Degrees of Freedom

In general, a control system has many different requirements. It should have good transient response to set-point changes, and it should reject load disturbances and measurement noise. For a system with error feedback only, an attempt is made to satisfy all demands with the same mechanism. Such systems are called one-degree-of-freedom systems.

The system shown in Figure 3.10 is said to have two degrees of freedom because the signal path from the set point to the control signal is different from the signal path from measured value to control signal.

There are many possible configurations of systems with two degrees of freedom. The system shown in Figure 3.10 is only one alternative. An extended use of structures with two degrees of freedom is a very natural extension of the PID controller. The key idea is to let the controller C be a PI or a PID controller but to use more flexible feedforward than the standard PID controller permits. This will be discussed more fully in Chapter 5.

3.5 Integrator Windup

Although many aspects of a control system can be understood based on linear theory, some nonlinear effects must be accounted for. All actuators have limitations: a motor has limited speed, a valve cannot be more than fully opened or fully closed, etc. For a control system with a wide range of operating conditions, it may happen that the control variable reaches the actuator limits. When this happens the feedback loop is broken and the system runs as an open loop because the actuator will remain at its limit independently of the process output. If a controller with integrating action is used, the error may continue to be integrated if the algorithm is not properly designed. This means that the integral term may become very large or, colloquially, it "winds up." It is then required that the error has opposite sign for a long period before things return to normal. The consequence is that any controller with integral action may give large transients when the actuator saturates.

EXAMPLE 3.1—ILLUSTRATION OF INTEGRATOR WINDUP
Figure 3.11 shows the control signal, the measurement signal, and the set point in a case where the control signal becomes saturated. After the first set-point change, the control signal increases to its upper limit $u_{\max}$. This control signal is not large enough to eliminate the control error. Therefore, the integral of the control error, and the integral part of the control signal, increases. Since the desired control signal u increases, there is a difference between the desired control signal and the true control signal u_{out}.

Figure 3.11 shows what happens when after a certain time the set point is lowered to a level where the controller is able to eliminate the control error. Since the sign of the control error becomes negative, the control signal starts to decrease, but since the desired control signal u is above the limit $u_{\max}$, the true control signal u_{out} is stuck at the limit for a while and the response becomes delayed. □

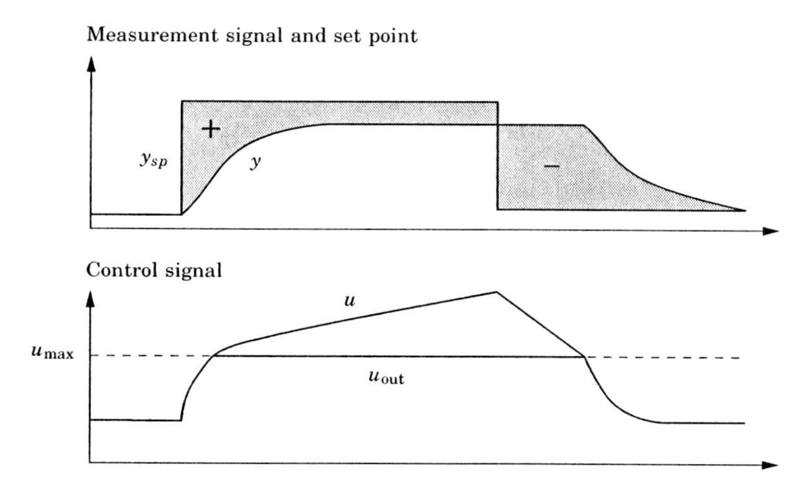

Figure 3.11 Illustration of integrator windup.

The following example shows some other effects that may occur due to integrator windup when the process is unstable.

EXAMPLE 3.2—ILLUSTRATION OF INTEGRATOR WINDUP
The windup phenomenon is illustrated in Figure 3.12, which shows control of an integrating process with a PI controller. The initial set-point change is so large that the actuator saturates at the high limit. The integral term increases initially because the error is positive; it reaches its largest value at time $t = 10$ when the error goes through zero. The output remains saturated at this point because of the large value of the integral term. It does not leave the saturation limit until the error has been negative for a sufficiently long time to let the integral part come down to a small level. Notice that the control signal bounces between its limits several times. The net effect is a large overshoot and a damped oscillation where the control signal flips from one extreme to the other. The output finally comes so close to the set point that the actuator does not saturate. The system then behaves linearly and settles. □

Integrator windup may occur in connection with large set-point changes, or it may be caused by large disturbances or equipment malfunctions. Windup can also occur when selectors are used so that several controllers are driving one actuator. In cascade control, windup may occur in the primary controller when the secondary controller is switched to manual mode, uses its local set point, or if its control signal saturates. See Section 12.4.

The phenomenon of windup was well known to manufacturers of analog controllers, who invented several tricks to avoid it. They were described under labels like preloading, batch unit, etc. Although the problem was well understood, there were often limits imposed because of the analog implementations. The ideas were often kept as trade secrets and not much spoken about. The problem of windup was rediscovered when controllers were implemented digitally and several methods to avoid windup were presented in the literature. In the following section we describe several of the ideas.

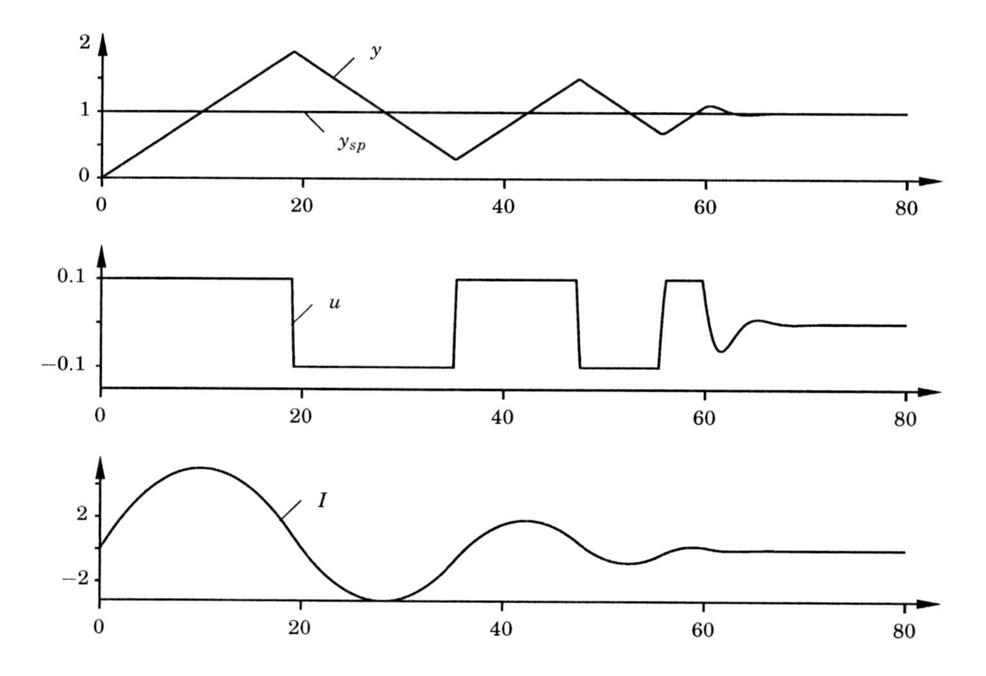

Figure 3.12 Illustration of integrator windup. The diagrams show process output y, set point y_{sp}, control signal u, and integral part I.

Set-Point Limitation

One way to try to avoid integrator windup is to introduce limiters on the set-point variations so that the controller output will never reach the actuator bounds. This often leads to conservative bounds and limitations on controller performance. Further, it does not avoid windup caused by disturbances.

Incremental Algorithms

In the early phases of feedback control, integral action was integrated with the actuator by having a motor drive the valve directly. In this case, windup is handled automatically because integration stops when the valve stops. When controllers were implemented by analog techniques, and later with computers, many manufacturers used a configuration that was an analog of the old mechanical design. This led to the so-called velocity algorithms, discussed in Chapter 13. In this algorithm the rate of change of the control signal is first computed and then fed to an integrator. In some cases this integrator is a motor directly connected to the actuator. In other cases the integrator is implemented internally in the controller. With this approach it is easy to handle mode changes and windup. Windup is avoided by inhibiting the integration whenever the output saturates. This method is equivalent to back-calculation, which is described below. If the actuator output is not measured, a model that computes the saturated output can be used. It is also easy to limit the rate of change of the control signal.

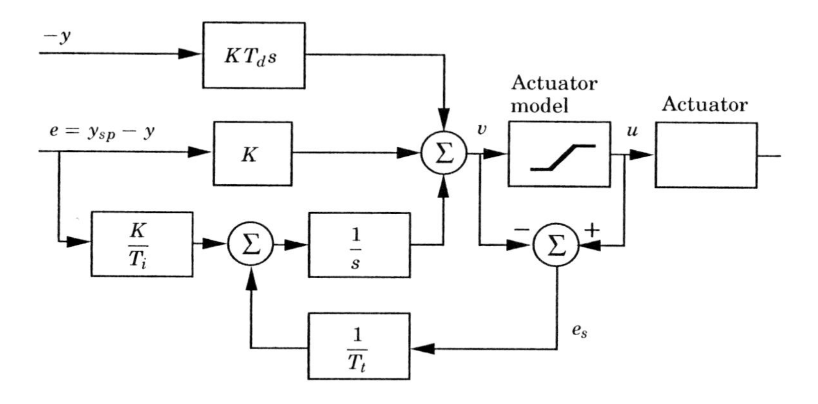

Figure 3.13 PID controller with anti-windup.

Back-Calculation and Tracking

Back-calculation works as follows. When the output saturates, the integral term in the controller is recomputed so that its new value gives an output at the saturation limit. It is advantageous not to reset the integrator instantaneously but dynamically with a time constant T_t.

Figure 3.13 shows a block diagram of a PID controller with anti-windup based on back-calculation. The system has an extra feedback path that is generated by measuring the actual actuator output, or the output of a mathematical model of the saturating actuator, and forming an error signal (e_s) as the difference between the output of the controller (v) and the actuator output (u). The signal e_s is fed to the input of the integrator through gain $1/T_t$. The signal is zero when there is no saturation. Thus, it will not have any effect on the normal operation when the actuator does not saturate. When the actuator saturates, the signal e_s is different from zero. The normal feedback path around the process is broken because the process input remains constant. There is, however, a feedback path around the integrator. Because of this, the integrator output is driven towards a value such that the integrator input becomes zero. The integrator input is

$$\frac{1}{T_t} e_s + \frac{K}{T_i} e,$$

where e is the control error. Hence,

$$e_s = -\frac{KT_t}{T_i} e$$

in steady state. Since $e_s = u - v$, it follows that

$$v = u_{\text{lim}} + \frac{KT_t}{T_i} e,$$

where u_{lim} is the saturating value of the control variable. Since the signals e and u_{lim} have the same sign, it follows that v is always larger than u_{lim} in

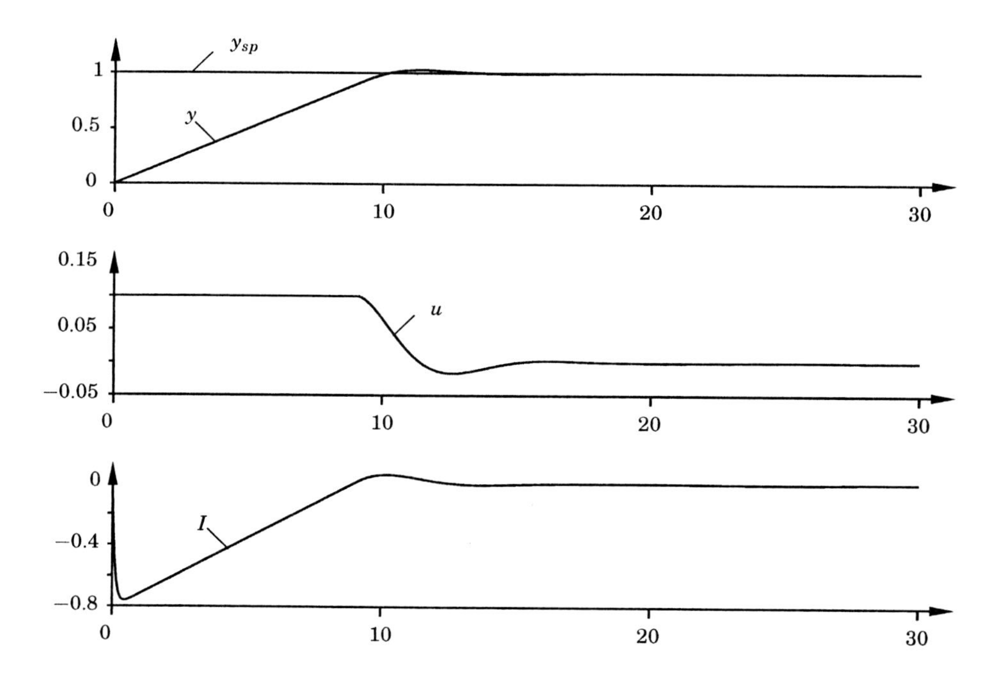

Figure 3.14 Controller with anti-windup applied to the system of Figure 3.12. The diagrams show process output y, set point y_{sp}, control signal u, and integral part I.

magnitude. This prevents the integrator from winding up. The rate at which the controller output is reset is governed by the feedback gain, $1/T_t$, where T_t can be interpreted as the time constant, which determines how quickly the integral is reset. We call this the tracking time constant.

Figure 3.14 shows what happens when a controller with anti-windup is applied to the system simulated in Figure 3.12. Notice that the output of the integrator is quickly reset to a value such that the controller output is at the saturation limit and the integral has a negative value during the initial phase when the actuator is saturated. This behavior is drastically different from that in Figure 3.12, where the integral has a positive value during the initial transient. Also notice the drastic improvement in performance compared to the ordinary PI controller used in Figure 3.12.

The effect of changing the values of the tracking time constant is illustrated in Figure 3.15. From this figure, it may thus seem advantageous to always choose a very small value of the time constant because the integrator is then reset quickly. However, some care must be exercised when introducing anti-windup in systems with derivative action. If the time constant chosen is too small, spurious errors can cause saturation of the output, which accidentally resets the integrator. The tracking time constant T_t should be larger than T_d and smaller than T_i. A rule of thumb that has been suggested is to choose $T_t = \sqrt{T_i T_d}$.

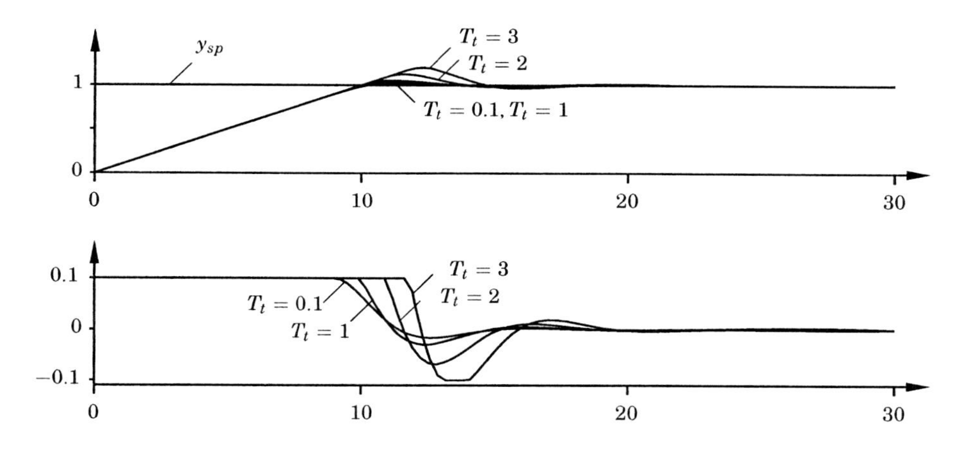

Figure 3.15 The step response of the system in Figure 3.12 for different values of the tracking time constant T_t. The upper curve shows process output y and set point y_{sp}, and the lower curve shows control signal u.

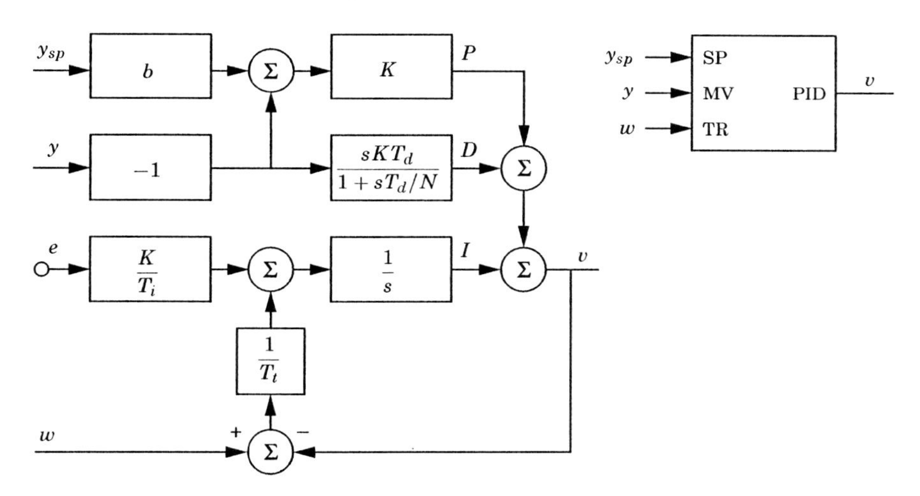

Figure 3.16 Block diagram and simplified representation of PID module with tracking signal.

Controllers with a Tracking Mode

A controller with back-calculation can be interpreted as having two modes: the normal *control mode*, when it operates like an ordinary controller, and a *tracking mode*, when the controller is tracking so that it matches given inputs and outputs. Since a controller with tracking can operate in two modes, we may expect that it is necessary to have a logical signal for mode switching. However, this is not necessary, because tracking is automatically inhibited when the tracking signal w is equal to the controller output. This can be used with great advantage when building up complex systems with selectors and cascade control.

Figure 3.16 shows a PID module with a tracking signal. The module has

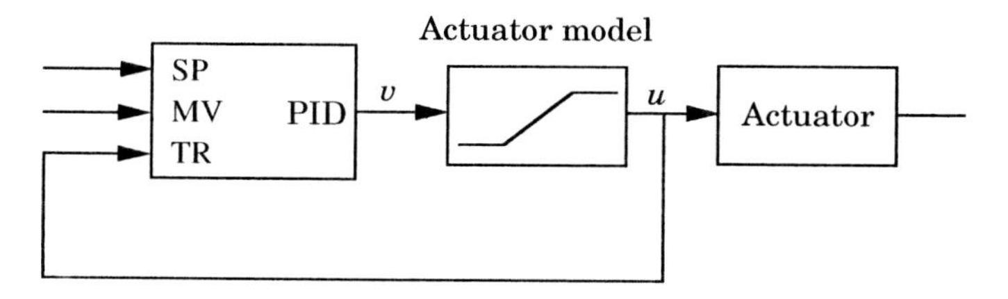

Figure 3.17 Representation of the controller with anti-windup in Figure 3.13 using the basic control module with tracking shown in Figure 3.16.

three inputs: the set point, the measured output, and a tracking signal. The new input TR is called a tracking signal because the controller output will follow this signal. Notice that tracking is inhibited when $w = v$. Using the module the system shown in Figure 3.13 can be presented as shown in Figure 3.17.

The Proportional Band

The notion of proportional band is useful in order to understand the windup effect and to explain schemes for anti-windup. The *proportional band* is an interval such that the actuator does not saturate if the instantaneous value of the process output or its predicted value is in the interval. For PID control without derivative gain limitation, the control signal is given by

$$u = K(by_{sp} - y) + I - KT_d \frac{dy}{dt}. \tag{3.21}$$

Solving for the predicted process output

$$y_p = y + T_d \frac{dy}{dt},$$

gives the proportional band (y_l, y_h) as

$$\begin{aligned} y_l &= by_{sp} + \frac{I - u_{\max}}{K} \\ y_h &= by_{sp} + \frac{I - u_{\min}}{K}. \end{aligned} \tag{3.22}$$

and $u_{\min}$, $u_{\max}$ are the values of the control signal for which the actuator saturates. The controller operates in the linear mode, if the predicted output is in the proportional band. The control signal saturates when the predicted output is outside the proportional band. Notice that the proportional band can be shifted by changing the integral term.

To illustrate that the proportional band is useful in understanding windup, we show the proportional band in Figure 3.18 for the system discussed in Example 3.2. (Compare with Figure 3.12.) The figure shows that the proportional band starts to move upwards because the integral term increases. This implies that the output does not reach the proportional band until it is much larger than the set point. When the proportional band is reached the control signal

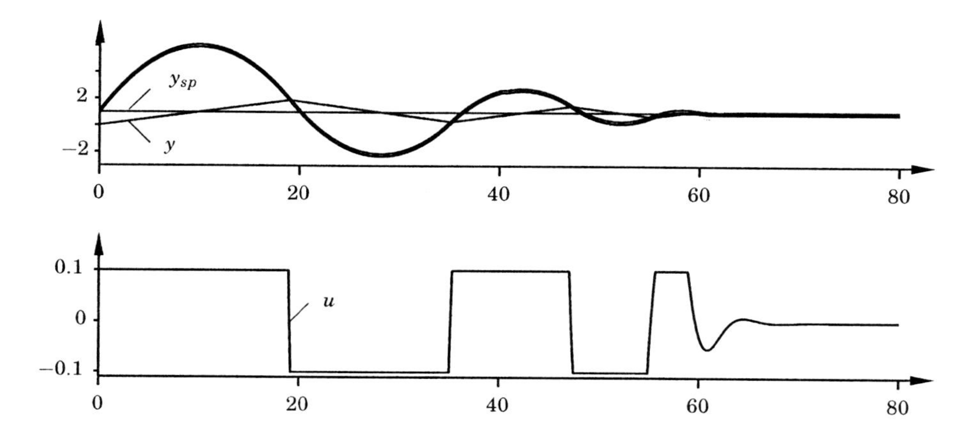

Figure 3.18 The proportional band for the system in Example 3.2. The upper diagram shows process output y and the proportional band. The lower diagram shows control signal u.

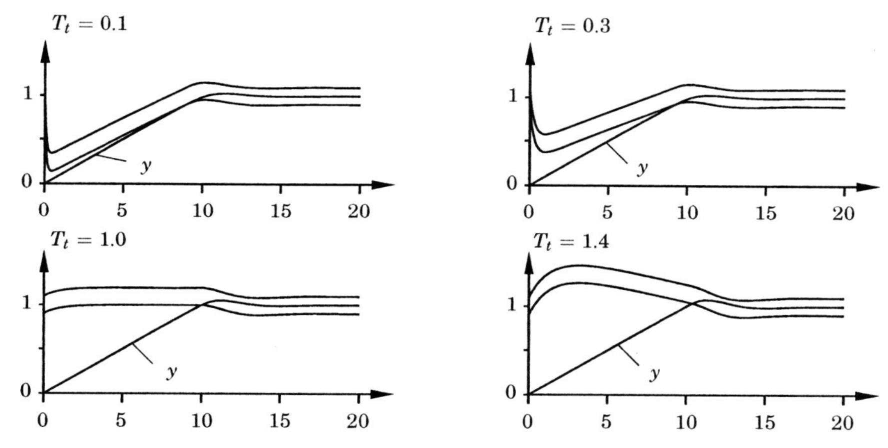

Figure 3.19 The proportional band and the process output y for a system with conditional integration and tracking with different tracking time constants T_t.

decreases rapidly. The proportional band changes so rapidly, however, that the output very quickly moves through the band, and this process repeats several times.

The notion of proportional band helps us to understand several schemes for anti-windup. Figure 3.19 shows the proportional band for the system with tracking for different values of the tracking time constant T_t. The figure shows that the tracking time constant has a significant influence on the proportional band. Because of the tracking, the proportional band is moved closer to the process output. How rapidly it does this is governed by the tracking time constant T_t. Notice that there may be a disadvantage in moving it too rapidly, since the predicted output may then move into the proportional band because of noise and cause the control signal to decrease unnecessarily.

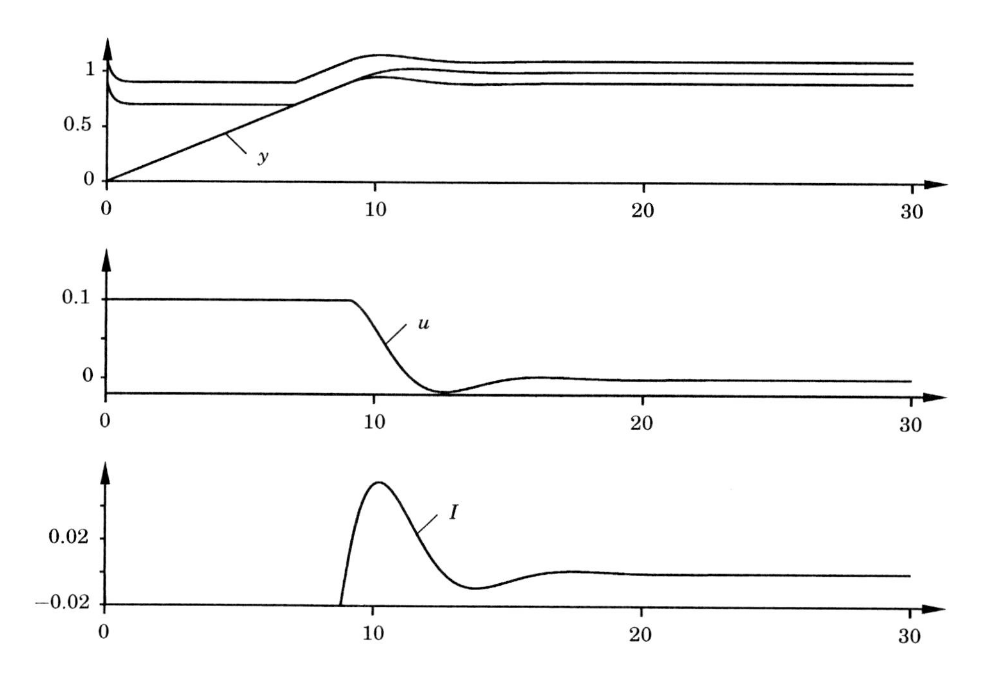

Figure 3.20 Simulation of the system in Example 3.2 with conditional integration. The diagrams show the proportional band, process output y, control signal u, and integral part I.

Conditional Integration

Conditional integration is an alternative to back-calculation or tracking. In this method integration is switched off when the control is far from steady state. Integral action is thus only used when certain conditions are fulfilled; otherwise the integral term is kept constant. The method is also called integrator clamping.

The conditions when integration is inhibited can be expressed in many different ways. Figure 3.20 shows a simulation of the system in Example 3.2 with conditional integration such that the integral term is kept constant during saturation. A comparison with Figure 3.19 shows that, in this particular case, there is very little difference in performance between conditional integration and tracking. The different wind-up schemes do, however, move the proportional bands differently.

A few different switching conditions are now considered. One simple approach is to switch off integration when the control error is large. Another approach is to switch off integration during saturation. Both these methods have the disadvantage that the controller may get stuck at a non-zero control error if the integral term has a large value at the time of switch-off.

A method without this disadvantage is the following. Integration is switched off when the controller is saturated *and* the integrator update is such that it causes the control signal to become more saturated. Suppose, for example, that the controller becomes saturated at the upper saturation. Integration is then

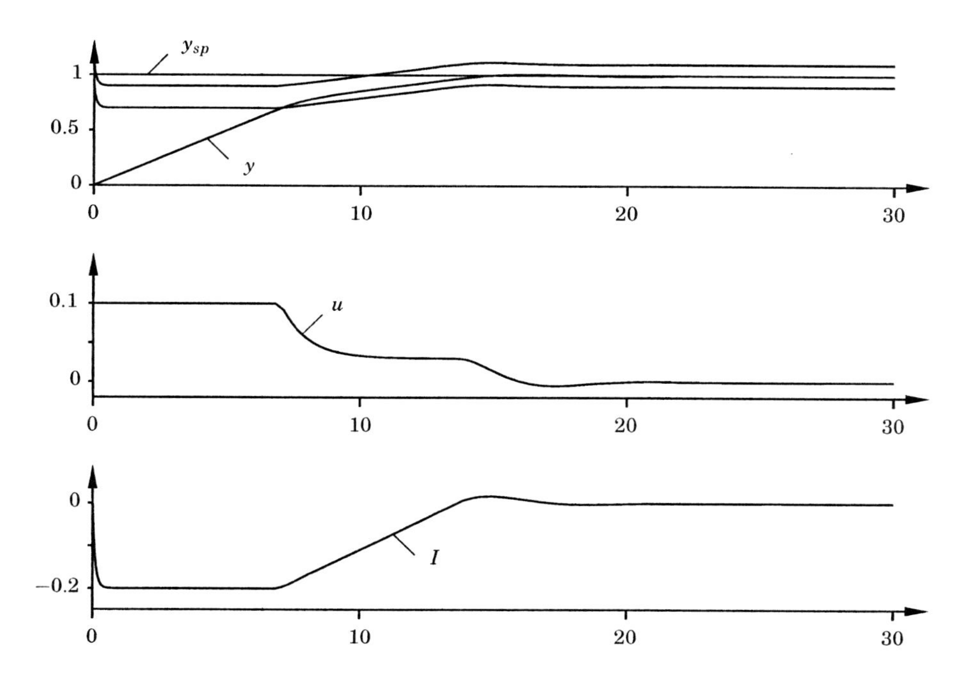

Figure 3.21 Adjustment of the proportional band using cut-back parameters. The diagrams show the proportional band, set point y_{sp}, process output y, control signal u, and integral part I.

switched off if the control error is positive, but not if it is negative.

Some conditional integration methods are intended mainly for startup of batch processes, when there may be large changes in the set point. One particular version, used in temperature control, sets the proportional band outside the set point when there are large control deviations. The offset can be used to adjust the transient response obtained during startup of the process. The parameters used are called cutback or preload (see Figure 3.21). In this system the proportional band is positioned with one end at the set point and the other end towards the measured value when there are large variations. These methods may give windup during disturbances.

Series Implementation

In Figure 3.3, we showed a special implementation of a controller in interacting form. To avoid windup in this controller we can incorporate a model of the saturation in the system as shown in Figure 3.22a. Notice that in this implementation the tracking time constant T_t is the same as the integration time T_i. This value of the tracking time constant is often too large.

In Figure 3.22a, the model of the saturation will limit the control signal directly. It is important, therefore, to have a good model of the physical saturation. Too hard a limitation will cause an unnecessary limitation of the control action. Too weak a limitation will cause windup.

More flexibility is provided if the saturation is positioned as in Figure 3.22b.

a)

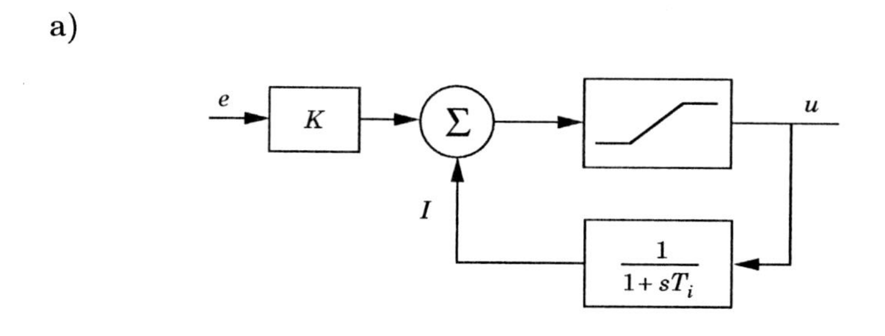

b)

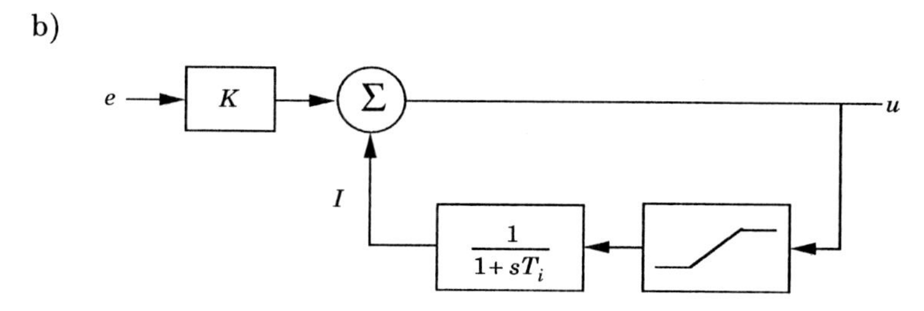

Figure 3.22 Two ways to provide anti-windup in the controller in Figure 3.3 where integral action is generated as automatic reset.

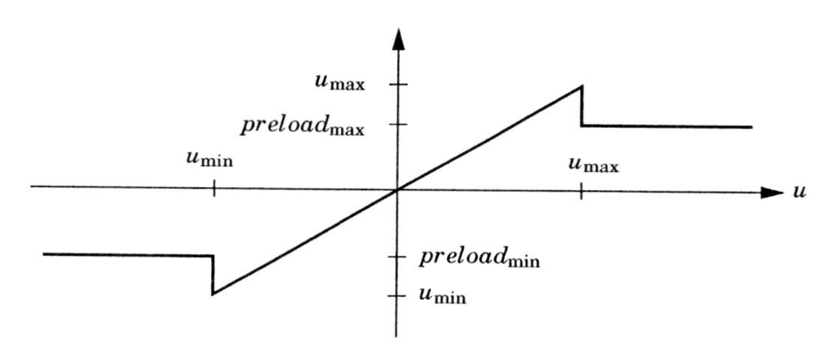

Figure 3.23 A "batch unit" used to provide anti-windup in the controller in Figure 3.3.

In this case, the saturation will not influence the proportional part of the controller. With this structure it is also possible to force the integral part to assume other preload values during saturation. This is achieved by replacing the saturation function by the nonlinearity shown in Figure 3.23. This anti-windup procedure is sometimes called a "batch unit" and may be regarded as a type of conditional integration. It is mainly used for adjusting the *overshoot* during startup when there is a large set point change. In early single-loop controllers the batch unit was supplied as a special add-on hardware.

Combined Schemes

Tracking and conditional integration can also be combined. In [Howes, 1986] it is suggested to manipulate the proportional band explicitly for batch control. This is done by introducing so-called *cutback points*. The high cutback is above the set point, and the low cutback is below. The integrator is clamped when the predicted process output is outside the cutback interval. Integration is performed with a specified tracking time constant when the process output is between the cutback points. The cutback points are considered as controller parameters that are adjusted to influence the response to large set-point changes. A similar method is proposed in [Dreinhofer, 1988], where conditional integration is combined with back-calculation. In [Shinskey, 1988], the integrator is given a prescribed value $i = i_0$ during saturation. The value of i_0 is tuned to give satisfactory overshoot at startup. This approach is also called preloading.

3.6 When Can PID Control Be Used?

There are many requirements on a controlled system. It should respond well to set-point changes, it should attenuate load disturbances, measurement noise should not give excessive control actions, and the system should be insensitive to process variations. Design of a control system also involves aspects of process dynamics, actuator saturation, and disturbance characteristics. It may seem surprising that a controller as simple as the PID controller can work so well. The general empirical observation is that most industrial processes can be controlled reasonably well with PID control provided that the demands on the performance of the control are not too high. In the following paragraphs we delve further into this issue by first considering cases where PID control is sufficient and then discussing some generic problems where more sophisticated control is advisable.

When Is PI Control Sufficient?

All stable processes can be controlled by an I controller if the performance requirements are modest. Proportional action gives additional performance enhancements. It is therefore not surprising that the PI controller is the most common controller. Disregarding saturations a process with first-order dynamics can be given any desired performance using a PI controller. PI control can also be used for processes with integral action.

Derivative action is frequently not used. It is an interesting observation that many industrial controllers only have PI action and that in others the derivative action can be (and frequently is) switched off. It can be shown that PI control is adequate for all processes where the dynamics are essentially of the first order (level controls in single tanks, stirred tank reactors with perfect mixing, etc). It is fairly easy to find out if this is the case by measuring the step response or the frequency response of the process. If the step response looks like that of a first-order system or, more precisely, if the Nyquist curve lies in the first and the fourth quadrants only, then PI control is sufficient. Another reason is that the process has been designed so that its operation does not require tight control. Then, even if the process has higher-order dynamics, what it needs is an integral action to provide zero steady-state offset and an adequate transient response by proportional action.

When Is Derivative Action Useful?

A double integrator cannot be controlled by a PI controller. The reason is that the process has a phase lag of 180° and that a PI controller also has a phase lag. Derivative action is needed for such a process. Disregarding saturations a process with second-order dynamics can be given any desired performance using a PID controller.

Similarly, PID control is sufficient for processes where the dominant dynamics are of the second order. For such processes there are no benefits gained by using a more complex controller. A typical case of derivative action improving the response is when the dynamics are characterized by time constants that differ in magnitude. Derivative action can then profitably be used to speed up the response. Temperature control is a typical case. Derivative control is also beneficial when tight control of a higher-order system is required. The higher-order dynamics would limit the amount of proportional gain for good control. With a derivative action, improved damping is provided, hence, a higher proportional gain can be used to speed up the transient response.

Many processes encountered in process control have dynamics with essentially monotone step responses, often with time delay. If the dynamics is delay dominated derivative action gives modest performance improvements compared with PI control, but derivative action gives significant improvements for processes that are lag dominated. This is discussed further in Chapter 7.

When Is More Sophisticated Control Needed?

The benefits of using a more sophisticated controller than the PID is demonstrated by some examples below.

Higher-Order Processes When the system is of an order higher than two, the control can be improved by using a more complex controller than the PID controller. This is illustrated by the following example.

EXAMPLE 3.3—SYSTEMS OF HIGH ORDER
Consider a third-order process described by the following transfer function

$$P(s) = \frac{1}{(s+1)^3}.$$

Figure 3.24 shows the control obtained using a PID controller and a more complex controller of higher order.

The PID controller has the parameters $K = 3.4$, $T_i = 2.0$, and $T_d = 0.6$. The PID controller is compared with a controller of the form

$$R(s)u(t) = -S(s)y(t) + T(s)y_{sp}(t),$$

with the following controller polynomials

$$\begin{aligned} R(s) &= s(s^2 + 11.5s + 57.5) \\ S(s) &= 144s^3 + 575s^2 + 870s + 512 \\ T(s) &= 8s^3 + 77s^2 + 309s + 512. \end{aligned}$$

The benefits of using a more complex controller in the case of higher-order dynamics is clearly demonstrated in the figure. □

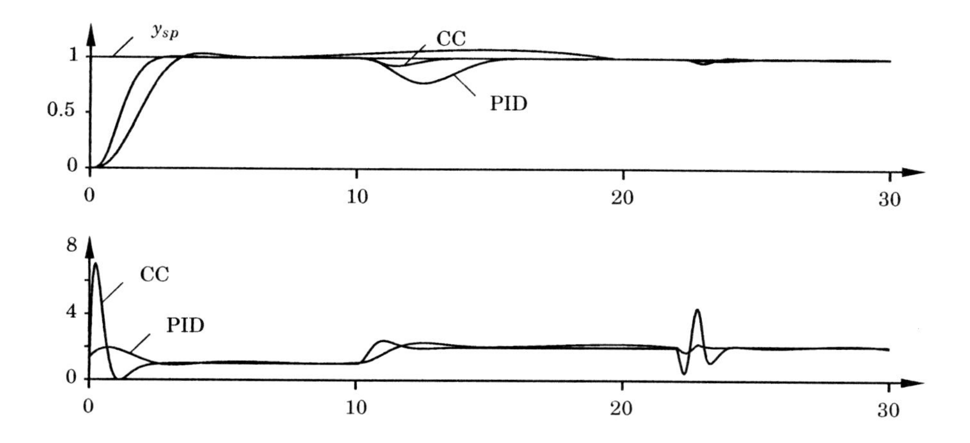

Figure 3.24 Control of the third-order system in Example 3.3 using a PID controller (PID) and a more complex controller (CC). The figure shows responses to a set-point change, a load disturbance, and finally a measurement disturbance. The upper diagram shows set point y_{sp} and measurement signal y, and the lower diagram shows control signal u.

Systems with Long Time Delay Control of systems with a dominant time delay are notoriously difficult. It is also a topic on which there are many different opinions concerning the merit of PID control. There seems to be general agreement that derivative action does not help much for processes with dominant time delays. For open-loop stable processes, the response to command signals can be improved substantially by introducing dead-time compensation. The load disturbance rejection can also be improved to some degree because a dead-time compensator allows a higher loop gain than a PID controller. Systems with dominant time delays are thus candidates for more sophisticated control.

EXAMPLE 3.4—COMPENSATION FOR TIME DELAYS
Consider a process with the transfer function

$$P(s) = \frac{1}{1+2s} e^{-4s},$$

which has a significant time delay. Figure 3.25 shows a simulation of the closed-loop system obtained with a PI controller with a gain $K = 0.2$ and an integral time $T_i = 2.5$. For comparison the figure also shows the performance with a Smith predictor, which is a special controller for a system with time delays. This controller will be discussed in detail in Chapter 8. The response to set-point changes is much better with the Smith predictor, but the improvement in response to load disturbances is less. □

Systems with Oscillatory Modes Systems with oscillatory modes occur in applications such as flexible robot arms, disk drives and optical memories,

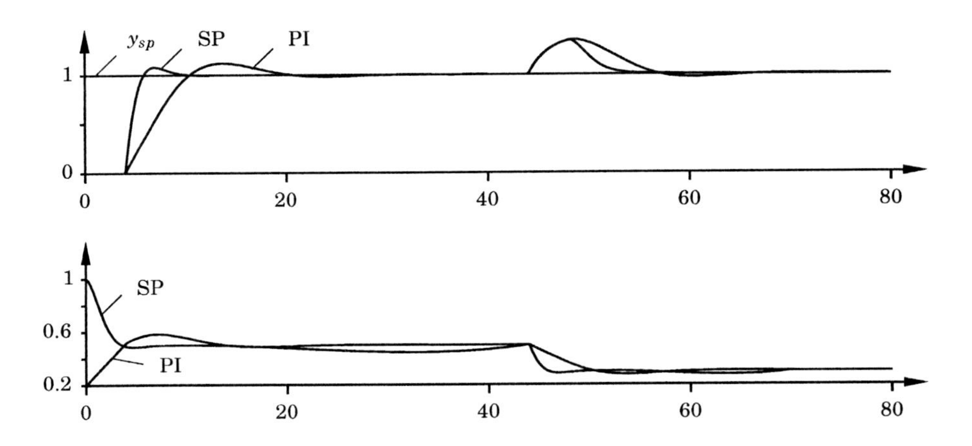

Figure 3.25 Control of the system in Example 3.4 with PI control (PI) and with a Smith predictor (SP). The upper diagram shows set point y_{sp} and measurement signal y, and the lower diagram shows control signal u.

atomic force microscopes (AFMs), micro-mechanical systems (MEMS), flexible space structures, and combustion systems. There are particular difficulties when the damping is very low so that the system is highly resonant. Typical applications are in micro-mechanical systems and in atomic force microscopes. Systems with resonant modes are not so common in process control applications. Derivative action can give drastic improvements for oscillatory systems as is illustrated by the following example.

EXAMPLE 3.5—AN OSCILLATORY SYSTEM WITH LOW DAMPING
Consider a system with the normalized transfer function

$$P(s) = \frac{a^2}{s^2 + 2\zeta as + a^2}.$$

We will consider systems with very low relative damping $\zeta = 0.005$. The performance obtained with a PI controller is severely limited by the low relative damping of the process. Since a PI controller cannot provide any phase lead the damping of the oscillatory modes of the closed-loop system will be smaller than those of the open-loop system. A key requirement is that the PI controller does not excite the high-frequency modes.

A pure integrating controller is a reasonable choice. The stability condition for such a controller is $k_i < 2\zeta a^3$, which implies that $k_i = \zeta a^3$ is a good value of the controller gain. With this choice the closed-loop system has the same settling time as the open-loop system. The response time can only be improved a little by adding proportional action. Figure 3.26 shows the input and the output for a step change in the set point for a controller with these parameters. The step response has a settling time of about 1500 s. The reason why the system has to be so slow is that the oscillatory motion cannot be damped by the PI controller, and it is therefore necessary to have a slow controller so that

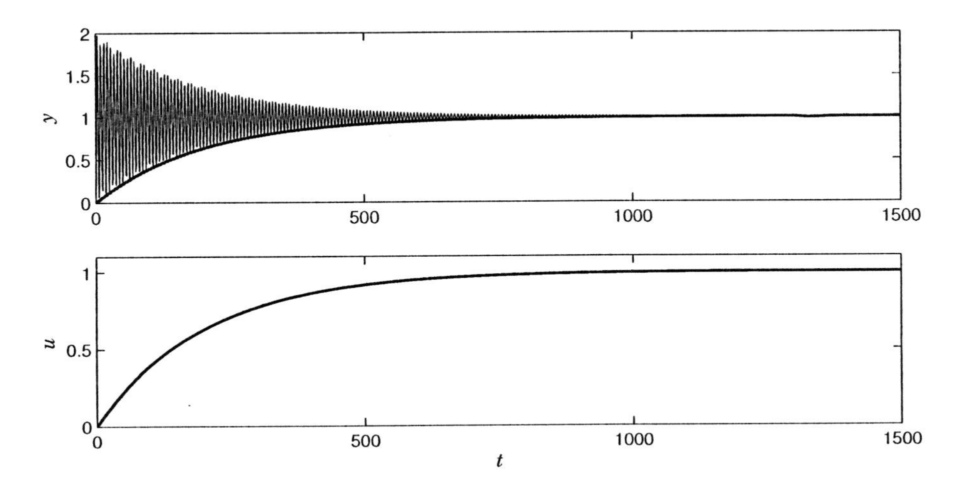

Figure 3.26 Output and control signals for PI control of the oscillatory system. The oscillating signal is the open-loop response.

the oscillatory modes are not excited by the controller. In the figure we have also shown the step response of the process.

The performance can be improved drastically by using a PID controller. One possibility is to use a PID controller with parameters

$$\begin{aligned} k &= (1 + 2\alpha_0\zeta_0)\frac{\omega_0^2}{a^2} - 1 \\ k_i &= \frac{\alpha_0\omega_0^3}{a^2} \\ k_d &= \frac{(\alpha_0 + 2\zeta_0)\omega_0 - 2\zeta a}{a^2}. \end{aligned} \tag{3.23}$$

Here ω_0, α_0, and ζ_0 are design parameters that give the properties of the closed-loop system. A reasonable choice is $\omega_0 = 3a$, $\alpha_0 = 1$, $\zeta_0 = 0.5$. The rationale for the formulas and the parameter choices will be given later in Section 6.4. Figure 3.27 shows the responses of the output and the controller to a step change in the set point. In this case the system has a settling time of about 2 s. This is about three orders of magnitude better than with PI control! The reason for this is that by using derivative action it is possible to damp the oscillations. This is indicated in the figure by showing the open-loop response of the process in dashed line. Also notice the drastic differences in the control signals for PI and PID control. It is also important to use set-point weighting with $b = 0$ to avoid a rapid change of the control variable. Such a change will excite the poorly damped oscillatory modes. □

Summary

When the dynamics of the process to be controlled are simple, a PID controller is sufficient. When the dynamics become more complicated, the performance

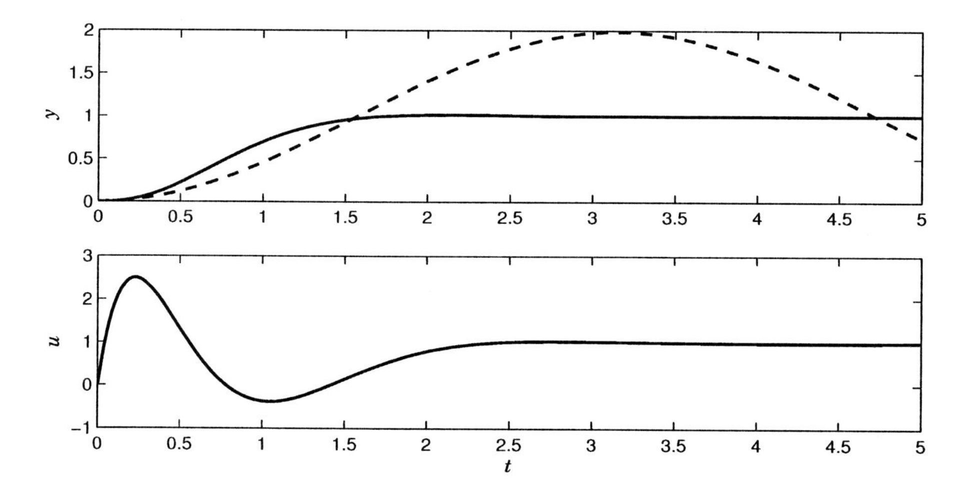

Figure 3.27 Output and control signals for PID control of the oscillatory system. The dashed curve is the open-loop response.

can be improved by using a more sophisticated controller structure than the PID. Examples of such processes have been given above.

For some systems with large parameter variations it is possible to design linear controllers that allow operation over a wide parameter range. Such controllers are, however, often of high order.

The control of process variables that are closely related to important quality variables may be of a significant economic value. In such control loops it is frequently necessary to select the controller with respect to the disturbance characteristics. This often leads to strategies that are not of the PID type. These problems are often associated with time delays.

A general controller attempts to model the disturbances acting on the system. Since a PID controller has limited complexity, it cannot model complex disturbance behavior in general nor periodic disturbances in particular.

3.7 Summary

A detailed presentation of the PID algorithm has been given. Several modifications of the "textbook" version must be made to obtain a practical, useful controller. Problems that must be handled are, for example, integral windup and introduction of set-point values. A discussion of the limitations of the PID algorithm and a characterization of processes in which the PID controller manages to perform the control have also been given.

3.8 Notes and References

An interesting summary of the development of the PID controller is given in [Bennett, 2000]. Proportional feedback in the form of a centrifugal governor was used to regulate the speed of windmills around 1750. James Watt used a similar system for speed control of steam engines in 1788. Integral action was discovered later by several authors. It is explained analytically by [Maxwell, 1868] and [Vyshnegradskii, 1876]. Feedback control with proportional and integral action was rediscovered many times after that. In the early stages, the development of controllers was closely related to development of sensors and actuators. Sensing, actuation, and control were often combined in the same device. There was also confusion about integral and derivative action because some controllers acted through motors that had integral action.

The PID controller in the form we know it today emerged in the period from 1915 to 1940. The major development was made in legendary instrument companies such as Bristol, Fisher, Foxboro, Honeywell, Leeds & Northrup, Mason-Neilan, and Taylor Instrument. Integral action was called automatic reset because it replaced a manual reset that was used in proportional controllers to obtain the correct steady-state value. The potential of a controller that could anticipate future control errors was discussed in the 1920s. However, it took some time before the idea could be implemented. A controller with derivative action was introduced by Ralph Clarridge of the Taylor Instrument Company in 1935. At that time the function was called "pre-act." An interesting overview of the early history of PID controllers is given by [Stock, 1988]. There is also much interesting material in publications from the instrument companies. The interview with Ziegler, who is one of the pioneers in our field, in [Blickley, 1990], gives a perspective on the early development; other interesting material is found in [Bennett, 1993].

It is interesting to observe that feedback was crucial for the construction of the controller itself. The early pneumatic systems used the idea that an essentially linear controller can be obtained by a feedback loop composed of linear passive components and a nonlinear amplifier, the flapper valve. Similar ideas were used in electronic controllers with electric motors and relays.

Many of the practically useful modifications of the controller first appeared as special hardware functions. They were not expressed in mathematical form. An early mathematical analysis of a steam engine with a governor was made independently by [Maxwell, 1868] and [Vyshnegradskii, 1876]. This analysis clearly showed the difference between proportional and integral control. The papers [Minorsky, 1922; Küpfmüller, 1928; Nyquist, 1932; Hazen, 1934] were available at the time when the PID controller was developed. However, there is little evidence that the engineers in the process control field knew about them. Process control therefore developed independently. Two of the early papers [Grebe *et al.*, 1933] and [Ivanoff, 1934] were written by engineers at the Dow Chemical Company. There were also contributions from university researchers [Callender *et al.*, 1936] and [Hartree *et al.*, 1937].

The PID controller has gone through an interesting development because of the drastic technology changes that have happened since 1930. The pneumatic controller improved drastically by making systematic use of the force balance principle. Pneumatics was replaced by electronics when the operational amplifier appeared in the 1950s. The emergence of computer control in the 1960s was an important development. In the early computer control systems the computer commanded the set points of analog controllers. The next stage of the development was direct digital control (DDC), where the computer was controlling the actuator directly; see [Webb, 1967]. A digital computer was then used to implement many PID controllers. This development led to a reconsideration of much of the fundamentals of PID control; see [Goff, 1966b], [L&N, 1968], [Moore *et al.*, 1970], and [Palmor and Shinnar, 1979]. The appearance of microprocessors in the 1970s made it possible to use digital control for single-loop controllers. It also led to the development of distributed control systems for process control, where the PID controller was a key element; see [Lukas, 1986]. As the computing power of the microprocessors increased it was possible to introduce tuning, adaptation, and diagnostics in the single-loop controllers. This development started in the 1980s and has accelerated in the 1990s; see [Åström *et al.*, 1993].

It is interesting to observe that many facts about PID control were rediscovered in connection with the shifts in technology. One reason being that many practical aspects of PID control were considered as proprietary information that was not easily accessible in public literature. Much useful information was also scattered in the literature.

Two different approaches were used to deal with set-point changes in early controllers. Some controllers used error feedback but others introduced the set point only in the integral part. The effect of this is that the overshoot that occurs with set-point changes can be reduced. The idea that a separation of the responses to set points and load disturbances can be accomplished by using a controller with two degrees of freedom was introduced in [Horowitz, 1963]. The application to PID control was introduced in [Araki, 1984]. An early industrial application is described in [Shigemasa *et al.*, 1987], see also [Araki and Taguchi, 1998; Taguchi and Araki, 2000]. Set-point weighting where an adjustable fraction of the set point is introduced in proportional and derivative parts is now a common feature of PID controllers.

The phenomenon of integral windup was well known in the early analog implementations. The controller structures used were often such that windup was avoided. The anti-windup schemes were rediscovered in connection with the development of direct digital control. This is discussed in [Fertik and Ross, 1967]. Much work on avoiding windup has been done since then, and windup has now made its way into some textbooks of control; see [Åström and Wittenmark, 1997]. There are many papers written on the windup phenomenon; see [Kramer and Jenkins, 1971; Glattfelder and Schaufelberger, 1983; Krikelis, 1984; Gallun *et al.*, 1985; Kapasouris and Athans, 1985; Glattfelder and Schaufelberger, 1986; Howes, 1986; Åström, 1987b; Hanus *et al.*, 1987; Chen and Wang, 1988; Glattfelder *et al.*, 1988; Hanus, 1988; Zhang and Evans, 1988; Åström and Rundqwist, 1989; Rundqwist, 1990; Walgama and Sternby, 1990]. A detailed treatment of the windup problem is given in the book [Glattfelder and Schaufelberger, 2003]. Mode switching is treated in the paper [Åström, 1987b].

4

Controller Design

4.1 Introduction

Control system design is a very rich field. There have been substantial advances over the past 50 years that have resulted in much insight and understanding as well as specific design methods. This development has been augmented by the advances in computing and the development of computer-based design tools. Broadly speaking, PID controllers have been designed using two different approaches; model-based control and direct tuning. The model based approaches start with a simple mathematical model of the process. Very simple models have been used, typically a first-order system with a time delay. In direct tuning a controller is applied to the process, and some simple experiments are performed to arrive at the controller parameters. Because of the simplicity of models and the controller special methods have been developed for PID control. From 1990 there has been a significant increase in the interest in design of PID controllers, partially motivated by the needs of automatic tuning devices for such controllers.

To develop design methods it is necessary to realize that there is a very wide range of different types of control problems even if the controller is restricted to PID. Some typical examples are:

- Design of a simple controller for a non-critical application.
- Design of a controller for a special process that minimizes fluctuations in important control variables.
- Development of a design technique that can be used in a universal auto-tuner for PID control.

There are also a number of important non-technical issues that should be considered: What is the time and effort required to apply the method? What is the knowledge level required of the user? A solution to the design problem should also give an understanding of when it is beneficial to add derivative action to a PI controller and when even more complex controllers should be considered.

This chapter gives an overview of ideas and concepts that are relevant for

PID control. It is attempted to bring design of PID controllers more into the mainstream of control design.

4.2 A Rich Variety of Control Problems

Before discussing specific tuning methods it is useful to realize that there is a wide range of control problems with very diverse goals. Some examples are: steady-state regulation, set-point tracking and path following, and control of buffers and surge tanks.

The goal of steady-state regulation is to keep process variables close to desired values. The key problems are caused by load disturbances, measurement noise, and process variations. Steady-state regulation is very common in process control.

In set-point tracking it is attempted to make process variables follow a given time function or a given curve. These problems typically occur in motion control and robotics. In some cases, for example, machine tool control or robotics, the demand on tracking precision is very severe. In other cases, for example, moving robots, the requirements are less demanding. There is a significant difference between tracking a given time curve and path following, which typically involves control of several variables.

Buffers are common in the industrial production. They are used to smooth variations between different production processes, both in process control and in discrete manufacturing. In process control they are often called *surge tanks*. Buffers are also common in computing systems. They are used in servers to smooth variations in demand of clients, and they are used in computer networks to smooth variations in the load. Buffers are also key elements in supply chains where effective buffer control has a major impact on profitability. The buffer levels should fluctuate; otherwise the buffer does not function. Ideally, no control should be applied unless there is a risk of over- or underflow. An integrating controller with low gain and a scheduling that gives higher gains at the buffer limits are commonly used.

The key issues in many of the control problems are attenuation of load disturbances, injection of measurement noise, robustness to process variations, and set-point following. The relative importance of these factors and the requirements vary from application to application, but all factors must be considered.

4.3 Feedback Fundamentals

A block diagram of a basic feedback loop with a controller having two degrees of freedom is shown in Figure 4.1. The process is represented by the block P. The controller is represented by the feedback block C and the feedforward

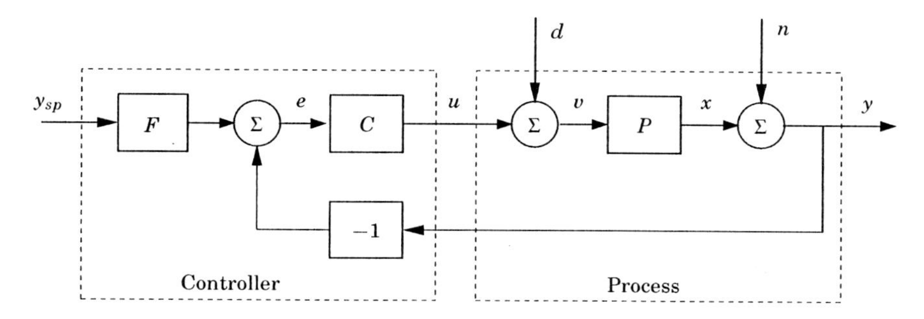

Figure 4.1 Block diagram of a basic feedback loop having two degrees of freedom.

part F. For an ideal PID controller with set-point weighting we have

$$
\begin{aligned}
C(s) &= K\Big(1 + \frac{1}{sT_i} + sT_d\Big) \\
F(s) &= \frac{b + \frac{1}{sT_i} + csT_d}{1 + \frac{1}{sT_i} + sT_d}.
\end{aligned}
\tag{4.1}
$$

Compare with (3.7) and (3.20). The signal u is the control signal, and the signal x is the real process variable. Information about x is obtained from the sensor signal y, which is corrupted by measurement noise n. The signal d represents load disturbances that drive the system away from its desired state. This signal can enter the process in different ways; in Figure 4.1 it is assumed that it acts on the process input.

The goal of control design is to determine the transfer functions C and F so that the process variable x is close to the set point y_{sp} in spite of load disturbances, measurement noise, and process uncertainties. The feedback can reduce the effect of load disturbances. Because of the feedback measurement noise is fed back into the system. It is essential to make sure that this does not cause large variations in the process variable. Since the process model is never accurate it is essential that the behavior of the closed-loop system is insensitive to variations in the process. The feedforward transfer function F is designed to give the desired response to set-point changes.

Fundamental Relations

The feedback loop is influenced by three external signals, the set point y_{sp}, the load disturbance d, and the measurement noise n. There are at least three signals x, y, and u that are of great interest for control. This means that there are nine relations between the input and the output signals. Since the system is linear these relations can be expressed in terms of the transfer functions. Let X, Y, U, D, N, and Y_{sp} be the Laplace transforms of x, y, u, d, n, and y_{sp}, respectively. The following relations are obtained from the block diagram in Figure 4.1:

$$
\begin{aligned}
X &= \frac{PCF}{1+PC}Y_{sp} + \frac{P}{1+PC}D - \frac{PC}{1+PC}N \\
Y &= \frac{PCF}{1+PC}Y_{sp} + \frac{P}{1+PC}D + \frac{1}{1+PC}N \\
U &= \frac{CF}{1+PC}Y_{sp} - \frac{PC}{1+PC}D - \frac{C}{1+PC}N.
\end{aligned}
\tag{4.2}
$$

There are several interesting conclusions we can draw from these equations. First, we can observe that several transfer functions are the same and that all relations are given by the following six transfer functions, called the *Gang of Six*.

$$
\begin{array}{ccc}
\dfrac{PCF}{1+PC} & \dfrac{PC}{1+PC} & \dfrac{P}{1+PC} \\
\dfrac{CF}{1+PC} & \dfrac{C}{1+PC} & \dfrac{1}{1+PC}.
\end{array}
\tag{4.3}
$$

The transfer functions in the first column give the response of process variable and control signal to the set point. The second column gives the same signals in the case of pure error feedback when $F = 1$. The transfer function $P/(1+PC)$ in the third column tells how the process variable reacts to load disturbances, and the transfer function $C/(1+PC)$ gives the response of the control signal to measurement noise.

Notice that only four transfer functions,

$$
\begin{array}{cc}
\dfrac{PC}{1+PC} & \dfrac{P}{1+PC} \\
\dfrac{C}{1+PC} & \dfrac{1}{1+PC},
\end{array}
\tag{4.4}
$$

are required to describe how the system reacts to load disturbance and the measurement noise. These transfer functions are called the *Gang of Four*. They also capture robustness, as will be discussed in Section 4.6. Two additional transfer functions are required to describe how the system responds to set-point changes.

The special case when $F = 1$ is called a system with (pure) error feedback. In this case, all control actions are based on feedback from the error only. In this case, the system is completely characterized by the Gang of Four (4.4).

We are often interested in the magnitude of the transfer functions given by Equation 4.4. It is important to be aware that the transfer functions $PC/(1+PC)$ and $1/(1+PC)$ are dimension free, but the transfer functions $P/(1+PC)$ and $C/(1+PC)$ are not. For practical purposes it is therefore important to normalize the signals, for example, by scaling process inputs and outputs to the interval 0 to 1 or -1 to 1.

A Practical Consequence

The fact that six relations are required to capture properties of the basic feedback loop is often neglected in literature, particularly in the papers on PID control. To describe the system properly it is thus necessary to show the response of all six transfer functions. The transfer functions can be represented

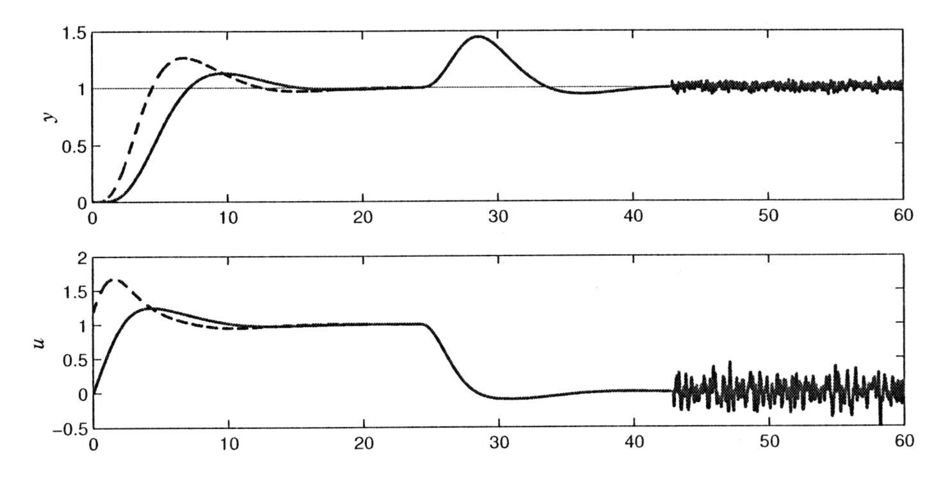

Figure 4.2 Representation of the properties of a basic feedback system by responses to a step in the reference, a step in the load disturbance, and measurement noise. The full lines are for set-point weight $b = 0$ and the dashed line is for set-point weight $b = 1$.

in different ways, by their step responses or by their frequency responses. Most papers on control only show the response of the process variable to set-point changes. Such a curve gives only partial information about the behavior of the system. To get a more complete representation of the system all six responses should be given, for example, as shown in Figure 4.2. This figure shows the responses in process variable and control signal to an experiment with a step change in set point followed by a step in the load disturbance, and measurement noise. The solid lines show the response when $F = 1$ and the dashed lines show the response when feedforward is used. Figure 4.2 thus gives a complete characterization of all six transfer functions in Equation 4.3.

Many Variations

The system shown in Figure 4.1 is a prototype problem. There are many variations of this problem. In Figure 4.1 the load disturbances act on the process input. In practice the disturbances can appear in many other places in the system. The measurement noise also acts at the process output. There may also be dynamics in the sensor, and the measured signal is often filtered. All these variations can be studied with minor modifications of the analysis based on Figure 4.1. As an illustration we will investigate the effects of a sensor filter. Figure 4.3 shows a block diagram of such a system. A typical example is a PID controller with set-point weighting and a second-order measurement filter. The transfer functions $F(s)$ and $C(s)$ in Figure 4.3 are given by (4.1) and the filter transfer function $G_f(s)$ is

$$G_f(s) = \frac{1}{1 + sT_f + s^2T_f^2/2}. \tag{4.5}$$

The relations between the input signals and output signals in Figure 4.3 are

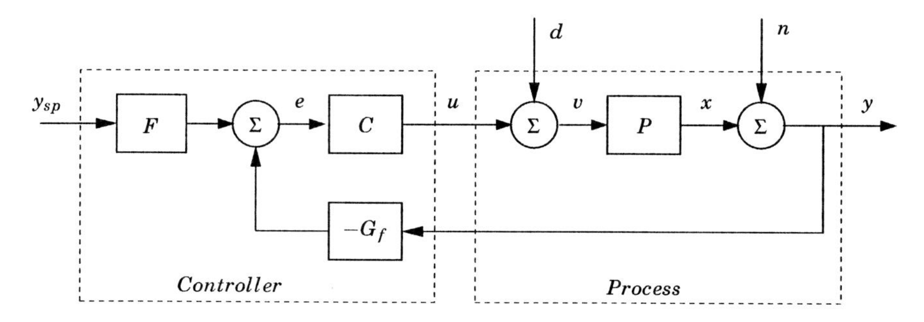

Figure 4.3 Block diagram of a basic feedback loop having two degrees of freedom and filtering of the measurement.

given by

$$
\begin{aligned}
X &= \frac{PCF}{1+PCG_f}Y_{sp} + \frac{P}{1+PCG_f}D - \frac{PCG_f}{1+PCG_f}N \\
Y &= \frac{PCF}{1+PCG_f}Y_{sp} + \frac{P}{1+PCG_f}D + \frac{1}{1+PCG_f}N \\
U &= \frac{CF}{1+PCG_f}Y_{sp} - \frac{PCG_f}{1+PCG_f}D - \frac{CG_f}{1+PCG_f}N.
\end{aligned}
\tag{4.6}
$$

Equation (4.6) is identical to (4.2) if the transfer function $C(s)$ and $F(s)$ are replaced by

$$
\bar{C}(s) = C(s)G_f(s), \quad \bar{F}(s) = \frac{F(s)}{G_f(s)}, \tag{4.7}
$$

The modifications required to deal with filtering are thus minor, and it suffices to develop the theory for the configuration in Figure 4.1.

Separation of Responses to Disturbances and Set Points

In early work on PID control it was a tradition to have two tuning rules, one for good set-point response and another for efficient attenuation of load disturbances. This practice still continues. A strong advantage of a controller with two degrees of freedom is that the responses to disturbances and set point can be designed separately. This follows from (4.2), which shows that the response to load disturbances and measurement noise is given by the $C(s)$, or from (4.6) by $\bar{C}(s) = C(s)G_f(s)$ when the measurement is filtered. A good design procedure is thus to determine $C(s)$ to account for robustness and disturbances. The feedforward transfer function $F(s)$ can then be chosen to give the desired set-point response. In general, this requires that the feedforward transfer function can be chosen freely. Simply choosing the set-point weights often give satisfactory results. Notice that there are some situations where only the error signal is available. The decoupling of the design problem then is not possible, and the design of the feedback then has to consider a trade-off between disturbances, robustness, and set-point response.

Fundamental Limitations

In any design problem it is important to be aware of the fundamental limitations. Typical sources of limitations are

- Process dynamics
- Nonlinearities
- Disturbances
- Process uncertainty

Process dynamics is often the limiting factor. Time delays and poles and zeros in the right half plane are relevant factors. It is important to be aware of these limitations. Time delays are the most common factor for PID control. It seems intuitively reasonable that it is impossible to have tight control of a system with a time delay. It can be shown that for a process with a time delay L the achievable gain crossover frequency ω_{gc}, which is defined in Section 4.4, is limited by

$$\omega_{gc}L < 1. \tag{4.8}$$

Since

$$e^{-sL} \approx \frac{1 - sL/2}{1 + sL/2},$$

it also seems reasonable that right-half plane zeros also limit the achievable performance. It can be shown that a right-half plane zero at $s = b$ limits the gain crossover frequency to

$$\omega_{gc} < 0.5b. \tag{4.9}$$

A right-half plane pole $s = a$ in the process limits the achievable gain crossover frequency ω_{gc} to

$$\omega_{gc} > 2a. \tag{4.10}$$

Notice that time delays and right-half plane zeros give an upper bound to the achievable gain crossover frequency while right-half plane poles give a lower bound.

Nonlinearities, saturation, and rate saturation are very common; they impose limitations on how much and how fast the process variables can change. Saturations combined with unstable process dynamics are particularly serious because they may lead to situations where it is not possible to recover stable operating conditions. Such situations are fortunately not common in process control.

Load disturbances and measurement noise limit how accurately a process variable can be controlled. The limitations often interact. The allowable controller gain is, for example, limited by a combination of measurement noise and actuator saturation. The effect of load disturbances depends critically on the achievable bandwidth.

Process models used for control are always approximations. Process dynamics may also change during operation. Insensitivity to model uncertainty is one of the essential properties of feedback. There is, however, a limit to the

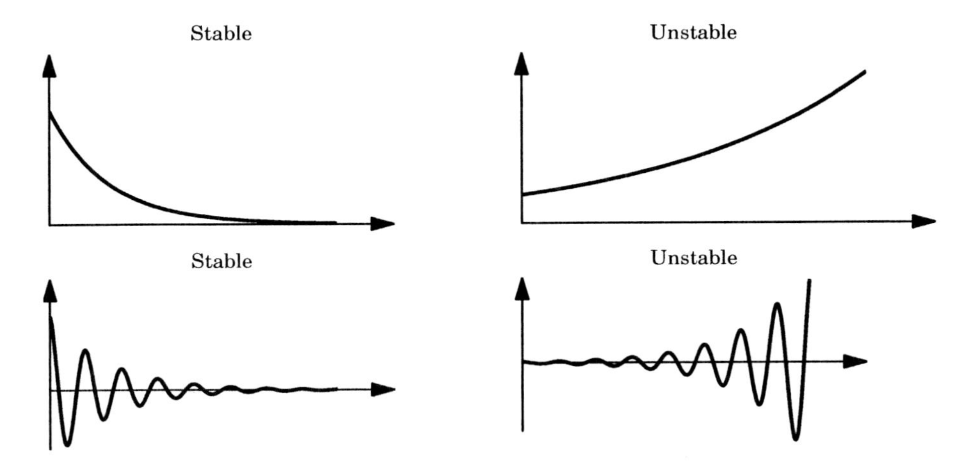

Figure 4.4 Illustration of different system behaviors used to define stability.

uncertainty that can be dealt with. Feedback cannot be active in frequency ranges where the uncertainty in the phase of the process is larger than $\pm 90^\circ$. To have reasonable control performance the uncertainty should be less than about $\pm 15^\circ$. If the process variations correlate well with some measured quantity it is possible to compensate for the uncertainties by changing the controller parameters. This technique, which is called gain scheduling, will be discussed in Section 9.3.

4.4 Stability

Feedback has many useful properties. The main drawback is that feedback may cause instability. It is therefore essential to have a good understanding of stability and the mechanisms that cause instability.

Stability Concepts

The notion of stability is intuitively very simple. It tells how a system behaves after a perturbation. Already in 1868 Maxwell classified the behavior as follows:

U1: The variable increases continuously

S1: The variable decreases continuously

U2: The variable increases in an oscillatory manner

S2: The variable decreases in an oscillatory manner

These behaviors are illustrated in Figure 4.4. Maxwell called the behaviors labeled S stable and the ones labeled U unstable. He also found that for linear time-invariant systems stability was related to properties of the roots of an algebraic equation.

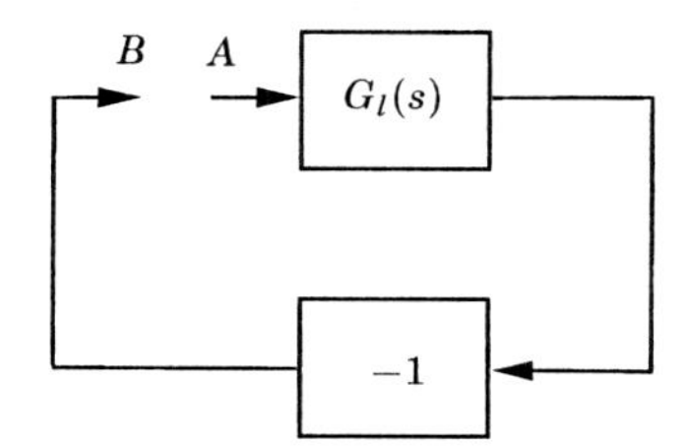

Figure 4.5 Block diagram of a simple feedback system.

Consider a system with the transfer function

$$G(s) = \frac{b(s)}{a(s)}, \tag{4.11}$$

where $a(s)$ and $b(s)$ are polynomials. Recall that the roots of the polynomial $a(s)$ are called the poles of the system. Since a pole s_i corresponds to a time function $e^{s_i t}$ the following relations are obtained between the behaviors and the roots of an algebraic equation:

U1: Corresponds to real poles with positive real part

S1: Corresponds to real poles with negative real part

U2: Corresponds to complex poles with positive real part

S2: Corresponds to complex poles with negative real part

The system (4.11) is stable if it has no poles in the right half plane. The equation

$$a(s) = 0 \tag{4.12}$$

is called the characteristic equation. A system is stable if the characteristic equation does not have any roots with positive real parts. It is common practice to label poles on the imaginary axis as unstable.

Nyquist's Stability Criterion

The algebraic definition of stability based on the roots of the characteristic equation is useful, but it also has some drawbacks. Consider, for example, the feedback system in Figure 4.5 where the transfer functions of the process and the controller have been combined into one block with the transfer function $G_l = PC$. The characteristic equation for this system is

$$1 + G_l(s) = 0. \tag{4.13}$$

The transfer function, which is the product of the transfer functions of the process and the controller, describes how signals propagate around the feedback loop and is called the *loop transfer function*. It is not easy to see how the roots of (4.13) are influenced by the transfer functions of the process and the controller. This can, however, be done by using a totally different view of stability, which was developed by Nyquist. He started by investigating the conditions for maintaining an oscillation in the system shown in Figure 4.5. Assume that the feedback loop is broken as is indicated in the figure and that the signal $u_A(t) = \sin \omega_0 t$ is injected at point A. After a transient the output at point B is then given by

$$u_B(t) = -|G_l(i\omega_0)| \sin\bigl(\omega_0 t + \arg G_l(i\omega)\bigr).$$

The signals $u_A(t)$ and $u_B(t)$ are identical if

$$G_l(i\omega_0) = -1, \tag{4.14}$$

and an oscillation will be maintained if the loop is closed by joining points A and B. Equation 4.14 thus gives the condition for oscillations in the system. It follows from (4.13) and (4.14) that the condition for oscillation implies that the characteristic equation of the system has a root $s = i\omega_0$. The frequencies where the system can maintain an oscillation can be determined by solving (4.14) for ω_0.

Nyquist developed a stability criterion based on the idea of how sinusoids propagate around the feedback loop. Nyquist argues as follows. He first investigated frequencies where the signals u_A and u_B are in phase, i.e., when $\arg G_l(i\omega_0) = \pi$. Intuitively it seems reasonable that the system is stable if $|G_l(i\omega_0)| < 1$ because the amplitude is then decreased when the signal traverses the loop. The situation is actually a little more complicated because the system may be stable even if $|G_l(i\omega_0)| > 1$. The precise result can be expressed in terms of the Nyquist curve introduced in Section 2.3. Recall that the Nyquist curve is a plot of $(\text{Re}G_l(i\omega), \text{Im}G_l(i\omega))$ for $0 \le \omega \le \infty$. When the loop transfer function does not have poles in the right half plane the condition for stability is that the critical point -1 is to the left of the Nyquist curve when it is traversed for increasing ω.

A nice property of the Nyquist's criterion is that it indicates how a system should be changed in order to move the Nyquist curve away from the critical point. Figure 6.4 shows that derivative action, which introduces phase lead, bends the curve away from the critical point. Integral action, which introduces phase lag, bends the curve towards the critical point. The idea is to modify the controller so that the curve is bent away from the critical point. This has led to a whole class of design methods called loop shaping.

Stability Margins

In practice it is not enough to require that the system is stable. There must also be some margins of stability. This means that the Nyquist curve should not be too close to the critical point. This is illustrated in Figure 4.6, which shows several stability margins. The gain margin g_m tells how much controller gain can be increased before reaching the stability limit. Let *phase crossover frequency* ω_{180} be the smallest frequency where the phase lag of the loop transfer function $G_l(s)$ is 180° and the gain margin be defined as

$$g_m = \frac{1}{|G_l(i\omega_{180})|}. \tag{4.15}$$

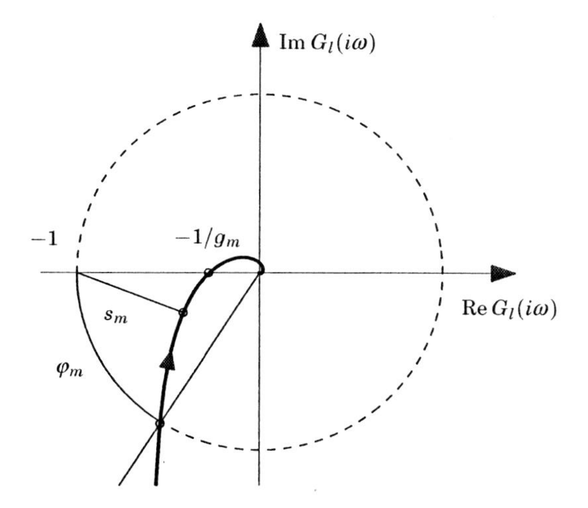

Figure 4.6 Nyquist plot of the loop transfer function G_l with gain margin g_m, phase margin φ_m and stability margin s_m.

The point where the Nyquist curve intersects the unit circle is another interesting point. This point can be characterized by the angle φ_m. This angle called *the phase margin* is also a measure of how close the Nyquist curve is to the critical point. The angle φ_m is the amount of phase lag required to reach the stability limit. The *gain crossover frequency* ω_{gc} is the lowest frequency where the loop transfer function $G_l(s)$ has unit magnitude. The phase margin is formally defined as

$$\varphi_m = \pi + \arg G_l(i\omega_{gc}). \tag{4.16}$$

Both gain and phase margin are classical measures of degrees of stability. Both values must be specified in order to ensure that the Nyquist curve is far from the critical point. They can be replaced by a single number, the shortest distance from the Nyquist curve to the critical point -1, which is called the *stability margin* s_m.

Reasonable values of the margins are phase margin $\varphi_m = 30^\circ - 60^\circ$, gain margin $g_m = 2 - 5$, stability margin $s_m = 0.5 - 0.8$.

The gain and phase margins were originally conceived for the case when the Nyquist curve only intersects the unit circle and the negative real axis once. For more complicated systems there may be many intersections, and it is then necessary to consider the intersections that are closest to the critical point. For more complicated systems there is also another number that is highly relevant, namely, *the delay margin*. The delay margin is defined as the smallest time delay required to make the system unstable. For loop transfer functions that decay quickly the delay margin is closely related to the phase margin, but for systems where the amplitude ratio of the loop transfer function has several peaks at high frequencies the delay margin is a much more relevant measure. This is particularly relevant for the Smith predictor that will be discussed in Chapter 8.

Internal Stability

So far we have only discussed the simple feedback system in Figure 4.5. For the more general system in Figure 4.1 which is characterized by six transfer functions, it is necessary to require that all four transfer functions,

$$\begin{array}{cc} \dfrac{PC}{1+PC} & \dfrac{P}{1+PC} \\ \dfrac{C}{1+PC} & \dfrac{1}{1+PC}, \end{array} \tag{4.17}$$

are stable; compare with (4.3). This is called internal stability. Notice that there may be cancellations of poles and zeros in the product PC.

Stability Regions

A primary requirement for a PID controller is that the parameters of the controller are chosen in such a way that the closed-loop system is stable. A PID controller of the form

$$C(s) = k + \frac{k_i}{s} + k_d s \tag{4.18}$$

has three parameters only, and the stability region can be represented by a volume in three dimensions. To describe this volume the process transfer function is represented as

$$P(i\omega) = r(\omega)e^{i\phi(\omega)} = r(\omega)(\cos(\omega) + i\sin(\omega)),$$

and the condition for oscillation (4.14) then becomes

$$P(i\omega)C(i\omega) = r(\omega)(\cos(\omega) + i\sin(\omega))\big(k - i\frac{k_i}{\omega} + ik_d\omega\big) = -1.$$

Separating the real and imaginary parts we find that the boundary of the stability region can be represented parametrically as

$$\begin{aligned} k &= -\frac{\cos\phi(\omega)}{r(\omega)} \\ k_i &= \omega^2 k_d - \frac{\omega \sin\phi(\omega)}{r(\omega)}. \end{aligned} \tag{4.19}$$

It is thus straightforward to determine the stability region for a constant value of k_d. Repeating the calculations for a set of k_d-values gives the stability region for the PID controller.

EXAMPLE 4.1—STABILITY REGION FOR $P(s) = 1/(s+1)^4$

Figure 4.7 shows the stability region for a process with the transfer function $P(s) = 1/(s+1)^4$. The value $k_d = 0$ corresponds to PI control. Integral gain k_i may be increased by adding derivative action. The integral gain has its maximum $k_i = 36$ at the boundary of the stability region for $k = 8$ and $k_d = 20$. The system is unstable for all values of k and k_i if $k_d > 20$. □

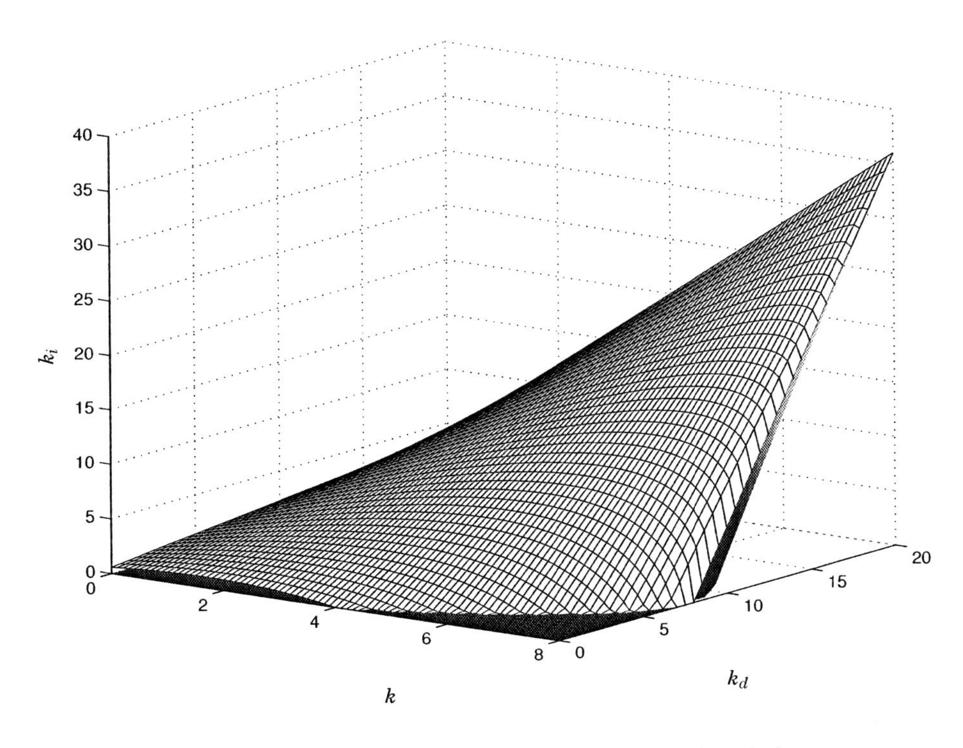

Figure 4.7 Stability region for the system $P(s) = (1+s)^{-4}$.

Some interesting conclusions can be drawn from Example 4.1. To have good disturbance rejection it is desirable to have a large value of k_i. This is shown in Section 4.9. With PI control, the largest value of k_i for a stable system is $k_i = 1$. Figure 4.7 shows that the value of k_i can be increased substantially by introducing derivative action. The highest value of k_i that can be obtained with a stable system is $k_i = 36$. This will, however, be a very fragile controller because the system can be made unstable by arbitrarily small changes in controller gains. For large values of k_d the curves have sharp corners at the point of maximum integral gain. This property of derivative action is one reason why tuning of controllers with derivative action is difficult. It will be discussed further in Chapters 6 and 7.

Constant Proportional Gain

The region of parameters where the system is stable is a subset of R^3. The calculations performed give the two-dimensional intersections with constant derivative gain. Additional insight can be obtained from another representation of the stability regions. To investigate the stability we will use the Nyquist criterion and plot the locus of the loop transfer function $G_l(s)$. With proportional control we have $G_l = kP$. For a fixed value of the proportional gain $k > 0$ we determine the frequency ω_n where the Nyquist curve of $kP(i\omega)$ intersects the circle with the line segment $(-1, 0)$ as a diameter; see Figure 4.8. We will first consider the case when the intersection of the Nyquist curve and the circle

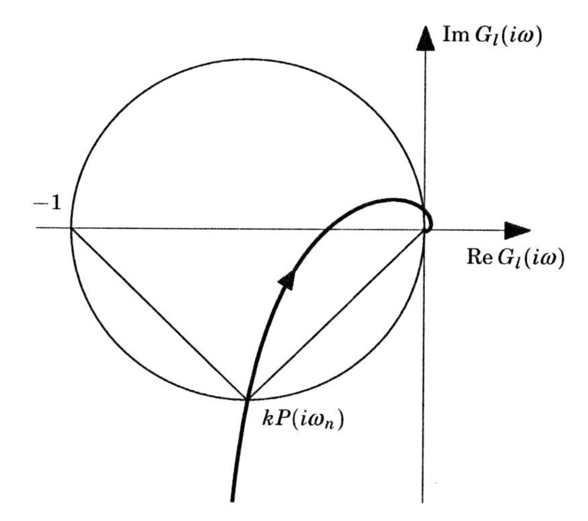

Figure 4.8 Nyquist curve of the loop transfer function $G_l(s) = kP(s)$.

occurs in the lower half plane as shown in Figure 4.8. The controller transfer function is

$$C(i\omega) = k + i\Big(-\frac{k_i}{\omega} + k_d\omega\Big) = k - i\Big(\frac{k_i}{\omega} - k_d\omega\Big),$$

hence,

$$G_l(i\omega_n) = P(i\omega_n)C(i\omega_n) = kP(i\omega_n) - i\Big(\frac{k_i}{\omega_n} - k_d\omega_n\Big)kP(i\omega_n).$$

If proportional gain k is fixed the point $kP(i\omega_n)$ moves to $G_l(i\omega_n)$ when proportional and integral gains are different from zero. To avoid reaching the critical point it must be required that

$$\Big(\frac{k_i}{\omega_n} - k_d\omega_n\Big)|P(i\omega_n)| < |1 + P(i\omega_n)|.$$

The same analysis can be made when the intersection of the Nyquist curve and the circle occurs in the upper half plane. Combining the inequalities we find that the stability regions are given by the conditions

$$\begin{aligned} &k_i > 0 \\ &k_i < \omega_n^2 k_d + \omega_n \frac{|1 + kP(i\omega_n)|}{|P(i\omega_n)|}, \text{ for Im } P(i\omega_n) < 0 \\ &k_i > \omega_n^2 k_d - \omega_n \frac{|1 + kP(i\omega_n)|}{|P(i\omega_n)|}, \text{ for Im } P(i\omega_n) > 0 \end{aligned} \tag{4.20}$$

which should hold for for all ω_n such that

$$\Big|kP(i\omega_n) + \frac{1}{2}\Big| = \frac{1}{2}. \tag{4.21}$$

We can thus conclude that for constant proportional gain the stability region is represented by several convex polygons in the k_i-k_d plane. In general, there may be several polygons, and each may have many surfaces. The number of surfaces of the polygons is determined by the number of roots of the Equation 4.21. In many cases, the polygons are also very simple, as is illustrated with the following example.

EXAMPLE 4.2—FOUR EQUAL POLES
To illustrate the results we consider a process with the transfer function

$$P(s) = \frac{1}{(s+1)^4} = \frac{1}{s^4 + 6s^2 + 1 + 4s(s^2+1)}.$$

In this case, Equation 4.21 becomes

$$\omega^4 - 6\omega^2 + 1 + k = 0.$$

This equation has only two positive solutions,

$$\omega^2 = 3 \pm \sqrt{8-k},$$

and it follows from (4.20) that the stability region is given by the inequalities

$$\begin{aligned} k_i &> 0 \\ k_i &< (3 - \sqrt{8-k})k_d + 4k - 56 + 20\sqrt{8-k} \\ k_i &> (3 + \sqrt{8-k})k_d + 4k - 56 - 20\sqrt{8-k}. \end{aligned} \tag{4.22}$$

The stability region is shown in Figure 4.7. The integral gain has its maximum $k_i = 36$ at the boundary of the stability region for $k = 8$ and $k_d = 20$. □

4.5 Closed-Loop Poles and Zeros

Many properties of a feedback system can be obtained from the closed-loop poles and zeros. For PID control the behavior is often characterized by a few dominant poles, typically those closest to the origin. Many properties of the closed-loop system can be deduced from the poles and the zeros of complementary sensitivity function

$$T(s) = \frac{PC(s)}{1 + PC(s)}.$$

With error feedback, $F = 1$ in Figure 4.1, the closed-loop zeros are the same as the zeros of loop transfer function $G_l(s)$, and the closed-loop poles are the roots of the equation

$$1 + G_l(s) = 0.$$

The pole-zero configurations of closed-loop systems may vary considerably. Many simple feedback loops, however, will have a configuration of the type

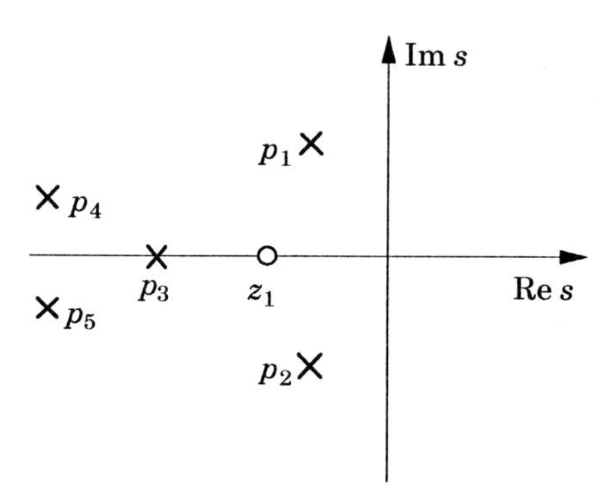

Figure 4.9 Pole-zero configuration of the transfer function from set point to output for a simple feedback system.

shown in Figure 4.9, where the principal characteristics of the response are given by a complex pair of poles, p_1 and p_2, called the *dominant poles*. The response is also influenced by real poles and zeros p_3 and z_1 close to the origin. The position of p_3 and z_1 may be reversed. There may also be more poles and zeros far from the origin, which typically are of less influence. Poles and zeros to the left of the dominant poles have little influence on the transient response if they are sufficiently far away from the dominant poles. The influence of a pole diminishes if there is a zero close to it.

Complex poles can be characterized in terms of their frequency ω_0, which is the distance from the origin, and their relative damping ζ. A first approximation of the response is obtained from the equivalent second-order system. The response is modified if there are poles and zeros close to the dominating poles. Classical control was very much concerned with closed-loop systems having the pole-zero configuration shown in Figure 4.9.

Even if many closed-loop systems have a pole-zero configuration similar to the one shown in Figure 4.9, there are, however, exceptions. For instance, systems with mechanical resonances, which may have poles and zeros close to the imaginary axis, are generic examples of systems that do not fit the pole-zero pattern of the figure. Another example is processes with a long dead time.

Design of PID controllers are typically based on low-order models, which gives closed-loop systems with a small number of poles and zeros.

Dominant Poles from the Loop Transfer Function

A simple method for approximate determination of the dominant poles from knowledge of the Nyquist curve of the loop transfer function will now be given. Consider the loop transfer function $G_l(s)$ as a mapping from the s-plane to the G_l-plane. The map of the imaginary axis in the s-plane is the Nyquist curve $G_l(i\omega)$, which is indicated in Figure 4.10. The closed-loop poles are the roots of the characteristic equation

$$1 + G_l(s) = 0.$$

The map of a straight vertical line through the dominant closed-loop poles in

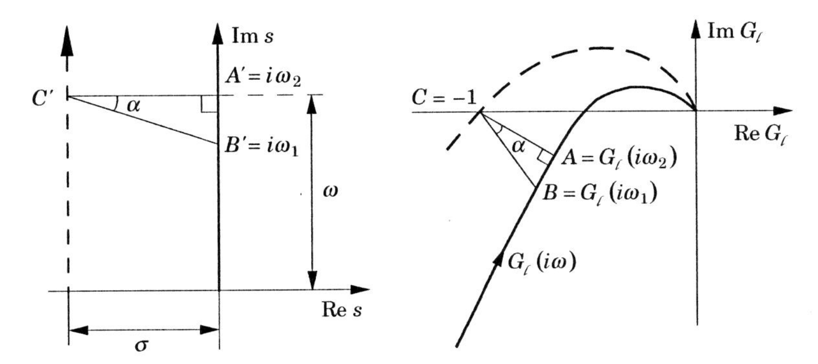

Figure 4.10 Representation of the loop transfer function $G_l(i\omega)$ as a map of complex planes.

the s-plane is thus a curve through the critical point $G_l = -1$ in the G_l-plane. This curve is shown by a dashed line in Figure 4.10. Since the map is conform, the straight line A′C′ is mapped on the curve AC, which intersects the Nyquist curve orthogonally. The triangle ABC is also mapped conformally to A′B′C′. If ABC can be approximated by a triangle, we have

$$\frac{G_l(i\omega_2) - G_l(i\omega_1)}{i\omega_2 - i\omega_1} \approx \frac{1 + G_l(i\omega_2)}{\sigma}.$$

When ω_1 is close to ω_2 this becomes

$$\sigma = (1 + G_l(i\omega_2)) \frac{i\omega_2 - i\omega_1}{G_l(i\omega_2) - G_l(i\omega_1)} \approx \frac{1 + G_l(i\omega_2)}{G_l'(i\omega_2)}, \tag{4.23}$$

where $G_l'(s) = dG_l(s)/ds$. To determine the dominant poles we first determine the point A on the Nyquist curve that is closest to the critical point -1. This point is characterized by the frequency ω_2. Then determine the derivative of the loop transfer function at ω_2. The dominant poles are then given by $s = -\sigma \pm i\omega_2$, where σ is given by Equation 4.23.

4.6 The Sensitivity Functions

Two of the transfer functions (4.3) are of particular interest, the sensitivity function S and the complementary sensitivity function T. These functions are defined by

$$S = \frac{1}{1 + PC} = \frac{1}{1 + G_l}, \qquad T = \frac{PC}{1 + PC} = \frac{G_l}{1 + G_l}. \tag{4.24}$$

The sensitivity functions are uniquely given by the loop transfer function $G_l(s) = P(s)C(s)$ and have the property $S + T = 1$. The transfer functions reflect many interesting properties of the closed-loop system, particularly robustness to process variations.

Small Process Variations—The Sensitivity Function

We will start by investigating how sensitive the response to set-point changes is to small process variations. It follows from (4.2) that the transfer function from set point to process variable is

$$G_{xy_{sp}} = G_{yy_{sp}} = \frac{PCF}{1+PC}.$$

Consider $G_{xy_{sp}}$ as a function of the process transfer function P. Differentiating with respect to P gives

$$\frac{dG_{xy_{sp}}}{dP} = \frac{CF}{1+PC} - \frac{PC^2F}{(1+PC)^2} = \frac{CF}{(1+PC)^2} = \frac{1}{1+PC}\frac{CF}{1+PC}.$$

Hence,

$$\frac{dG_{xy_{sp}}}{G_{xy_{sp}}} = \frac{1}{1+PC}\frac{dP}{P} = S\frac{dP}{P}. \tag{4.25}$$

Notice that the quantity dG/G can be interpreted as the relative variation in the transfer function G. Equation 4.25 thus implies that the relative error in the closed-loop transfer function $G_{yy_{sp}}$ is equal to the product of the sensitivity function and the relative error in the process. For frequencies where the sensitivity function is small it thus follows that the closed-loop system is very insensitive to variations in the process. This is actually one of the key reasons for using feedback. The formula (4.25) is one of the reasons why S is called the sensitivity function. The sensitivity function also has other interesting properties.

Disturbance Attenuation

A very fundamental question is how much the fluctuations in the process variable are influenced by feedback. Consider the situation shown in Figure 4.11 where the same load disturbance acts on a process P in open loop and on the process P in a closed loop with the controller C. Let y_{ol} be the output of the open-loop system and y_{cl} the output of the closed-loop system. We have the following relation between the Laplace transforms of the signals,

$$\frac{Y_{cl}(s)}{Y_{ol}(s)} = \frac{1}{1+P(s)C(s)} = S(s). \tag{4.26}$$

Disturbances with frequencies ω such that $|S(i\omega)| < 1$ are thus attenuated by feedback, but disturbances such that $|S(i\omega)| > 1$ are amplified by the feedback. A plot of the amplitude ratio of S thus immediately tells the effect of feedback.

Since the sensitivity only depends on the loop transfer function it can be visualized graphically in the Nyquist plot of the loop transfer function. This is illustrated in Figure 4.12. The complex number $1 + G_l(i\omega)$ can be represented as the vector from the point -1 to the point $G_l(i\omega)$ on the Nyquist curve. The sensitivity is thus less than one for all points outside a circle with radius 1 and center at -1. Disturbances of these frequencies are attenuated by the feedback.

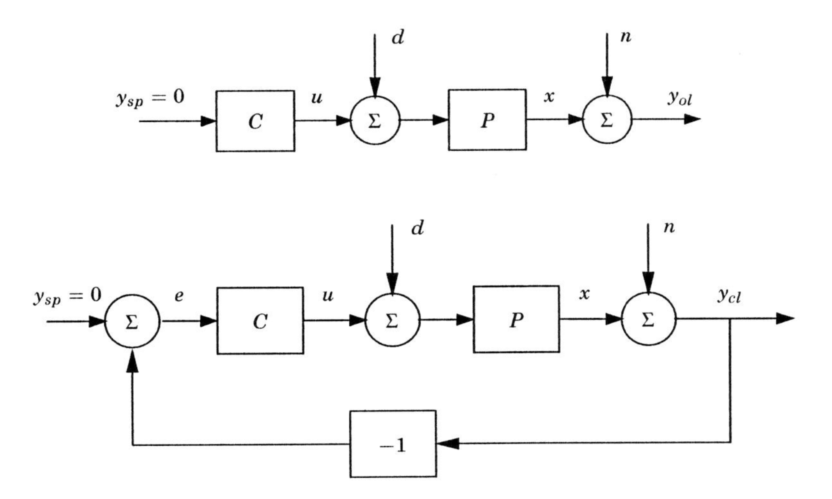

Figure 4.11 Block diagrams of open- and closed-loop systems subject to the same disturbances.

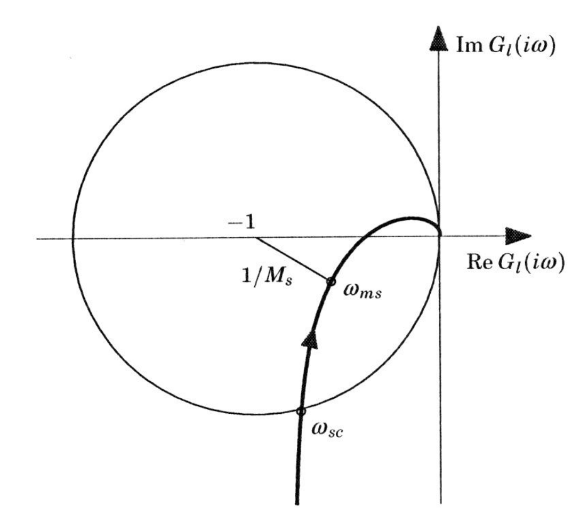

Figure 4.12 Nyquist curve of loop transfer function showing graphical interpretation of maximum sensitivity. The sensitivity crossover frequency ω_{sc}, and the frequency ω_{ms} where the sensitivity has its largest value are indicated in the figure. All points inside the circle with center at the -1 have sensitivities greater than 1.

The lowest frequency where the sensitivity function has magnitude 1 is called the *sensitivity crossover frequency* ω_{sc}. The value

$$M_s = \max_{\omega} |S(i\omega)| = \max_{\omega} \left| \frac{1}{1 + P(i\omega)C(i\omega)} \right| = \max_{\omega} \left| \frac{1}{1 + G_l(i\omega)} \right|, \tag{4.27}$$

which is called the maximum sensitivity, tells the worst-case amplification of the disturbances.

The sensitivity cannot be made arbitrarily small. The following relation holds under reasonably general conditions for stable systems

$$\int_0^\infty \log |S(i\omega)| d\omega = 0. \tag{4.28}$$

This very important relation is called Bode's integral. It says that if the sensitivity is reduced for one frequency it increases at another frequency. Feedback can thus redistribute the attenuation of disturbances for different frequencies, but it cannot reduce the effect of disturbances for all frequencies.

In Section 2.6 it was mentioned that random fluctuations can be modeled by a power spectral density. If the spectral density is $\phi(\omega)$ for a system without control it becomes $|S(i\omega)|^2\phi(\omega)$ for a system with control. The rations of the variances under open and closed loop are thus

$$\frac{\sigma_{cl}^2}{\sigma_{ol}^2} = \frac{\int_{-\infty}^{\infty} |S(i\omega)|^2 \phi(\omega) d\omega}{\int_{-\infty}^{\infty} \phi(\omega) d\omega}. \tag{4.29}$$

Stability Margins and Maximum Sensitivity

Notice that $|1+G_l(i\omega)|$ is the distance from a point on the Nyquist curve of the loop transfer function to the point -1. See Figure 4.12. The shortest distance from the Nyquist curve of the loop transfer function to the critical point -1 is thus $1/M_s$, which is equal to the stability margin s_m. Compare Figures 4.12 and 4.6. The maximum sensitivity can thus also serve as a stability margin. A requirement on M_s gives the following bounds for gain and phase margins

$$\begin{aligned} g_m &\geq \frac{M_s}{M_s - 1} \\ \varphi_m &\geq 2 \arcsin\left(\frac{1}{2M_s}\right). \end{aligned}$$

The requirement $M_s = 2$ implies that $g_m \geq 2$ and $\varphi_m \geq 29°$ and $M_s = 1.4$ implies that $g_m \geq 3.5$ and $\varphi_m \geq 41°$.

Nonlinearities in the Loop

The condition that the Nyquist curve of the loop transfer function is outside a circle at the critical point with radius $1/M_s$ has strong implications. It follows from Nyquist's stability criterion that the system remains stable even if the gain is increased by the factor $M_s/(M_s - 1)$ or if it is decreased by the factor $M_s/(M_s + 1)$. More surprising is that the closed loop is stable even if a static nonlinearity f is inserted in the loop, provided that

$$\frac{M_s}{M_s + 1} < \frac{f(x)}{x} < \frac{M_s}{M_s - 1}. \tag{4.30}$$

A small value of M_s thus ensures that the system will remain stable in spite of nonlinear actuator characteristics. With $M_s = 2$ the function lies in a sector limited by straight lines through the origin with slopes $2/3$ and 2. With $M_s = 1.4$ the slopes are between 0.28 and 3.5.

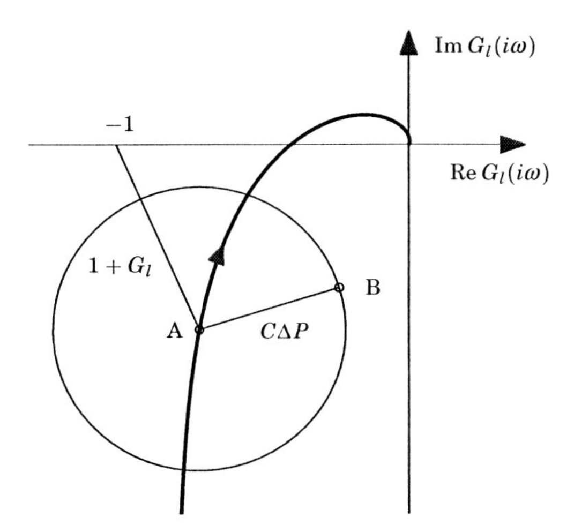

Figure 4.13 Nyquist curve of a nominal loop transfer function and its uncertainty caused by process variations ΔP.

Large Variations

We will now investigate conditions for the system to remain stable under large variations in the process transfer function. Assume that the process transfer function changes from P to $P + \Delta P$, where ΔP is a stable transfer function. Consider a point A on the the Nyquist curve of the loop transfer function; see Figure 4.13. This point then changes from A to B in the figure. The distance from the critical point -1 to the point A is $|1 + G_l|$. This means that the perturbed Nyquist curve will not reach the critical point -1 provided that

$$|C\Delta P| < |1 + G_l|,$$

which implies

$$|\Delta P| < \left|\frac{1 + G_l}{C}\right|. \tag{4.31}$$

Notice that the condition is conservative because it follows from Figure 4.13 that much larger changes can be made in directions from the critical point. The condition (4.31) must be valid for all points on the Nyquist curve, i.e, pointwise for all frequencies. The condition (4.31) for stability can then be written as

$$\left|\frac{\Delta P(i\omega)}{P(i\omega)}\right| < \frac{1}{|T(i\omega)|}, \tag{4.32}$$

where T is the complementary transfer function. The inequality (4.32) tells that large relative perturbations are permitted as long as T is small. A simple conservative estimate of the permissible relative error in the process transfer function is $1/M_t$ where

$$M_t = \max_{\omega} |T(i\omega)| = \max_{\omega}\left|\frac{P(i\omega)C(i\omega)}{1 + P(i\omega)C(i\omega)}\right| = \max_{\omega}\left|\frac{G_l(i\omega)}{1 + G_l(i\omega)}\right|, \tag{4.33}$$

is the largest magnitude of $|T|$. Notice that M_t is also the largest gain of the transfer function from set point to output for a system with error feedback.

Equation 4.32 can also be written as

$$|\Delta P(i\omega)| < \frac{|P(i\omega)|}{|T(i\omega)|}. \tag{4.34}$$

It follows from this equation that the magnitude of the permissible error $|\Delta P(i\omega)|$ is small when $|P(i\omega)|$ is less than $|T(i\omega)|$. High model precision is thus required for frequencies where the gain of the closed-loop system is larger than the gain of the open-loop system.

Graphical Interpretation of Constraint on Sensitivities

The requirements that the sensitivities are less than given values have nice geometric interpretations in the Nyquist plot. Since the sensitivity is defined by

$$S(i\omega) = \frac{1}{1 + G_l(i\omega)},$$

it follows that the sensitivity has constant magnitude on circles with center at the critical point -1. The condition that the largest sensitivity is less than M_s is equivalent to the condition that the Nyquist curve of the loop transfer function is outside a circle with center at -1 and radius $1/M_s$.

There is a similar interpretation of the complementary sensitivity

$$T = \frac{G_l(i\omega)}{1 + G_l(i\omega)}.$$

Introducing

$$G_l(i\omega) = \mathrm{Re}G_l(i\omega) + i\mathrm{Im}G_l(i\omega) = x + iy,$$

we find that the magnitude of T is given by

$$|T| = \frac{\sqrt{x^2 + y^2}}{\sqrt{(1 + x)^2 + y^2}}.$$

The magnitude of the complementary sensitivity function is constant if

$$x^2 + y^2 = M_t^2((1 + x)^2 + y^2) = M_t^2(1 + 2x + x^2 + y^2),$$

or

$$x^2 \frac{M_t^2 - 1}{M_t^2} + 2x + y^2 \frac{M_t^2 - 1}{M_t^2} + 1 = 0.$$

This condition can be written as

$$\begin{aligned} &x^2 + 2\frac{M_t^2}{M_t^2 - 1}x + y^2 + \frac{M_t^2}{M_t^2 - 1} \\ &= \left(x + \frac{M_t^2}{M_t^2 - 1}\right)^2 + y^2 + \frac{M_t^2}{M_t^2 - 1} - \left(\frac{M_t^2}{M_t^2 - 1}\right)^2 \\ &= \left(x + \frac{M_t^2}{M_t^2 - 1}\right)^2 + y^2 - \frac{M_t^2}{(M_t^2 - 1)^2} = 0. \end{aligned}$$

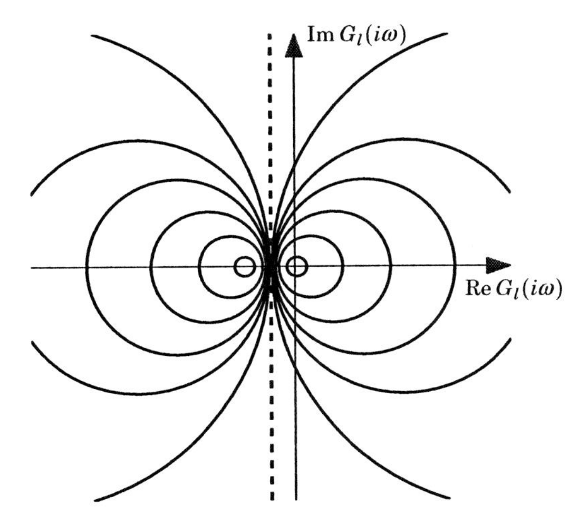

Figure 4.14 Loci where the complementary sensitivity function has constant magnitude. The solid lines show points where the magnitude of the sensitivity function is $M_t = 1.1$, 1.2, 1.4, 1.5, 2, and 5 and the inverses of these values. The dashed line corresponds to $M_t = 1$.

This is a circle with center at $x = -M_t^2/(M_t^2 - 1)$ and $y = 0$, and with radius $r = M_t/(M_t^2 - 1)$. For $M_t = 1$ the circle degenerates to the straight line with $x = -0.5$. The requirement that the complementary sensitivity function is less than M_t thus implies that the Nyquist curve is outside the corresponding circle. The loci of constant gain of the complementary sensitivity function G_l are shown in Figure 4.14. Notice that the circles enclose the critical point -1. Notice also that the closed-loop transfer function is insensitive to variations at frequencies where the loop transfer function is far from the origin, particularly if the Nyquist curve is close to the straight line $\mathrm{Re}G_l(i\omega) = -0.5$. This implies that controllers with the property

$$T_i \approx T_{ar} \frac{2KK_p}{1 + 2KK_p} \tag{4.35}$$

are very robust. Compare with Section 6.3.

Combined Sensitivities

The requirements that the maximum sensitivity is less than M_s and the complementary sensitivity is less than M_t imply that the Nyquist curve should be outside the corresponding circles. It is possible to find a slightly more conservative condition by determining a circle that encloses both circles as is illustrated in Figure 4.15. The radii and the centers of the circles are given in Table 4.1. In that table we have also given the circles that guarantee that both M_t and M_s are smaller than specified values. A particular simple criteria is obtained if it is required that $M_s = M_t$.

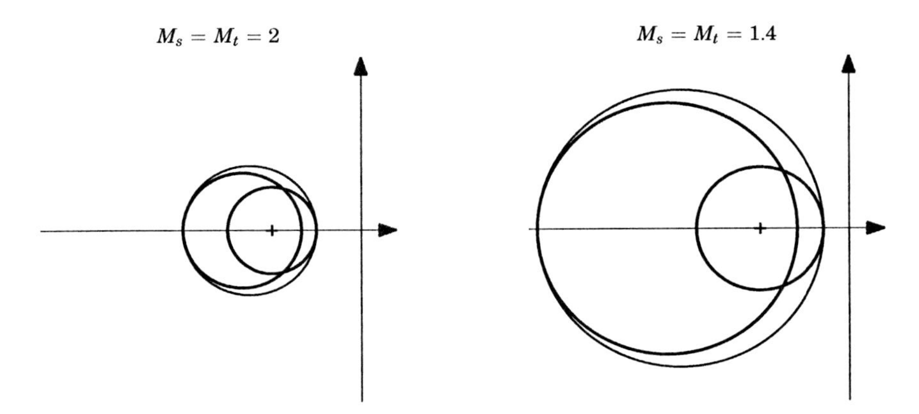

Figure 4.15 Curves for constant sensitivity, constant complementary sensitivity, and constant combined sensitivity.

Table 4.1 Center and radius of circles defining locus for constant sensitivity M_s, constant complementary sensitivity M_t, constant mixed sensitivity, and equal sensitivities $M = M_s = M_t$.

Contour	Center	Radius
M_s	-1	$1/M_s$
M_t	$-\dfrac{M_t^2}{M_t^2 - 1}$	$\dfrac{M_t}{M_t^2 - 1}$
M_s, M_t	$-\dfrac{M_s(2M_t - 1) - M_t + 1}{2M_s(M_t - 1)}$	$\dfrac{M_s + M_t - 1}{2M_s(M_t - 1)}$
$M = M_s = M_t$	$-\dfrac{2M^2 - 2M + 1}{2M(M - 1)}$	$\dfrac{2M - 1}{2M(M - 1)}$

4.7 Robustness to Process Variations

Robustness to process variations is a key issue in control systems design. Process parameters can change for many reasons; they typically depend on operating conditions. Time delays and time constants often change with production levels. Parameters can also change because of aging of equipment. One of the key reasons for using feedback is that it is possible to obtain closed-loop systems that are insensitive to variations in the process.

The analysis of the sensitivity functions in Section 4.6 gives insight into the effects of process variations. Equation 4.25 shows the effect of small process variations on the closed-loop system. In particular it tells that a closed-loop system is insensitive to small process variations for frequencies where the sensitivity function is small.

The robustness inequality given by (4.32) tells that a closed-loop system will remain stable when the process is perturbed from $P(s)$ to $P(s) + \Delta P(s)$,

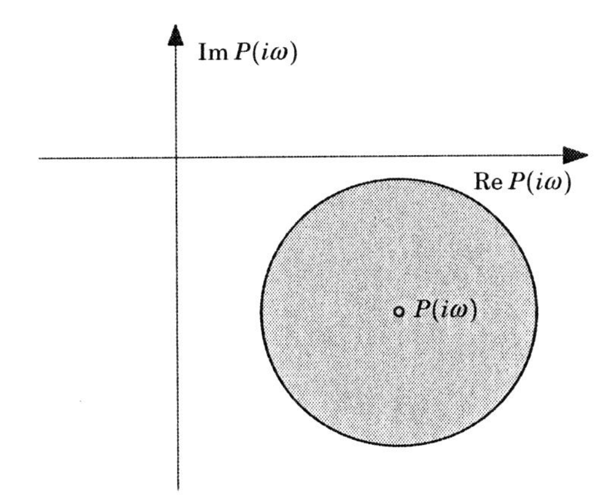

Figure 4.16 Shaded circle shows permissible values of $P(i\omega) + \Delta P(i\omega)$ given by the inequality (4.32). The circle is drawn for $M_t = 2$.

where $\Delta P(i\omega)$ is a stable transfer function, if the perturbations are bounded by

$$\frac{|\Delta P(i\omega)|}{|P(i\omega)|} < \frac{1}{|T(i\omega)|}.$$

This equation is one of the reasons why feedback systems work so well in practice. The mathematical models used to design control system are often strongly simplified. There may be model errors and the properties of a process may change during operation.

Equation (4.32) implies that the closed-loop system will be stable for substantial variations in the process dynamics. The closed-loop system is stable if, for all ω, the perturbed process transfer function $P(i\omega) + \Delta P(i\omega)$ lies in a circle with center at $P(i\omega)$ and radius $1/|T(i\omega)|$, see Figure 4.16. For a system designed with $M_t = 2$ it is possible to change the process gain by factors in the range 0.5 to 1.5 and the phase can be changed by 60°. For a system with $M_t = 1.414$ the gain can be changed by factors in the range 0.3 to 1.7, and the phase can be changed by 45°.

The Cancellation Problem

The sensitivities depend on the loop transfer function $G_l = PC$. Robustness criteria based on sensitivities can give misleading results when there are factors in the process and controller transfer functions that cancel each other. We will illustrate what happens with an example.

EXAMPLE 4.3—CANCELLATIONS
Consider a process with the transfer function

$$P(s) = \frac{1}{s^2 + 2\zeta a s + a^2},$$

and a controller with the transfer function

$$C(s) = \frac{50(s^2 + 2\zeta as + a^2)}{s(s^2 + 10s + 50)}.$$

This controller is a combination of a PID controller with a filter to provide high-frequency roll-off and a notch filter to reduce the excitation of the low-frequency oscillatory mode. The loop transfer function is

$$G_l(s) = \frac{50}{s(s^2 + 10s + 50)}.$$

Notice that the oscillatory modes vanish because the same factor appears both in the controller and the process. The sensitivity functions are

$$\begin{aligned} S(s) &= \frac{s(s+5)^2}{s^3 + 10s^2 + 50s + 50} \\ T(s) &= \frac{1}{s^3 + 10s^2 + 50s + 50}. \end{aligned}$$

With the numerical values $a = 0.5$ and $\zeta = 0.02$ we get $M_s = 1.2$ and $M_t = 1$. A casual application of the robustness inequality (4.32) may lead us to believe that the closed-loop system is robust. However, if a controller is designed based on the nominal value $a = 0.5$ and if the process parameter is changed by 5 percent to $a = 0.4775$ the system becomes unstable. The reason is that if we interpret the parameter variation as an additive disturbance in the process model the small perturbation in the process parameter a translates as a much larger additive disturbance because it is associated with a resonant mode with a very small relative damping. □

The controller in the example is not a good design because it is bad practice to cancel slow process poles.

Other Robustness Measures

There are other robustness results that permit more realistic process variations than the stable additive perturbation used in the robustness inequality (4.32). One result represents the process transfer function as

$$P(s) = \frac{N(s)}{D(s)}$$

where $N(s)$ and $D(s)$ are stable transfer functions. The results state that the system is stable for variations ΔN and ΔD such that

$$\begin{aligned} \max(|N(i\omega)|, |D(i\omega)|) &= \bar{\sigma} \begin{pmatrix} \dfrac{1}{1 + P(i\omega)C(i\omega)} & \dfrac{P(i\omega)}{1 + P(i\omega)C(i\omega)} \\ \dfrac{C(i\omega)}{1 + P(i\omega)C(i\omega)} & \dfrac{P(i\omega)C(i\omega)}{1 + P(i\omega)C(i\omega)} \end{pmatrix} \quad (4.36) \\ &= \frac{\sqrt{(1 + |P(i\omega)|^2)(1 + |C(i\omega)|^2)}}{|1 + P(i\omega)C(i\omega)|} = \Sigma(\omega), \end{aligned}$$

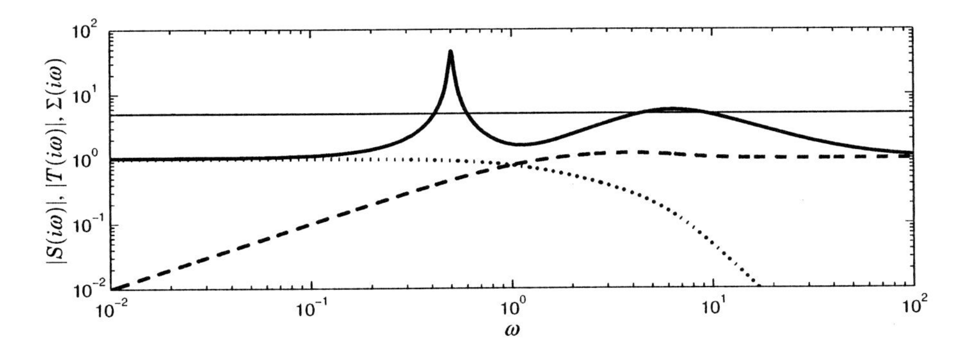

Figure 4.17 The magnitudes of the sensitivity function $|S(i\omega)|$ (dotted), the complementary sensitivity function $|S(i\omega)|$ (dashed) and the largest singular value $\Sigma(i\omega)$ (solid) for the system in Example 4.3.

where $\bar{\sigma}$ is the largest singular value. The parameter

$$M_\sigma = max_\omega \Sigma(\omega)$$

is a robustness measure. The robustness condition (4.32) requires that the process perturbation $\Delta P(s)$ is a stable transfer function. Criteria based on M_σ do not have this limitation because it permits more general perturbations of the process, for example, changing a small stable pole, an integrator, or an unstable pole. It also covers the situation when there are cancellations of poles and zeros. To have good robustness the parameter M_σ should be less than 3 to 5. Notice that M_σ is larger than both M_s and M_t.

To illustrate the effectiveness of M_σ we apply it to Example 4.3. Figure 4.17 shows $|S(i\omega)|$, $|T(i\omega)|$, and $\Sigma(\omega)$ for the nominal system in Example 4.3. We have $M_\sigma = 46$; since this is much larger than 5 it follows that the closed-loop system has very poor robustness.

Another way of investigating robustness is to explore variations in process parameters required to make the closed-loop system unstable. Changes in gain and time constants can be captured by replacing $P(s)$ by $\kappa P(\alpha s)$. Process variations that make the system unstable are given by

$$\kappa P(i\alpha\omega)C(i\omega) + 1 = 0.$$

Solving for α and κ for all ω gives the functions $\kappa(\omega)$ and $\alpha(\omega)$. Peter Hansen has suggested the following robustness index

$$R_{ph} = \min_\omega(\log|\kappa(\omega)| + \log|\alpha(\omega)|). \tag{4.37}$$

This measure is a generalization of gain margin and delay margin.

The largest singular value M_σ and the robustness measure R_{ph} are more complicated than M_s and M_t, and we will therefore mostly use M_s and M_t. It should, however, be kept in mind that evaluating robustness requires some care, particularly when there are cancellations and when the loop transfer function has high peaks above the gain crossover frequency. This is typically the cases for motion control with systems having mechanical resonances and for predictive controllers investigated in Chapter 8.

4.8 Quantifying the Requirements

Having understood the fundamental properties of the basic feedback loop we will now quantify the requirements on a typical control system. To do this it is necessary to have a clear understanding of the primary goal of control. Control problems are very rich as was discussed in Section 4.2. In general, we have to consider

- Load disturbance attenuation
- Measurement noise response
- Robustness to process uncertainties
- Set-point response

The emphasis on the different factors depends on the particular problem. Robustness is important for all applications. Set-point following is the major issue in motion control, where it is desired that the system follows commanded trajectories. In process control, the set point is normally kept constant most of the time; changes are typically made only when production is altered. Rejection of load disturbances is instead the key issue in process control. There are also situations where the purpose of control is not to keep the process variables at specified values. Level control in buffer tanks is a typical example. The reason for using a buffer tank is to smooth flow variations. In such a case the tank level should fluctuate within some limits. A good strategy is to take no control actions as long as the tank level is within certain limits and only apply control when the level is close to the limits. This is called averaging control or surge tank control. There are special strategies developed for dealing with such problems, techniques such as gain scheduling have also been applied. This is discussed in Section 9.3.

The linear behavior of the system is completely determined by the *Gang of Six* (4.3). Neglecting set-point response it is sufficient to consider the *Gang of Four* (4.4). Specifications can be expressed in terms of these transfer functions.

A significant advantage with a structure having two degrees of freedom, or set-point weighting, is that the problem of set-point response can be decoupled from the response to load disturbances and measurement noise. The design procedure can then be divided into two independent steps.

- First design the feedback controller C that reduces the effects of load disturbances and the sensitivity to process variations without introducing too much measurement noise into the system.
- Then design the feedforward F to give the desired response to set points.

We will now discuss how specifications can be expressed in terms properties of the transfer functions (4.4).

Response to Load Disturbances

An estimate of the effectiveness of a control system to reject disturbances is given by (4.26), which compares the outputs of a closed- and an open-loop system when the disturbances are the same. The analysis shows that disturbances with frequencies less than the sensitivity crossover frequency ω_{sc} are attenuated by feedback and that the largest amplification of disturbances is the maximum sensitivity M_s.

We will now turn specifically to load disturbances which are disturbances that drive the process variables away from their desired values. Attenuation of load disturbances is a primary concern for process control. This is particularly the case for regulation problems where the processes are running in steady state with constant set point. Load disturbances are often dominated by low frequencies. Step signals are therefore used as prototype disturbances. The disturbances may enter the system in many different ways. If nothing else is known, it is often assumed that the disturbances enter at the process input. The response of the process variable is then given by the transfer function

$$G_{xd} = \frac{P}{1+PC} = PS = \frac{T}{C}. \tag{4.38}$$

Since load disturbances typically have low frequencies it is natural that the criterion emphasizes the behavior of the transfer function at low frequencies. Filtering of the measurement signal has only marginal effect on the attenuation of load disturbances because the filter only attenuates high frequencies. For a system with $P(0) \neq 0$ and a controller with integral action control the controller gain goes to infinity for small frequencies, and we have the following approximation for small s;

$$G_{xd} = \frac{T}{C} \approx \frac{1}{C} \approx \frac{s}{k_i}. \tag{4.39}$$

Since load disturbances typically have low frequencies this equation implies that integral gain k_i is a good measure of load disturbance rejection.

EXAMPLE 4.4—LOAD DISTURBANCE ATTENUATION

Consider a process with the transfer function $P = (s+1)^{-4}$ and a PI controller with $k = 0.5$ and $k_i = 0.25$. The system has $M_s = 1.56$ and $\omega_{ms} = 0.494$. Figure 4.18 shows the magnitude curve of the transfer function (4.38). The figure shows clearly that feedback reduces the low-frequency gain significantly compared with the open-loop system. The dashed-dotted line in the figure shows the gain curve for the transfer function s/k_i. The figure shows clearly that this is a very good approximation of G_{xd} for low frequencies, approximately up to ω_{ms}. Integral gain k_i is a good measure of load frequency disturbance attenuation. For high frequencies the load disturbance rejection is given by the process dynamics; feedback has no influence. The sensitivity crossover frequency is $\omega_{sc} = 0.25$, which is close to k_i.

Attenuation of load disturbances can also be characterized in the time domain by showing the time response due to a representative disturbance. This

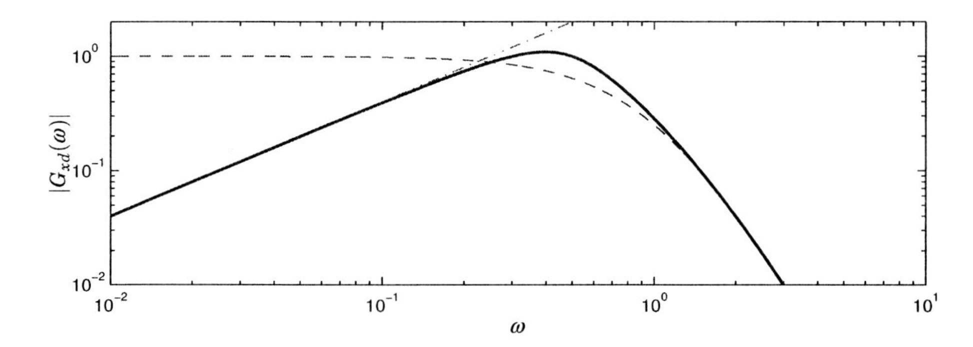

Figure 4.18 The gain of the transfer function G_{xd} from load disturbance to process variable for PI control ($k = 0.5$, $T_i = 2.0$) of the process $P = (s+1)^{-4}$. The dashed dotted curve shows the gain of s/k_i, and the dashed curve shows gain of the process transfer function P.

is illustrated in Figure 4.19, which shows the response of the process output to a unit step disturbance at the process input. The output has its maximum $y_{max} = 0.66$ for $t_{max} = 5.62$. Furthermore, $t_{max}\omega_{ms} = 2.76$, integrated error $IE = 4.00$ and integrated absolute error $IAE = 4.26$. □

The steady-state error caused by a unit step load disturbance for proportional control is

$$e_{ss} = \frac{P(0)}{1 + kP(0)}, \tag{4.40}$$

where k is the proportional gain of the controller. As indicated in Figure 4.19, the steady-state error for proportional control can be used as an approximation of the largest error for PID control. For the system in Example 4.4 we have $P(0) = 1$ and $k = 0.5$ and (4.40) gives the estimate $e_{max} \approx e_{ss} = 1/1.5 = 0.67$ which is close to the correct value 0.66.

Response to Measurement Noise

An inevitable consequence of using feedback is that measurement noise is fed back into the system. Measurement noise, which typically has high frequencies, generates undesirable control actions and variations in the process variable. Rapid variations in the control variable are detrimental because they cause wear in valves and motors and they even saturate the actuator. It is important to keep these variations at a reasonable level. A typical requirement is that the variations are only a fraction of the span of the control signal. The variations can be influenced by filtering and by proper design of the high-frequency properties of the controller.

The effects of measurement noise are thus captured by the transfer function from measurement noise to the control signal

$$G_{un} = \frac{C}{1 + PC} = CS = \frac{T}{P}. \tag{4.41}$$

For low frequencies (small s) the transfer function approaches $1/P(0)$ and for

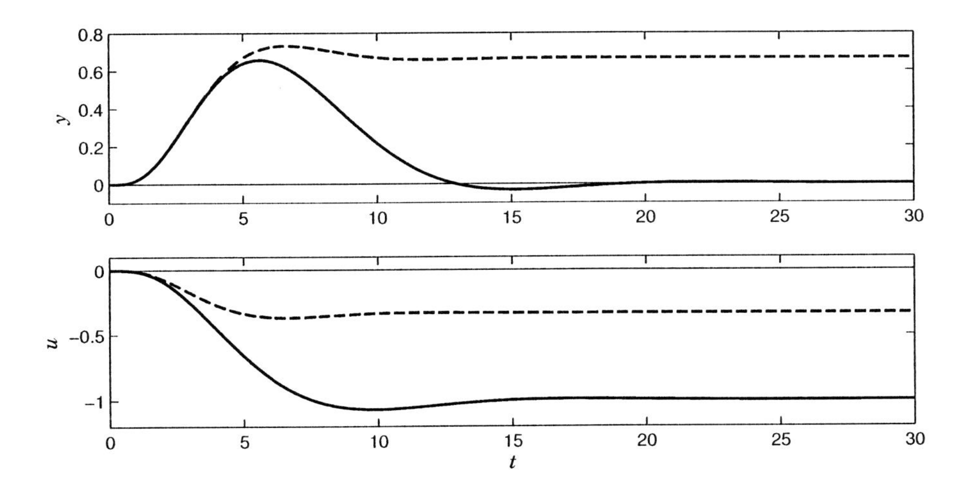

Figure 4.19 Response to a load disturbance in the form of a unit step with a PI controller having parameters $k = 0.5$ and $k_i = 0.25$ and the process $P = (s+1)^{-4}$. The dashed curve shows the response to a proportional controller with gain $k = 0.5$.

high frequencies (large s) we have approximately

$$G_{un} \approx C.$$

For an ideal PID controller the transfer function G_{un} becomes infinite for large s which clearly indicates the necessity to filter the derivative, as discussed in Section 3.3. We illustrate with an example.

EXAMPLE 4.5—EFFECT OF FILTERING

Figure 4.20 shows the gain curve of the transfer function (4.41) for PID control of the process $P = (s+1)^{-4}$. The dashed line is for a controller with a first-order filter of the derivative and the full line for a controller with a second-order filter of the measured signal. The significant differences in the transfer functions for high frequencies is a good motivation for preferring the controller with filtering of the measurement signal. For low frequencies (small s) the transfer function approaches $1/P(0)$. □

A simple measure of the effect of measurement noise is the largest gain of the transfer function G_{un},

$$M_{un} = \max_{\omega} |G_{un}(i\omega)|. \tag{4.42}$$

For PI control the gain of the transfer function G_{un} has a peak close to the peak of the sensitivity function and we have approximately

$$M_{un} \approx M_s K. \tag{4.43}$$

For PID control the gain of the transfer function G_{un} typically has two local maxima, one is close to the maximum of the sensitivity function. The other peak is larger

$$M_{un} \approx k_d/T_d, \tag{4.44}$$

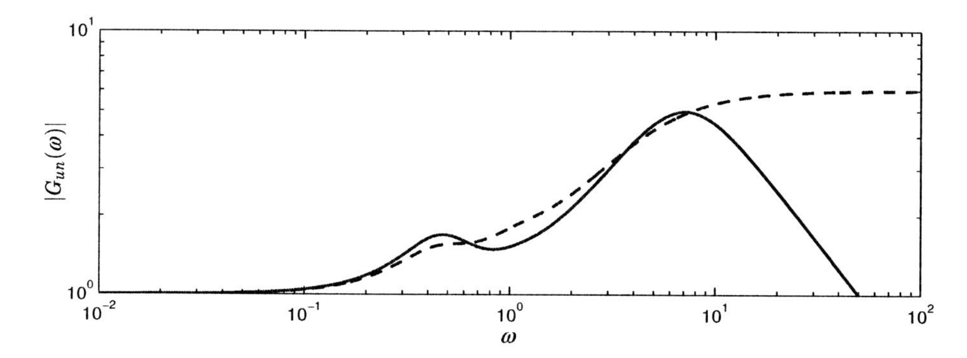

Figure 4.20 The magnitude of the transfer function $G_{un} = CS$ for PID control ($k = 1$, $T_i = 2$, $T_d = 1$, $T_f = 0.2$) of the process $P = (s+1)^{-4}$. The solid line represents a controller with a second-order noise filter of the measured signal (3.16) and the dashed line a controller with a first-order filter of the derivative (3.15).

and it occurs close to the frequency $1/T_f$.

If the standard deviation of the measurement noise is σ_n, a crude estimate of the variations in the control signal is $M_{un}\sigma_n$. More accurate assessment can be made if the power spectrum ϕ_n of the measurement noise is known. The standard deviation of the control signal is then given by

$$\sigma_u^2 = \int_{-\infty}^{\infty} |G_{un}(i\omega)|^2 \phi_n(\omega) d\omega. \tag{4.45}$$

However, it is rare that such detailed information is rarely available for typical applications.

Robustness to Process Variations

The inverse of the maximum sensitivity is the shortest distance from the critical point -1 to the Nyquist curve of the loop transfer function.

The sensitivity to small variations in process dynamics is captured by the sensitivity function. We have

$$\frac{dT}{T} = S\frac{dP}{P}.$$

Variations in process dynamics thus have small influence on the closed-loop system for frequencies where the sensitivity function is small.

Variations in process dynamics may also lead to instability. The condition

$$\frac{|\Delta P(i\omega)|}{|P(i\omega)|} < \frac{1}{|T(i\omega)|}$$

guarantees that a variation $\Delta P(i\omega)$ in the process transfer function does not make the system unstable. Robustness to process variations is thus captured by the sensitivity and the complementary sensitivity functions. Simple measures are the maximum sensitivity M_s, the maximum of the complementary sensitivity M_t, or the largest combined sensitivity M. Typical values of the sensitivities are in the range of $1.2 - 2.0$.

Other measures are the gain margin g_m (typically 2 to 8), the phase margin φ_m (typically 30° to 60°), or the stability margin $s_m = 1/M_s$ (typically 0.5 to 0.8). Compare with Section 4.4.

Trade-offs

Load disturbance attenuation is captured by integral gain k_i. It follows from (4.39) that attenuation of low-frequency disturbances is approximately inversely proportional to k_i. Injection of measurement noise is captured by the noise gain M_{un}. It follows from (4.42) that M_{un} gives the gain from measurement noise to control variable. The trade-off between load disturbance attenuation and injection of measurement noise can thus be achieved by balancing k_i and M_{un}.

Set-Point Response

By using a controller with two degrees of freedom it is possible to obtain any desired response to set-point changes. This will be discussed further in Chapter 5. The limitations are given by the permissible magnitude of the control signal. In some cases only the control error is measured. A controller with two degrees of freedom then cannot be used and the response to set points has to be handled by proper choosing of the controller transfer function. Large overshoots can be avoided by requiring low values of M_t.

Summary

Summarizing we find that the behavior of the system can be characterized in the following way. The transfer function from load disturbance to process variable is

$$G_{yd} = \frac{P}{1+PC} = PS \approx \frac{s}{k_i}, \tag{4.46}$$

where the approximation holds for low frequencies.

The effect of measurement noise can be captured by the noise gain

$$M_{un} = \max_{\omega} |G_{un}(i\omega)| \approx \begin{cases} kM_s & \text{for PI control} \\ k_d/T_f & \text{for the PID controller (3.16),} \end{cases} \tag{4.47}$$

which strongly depends on the filtering of measurement noise.

Stability and robustness to process uncertainties can be expressed by the sensitivity function and the complementary sensitivity function

$$S = \frac{1}{1+PC}, \qquad T = \frac{PC}{1+PC},$$

where the largest values of the sensitivity functions M_s and M_t are good quantitative measures. The parameter $1/M_s$ is the shortest distance from the critical point to the Nyquist curve of the loop transfer function.

Essential features of load disturbance attenuation, measurement noise injection, and robustness can thus be captured by four parameters k_i, M_{un}, M_t, and M_s. An attractive feature of this choice of parameters is that k_i and M_{un} are directly related to the controller parameters and that there are good design methods that can guarantee given M_s and M_t.

4.9 Classical Specifications

The specifications we have given have the advantage that they capture robustness as well as the responses to load disturbances, measurement noise, and set points with only four parameters. Unfortunately, it has been the tradition in PID to judge a system based on one response only, typically the response of the output to a step change in the set point. This can be highly misleading as we have discussed previously. A large number of different parameters have also been used to characterize the responses. For completeness and to connect with classical literature on PID control some of the classical specifications will be summarized in this section.

Criteria Based on Time Responses

Many criteria are related to time responses, for example, the step response to set-point changes or the step response to load disturbances. It is common to use some feature of the error typically extrema, asymptotes, areas, etc.

The maximum error $e_{\max}$ is defined as

$$\begin{aligned} e_{\max} &= \max_{0\leq t<\infty} |e(t)| \\ T_{\max} &= \arg\max |e(t)|. \end{aligned} \tag{4.48}$$

The time $T_{\max}$ where the maximum occurs is a measure of the response time of the system. The integrated absolute error (IAE) is defined as

$$IAE = \int_0^\infty |e(t)|dt. \tag{4.49}$$

A related error is integrated error (IE), defined as

$$IE = \int_0^\infty e(t)dt. \tag{4.50}$$

The criteria IE and IAE are the same if the error does not change sign. Notice that IE can be very small even if the error is not. For IE to be relevant it is necessary to add conditions that ensure that the error is not too oscillatory. The criterion IE is a natural choice for control of quality variables for a process where the product is sent to a mixing tank. The criterion may be strongly misleading, however, in other situations. It will be zero for an oscillatory system with no damping. It will also be zero for a control loop with two integrators.

There are many other criteria, for example, the time multiplied absolute error, defined by

$$ITNAE = \int_0^\infty t^n|e(t)|dt. \tag{4.51}$$

The integrated squared error (ISE) is defined as

$$ISE = \int_0^\infty e(t)^2 dt. \tag{4.52}$$

There are other criteria that take account of both input and output signals, for example, the quadratic criterion

$$QE = \int_0^\infty (e^2(t) + \rho u^2(t))dt, \tag{4.53}$$

where ρ is a weighting factor. The criteria IE and QE can easily be computed analytically, simulations are, however, required to determine IAE.

One reason for using IE is that its value is directly related to the parameter k_i of the PID controller, as is illustrated by the following example.

EXAMPLE 4.6—INTEGRAL GAIN AND IE FOR LOAD DISTURBANCES
Consider the control law

$$u(t) = ke(t) + k_i \int_0^t e(t)dt - k_d \frac{dy}{dt}.$$

Assume that this controller gives a stable closed-loop system. Furthermore, assume that the error is zero initially and that a unit step load disturbance is applied at the process input. Since the closed-loop system is stable and has integral action the control error will go to zero. We thus find

$$u(\infty) - u(0) = k_i \int_0^\infty e(t)dt.$$

Since the disturbance is applied at the process input, the change in control signal is equal to the change of the disturbance. Hence, $u(\infty) - u(0) = 1$ and we get

$$IE = \int_0^\infty e(t)dt = \frac{1}{k_i} = \frac{T_i}{K}. \tag{4.54}$$

□

Integral gain k_i is thus inversely proportional to the integrated error caused by a unit step load disturbance applied to the process input.

Set-Point Response

Specifications on set-point following are typically expressed in the time domain. They may include requirements on rise time, settling time, decay ratio, overshoot, and steady-state offset for step changes in set point. These quantities are defined as follows, see Figure 4.21.

- The *rise time* T_r is defined either as the inverse of the largest slope of the step response or the time it takes for the step response to change from 10 percent to 90 percent of its steady-state value.

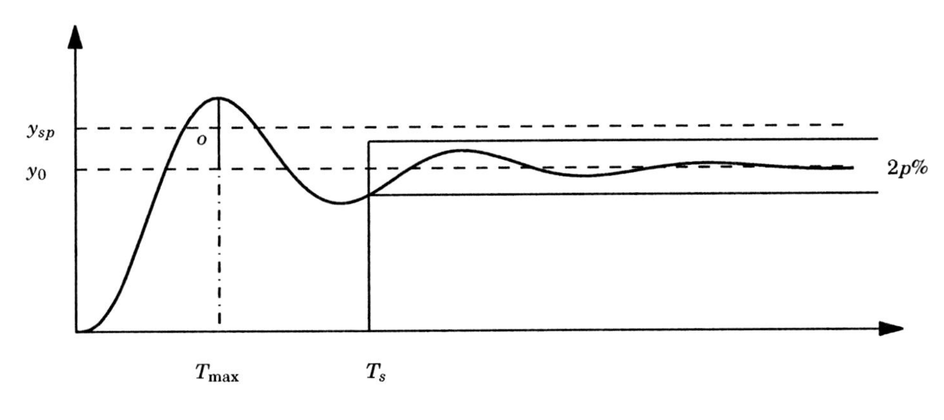

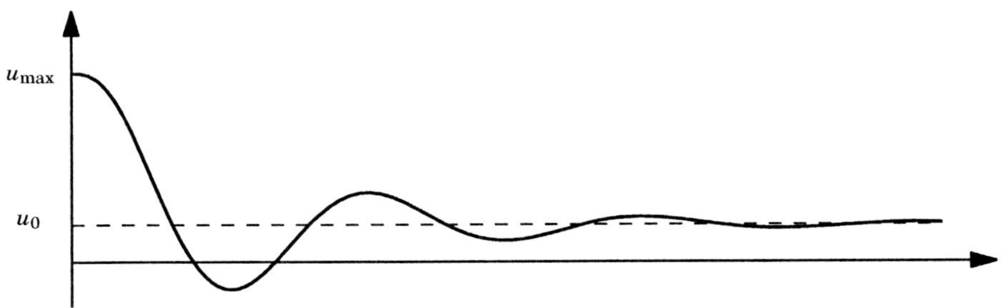

Figure 4.21 Specifications on set-point following based on the time response to a unit step in the set point. The upper curve shows the response of the output, and the lower curve shows the corresponding control signal.

- The *settling time* T_s is the time it takes before the step response remains within p percent of its steady-state value. The values $p = 1, 2$, and 5 percent of the steady-state value are commonly used.
- The *decay ratio* d is the ratio between two consecutive maxima of the error for a step change in set point or load; see Figure 2.35. The value $d = 1/4$, which is called quarter amplitude damping, has been used traditionally. This value is, however, normally too high, as will be shown later.
- The *overshoot* o is the ratio between the difference between the first peak and the steady-state value of the step response. It is often given in percent. In industrial control applications it is common to specify a maximum overshoot of 8 to 10 percent. In many situations it is desirable, however, to have an over-damped response with no overshoot.
- The *steady-state error* $e_{ss} = y_{sp} - y_0$ is the steady-state control error e. This is always zero for a controller with integral action.

Actuators may have rate limitations, which means that step changes in the control signal will not appear instantaneously. In motion control systems it is often more relevant to consider responses to ramp signals instead of step signals.

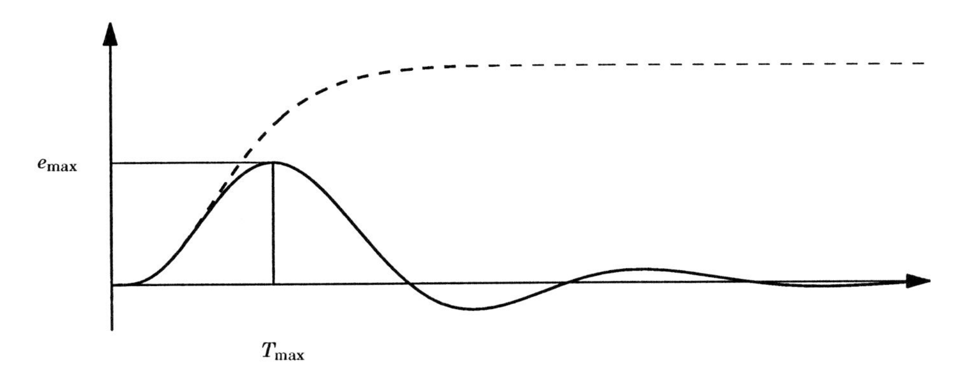

Figure 4.22 The error due to a unit step load disturbance at the process input and some features used to characterize attenuation of load disturbances. The dashed curve show the open-loop error.

Response to Load Disturbances

The response to load disturbances is of primary importance in process control. Figure 4.22 shows the output for a step disturbance in a load applied at the process input and some features that are used to characterize the response. The figure shows the *maximum error* e_{max}, the time it takes to reach the maximum T_{max}, and the settling time T_s. In addition to these numbers the integrated error (IE) or the integrated absolute error (IAE) are also commonly used to characterize the load disturbance response. The maximum error for a unit step and the time where this is reached can be approximated by

$$\begin{aligned} e_{max} &= \frac{1}{1+kP(0)} \\ T_{max} &\approx \frac{3}{\omega_{ms}}. \end{aligned} \tag{4.55}$$

We illustrate these estimates by an example.

EXAMPLE 4.7—ESTIMATING THE MAXIMUM ERROR

When a process with the transfer function $P(s) = (s+1)^{-4}$ is controlled by a PI controller having parameters $k = 0.78$ and $k_i = 0.38$, we have $\omega_{ms} = 0.559$, $e_{max} = 0.59$, and $T_{max} = 5.15$. The estimates above give $e_{max} \approx 0.56$, $T_{max} = 5.6$. □

Criteria Based on Frequency Responses

Specifications can also be related to frequency responses. Since specifications were originally focused on set-point response it was natural to consider the transfer function from set point to output. A typical gain curve for this response is shown in Figure 4.23. It is natural to require that the steady-state gain is one. Typical specifications are then as follows:

- The *resonance peak* M_p is the largest value of the frequency response.

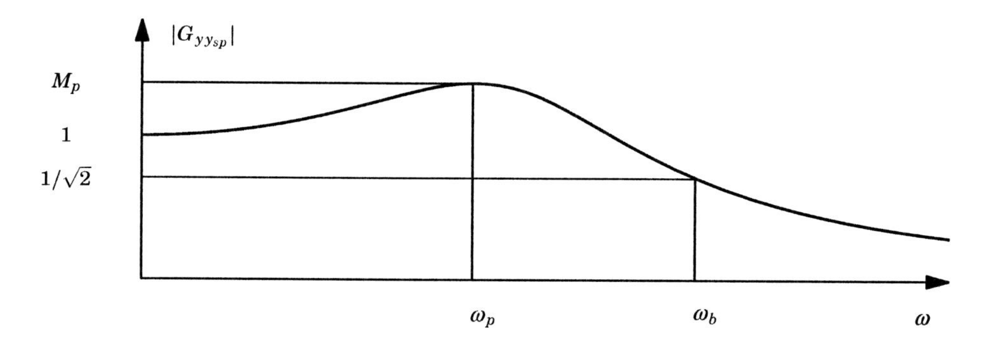

Figure 4.23 Gain curve for transfer function from set point to output.

- The *peak frequency* ω_p is the frequency where the maximum occurs.
- The *bandwidth* ω_b is the frequency where the gain has decreased to $1/\sqrt{2}$.

For a system with error feedback the transfer function from set point to output is equal to the complementary transfer function, and we have $M_p = M_t$.

Specifications can also be related to the loop transfer function. Useful features that have been discussed previously are

- Gain crossover frequency ω_{gc}.
- Gain margin g_m.
- Phase margin φ_m.
- Maximum sensitivity M_s.
- Frequency where the sensitivity function has its maximum ω_{ms}.
- Sensitivity crossover frequency ω_{sc}.
- Maximum complementary sensitivity M_t.
- Frequency where the complementary sensitivity function has its maximum ω_{mt}.

Relations between Time and Frequency Domain Specifications

There are approximate relations between specifications in the time and frequency domain. Let $G(s)$ be the transfer function from set point to output. In the time domain the response speed can be characterized by the rise time T_r, the average residence time T_{ar}, or the settling time T_s. In the frequency domain the response time can be characterized by the closed-loop bandwidth ω_b, the gain crossover frequency ω_{gc}, and the sensitivity frequency ω_{ms}. The product of bandwidth and rise time is approximately constant, and we have

$$T_r\omega_b \approx 2. \tag{4.56}$$

It has previously been shown that

$$T_{ar} = -\frac{G'(0)}{G(0)};$$

see (2.16).

The overshoot of the step response o is related to the peak M_p of the frequency response in the sense that a larger peak normally implies a larger overshoot. Unfortunately, there are no simple relations because the overshoot also depends on how quickly the frequency response decays. For $M_p < 1.2$ the overshoot o in the step response is often close to $M_p - 1$. For larger values of M_p the overshoot is typically less than $M_p - 1$. These relations do not hold for all systems; there are systems with $M_p = 1$ that have a positive overshoot. These systems have transfer functions that decay rapidly around the bandwidth.

To avoid overshoots in systems with error feedback it is therefore advisable to require that the maximum of the complementary sensitivity function is small, say, $M_t = 1.1 - 1.2$ in order to avoid too large overshoot in the step response to command signals.

Performance Assessment

Before designing a controller it is useful to make a preliminary assessment of achievable performance. It is interesting to know if a PID controller is sufficient or if the performance can be increased substantially by using a more complex controller. It is also interesting to know if a PI controller is sufficient or if derivative action gives significant improvements. To make the assessment we need some measure of performance. In this section we will use the gain crossover frequency ω_{gc} as a yardstick. When the phase margin is 60° this frequency is equal to the sensitivity crossover frequency ω_{sc}. Recall from Section 4.6 that disturbances with frequencies lower than ω_{sc} are reduced by feedback. For phase margins lower than 60° we have $\omega_{sc} < \omega_{gc}$, and for larger phase margins we have $\omega_{sc} > \omega_{gc}$. Attenuation of load disturbances is thus improved with increasing gain crossover frequencies.

Process dynamics with non-minimum phase properties like a time delay imposes fundamental limitations on the achievable performance which can be expressed by the inequality

$$\omega_{gc} L < a, \tag{4.57}$$

where a is a number less than 1. Since the true time delay L is rarely known it can be approximated by the apparent time delay L_a. Figure 4.24 shows the product $\omega_{gc} L_a$ for a large batch of systems under robust PID control. The circles which represents FOTD systems show that the product is 0.5 for FOTD systems. For high order systems with lag dominated dynamics the product $\omega_{gc} L_a s$ is larger than 0.5 because the apparent time delay of the approximating FOTD model is larger than the true time delay of the system.

Consider a closed-loop system with a process having transfer function $P(s)$ and a controller with transfer function $C(s)$. The gain crossover frequency is defined by

$$\arg P(i\omega_{gc}) + \arg C(i\omega_{gc}) = -\pi + \varphi_m. \tag{4.58}$$

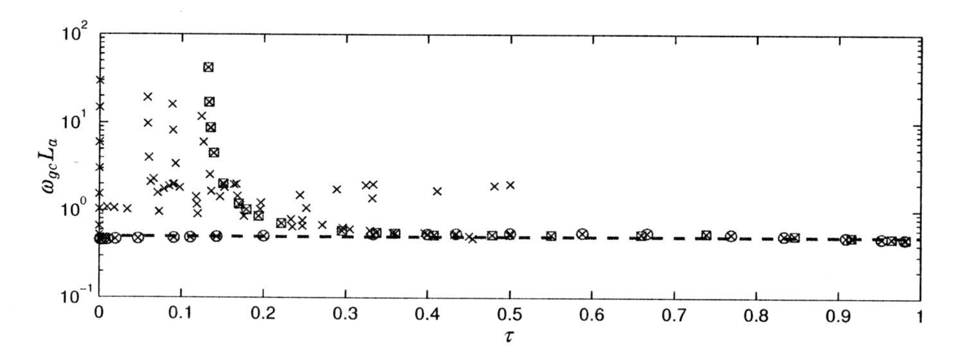

Figure 4.24 The product of gain crossover frequency ω_{gc} and apparent time delay L_a as a function of normalized dead time, for a large set of systems under PID control. The circles show results for FOTD systems and the squares for SOTD systems. The PID controllers are designed to give a combined sensitivity $M = 1.4$. All systems are described in Section 7.2. The dashed line gives the relation $\omega_{gc}La = 0.5$.

Notice that the units radians is used in this equation. A PD controller has a maximum phase lead of about 60° ($\pi/3$ rad), a proportional controller has zero phase lag, a PI controller has a phase lag of about 45° ($\pi/4$ rad), and a PID controller can have a phase lead of about 45°. If a phase margin of 45° is desired it follows from Equation 4.58 that crossover frequencies for PI, PID, and PD control are the frequencies where the process has phase-lags of 90°, 135°, and 195°, respectively. These frequencies are denoted as ω_{90}, ω_{135}, and ω_{195}. An estimate of the controller gains required can be obtained by computing the process gains at the corresponding frequencies. Notice that this assessment only requires the process transfer function. We illustrate by two examples.

EXAMPLE 4.8—MULTI-LAG PROCESS
Consider a process with the transfer function

$$P(s) = \frac{1}{(s+1)^4}.$$

We have $\omega_{90} = 0.41$ and $K_{90} = 0.73$, where K_{90} denotes the process gain at ω_{90}. Furthermore, we have $\omega_{135} = 0.67$, $K_{135} = 0.48$, and $\omega_{195} = 1.14$, $K_{195} = 0.19$. We can thus expect that disturbances with frequencies lower than 0.4 rad/s can be reduced by PI control. Since ω_{135} is moderately larger than ω_{90} we can expect that a PI controller can be improved somewhat by introducing derivative action. The gain of a PID controller can be expected to be about twice as large as for PI control. Also notice that the apparent time delay is $L = 2.14$ and that $\omega_{gc} = 0.47$ which is in good agreement with (4.24). □

EXAMPLE 4.9—A LAG-DOMINATED PROCESS
Consider a process with the transfer function

$$P(s) = \frac{1}{(s+1)(0.1s+1)(0.01s+1)(0.001s+1)}.$$

Table 4.2 Parameters of PI controllers for the process $P(s) = (s+1)^{-3}$ designed with different M_s.

M_s	k	k_i	M_{un}	b	ω_{ms}	IAE	T_s	M_t
1.2	0.355	0.171	0.426	1.00	0.671			1.00
1.4	0.633	0.325	0.866	1.00	0.74	3.07	10.3	1.00
1.6	0.862	0.461	1.379	0.93	0.79	2.28	7.87	1.05
1.8	1.056	0.580	1.901	0.70	0.83	2.00	6.77	1.24
2.0	1.222	0.685	2.444	0.50	0.86	1.89	6.27	1.45

We have $\omega_{90} = 3.0$, $K_{90} = 0.3$, $\omega_{135} = 9.9$, $K_{135} = 0.07$, and $\omega_{195} = 47.5$, $K_{195} = 0.004$. We can thus expect that disturbances with frequencies lower than 3 rad/s can be reduced by PI control. With a PID controller disturbances with frequencies up to 9.9 rad/s can be reduced. In this case, there are significant performance benefits from using derivative action. Since ω_{195} is much larger than ω_{135} there may be substantial benefits by using more complex controllers. Since the process gain K_{135} is so low the improved benefits require controllers with high gain, and the benefits may be not be realizable unless sensor noise if very low. □

Design Parameters

In control system design and implementation it is convenient to have a parameter that can be changed to influence the key trade-offs in a design problem. Performance expressed by fast response time and good attenuation of load disturbances can be obtained, but large control signals may be required. Stricter requirements on robustness may lead to poorer performance.

The trade-off between performance and robustness varies between different control problems. Therefore, it is desirable to have a design parameter to change the properties of the closed-loop system. Ideally, the parameter should be directly related to the performance or the robustness of the system; it should not be process oriented. There should be good default values so a user is not forced to select some value. This is of special importance when the design procedure is used for automatic tuning. The design parameter should also have a good physical interpretation and natural limits to simplify its adjustment.

The behavior of a system can often be characterized by a few dominant poles that are close to the origin. When there is one real dominant pole this pole can be used as a design parameter. This is used, for example, in the design method Lambda Tuning, which will be discussed in Section 6.5. When the dominant poles are complex the distance from the origin of the poles ω_0 and their relative damping ζ are good design parameters. This applies to controllers based on pole placement design, which will be discussed in Section 6.4. The maximum sensitivity M_s or the combined sensitivity M are good design variables for regulation problems. This is illustrated in Figure 4.25, which shows the effect of M_s on time and frequency responses for a PI controller, and in Table 4.2,

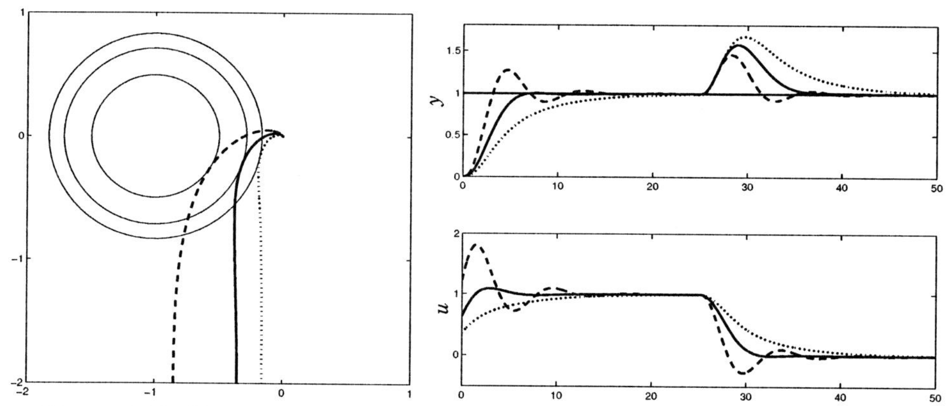

Figure 4.25 Illustrates the effects of using M_s as a design parameter. The left curves show the Nyquist plots of the loop transfer functions together with the circles of constant M_s=1.2 (dotted), 1.4, 2.0 (dashed). The curves on the right show process outputs and control signals for the different values of the design parameters.

which gives numerical values of controller parameters and various criteria. The step response for $M_s = 1.2$ has no overshoot and relatively long settling time. The settling time decreases and the overshoot increases as the value of M_s is increased. Notice that set-point weighting is used for larger values of M_s to reduce the overshoot. The performance also increases with increasing M_s. The values of IAE decrease with about a factor of 2. Apart from use in design it is also possible to implement systems where the user can adjust the design parameters on line.

4.10 Summary

In this section we have summarized some important issues for the design of control systems, with particular emphasis for PID control. A discussion of the basic feedback loop showed that it is necessary to consider six transfer functions (the Gang of Six) to determine the properties of a feedback loop. This is severely neglected in most elementary texts in control and in the literature of PID controllers. The notion of stability was then discussed. This is important because the risk for instability is the main disadvantage of feedback. Stability criteria and stability margins were also introduced. The stability criteria also made it possible to obtain the parameter regions which give a stable closed-loop system under PID control. Characterization of a closed-loop system by its poles and zeros give very valuable insight, and it is also closely related to many design methods. The sensitivity function and the complementary sensitivity functions, which are useful to express the robustness to parameter variations, were also introduced. The problem of controller design was then discussed, and a number of criteria used to give specifications on a control system were also introduced. The key factors are load disturbances, measurement noise, robustness, and set-point response. A nice result is that for systems having two degrees of freedom it is possible to design for disturbances and robustness. The desired set-point response can then be obtained by using feedforward. For PID control set-point weighting is a special form of controller with two degrees of freedom that often is sufficient. It is also shown that the key requirements can be parameterized in a simple way. Load disturbance response is captured by integral gain of the controller k_i. Effects of measurement noise are captured by the noise gain M_{un}, which has a simple relation to controller parameters. Robustness is captured by the maximum sensitivities M_s and M_t.

4.11 Notes and References

Control system design is complicated because many factors have to be considered and trade-offs have to be made. It is therefore natural that it took time before a good understanding was developed. Early work on control design was based on the differential equations describing the closed-loop system. A typical approach was to adjust the controller parameters so that the dominant closed-loop poles had desired properties. Systematic methods for control system design appeared in the 1940s when the field of control emerged. The design methods were based on frequency response, computations were based on graphics, and modeling was often done experimentally by perturbing the system with sinusoidal signals; see [Bode, 1945; James *et al.*, 1947; Brown and Campbell, 1948; Chestnut and Mayer, 1959]. It is noteworthy that particular emphasis was given to robustness to process variations. The insightful book [Horowitz, 1963] gives a mature account. This book also emphasizes the important concept of controllers that have two degrees of freedom. Such controllers admit a decoupling of the responses to set points and load disturbances.

There was a paradigm shift in the 1960s when differential equations reappeared in the name of state-space systems; see [Zadeh and Desoer, 1963]. This coincided with the appearance of digital computers, which permitted efficient numerical computations. The important ideas of optimal control and Kalman filtering are key contributions; see [Bellman, 1957; Kalman, 1960; Kalman and Bucy, 1961; Kalman, 1961; Pontryagin *et al.*, 1962; Athans and Falb, 1966; Bryson and Ho, 1969].

There was a very dynamic development of theory, many design methods and efficient computational techniques were also developed; see [Boyd and Barratt, 1991].

The robustness issue was unfortunately neglected for a long period. This was remedied with the emergence of the so called $\mathcal{H}_\infty$ theory, which led to a reconciliation with the classical frequency response methods. The books [Doyle *et al.*, 1992; Zhou *et al.*, 1996; Skogestad and Postlethwaite, 1996] give a balanced perspective. The robustness criteria M_s, M_t, and M_σ are results of robust control theory. An interesting novel robustness criterion which focuses on variations in the process parameters has been suggested in [Hansen, 2000] and [Hansen, 2003]. The question of fundamental limitations is closely related to robustness as is discussed by [Åström, 2000]. For process control the true time delay is a key limiting factor. Notice that the true time delay can be different from the apparent time delay obtained when fitting FOTD models.

Many practitioners of control have been fully aware of the importance of the compromise between performance and robustness; see [Shinskey, 1990], and it is now pleasing to see that robust control theory has made it possible to merge theory and practice; see [Panagopoulos and Åström, 2000].

In the literature on PID control there has been a long discussion, whether tuning should be based on response to set-point changes or load disturbances. It is surprising that so many papers just show the response of process output to a step change in the set point. Since steady-state regulation is the essential problem in process control, load-disturbance responses are more important than responses to set points as has been emphasized many times by Shinskey; see for example [Shinskey, 1996]. One of the useful conclusions of robust control theory is that six responses are required to get a complete understanding of a closed loop system,

Another lesson from robust control theory is that high-frequency roll-off improves robustness. This is a good reason to use effective filtering in PID control.

5

Feedforward Design

5.1 Introduction

Feedforward is a simple and powerful technique that complements feedback. Feedforward can be used both to improve the set-point responses and to reduce the effect of measurable disturbances. Use of feedforward to improve set-point response has already been discussed in connection with set-point weighting in Section 3.4. We will now give a systematic treatment of design of feedforward control and also discuss design of model-following systems. The special case of set-point weighting will be discussed in detail, and we will present methods for determining the set-point weights. We will also show how feedforward can be used to reduce the effect of disturbances that can be measured.

5.2 Improved Set-Point Response

Feedforward can be used very effectively to improve the set-point response of the system. By using feedforward it is also possible to separate the design problem into two parts. The feedback controller is first designed to give robustness and good disturbance rejection and the feedforward is then designed to give a good response to set-point changes.

Effective use of feedforward requires a system structure that has two degrees of freedom. An example of such a system is shown in Figure 3.10. It is first assumed that the system has the structure shown in Figure 5.1. Let the process have the transfer function $P(s)$. We assume that a feedback controller $C(s)$, which gives good rejection of disturbances and good robustness, has been designed, and we will consider the problem of designing a feedforward compensator that gives a good response to set-point changes.

The feedforward compensator is characterized by the transfer functions $M_u(s)$ and $M_y(s)$, where $M_y(s)$ gives the desired set-point response. The system works as follows. When the set point is changed the transfer function $M_u(s)$ generates the signal u_{ff}, which gives the desired output when applied as input to the process. The desired output y_m is generated by $M_y(s)$. Under ideal conditions this signal is equal to the process output y. The control error e

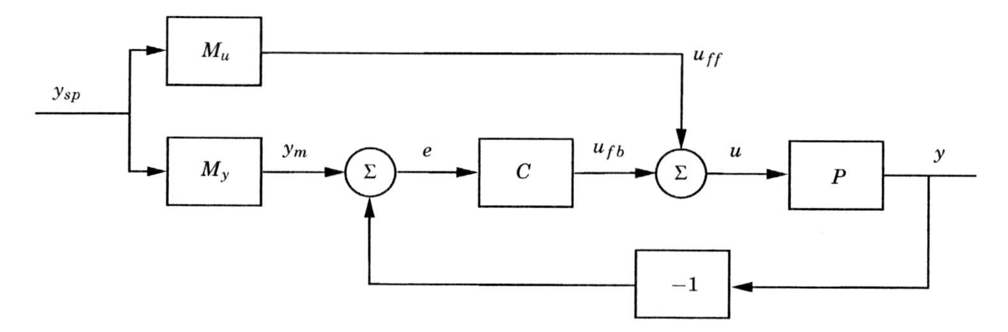

Figure 5.1 Block diagram of a system with two degrees of freedom.

is zero, and the feedback signal u_{fb} remains constant. If there are disturbances or modeling errors the signal y_m and y will differ. The feedback then attempts to bring the error to zero. The transfer function from set point to process output is

$$G_{yy_{sp}}(s) = \frac{P(CM_y + M_u)}{1+PC} = M_y + \frac{PM_u - M_y}{1+PC}. \tag{5.1}$$

The first term represents the desired transfer function. The second term can be made small in two ways. Feedforward compensation can be used to make $PM_u - M_y$ small, or feedback compensation can be used to make the error small by making the loop gain PC large. The condition for ideal feedforward is

$$M_y = PM_u. \tag{5.2}$$

Notice the different character of feedback and feedforward. With feedforward it is attempted to match two transfer functions, and with feedback it is attempted to make the error small by dividing it by a large number. With a controller having integral action the loop gain is very large for small frequencies. It is thus sufficient to make sure that the condition for ideal feedforward holds at higher frequencies. This is easier than to satisfy the condition (5.2) for all frequencies.

System Inverses

From (5.2) the feedforward compensator M_u is

$$M_u = P^{-1}M_y, \tag{5.3}$$

which means that it contains an inverse of the process model P. A key issue in design of feedforward compensators is thus to find inverse dynamics. It is easy to compute the inverse formally. There are, however, severe fundamental problems in system inversion, which are illustrated by the following examples.

EXAMPLE 5.1—INVERSE OF FOTD SYSTEM
The system

$$P(s) = \frac{1}{1+sT}\, e^{-sL}$$

has the formal inverse

$$P^{-1}(s) = (1 + sT)e^{sL}.$$

This system is not a causal dynamical system because the term e^{sL} represents a prediction. The term $(1 + sT)$ requires an ideal derivative, which also is problematic as was discussed in Section 3.3. Implementation of feedforward thus requires approximations. □

EXAMPLE 5.2—INVERSE OF SYSTEM WITH RHP ZERO
The system

$$P(s) = \frac{s-1}{s+2}$$

has the inverse

$$P^{-1}(s) = \frac{s+2}{s-1}.$$

Notice that this inverse is an unstable system. □

It follows from (5.2) that there will be pole-zero cancellations when designing feedforward. The canceled poles and zeros must be stable and sufficiently fast; otherwise, there will be signals in the system that will grow exponentially or decay very slowly.

The difficulties in computing inverses can be avoided by restricting the choice of M_y. Since $M_u = P^{-1}M_y$ we can require that the transfer function M_y has a time delay that is at least as long as the time delay of P. Further, M_y and P must have the same zeros in the right half plane. To avoid differentiation, the pole excess in M_y must be at least as large as the pole excess in P. One possibility is to approximate process dynamics by a simple model and to choose M_y as a model having the same structure. To design feedforward we thus have to compute approximate system inverses with suitable properties.

Approximate Inverses

Different ways to find approximate process models were discussed in Section 2.8. Here we will give an additional method that is tailored for design of feedforward control.

Let $P^{\dagger}$ denote the approximate inverse of the transfer function P. A common approximation in process control is to neglect all dynamics and simply take the inverse of the static gain, i.e.;

$$P^{\dagger}(s) = P(0)^{-1}.$$

A number of results on more accurate system inverses have been derived in system theory. Some of these will be shown here. Note that the inverse transfer function only has to be small for those frequencies where the sensitivity function is large.

EXAMPLE 5.3—APPROXIMATE INVERSE OF FOTD SYSTEM
The system

$$P(s) = \frac{1}{1+sT} e^{-sL}$$

has the approximate inverse

$$P^\dagger(s) = \frac{1+sT}{1+sT/N},$$

where N gives the frequency range where inversion is valid. □

EXAMPLE 5.4—APPROXIMATE INVERSE OF SYSTEM WITH RHP ZERO
The system

$$P(s) = \frac{s-1}{s+2}$$

has the inverse

$$P^\dagger(s) = \frac{s+2}{s+1}.$$

Notice that the unstable zero in P gives rise to a pole in $P^\dagger$ that is the mirror image of the unstable zero. □

A simple model for systems with monotone step responses has the transfer function

$$P(s) = \frac{K}{(1+sT)^n} e^{-sL}. \tag{5.4}$$

We call this the NOTD model because it has one time delay and n equal lags. The approximation can be made by fitting the transfer functions at a few relevant frequencies. Assuming that we want a perfect fit at $\omega = 0$ and $\omega = \omega_0$ we find that

$$\begin{aligned} P(0) &= K \\ |P(i\omega_0)| &= \frac{1}{(1+(\omega_0 T)^2)^{n/2}} \\ \arg P(i\omega_0) &= -n \arctan \omega_0 T - \omega_0 L. \end{aligned}$$

Solving these equations we find

$$\begin{aligned} K &= P(0) \\ T &= \frac{\sqrt{|P(i\omega_0|^{-n/2} - 1}}{\omega_0} \\ L &= -\frac{\arg P(i\omega_0) + n \arctan \omega_0 T}{\omega_0}. \end{aligned} \tag{5.5}$$

A good fit is required at the frequency ω_{ms} of maximum sensitivity. Since this frequency is known when the feedback controller C has been designed it is natural to choose $\omega_0 = \omega_{ms}$.

We will give an example to illustrate the accuracy of the approximation.

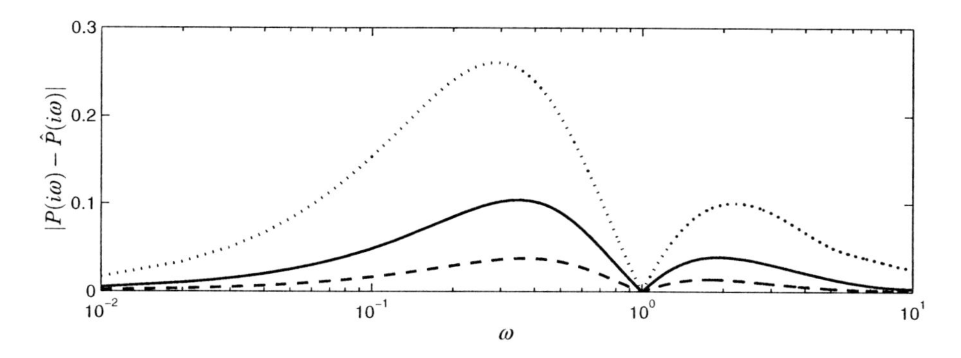

Figure 5.2 Error when fitting NOTD models of different orders to the transfer function $P(s) = 1/(s+1)^4$ for $n = 1$ (dotted), $n = 2$ (solid), and $n = 3$ (dashed).

Table 5.1 Parameters and maximum errors when fitting NOTD models of different orders to the transfer function $P(s) = 1/(s+1)^4$.

n	ω	K	L	T	$e_{\max}$	$\omega_{\max}$
1	0.5	1	1.9566	2.4012	0.1828	1.7400
2	0.5	1	1.1352	1.5000	0.0710	1.4500
3	0.5	1	0.5169	1.1773	0.0255	1.3300
1	1.0	1	1.8235	3.8730	0.2603	0.2800
2	1.0	1	1.0472	1.7321	0.1043	0.3400
3	1.0	1	0.4737	1.2328	0.0378	0.3600

EXAMPLE 5.5—FOUR EQUAL LAGS

Consider a process with the transfer function

$$P(s) = \frac{1}{(s+1)^4}.$$

In Figure 5.2 we show the error $|P(i\omega) - \hat{P}(i\omega)|$ for NOTD models with different n, and in Table 5.1 we give the parameters and the maximum error for different fits. Notice that relatively large errors, 20 to 30 percent, are obtained for first-order models, and significant reductions are obtained by increasing the model order. □

For a process given by (5.4) it is reasonable to choose the response model as

$$M_y = \frac{1}{(1+sT_m)^n} e^{-sL}.$$

It then follows from (5.2) that the feedforward compensator is given by

$$M_u = \frac{1}{K}\left(\frac{1+sT}{1+sT_m}\right)^n. \tag{5.6}$$

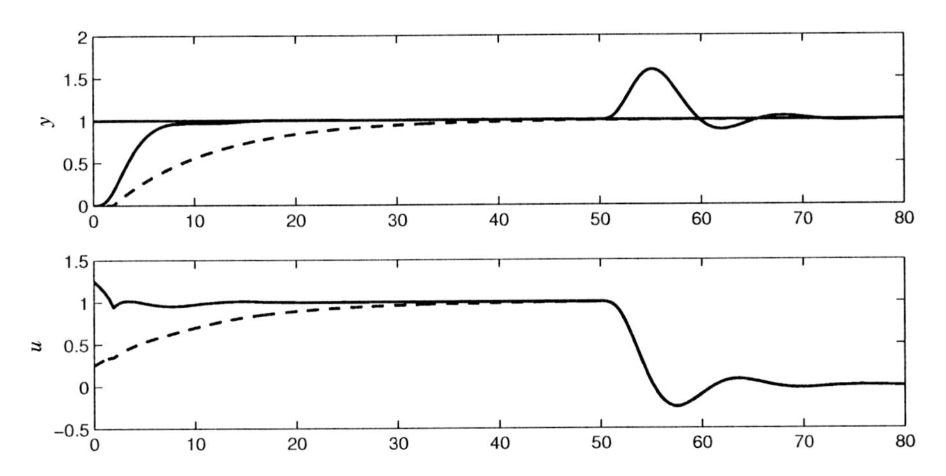

Figure 5.3 Responses to set points and load disturbances of the process $P(s) = 1/(s+1)^4$ with a PI controller and a feedforward based on the FOTD model for desired response times $T_r = 10$ (dashed line) and $T_r = 2$ (solid line).

In this particular case the feedforward compensator thus consists of a process model and a lead-lag or lag-lead network.

There are situations where it is desired that a feedback loop should have a set-point response with specified response time. A typical case is when several substances coming from different sources are mixed. When making production changes it is highly desirable that all systems react to production changes in the same manner. It is very easy to accomplish this when the required process dynamics are slow in comparison to the bandwidth of the feedback, because it follows from (5.1) that the set-point response is not very sensitive to the process model. We illustrate this with an example.

EXAMPLE 5.6—SLOW SET-POINT RESPONSE

Consider a process with the transfer function

$$P(s) = \frac{1}{(s+1)^4},$$

controlled with a PI controller with $K = 0.775$ and $T_i = 2.05$. This gives $M_s = 2$ and $\omega_{ms} = 0.559$. Approximating the process model with a first-order FOTD model gives the parameters $K_p = 1$, $T = 2.51$, and $L = 1.94$, see (5.5). Assume that the desired set-point response is given by

$$M_y(s) = \frac{1}{1 + sT_r}.$$

Figure 5.3 shows set-point responses for different values of T_r. The figure shows that the load disturbance response is the same in both cases and that the set-point response has the expected behavior. Notice the distortions of the curves for $T_r = 2$; they are due to the fact that the model does not fit so well for high frequencies. A rule of thumb is that the first-order model is reasonable for $\omega_{ms} T_r > 2$. In this case this gives $T_r > 3.6$. More accurate models are required to get the desired behavior for $T_r = 2$. □

The advantage by using a controller with two degrees of freedom is that the good disturbance attenuation can be maintained while making the set-point response slower.

5.3 Set-Point Weighting

For simple PID controllers it may not be necessary to use a complete system with two degrees of freedom. The desired set-point response can often be maintained simply by adjusting the set-point weights; see Section 3.4. To determine the set-point weights we consider the transfer function from set point to process output, and we choose set-point parameters so that the largest gain of this transfer function is one or close to one. This gives a set-point response without overshoot for most systems.

It follows from Figure 3.10 and Equation 3.20 that the transfer function from set point to process output is

$$G_{yy_{sp}}(s) = \frac{k_i + bks + ck_d s^2}{k_i + ks + k_d s^2} \frac{P(s)C(s)}{1 + P(s)C(s)} = \frac{k_i + bks + ck_d s^2}{k_i + ks + k_d s^2} T(s). \tag{5.7}$$

One possibility to achieve the largest gain close to one is to specify that the maximum sensitivity M_t is close to one. In such a case it may not be necessary to use set-point weighting. For designs with larger values of M_t we can simply compute maximum of $|G_{yy_{sp}}(i\omega)|$ and adjust the values of b and c that give a value close to one. The weight c is often set to zero. In that case, there is only one parameter to choose. If $|G_{yy_{sp}}(i\omega)|$ is larger than one for $b = 0$, a low-pass filtering of the set point may be used to reduce the magnitude of $|G_{yy_{sp}}(i\omega)|$ further. The set-point filter $F_{sp}(s)$ can be determined in the following way. Let m_s be the maximum of the transfer function (5.7) with $b = c = 0$, and let ω_{sp} be the frequency where the maximum occurs. A first-order filter

$$F_{sp} = \frac{1}{1 + sT_{sp}},$$

has the magnitude $1/m_s$ at the frequency ω_{sp} if the time constant is

$$T_{sp} = \frac{1}{\omega_{sp}} \sqrt{m_s^2 - 1}.$$

Feeding the set point through a low-pass filter designed in this way will reduce the magnitude at the frequency ω_{sp} to one.

A drawback with set-point weighting and filtering is that the set-point response may be unnecessarily slow.

5.4 Neutral Feedforward

A very simple choice of feedforward control for systems with monotone step responses that satisfies (5.2) is given by

$$\begin{aligned} M_y &= \frac{P}{P(0)} = \frac{P}{K_p} \\ M_u &= \frac{1}{K_p}. \end{aligned} \tag{5.8}$$

This means that the desired set-point response is the normalized open-loop response of the system. Since $M_u = 1/K_p$ the control signal is proportional to the set point. At a step change in the set point the control signal thus changes stepwise to the constant value that gives the desired steady-state, and remains at that value. The design of a neutral feedforward is thus very simple.

A complicated process model can be replaced by an approximate model. For PID control it is natural to base design of feedforward on the NOTD model. One way to determine appropriate parameters is to match the model at the frequency ω_{ms} where the sensitivity function has its largest value. We illustrate the design procedure with an example.

EXAMPLE 5.7—FOUR EQUAL LAGS
Consider a process with the transfer function

$$P(s) = \frac{1}{(s+1)^4}.$$

A PI controller with a specification on $M_s = 2$ for this system gives the parameters $K = 0.775$, $T_i = 2.05$, and $\omega_{ms} = 0.56$ of an approximate model. Equation 5.5 gives the parameters $K = 1$, $L = 1.94$, and $T = 2.50$. Figure 5.4 shows the response of the system to step and load disturbances. Notice that there is a dip in the control signal around time $t = 2$. The reason is the mismatch between the process and the model used to design the feedforward. This is illustrated in Figure 5.5, which shows the initial responses of the process and the model. Notice that the process responds faster than the model initially. There is then an error, which is compensated for by the feedback.

The set-point response can be improved by using a better approximation of the process model. One possibility is to fit a second-order NOTD model. Such a model has the parameters $K = 1$, $T = 1.52$, and $L = 1.13$. Figure 5.6 shows the responses of the system to step and load disturbances. Compared with Figure 5.4 the control signal is closer to the ideal value $u = 1$ and the set-point response is a little better. Figure 5.7 shows the comparison of the model output y_m and the process output. A comparison with Figure 5.5 shows that the second-order model gives a better fit. A comparison of Figure 5.4 with Figure 5.6 also illustrates that feedforward requires good modeling. □

In temperature control it is often desirable to have a controller without overshoot to step responses. The next example illustrates how neutral feedforward can be used to accomplish this.

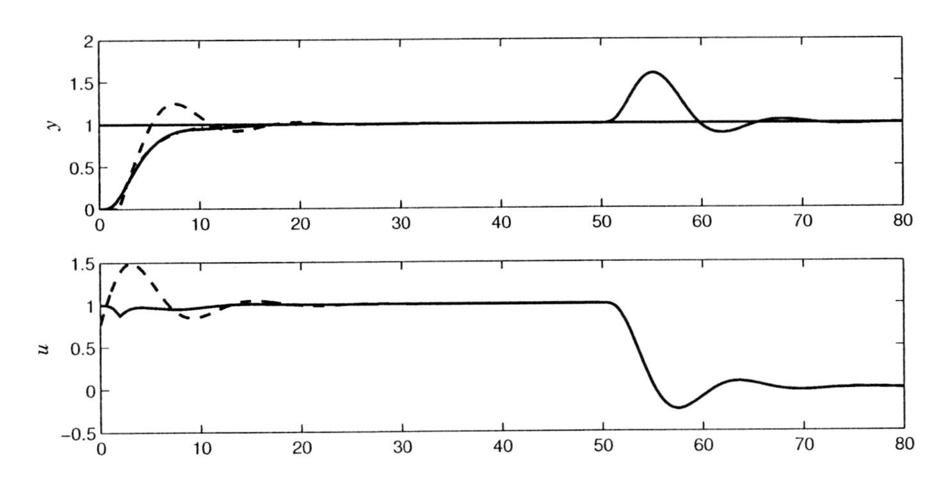

Figure 5.4 Responses to set point and load disturbances of the process $P(s) = 1/(s+1)^4$ with a PI controller (dashed line) and feedforward based on the FOTD model (solid line).

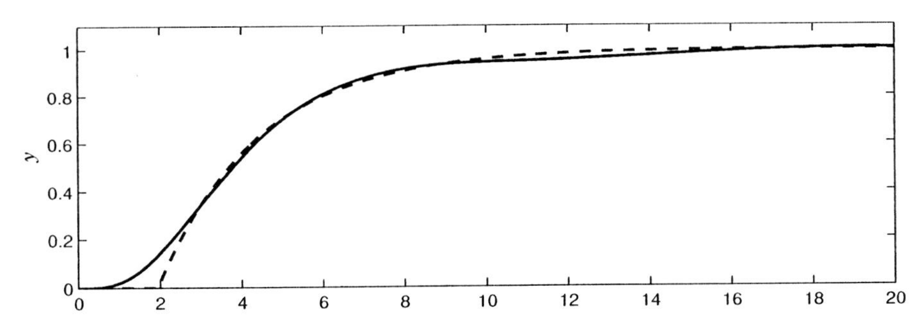

Figure 5.5 Step responses of the process P (solid line) and the model used to design the feedforward (dashed line).

EXAMPLE 5.8—DISTRIBUTED LAGS

Consider a process with the transfer function

$$P(s) = \frac{1}{\cosh \sqrt{s}}.$$

An aggressive PI controller with $M_s = 2$ has $K = 2.66$, $T_i = 0.197$, and $\omega_{ms} = 9.68$. Even with $b = 0$ this controller gives an overshoot as is shown by the dashed curve in Figure 5.8. Fitting a FOTD model at the frequencies 0 and ω_{ms} gives $K = 1$, $T = 0.408$, and $L = 0.0917$. The error in the transfer function is less than 5 percent. Figure 5.8 shows a simulation of the system with neutral feedforward based on that model. The figure shows that neutral feedforward achieves the desired response. □

Oscillatory System

PID control is not the best strategy for oscillatory systems because much better performance can be obtained with more complex controllers. PID control is,

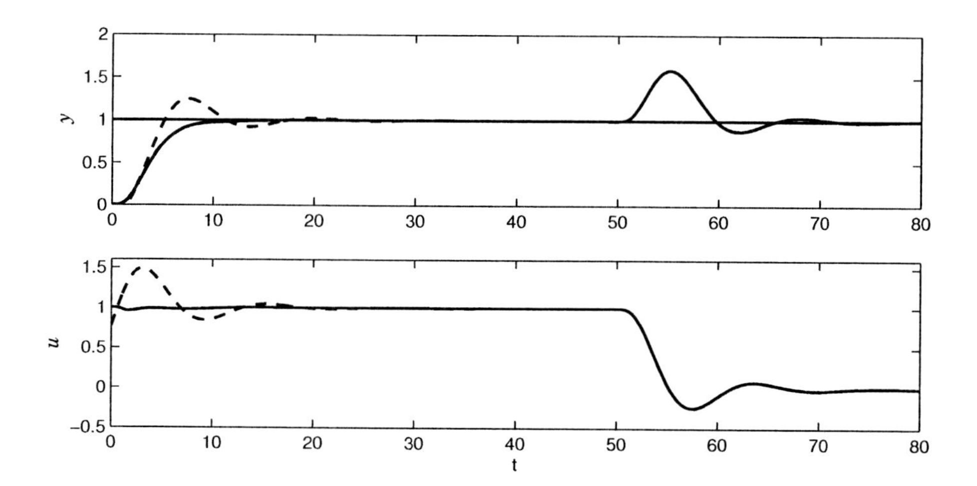

Figure 5.6 Responses to set points and load disturbances of the process $P(s) = 1/(s+1)^4$ with a PI controller (dashed line) and feedforward based on a SOTD model (solid line).

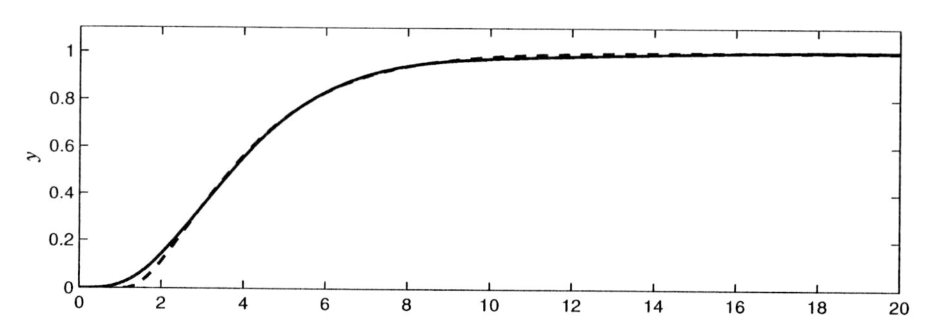

Figure 5.7 Step responses of the process P (solid line) and the model used to design the feedforward (dashed line).

however, sometimes used for such systems, and the performance of a conventional PID controller can often be improved by feedforward. Neutral feedforward, which gives a response similar to the uncontrolled system, can, however, not be used because it will give a response that is too oscillatory. We will illustrate how feedforward can be used by an example.

EXAMPLE 5.9—OSCILLATORY SYSTEM
Consider a system with the transfer function

$$P(s) = \frac{9}{(s+1)(s^2+0.1s+9)}.$$

The oscillatory mode has a relative damping $\zeta = 0.03$, which is quite low.

Reasonable PI controller parameters for the system are $K = -0.167$ and $T_i = -0.210$. Since the controller has negative gain, set-point weighting with $b = 0$ must be used to get a reasonable response. The overshoot is, however,

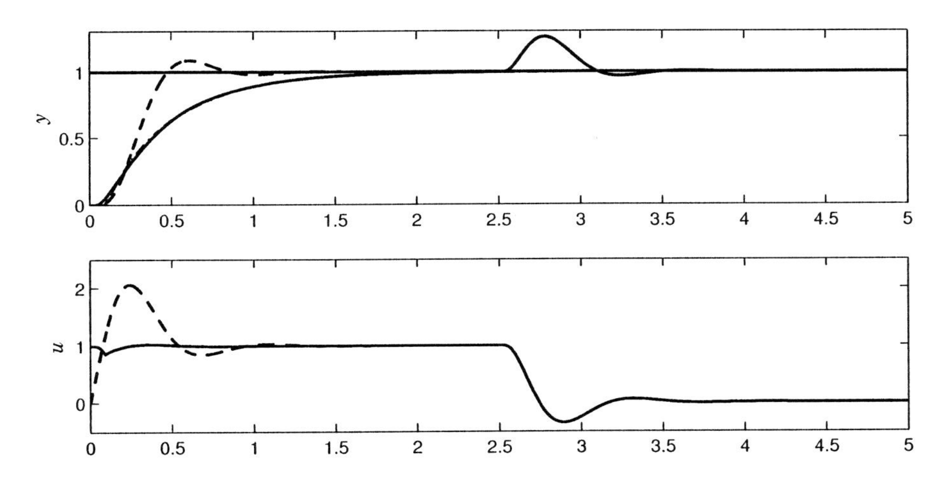

Figure 5.8 Responses to set points and load disturbances of the process $P(s) = 1/\cosh\sqrt{s}$ with a PI controller (dashed lines) and a neutral feedforward based on a first order FOTD model (solid lines).

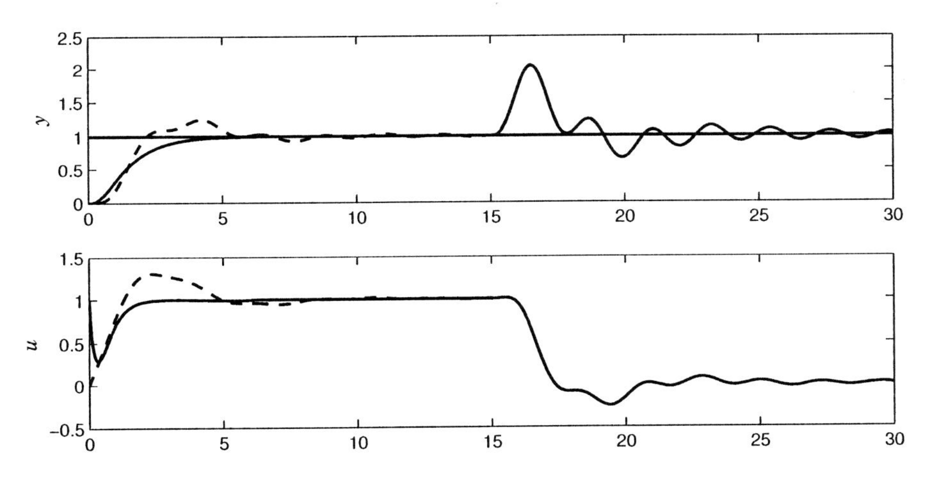

Figure 5.9 Responses to set points and load disturbances of the process $P(s) = 9/(s+1)(s^2+0.2s+9)$ with a PI controller (dashed lines) and a feedforward (solid lines).

still substantial as is seen by the dashed curve in Figure 5.9. To design the feedforward we choose a desired response given by the transfer function

$$M_y = \frac{9}{(s+1)(s^2+6s+9)}.$$

The dynamics of this system is essentially the same as for the process, but the complex poles now have critical damping. It follows from (5.2) that

$$M_u = \frac{s^2+0.1s+9}{s^2+6s+9}.$$

This transfer is close to one for all frequencies except those corresponding to the oscillatory modes where it has low gain. The transfer function thus blocks signals that can excite the oscillatory modes. Figure 5.9 shows the response of the system to set points and load disturbances. It is clear that the set-point response is improved substantially by the use of feedforward. The load disturbance response is still quite poor, which reflects the fact that PI control is not appropriate for a highly oscillatory system. □

5.5 Fast Set-Point Response

With neutral feedforward there is no overshoot in the control signal. It is possible to obtain more aggressive responses if we allow the control signal to overshoot. This is accomplished simply by requiring a faster response. To do this the model must also be accurate over a wider frequency range. The overshoot in the control signal may, however, increase very rapidly with increases in response time as is illustrated by the following example.

EXAMPLE 5.10—FAST SET-POINT RESPONSE
Consider the system

$$P(s) = \frac{1}{(s+1)^4}.$$

Assume that it is desired to have a set-point response given by

$$M_y = \frac{1}{(sT_m+1)^4}.$$

It follows from (5.3) that

$$M_u = \frac{(s+1)^4}{(sT_m+1)^4}.$$

For neutral feedforward we have $T_m = 1$, which gives $M_u = 1$. In general, we have

$$M_u(0) = T_m^{-4}.$$

The controller gain thus increases very rapidly with decreasing values of T_m. This is illustrated in the simulation shown in Figure 5.10, which shows the response for $T_m = 1$ (neutral feedforward), $T_m = 0.5$, and $T_m = 0.2$. The initial values of the control signal are 1, 16, and 625, respectively. Notice that the power 4 in the expressions is due to the fact that the process has a pole excess of 4. In practice, saturation of the actuator determines what can be achieved. □

Time Optimal Control

The example clearly illustrates that feedforward can be used to obtain fast set-point responses but that it requires models that are valid over a wide frequency range and that very large control signals may be required. The size of the

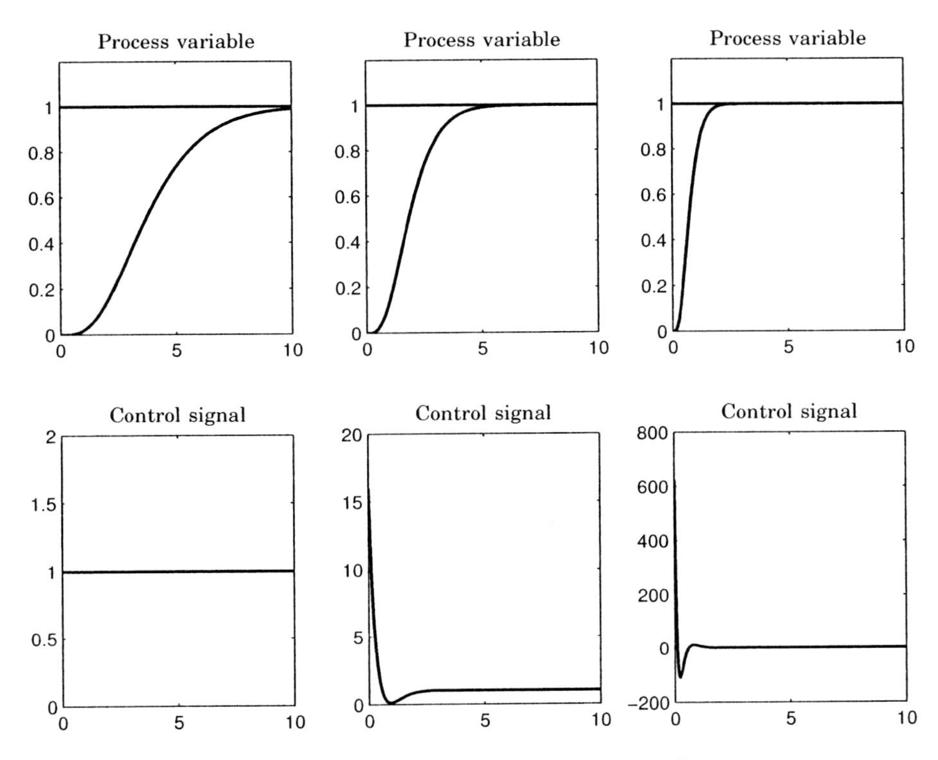

Figure 5.10 Set-point responses of the process $P(s) = (s+1)^{-4}$ with feedforward compensators designed to give $M_y(s) = (sT_m + 1)^{-4}$ for $T_m = 1$ (left), 0.5 and 0.2 (right).

control signal depends critically on the pole excess of the process. In practice, it is also necessary to take account of the fact that the control signals have limited range. It is therefore very natural to look for strategies that bring the process output from one set point to another in minimum time. This problem is solved by optimal control theory. It is known that for linear systems the solution is bang-bang control which means the control signal switches between its extreme values. An example is given in Figure 5.11, which shows the minimum time solution for the process $P = (s+1)^{-2}$ when the control signal is limited to values between 0 and 2. The control is very simple in this case. There can, however, be a large number of switches for high-order systems or for oscillatory systems. Because of its complexity it is not feasible to use optimal control except in very special situations. Approximate methods will therefore be developed.

Pulse Step Control

For stable systems with monotone step responses fast set-point responses can often be achieved with control signals that have the shape shown in Figure 5.11. This means that the maximum control signal is used initially. The control signal is then switched to its lowest value, and the control signal is finally given the value that gives the desired steady state. If the initial pulse is approximated with an impulse we obtain the situation shown in Figure 5.12.

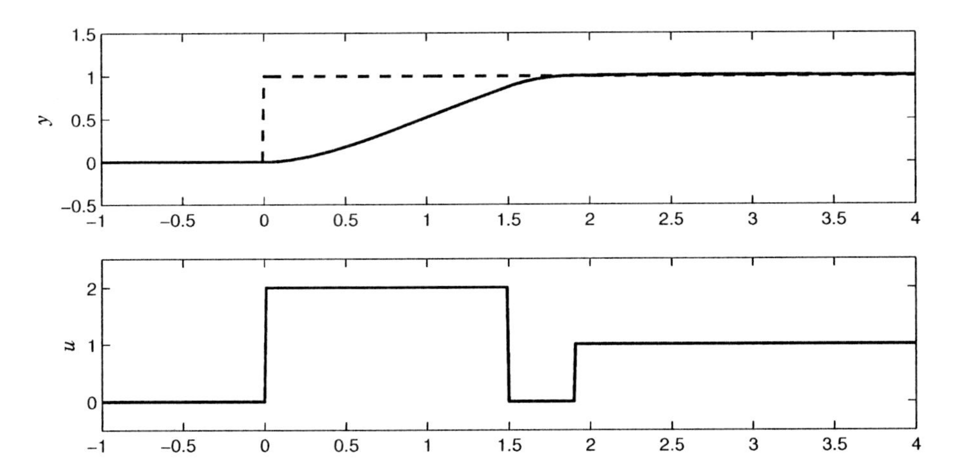

Figure 5.11 Time optimal set-point change for the process $P = (s+1)^{-2}$.

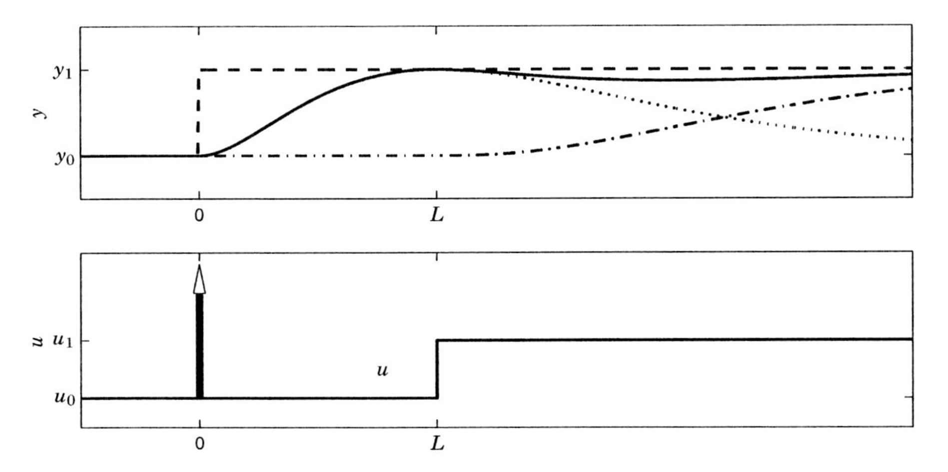

Figure 5.12 Response to an impulse (dotted line) and a delayed step (dashed-dotted line) for a system with monotone step response. The dashed line is the set point, and the solid line is the process output, composed of the sum of the impulse and step responses.

Assuming that the system is initially at rest its output is then given by

$$y(t) = ag(t) + bh(t - L),$$

where h is the step response and g the impulse response of the system. The parameters a, b, and L should be chosen so that the response matches the desired response as closely as possible. To do this the parameter a should be chosen as $y_{sp}/g_{\max}$, where $g_{\max}$ is the maximum of the impulse response $g(t)$. Parameter b should be chosen so that the desired steady state is obtained. Hence, $b = y_{sp}/K_p$ where K_p is the steady-state process gain. The parameter L should be adjusted to keep the output as close to the set point as possible. These choices imply that the settling time of the system is equal to the time

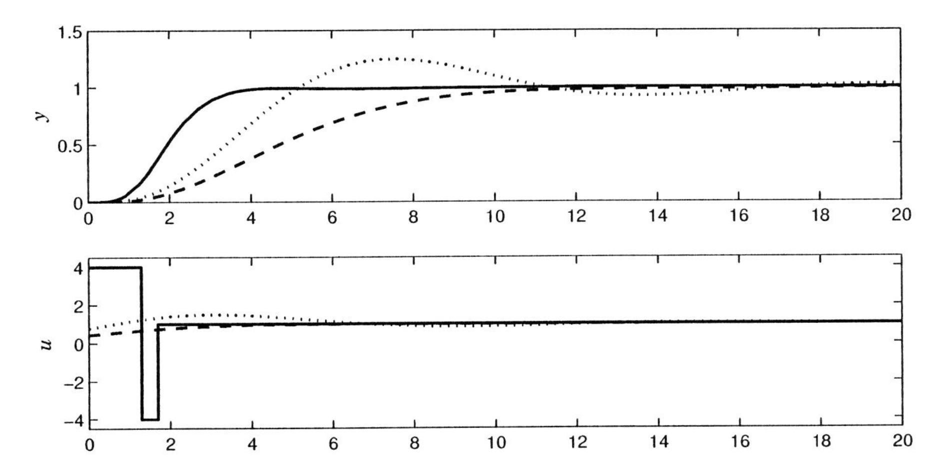

Figure 5.13 Comparison between the fast set-point response strategy (solid) and PI control with $M_s = 1.4$ (dashed) and $M_s = 2.0$ (dotted) for $P(s) = 1/(s+1)^4$.

where the impulse response has its maximum. The closed-loop settling time is thus matched to the natural response time of the system. It is of course not possible to have an impulse as an input. The impulse is therefore approximated by a pulse with an amplitude that corresponds to the maximum value of the control signal. The duration is chosen so that the area under the pulse equals a. The parameters given above can be fine-tuned by optimization. We illustrate the procedure by an example.

EXAMPLE 5.11—FAST SET-POINT RESPONSE
Consider a system with the transfer function

$$P(s) = \frac{1}{(s+1)^4}.$$

Figure 5.13 compares the fast set-point response method with regular PI control with two parameter settings. The fast set-point response has been computed with $u_{max} = 4$ and $u_{min} = -4$, and the resulting rise time and settling time are approximately 4 time units. The controllers have been designed with loop shaping for maximum sensitivities $M_s = 1.4$ and $M_s = 2.0$. The corresponding controller parameters are $K = 0.43$, $T_i = 2.25$, and $b = 1$ for $M_s = 1.4$, and $K = 0.78$, $T_i = 2.05$, and $b = 0.23$ for $M_s = 2.0$. Both PI designs are clearly outperformed by the pulse-step method. The rise times are a factor 2–3 longer, and the settling times approximately 3 times longer. The reason is, of course, that much less of the available control authority is used. If set-point weight b and/or M_s is increased, the size of the control signal will increase. This leads to a faster rise time, but at the expense of larger overshoot, so the settling time may actually be even higher. □

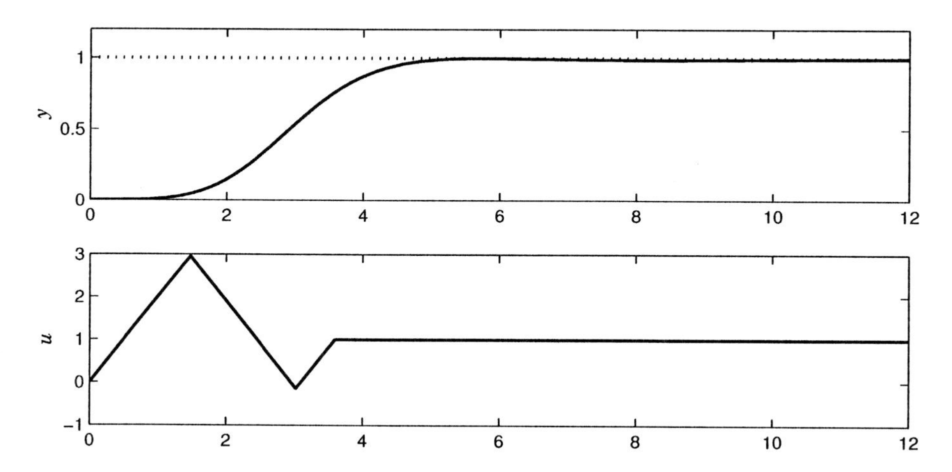

Figure 5.14 Process output and control signal for fast set-point changes with rate limitations $|du/dt| < 2$ for the process $P(s) = 1/(s+1)^4$.

Rate Limitations

The idea of fast set-point response can also be applied to the case when there are rate limitations. This is illustrated in Figure 5.14, which shows a simulation of the process with the transfer function $P = 1/(s+1)^4$ when there are rate limitations $|du/dt| < 2$. It is also possible to combine rate and level limitations.

5.6 Disturbance Attenuation

Disturbances can be eliminated by feedback. With a feedback system it is, however, necessary that there be an error before the controller can take actions to eliminate disturbances. In some situations, it is possible to measure disturbances before they have influenced the processes. It is then natural to try to eliminate the effects of the disturbances before they have created control errors. This control paradigm is called *feedforward*. The principle is illustrated in Figure 5.15.

In Figure 5.15 process transfer function P is composed of two factors, $P = P_1P_2$. A measured disturbance d enters at the input of process section P_2. The measured disturbance is fed to the process input via the feedforward transfer function G_{ff}. The transfer function from load disturbance to process output is

$$G_{yd}(s) = \frac{P_2(1 - P_1G_{ff})}{1 + PC} = P_2(1 - P_1G_{ff})S, \tag{5.9}$$

where $S = 1/(1+PC)$ is the sensitivity function. This equation shows that there are two ways of reducing the disturbance. We can try to make $1 - P_1G_{ff}$ small by a proper choice of the feedforward transfer function G_{ff}, or we can make the loop transfer function PC large by feedback. Feedforward and feedback can also be combined.

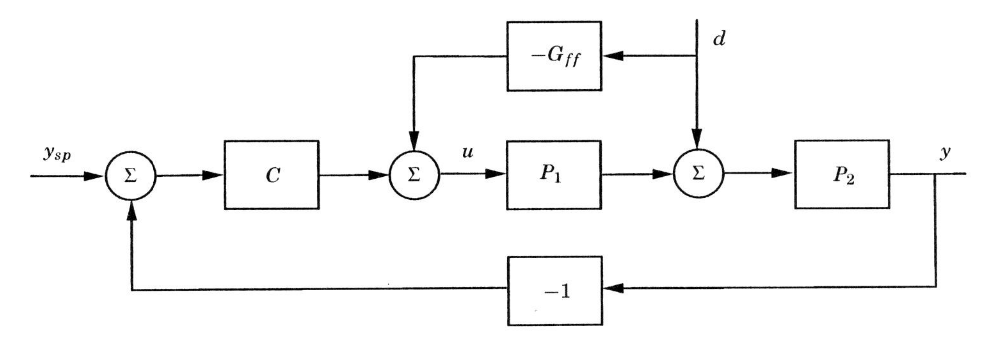

Figure 5.15 Block diagram of a system where a measured disturbance d is reduced by a combination of feedback and feedforward.

Notice that with feedforward we are trying to make the difference between two terms small, but with feedback we simply multiply with a small number. An immediate consequence is that feedforward is more sensitive than feedback. With feedback there is risk of instability; there is no such risk with feedforward. Feedback and feedforward are therefore complementary, and it is useful to combine them.

An ideal feedforward compensator is given by

$$G_{ff} = P_1^{-1} = \frac{P_{yd}}{P_{yu}}, \tag{5.10}$$

where P_{yd} is the transfer function from d to y and $P_{yu} = P$ is the transfer function from u to y. The ideal feedforward compensator is formed by taking the inverse of the process dynamics P_1. This inverse is often not realizable, but approximations have to be used.

Feedforward is most effective when the disturbance d enters early in the process. This occurs when most of the dynamics are in process section P_2. When $P_1 = 1$, and therefore $P_2 = P$, the ideal feedforward compensator is realizable, and the effects of the disturbance can be eliminated from the process output y. On the other hand, when the dynamics enter late in the process, so that $P_1 \approx P$, the effects of the disturbance are seen in the process output y at the same time as they are seen in the feedforward signal. In this case, there is no advantage of using feedforward compared to feedback.

Applications

In many process control applications there are several processes in series. In such cases, it is often easy to measure disturbances and use feedforward. Typical applications of feedforward control are drum-level control in steam boilers, control of distillation columns, and rolling mills. An application of combined feedback and feedforward control follows.

EXAMPLE 5.12—DRUM LEVEL CONTROL

A simplified diagram of a steam boiler is shown in Figure 5.16. The water in the raiser is heated by the burners. The steam generated in the raiser, which

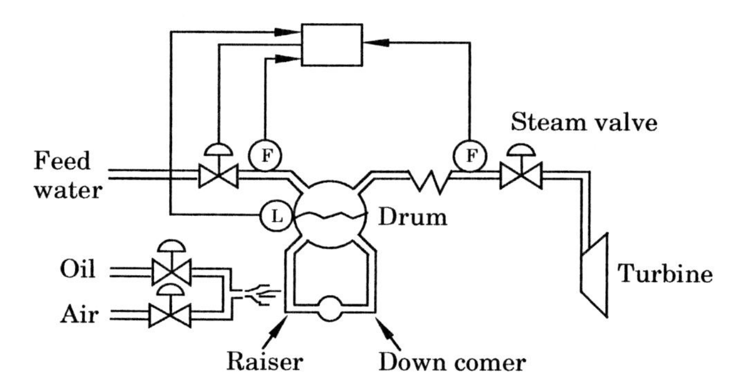

Figure 5.16 Schematic diagram of a drum boiler with level control.

is lighter than the water, rises toward the drum. This causes a circulation around the loop consisting of the raisers, the drum, and the down comers. The steam is separated from the water in the drum. The steam flow to the turbine is controlled by the steam valve.

It is important to keep the water level in the drum constant. Too low a water level gives insufficient cooling of the raisers, and there is a risk of burning. With too high a water level, water may move into the turbines, which may cause damage. There is a control system for keeping the level constant. The control problem is difficult because of the so-called *shrink and swell effect.* It can be explained as follows. Assume that the system is in equilibrium with a constant drum level. If the steam flow is increased by opening the turbine valve, the pressure in the drum will drop. The decreased pressure causes generation of extra bubbles in the drum and in the raisers. As a result, the drum level will initially increase. Since more steam is taken out of the drum, the drum level will of course finally decrease. This phenomenon, which is called the *shrink and swell effect,* causes severe difficulties in the control of the drum level. Mathematically, it also gives rise to right-half plane zero in the transfer function.

The problem can be solved by introducing the control strategy shown in Figure 5.16. It consists of a combination of feedback and feedforward. There is a feedback from the drum level to the controller, but there is also a feedforward from the difference between steam flow and feed-water flow so that the feed-water flow is quickly matched to the steam flow. □

5.7 Summary

Design of feedforward has been discussed in this chapter. Feedforward can be used to reduce the effect of measurable disturbances. Design of feedforward is essentially a matter of finding inverse process models. Different techniques to do this have been discussed. The major part of the chapter has been devoted to set-point response. A structure with two degrees of freedom has been used. This gives a clean separation of regulation and set-point response and of feedback and feedforward. It has been assumed that the feedback controller has been designed. A simple way to modify the set-point response is to use set-point weighting. If the desired results cannot be obtained by zero set-point weighting a full-fledged two-degree-of-freedom can be used. This makes is possible to make a complete separation between load disturbance response and set-point response. The crucial design issue is to decide the achievable response speed. For systems with monotone set-point responses the notion of neutral feedforward has been proposed. Many other variants have also been discussed. Finally, it has been demonstrated that very fast set-point responses can be obtained by using nonlinear methods.

Special care must be taken when implementing feedforward control, otherwise integrator windup may occur. Implementation of feedforward control is discussed in Section 13.4.

5.8 Notes and References

Feedforward is a useful complement to feedback. It was used in electronic amplifiers even before the feedback amplifier emerged as discussed in [Black, 1977]. Use of feedforward in process control was pioneered in [Shinskey, 1963]. The effectiveness of feedforward to improve set-point response using a system structure with two degrees of freedom (2DOF) was introduced in [Horowitz, 1963]. Set-point weighting, which is a simple form of 2DOF, has been used to a limited extent in early PID controllers where the weights have been 0 or 1. The use of continuously adjustable weights appeared in the 1980s. Use of feedforward to reduce the effect of measured disturbances is cumbersome to apply in the process control systems built on separate components but very easy in modern distributed control system; see [Bialkowski, 1995] and [ABB, 2002]. Applications of feedforward are gaining in popularity. Methods for assessment of potential improvements by using feedforward are also emerging; see [Petersson *et al.*, 2001; Petersson *et al.*, 2002; Petersson *et al.*, 2003].

6

PID Design

6.1 Introduction

This chapter describes methods for finding parameters of a PID controller, which is a special case of the problem of control system design that was discussed in Chapter 4. Design of PID controllers differs from the general design problem because the controller complexity is restricted. The general design methods give a controller with a complexity that matches the process model. To obtain a controller with restricted complexity we can either simplify the process models so that the design gives a PID controller, or we can design a controller for a complex model and approximate it with a PID controller. Another reason why special design methods for PID controllers emerged is the desire to have simple design methods that can be used by persons with poor knowledge of control. The situation has changed substantially with the advent of tuning tools and automatic tuners, which have made it possible to improve the process knowledge and permitted the use of more extensive calculations. This has brought design of PID controllers closer to the mainstream of control systems design.

In this chapter it has been attempted to strike a balance by providing both a historical perspective and to present powerful methods. Section 6.2 describes the methods developed by Ziegler and Nichols, which have had a major impact on the practice of PID control even if they do not result in good tuning. Some extensions of the Ziegler-Nichols methods are also discussed.

It is often necessary to complement the design methods with manual fine-tuning to obtain the desired goals of the closed-loop dynamics. These manual tuning rules are discussed in Section 6.3.

Section 6.4 presents the pole placement method, which is one of the main stream methods in control system design. To apply this method it is necessary to approximate process dynamics by a first order model for PI control and a second order model for PID control. Instead of attempting to position all closed-loop poles, it can be attempted to assign only a few dominating poles. Such methods are discussed in Section 6.4. The most common dominant pole placement design method is the lambda tuning method, presented in Section 6.5.

In Section 6.6, algebraic tuning methods are presented. In these methods,

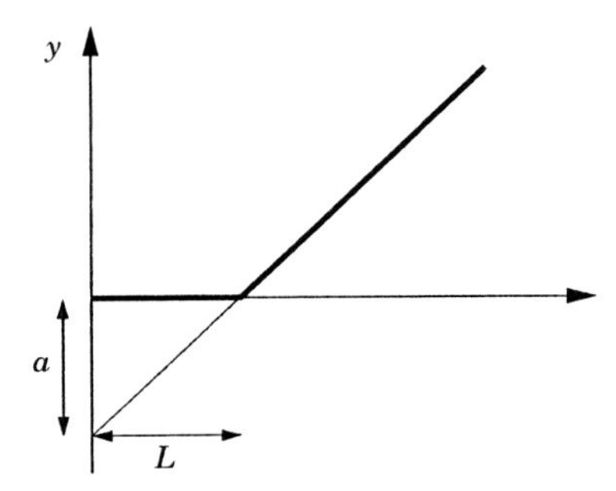

Figure 6.1 Characterization of a step response in the Ziegler-Nichols step response method.

the controller parameters are obtained from the specifications by a direct algebraic calculation. In these techniques it is also necessary to approximate process dynamics by low order models.

Many techniques for control system design are based on optimization. This gives a very flexible way of balancing conflicting design criteria. It is also possible to apply directly to controllers having restricted complexity. A number of uses of optimization for PID control are discussed in Section 6.7.

Loop shaping is another well-known technique for control system design. In Section 6.8 it is shown how this can be used for PID control. This gives a very flexible design method, which allows a nice trade-off between performance and robustness. An analysis of the method also gives useful insight into the difficulties with derivative action.

Conclusions and references are given in Sections 6.9 and 6.10.

6.2 Ziegler-Nichols and Related Methods

Two classical methods for determining the parameters of PID controllers were presented by Ziegler and Nichols in 1942. These methods are still widely used, either in their original form or in some modification. They often form the basis for tuning procedures used by controller manufacturers and the process industry. The methods are based on determination of some features of process dynamics. The controller parameters are then expressed in terms of the features by simple formulas. It is surprising that the methods are so widely referenced because they give moderately good tuning only in restricted situations. Plausible explanations may be the simplicity of the methods and the fact that they can be used for simple student exercises in basic control courses.

The Step Response Method

The first design method presented by Ziegler and Nichols is based on process information in the form of the open-loop step response. This method can be viewed as a traditional method based on modeling and control where a very simple process model is used. The step response is characterized by only two parameters a and L, as shown in Figure 6.1. Compare also with Figure 2.32.

Table 6.1 Controller parameters for the Ziegler-Nichols step response method.

Controller	aK	T_i/L	T_d/L	T_p/L
P	1			4
PI	0.9	3		5.7
PID	1.2	2	L/2	3.4

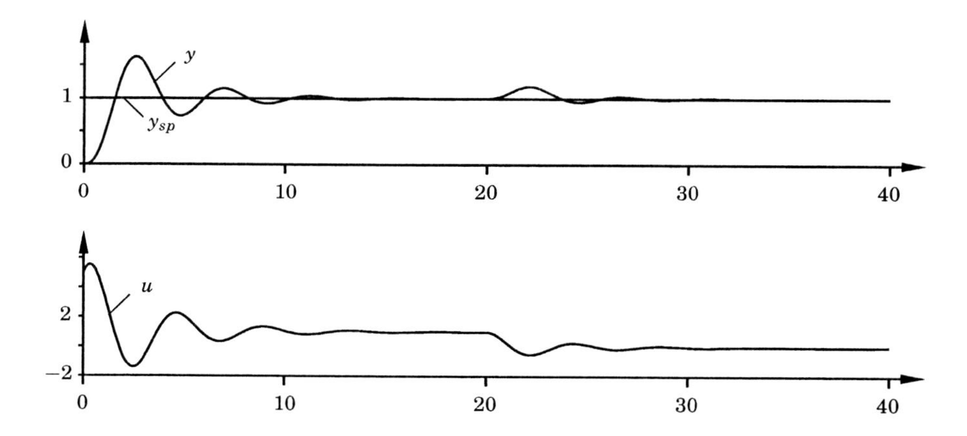

Figure 6.2 Set-point and load disturbance response of a process with transfer function $1/(s+1)^3$ controlled by a PID controller tuned with the Ziegler-Nichols step response method. The diagrams show set point y_{sp}, process output y, and control signal u.

The point where the slope of the step response has its maximum is first determined, and the tangent at this point is drawn. The intersections between the tangent and the coordinate axes give the parameters a and L. In Chapter 2, a model of the process to be controlled was derived from these parameters. This corresponds to modeling a process by an integrator and a time delay. Ziegler and Nichols have given PID parameters directly as functions of a and L. These are given in Table 6.1. An estimate of the period T_p of the closed-loop system is also given in the table.

EXAMPLE 6.1—ZIEGLER-NICHOLS STEP RESPONSE METHOD

Ziegler-Nichols' method will be applied to a process with the transfer function

$$P(s) = \frac{1}{(s+1)^3}. \tag{6.1}$$

Measurements on the step response give the parameters $a = 0.218$ and $L = 0.806$. The controller parameters can now be determined from Table 6.1. The parameters of a PID controller are $K = 5.50$, $T_i = 1.61$, and $T_d = 0.403$. The response of the closed-loop systems to a step change in set point followed by a step change in the load is shown in Figure 6.2. The behavior of the controller is as can be expected. The decay ratio for the step response is close to one quarter.

Table 6.2 Controller parameters for the Ziegler-Nichols frequency response method.

Controller	K/K_u	T_i/T_u	T_d/T_u	T_p/T_u
P	0.5			1.0
PI	0.4	0.8		1.4
PID	0.6	0.5	0.125	0.85

It is smaller for the load disturbance. The overshoot in the set-point response is too large. This can be improved by the set-point weighting b. Compare with Section 3.4. □

The Frequency Response Method

This method is also based on a simple characterization of the process dynamics. The design is based on knowledge of the point on the Nyquist curve of the process transfer function $P(s)$ where the Nyquist curve intersects the negative real axis. In Section 2.4 this point was characterized by K_{180} and ω_{180}. For historical reasons the point has been referred to as the ultimate point and characterized by the parameters $K_u = 1/K_{180}$ and $T_u = 2\pi/\omega_{180}$, which are called the *ultimate gain* and the *ultimate period.* These parameters can be determined in the following way. Connect a controller to the process, and set the parameters so that control action is proportional, i.e., $T_i = \infty$ and $T_d = 0$. Increase the gain slowly until the process starts to oscillate. The gain when this occurs is K_u, and the period of the oscillation is T_u. We have $K_u = 1/K_{180}$ and $T_u = 2\pi/\omega_u$. The parameters can also be determined approximately by relay feedback as is discussed in Section 2.7.

Ziegler-Nichols have given simple formulas for the parameters of the controller in terms of the ultimate gain and the ultimate period shown in Table 6.2. An estimate of the period T_p of the dominant dynamics of the closed-loop system is also given in the table.

The frequency response methods can also be viewed as an empirical tuning procedure where the controller parameters are obtained by direct experiments on the process combined with some simple rules. For a proportional controller the rule is simply to increase the gain until the process oscillates and then to reduce the gain by 50 percent.

We illustrate the design procedure with an example.

EXAMPLE 6.2—THE ZIEGLER-NICHOLS FREQUENCY RESPONSE METHOD
Consider the same process as in Example 6.1. The process given by (6.1) has the ultimate gain $K_u = 8$ and the ultimate period $T_u = 2\pi/\sqrt{3} = 3.63$. Table 6.2 gives the parameters $K = 4.8$, $T_i = 1.81$, and $T_d = 0.44$ for a PID controller. The closed-loop set-point and load disturbance responses when the controller is applied to the process given by (6.1) are shown in Figure 6.3.

The parameters and the performance of the controllers obtained with the frequency response method are close to those obtained by the step response method. The responses are slightly better damped. □

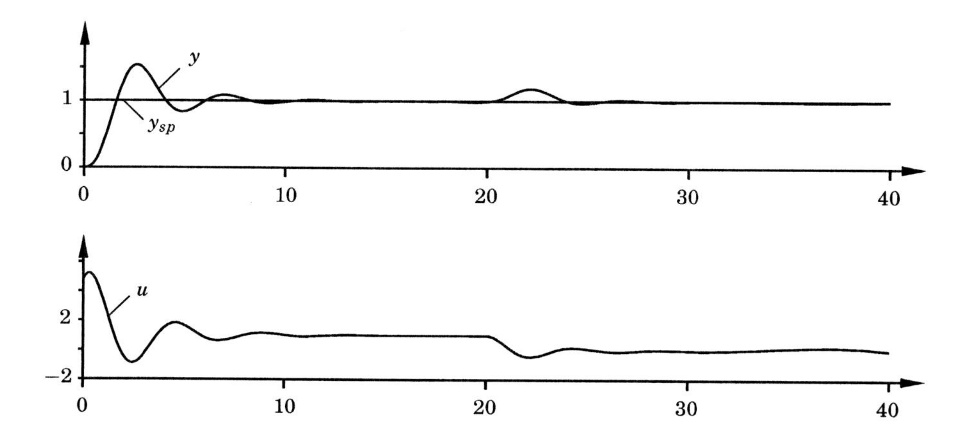

Figure 6.3 Set-point and load disturbance response of a process with the transfer function $1/(s+1)^3$ controlled by a PID controller that is tuned with the Ziegler-Nichols frequency response method. The diagrams show set point y_{sp}, process output y, and control signal u.

The Ziegler-Nichols tuning rules were originally designed to give systems with good responses to load disturbances. They were obtained by extensive simulations of many different systems with manual assessment of the results. The design criterion was quarter amplitude decay ratio, which is often too large, as is seen in the examples. For this reason the Ziegler-Nichols method often requires modification or re-tuning. Since the primary design objective was to reduce load disturbances, it is often necessary to choose set-point weighting carefully in order to obtain a satisfactory set-point response.

An Interpretation of the Frequency Response Method

The frequency response method can be interpreted as a method where one point of the Nyquist curve is positioned. With PI or PID control, it is possible to move a given point on the Nyquist curve of the process transfer function to an arbitrary position in the complex plane, as indicated in Figure 6.4. By changing the gain, a point on the Nyquist curve is moved radially from the origin. The point can be moved in the orthogonal direction by changing integral or derivative gain. Notice that with positive controller parameters the point can be moved to a quarter plane with PI or PD control and to a half plane with PID control. From this point of view the Ziegler-Nichols method can be interpreted as a primitive loop-shaping method where one point of the loop transfer function is moved to a desired point.

The frequency response method starts with determination of the point $(-1/K_u, 0)$ where the Nyquist curve of the open-loop transfer function intersects the negative real axis.

Let us now investigate how the ultimate point is changed by the controller. For a PI controller with Ziegler-Nichols tuning we have $K = 0.4K_u$ and $\omega_u T_i = (2\pi/T_u)0.8T_u = 5.02$. Therefore, the transfer function of the PI controller at

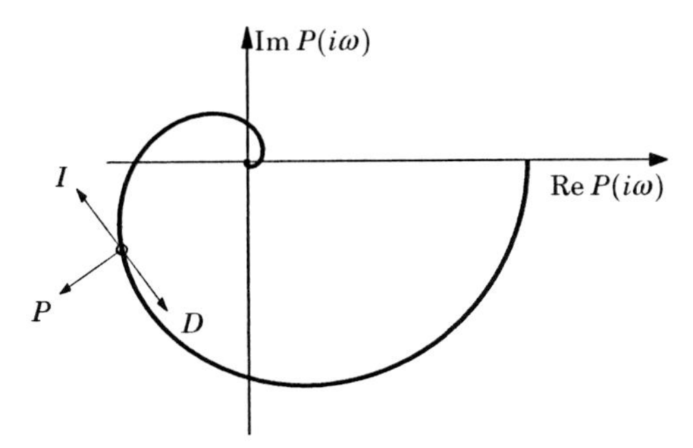

Figure 6.4 Illustrates that a point on the Nyquist curve of the process transfer function may be moved to another position by PID control. The point marked with a circle may be moved in the directions $P(i\omega)$, $-iP(i\omega)$, and $iP(i\omega)$ by changing the proportional, integral, and derivative gain, respectively.

the ultimate frequency is

$$C(i\omega_u) = K\left(1 + \frac{1}{i\omega_u T_i}\right) = 0.4K_u(1 - i/5.02) = K_u(0.4 - 0.08i).$$

The ultimate point is thus moved to $-0.4 + 0.08i$. This means that a lag of 11.2° is introduced at the ultimate frequency.

For a PID controller we have $K = 0.6K_u$, $\omega_u T_i = \pi$, and $\omega_u T_d = \pi/4$. The frequency response of the controller at frequency ω_u is

$$C(i\omega_u) = K\left(1 + i\left(\omega_u T_d - \frac{1}{\omega_u T_i}\right)\right) \approx 0.6K_u(1 + 0.467i).$$

This controller gives a phase advance of 25° at the ultimate frequency. The loop transfer function is

$$G_\ell(i\omega_u) = P(i\omega_u)C(i\omega_u) = -0.6(1 + 0.467i) = -0.6 - 0.28i.$$

The Ziegler-Nichols frequency response method for a PID controller thus moves the ultimate point $(-1/K_u, 0)$ to the point $-0.6 - 0.28i$. The distance from this point to the critical point is 0.5. This means that the method gives a sensitivity that is always greater than 2.

It has been suggested by Pessen to move the ultimate point to $-0.2 - 0.36i$ or $-0.2 - 0.21i$. Suda used approximations to obtain $M_\ell = 1.3$ by moving the critical point to $-0.628 - 0.483i$.

Design of PI Controller with a Given Phase Margin

Using the idea that the PI controller can be interpreted as moving a point on the loop transfer function it is easy to develop a design method that gives a closed-loop system with a given phase margin. Let the process transfer function be

$$P(i\omega) = \alpha(\omega) + i\beta(\omega) = \rho(\omega)e^{i\psi(\omega)}.$$

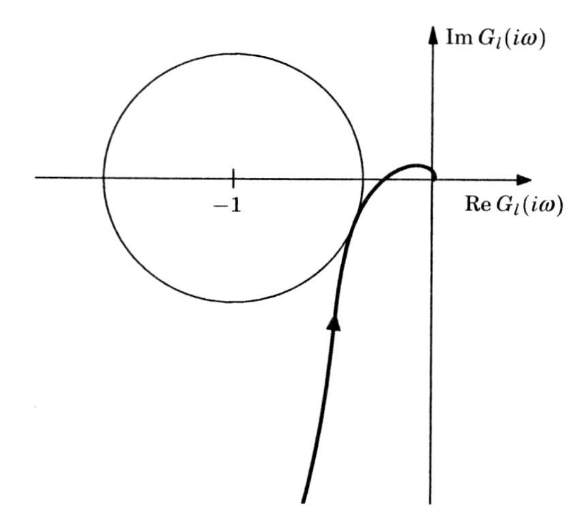

Figure 6.5 Nyquist plot for the loop transfer function G_l for PI control of the process $P(s) = e^{-\sqrt{s}}$. The controller was designed to give the phase margin of 60°.

With PI control the loop transfer function becomes

$$G_l(i\omega) = \Big(k - i\frac{k_i}{\omega}\Big)\big(\alpha(\omega) + i\beta(\omega)\big) = \alpha(\omega)k + \frac{\beta(\omega)k_i}{\omega} + i\Big(\beta(\omega)k - \frac{\alpha(\omega)k_i}{\omega}\Big).$$

Let ω_{gc} be the gain crossover frequency; requiring that the system has a phase margin φ_m it follows that

$$G_l(i\omega_{gc}) = -\cos(\varphi_m) - i\sin(\varphi_m),$$

which implies that

$$\begin{aligned} \alpha(\omega_{gc})k + \frac{\beta(\omega_{gc})k_i}{\omega_{gc}} &= -\cos(\varphi_m) \\ \beta(\omega_{gc})k - \frac{\alpha(\omega_{gc})k_i}{\omega_{gc}} &= -\sin(\varphi_m). \end{aligned}$$

Solving this equation for k and k_i gives

$$\begin{aligned} k &= -\frac{\alpha(\omega_{gc})\cos\varphi_m + \beta(\omega_{gc})\sin\varphi_m}{\alpha^2(\omega_{gc}) + \beta^2(\omega_{gc})} = -\frac{1}{\rho(\omega_{gc})}\cos\big(\varphi_m - \psi(\omega_{gc})\big) \\ k_i &= \omega_{gc}\frac{\alpha(\omega_{gc})\sin\varphi_m - \beta(\omega_{gc})\cos\varphi_m}{\alpha^2(\omega_{gc}) + \beta^2(\omega_{gc})} = \frac{\omega_{gc}}{\rho(\omega_{gc})}\sin\big(\varphi_m - \psi(\omega_{gc})\big). \end{aligned} \tag{6.2}$$

It is thus straightforward to compute the controller gains when the gain crossover frequency is given. Reasonable values of the gain crossover frequency are in the range $\omega_{90} \le \omega_{gc} \le \omega_{180-\varphi_m}$. The method can be improved by sweeping over ω_{gc} to maximize integral gain. Applying the method to design a PI controller for the process $P(s) = e^{-\sqrt{s}}$ with a phase margin of 60° gives $\omega_{gc} = 5.527$ $K = 4.79$ and $T_i = 0.392$ and $M_s = 1.53$. The Nyquist plot of the loop transfer function is shown in Figure 6.5.

Relations Between the Ziegler-Nichols Tuning Methods

The step response method and the frequency response method do not give the same values of the controller parameters. Comparing Examples 6.1 and 6.2 we find that the controller gains are 5.5 and 4.8 and that the integral times are 1.61 and 1.81. The step response method will in general give larger gains and smaller integral times. This is further illustrated in the following example.

EXAMPLE 6.3—PROCESS WITH INTEGRATION AND DELAY
Consider a process with the transfer function

$$P(s) = \frac{K_v}{s} e^{-sL},$$

which is the model originally used by Ziegler and Nichols to derive their tuning rules for the step response method. For this process we have $a = K_v L$. The ultimate frequency is $\omega_u = \pi/2L$, which gives the ultimate period $T_u = 4L$, and the ultimate gain is $K_u = \pi/2K_v L$.

For PI control the step response method gives the following parameters:

$$K = \frac{0.9}{K_v L}, \quad T_i = 3L.$$

This can be compared with the parameters

$$K = \frac{0.63}{K_v L}, \quad T_i = 3.2L$$

obtained for the frequency response method. Notice that the integral times are within 10 percent, but that the step response method gives a gain that is about 40 percent higher.

The PID parameters obtained from the step response method are

$$K = \frac{1.2}{bL}, \quad T_i = 2L \quad \text{and} \quad T_d = \frac{L}{2},$$

and those given by the frequency response methods are

$$K = \frac{0.94}{bL}, \quad T_i = 2L \quad \text{and} \quad T_d = \frac{L}{2}.$$

Both methods give the same values of integral and derivative times, but the step response method gives a gain that is about 25 percent higher than the frequency response method. □

EXAMPLE 6.4—PROCESS WITH PURE DELAY
Consider a process with the transfer

$$P(s) = K_p e^{-sL}.$$

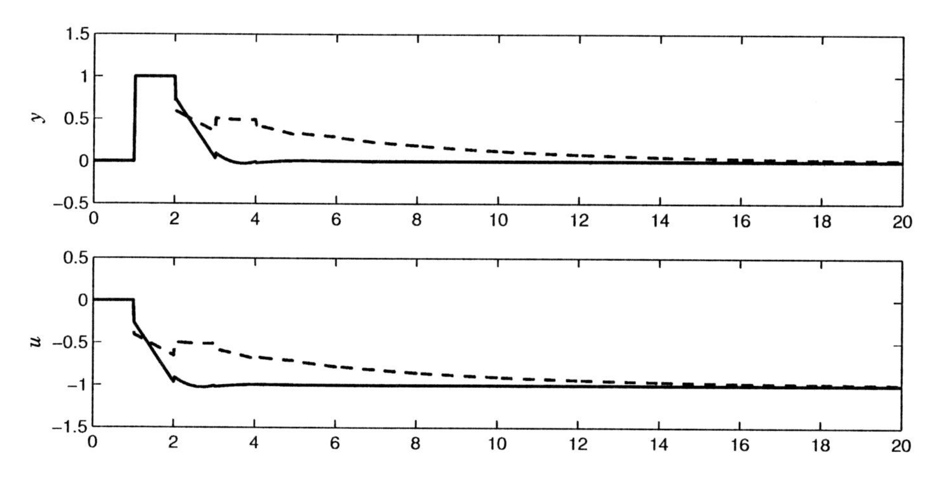

Figure 6.6 Responses to a load disturbance for a process with pure delay ($L = 1$) with PI controllers tuned by Ziegler-Nichols frequency response method (dashed) and a proper method (solid).

In this case we find that $a = \infty$! The step response method thus gives zero controller gain for PI and PID control.

The ultimate period is $T_u = 2L$, and the ultimate gain is $K_u = 1/K_p$. Using the frequency response method it follows from Table 6.2 that $KK_p = 0.4$ and $T_i = 1.6L$ for PI control. The PI controller gives a very poor result as is illustrated in Figure 6.6. The integral action is too small, which implies that it takes a very long time for the error to approach zero. For comparison we also show the response with a PI controller having $KK_p = 0.25$ and $T_i = 0.35$. This controller has a much better response to load disturbances.

For PID control the frequency response method gives $KK_p = 0.6$, $T_i = L$ and $T_d = 0.25$, which results in an unstable closed-loop system. □

These examples show that there can be considerable differences between the controller parameters obtained by the step response and the frequency response methods.

The Chien, Hrones, and Reswick Method

There have been many suggestions for modifications of the Ziegler-Nichols methods. There are methods that use the same information about the process as the Ziegler-Nichols methods, but the coefficients in Tables 6.1 and 6.2 are modified. Many methods of this type are used by controller manufacturers. There are also other methods that use more process data. Many methods are based on the idea that the process is approximated with the FOTD model

$$P(s) = \frac{K_p}{1 + sT} e^{-sL}.$$

As an illustration we will describe a method developed by Chien, Hrones, and Reswick (CHR). Their method gives closed-loop systems with slightly better

Table 6.3 Controller parameters obtained from the Chien, Hrones and Reswick load disturbance response method.

	No overshoot			20% overshoot		
Controller	aK	T_i/L	T_d/L	aK	T_i/L	T_d/L
P	0.3			0.7		
PI	0.6	4.0		0.7	2.3	
PID	0.95	2.4	0.42	1.2	2.0	0.42

Table 6.4 Controller parameters obtained from the Chien, Hrones and Reswick set-point response method.

	No overshoot			20% overshoot		
Controller	aK	T_i/L	T_d/L	aK	T_i/L	T_d/L
P	0.3			0.7		
PI	0.35	1.2		0.6	1.0	
PID	0.6	1.0	0.5	0.95	1.4	0.47

robustness than the Ziegler-Nichols method. The design criteria used were "quickest response without overshoot" or "quickest response with 20 percent overshoot." They proposed different tuning rules for load disturbances and set-point response.

To tune the controller according to the CHR method, the parameters a and L of the process model are first determined in the same way as for the Ziegler-Nichols step response method. The controller parameters are then given as functions of these two parameters. The tuning rule for load disturbance response are given in Table 6.3. The tuning rules in Table 6.3 have in general lower gains than the corresponding Ziegler-Nichols rule in Table 6.1.

Chien, Hrones, and Reswick found that tuning for set-point response was different than tuning for load disturbances. At that time the advantages of set-point weighting and systems with two degrees of freedom were not known. An additional parameter, time constant T, was required, and the controller gains were in general lower; see Table 6.4.

The Cohen-Coon Method

The Cohen-Coon method is also based on the FOTD process model

$$P(s) = \frac{K_p}{1 + sT} e^{-sL}.$$

The main design criterion is rejection of load disturbances. It attempts to position dominant poles that give a quarter amplitude decay ratio. For P and PD

Table 6.5 Controller parameters from the Cohen-Coon method.

Controller	aK	T_i/L	T_d/L
P	$1+\dfrac{0.35\tau}{1-\tau}$		
PI	$0.9\left(1+\dfrac{0.092\tau}{1-\tau}\right)$	$\dfrac{3.3-3.0\tau}{1+1.2\tau}$	
PD	$1.24\left(1+\dfrac{0.13\tau}{1-\tau}\right)$		$\dfrac{0.27-0.36\tau}{1-0.87\tau}$
PID	$1.35\left(1+\dfrac{0.18\tau}{1-\tau}\right)$	$\dfrac{2.5-2.0\tau}{1-0.39\tau}$	$\dfrac{0.37-0.37\tau}{1-0.81\tau}$

controllers the poles are adjusted to give maximum controller gain, subject to the constraint on the decay ratio. This minimizes the steady-state error due to load disturbances. For PI and PID control the integral gain $k_i = K/T_i$ is maximized. This corresponds to minimization of IE, the integral error due to a unit step load disturbance. For PID controllers three closed-loop poles are assigned; two poles are complex, and the third real pole is positioned at the same distance from the origin as the other poles. The pole pattern is adjusted to give quarter amplitude decay ratio, and the distance of the poles to the origin are adjusted to minimize IE.

Since the process is characterized by three parameters (K_p, L, and T), it is possible to give tuning formulas where controller parameters are expressed in terms of these parameters. Such formulas were derived by Cohen and Coon based on analytical and numerical computations. The formulas are given in Table 6.5. The parameters $a = K_pL/T$ and $\tau = L/(L+T)$ are used in the table to facilitate comparisons with Ziegler-Nichols tuning. A comparison with Table 6.1 shows that the controller parameters are close to those obtained by the Ziegler-Nichols step response method for small τ. Also notice that the integral time decreases for increasing τ, which is desirable as was found in Section 6.2. A peculiarity is that the gains go to infinity when τ goes to 1, which is not correct. The method does also suffer from the decay ratio being too large, which means that the closed-loop systems obtained have poor damping and high sensitivity.

Commentary

The Ziegler-Nichols tuning rules are simple and intuitive. They require little process knowledge, and they can be applied with modest effort. The process is characterized by two parameters that can be determined by simple experiments. The frequency response method has the advantage that parameters K_u and T_u are easier to determine accurately than the parameters a and L, which are used by the step response method.

The methods are still widely used even if they give closed-loop systems that are not robust. The rules are often combined with manual tuning, which will be discussed in Section 6.3. The main drawbacks with the methods are that too little process information is used and the design criterion quarter amplitude damping gives closed-loop systems with poor robustness. It is not clear why this design criterion was used. The load disturbance responses look quite reasonable, but without analysis or sensitivity studies it is not obvious that the closed-loop systems are not robust. The simulations shown in Figure 6.2 and Figure 6.3 indicate that the methods give reasonable control. Repeated simulations with perturbations in controller parameters reveal very clearly that the closed-loop system is not robust. Systems like the ones shown in Examples 6.3 and 6.4 also illustrate that it is not sufficient to characterize the process by two parameters only.

A very large number of variations of the Ziegler-Nichols methods have been proposed. Here we have chosen to discuss two methods. The modifications of the Chien-Hrones-Reswick method give systems with somewhat better robustness, but it still uses too little process information. The Cohen-Coon method uses three parameters to characterize the process, but it still uses quarter amplitude damping as a design criterion.

In Chapter 7 we will develop new methods that address the major shortcomings of the Ziegler-Nichols methods while retaining their simplicity.

6.3 Rule-Based Empirical Tuning

Since the Ziegler-Nichols methods only give "ball-park" values, it is necessary to complement the methods by manual tuning to obtain reasonable closed-loop properties. Manual tuning is typically performed by experiments on the process in closed loop. A perturbation is introduced either as a set-point change or as a change in the control variable. The closed-loop response is observed, and the controller parameters are adjusted. The adjustments are based on simple rules, which give guidelines for changing the parameters. The rules were developed by extensive experimentation. The following is a simple set of rules:

- Increasing proportional gain decreases stability
- Error decays more rapidly if integration time is decreased
- Decreasing integration time decreases stability
- Increasing derivative time improves stability

Lately, the tuning rules have also been formalized in various types of formal rule-based systems such as expert systems or fuzzy logic.

Tuning maps are one way to express the tuning rules. The purpose of these maps is to provide intuition about how changes in controller parameters influence the behavior of the closed-loop system. The tuning maps are simply arrays of transient or frequency responses corresponding to systematic variations in controller parameters. An example of a tuning map is given in Figure 6.7.

The figure illustrates how the load disturbance response is influenced by changes in gain and integral time. The process model

$$P(s) = \frac{1}{(s+1)^8}$$

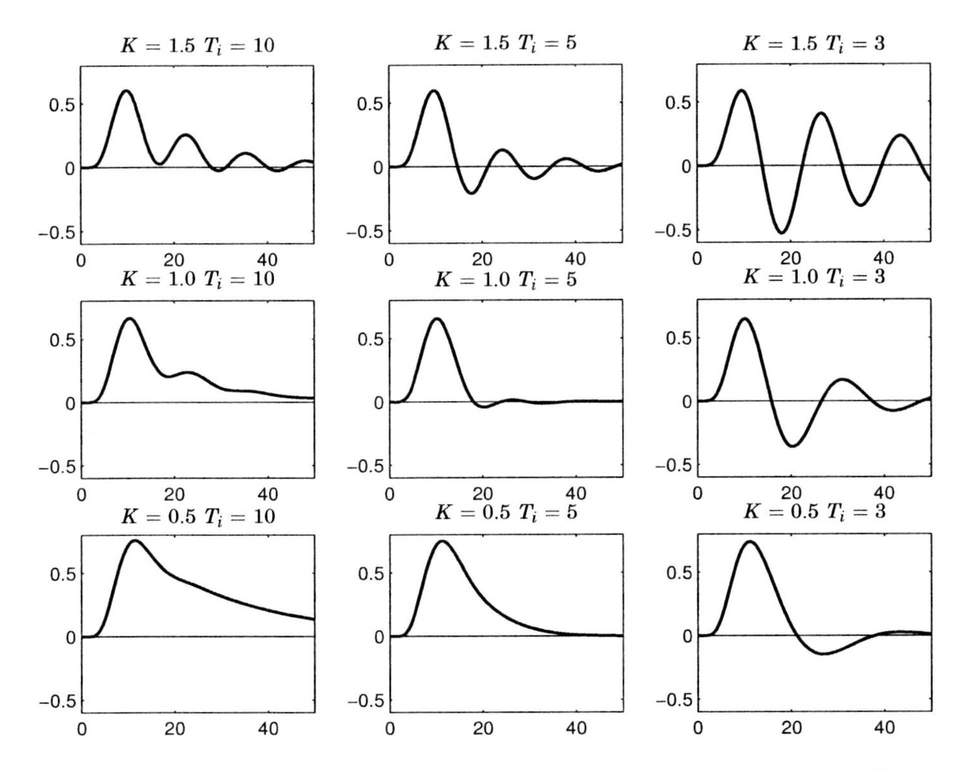

Figure 6.7 Tuning map for PID control of a process with the transfer function $P(s) = (s+1)^{-8}$. The figure shows the responses to a unit step disturbance at the process input. Parameter T_d has the value 1.9.

has been used in the example. The Ziegler-Nichols frequency response method gives the controller parameters $K = 1.13$, $T_i = 7.58$, and $T_d = 1.9$. The figure shows clearly the benefits of having a smaller value of T_i. Judging from the figure, the values $K = 1$ and $T_i = 5.0$ appear reasonable. The figure also shows that the choice of T_i is fairly critical. Also notice that controllers with $T_i < 7.6$ cannot be implemented on series form (compare with Section 3.4).

A different type of tuning map is shown in Figure 6.8, which shows the Nyquist curves of the loop transfer function. The figure shows that several of the Nyquist curves bend over too much to the right at low frequencies; see the figures in the left positions with $T_i = 10$. This means that the controller introduces too much phase lead. This is reduced by reducing parameter T_i.

A comparative study of curves like Figure 6.7 and Figure 6.8 is a good way to develop intuition for the relations between the time and frequency responses. An even better way is to use the interactive software that is now emerging.

Counter-Intuitive Behavior

Common rules for manual tuning says that the system becomes less oscillatory if the gain is reduced, if the integral time is increased, and if the derivative time is increased. Compare with Figure 6.4. These rules hold for the system

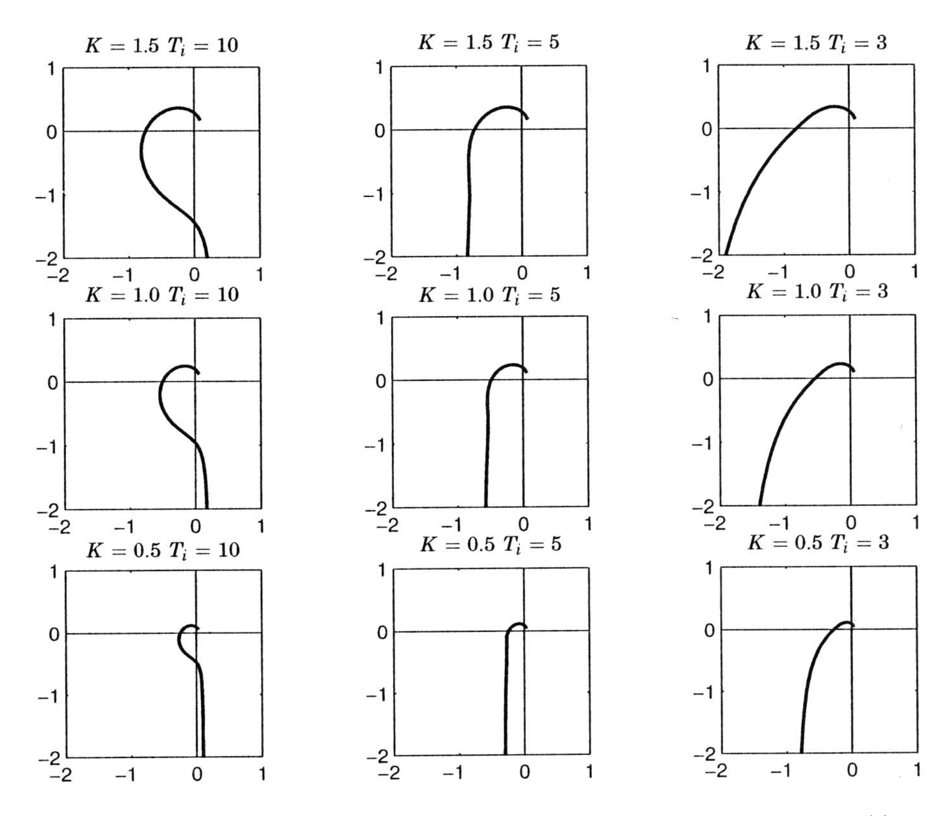

Figure 6.8 Tuning map for PID control of a process with the transfer function $P(s) = (s+1)^{-8}$. The figure shows the Nyquist plots of the loop transfer functions. Parameter T_d has the value 1.9.

shown in Figure 6.7 and Figure 6.8. There are, however, situations where these rules do not hold. The following is a simple common example.

EXAMPLE 6.5—PI CONTROL OF AN INTEGRATOR
Consider a process with the transfer function

$$P(s) = \frac{1}{s},$$

and a PI controller with the transfer function

$$C(s) = K(1 + \frac{1}{sT_i}).$$

The loop transfer function is

$$G_l(s) = P(s)C(s) = K\frac{1+sT_i}{s^2T_i};$$

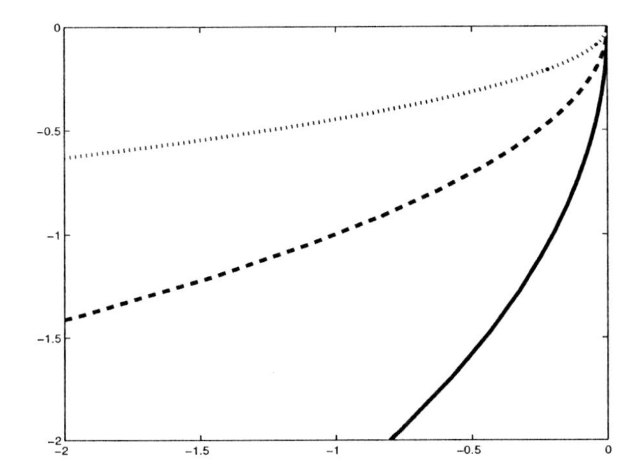

Figure 6.9 Nyquist curves for the loop transfer functions for an integrator with PI control. Integration time T_i is constant, and the gain has the values $K = 0.2$ (dotted), 1 (dashed), and 5 (solid). Notice the counterintuitive behavior that phase margin increases with increasing controller gain.

and the characteristic equation is

$$s^2 + Ks + \frac{K}{T_i} = 0.$$

Identifying this with a standard second-order system $s^2 + 2\zeta\omega_0 + \omega_0^2$ we find that

$$\zeta = \sqrt{\frac{K}{2T_i}}.$$

It follows from this equation that the damping increases when the controller gain is increased contrary to the intuition developed for the simple systems. This is also illustrated by the Nyquist curves in Figure 6.9. Notice that the Nyquist curve moves away from the critical point -1 as the gain increases. The reason for this is that the Nyquist curve is very close to the negative imaginary axis for large ω. Notice that a small time delay or a small lag will destroy this property. □

Situations like this make it difficult to form efficient rules that cover a wide range of conditions.

An Inequality for the Integration Time

It is useful to have a simple way to judge if the integral action of a controller is too weak, as in the three left and the lower middle examples in Figure 6.7 and Figure 6.8. Such a criterion can be based on a calculation of the asymptotic behavior of the loop transfer function for low frequencies. For a process with transfer function P and a PI controller with transfer function C we have

$$G_\ell(s) = P(s)C(s) \approx (P(0) + sP'(0))\, K\left(1 + \frac{1}{sT_i}\right)$$
$$= \frac{KP(0)}{sT_i} + KP(0) + \frac{KP'(0)}{T_i} + KP'(0)s.$$

Thus, for low frequencies the asymptote of the Nyquist curve is parallel to the imaginary axis with the real part equal to

$$KP(0) + \frac{KP'(0)}{T_i} = KK_p\left(1 - \frac{T_{ar}}{T_i}\right),$$

where $K_p = G(0)$ is the static process gain, and T_{ar} is the average residence time. It is reasonable to require that the real part of the asymptote be less than -0.5. This gives

$$T_i < T_{ar}\frac{2KK_p}{1 + 2KK_p} < T_{ar}. \tag{6.3}$$

For the system in Figure 6.7 and Figure 6.8, we get the requirement $T_i < 6.0$ for the systems in the upper row, $T_i < 5.3$ for the systems in the middle row, and $T_i < 4.0$ for the systems in the lower row. This means that condition (6.3) excludes the three left and the lower middle examples in Figure 6.7 and Figure 6.8.

The inequality for the integration time given by (6.3) can be used to give insight into the limitations of the Ziegler-Nichols rules for systems with large time delays. Consider a process with the transfer function

$$P(s) = K_p\frac{e^{-sL}}{1 + sT}.$$

For this system we have $T_{ar} = L + T$. Consider a PI controller tuned by the Ziegler-Nichols step response method. It follows from Table 6.1 that $KK_p = 0.9T/L$ and $T_i = 3L$. Equation 6.3 then gives

$$3L < (L + T)\frac{1.8T}{L + 1.8T},$$

which implies that $L < 0.38T$. This means that the Ziegler-Nichols step response method for PI control will not give good control unless the time delay is sufficiently small. Compare with Example 6.4.

Commentary

Manual tuning was used before any systematic tuning methods were available. It became a necessary complement to the Ziegler-Nichols method. It is essential for all practitioners of control to gain experience in judging the properties of closed-loop systems and to change controller parameters to modify the behavior. The assessment can be based on simple bump tests where set points or controller output is perturbed or by more elaborate frequency response measurements of the transfer function. It is necessary to be aware of the counter-intuitive behavior of processes with integral action illustrated in Example 6.5. The rule-based systems have been formalized when automatic tuners based on expert systems and fuzzy logic were developed. In Section 6.7 we will present systematic methods for improving the tuning based on optimization.

6.4 Pole Placement

Many properties of a closed-loop system are expressed by its poles. The idea with pole placement is to design a controller that gives a closed-loop system with desired closed-loop poles. The method requires a complete model of the process. Subject to some technical conditions it is possible to find a controller that gives the desired closed-loop poles, provided that the controller is sufficiently complex. To use the method for PID control it is necessary to restrict the complexity of the model by various approximation methods. The selected poles must then be chosen with care in order to ensure that the approximated model is valid for frequencies that correspond to the chosen poles.

A refinement of the procedure is to consider also the zeros of the transfer functions. This is particularly relevant for the set-point response. The zeros of the transfer function originating from the controller can be influenced by set-point weighting.

EXAMPLE 6.6—PI CONTROL OF A FIRST-ORDER SYSTEM
Suppose that the process can be described by the following first-order model

$$P(s) = \frac{K_p}{1+sT},$$

which has only two parameters, process gain K_p and time constant T. Let the process be controlled by a standard PI controller with set-point weighting,

$$C(s) = K\big(1 + \frac{1}{sT_i}\big)$$
$$C_{ff}(s) = K\big(b + \frac{1}{sT_i}\big).$$

The closed-loop system is of second order. The loop transfer function is

$$G_\ell(s) = P(s)C(s) = \frac{K_pK(1+sT_i)}{sT_i(1+sT)} = \frac{K_pK(s+1/T_i)}{T(s+1/T)},$$

and the characteristic polynomial

$$s^2 + \frac{1+K_pK}{T}s + \frac{K_pK}{TT_i}. \tag{6.4}$$

The closed-loop system has two poles that can be given arbitrary values by a suitable choice of gain K and integral time T_i of the controller. Now suppose that the desired closed-loop poles are characterized by their relative damping ζ and their frequency ω_0. The desired characteristic polynomial then becomes

$$s^2 + 2\zeta\omega_0 s + \omega_0^2. \tag{6.5}$$

Identifying coefficients of equal powers of s in (6.4) and (6.5) we get

$$\begin{aligned} K &= \frac{2\zeta\omega_0 T - 1}{K_p} \\ T_i &= \frac{2\zeta\omega_0 T - 1}{\omega_0^2 T} \\ k_i &= \frac{K}{T_i} = \frac{\omega_0^2 T}{K_p}. \end{aligned} \tag{6.6}$$

It is convenient to use the parameters ω_0 and ζ as design parameters; ω_0 determines the response speed and ζ determines the shape of the response.

With controller parameters given by (6.6) the closed-loop system is characterized by the *Gang of six*, see Equation (4.2).

$$\begin{aligned} \frac{PC}{1+PC} &= \frac{(2\zeta\omega_0 - 1/T)s + \omega_0^2}{s^2 + 2\zeta\omega_0 s + \omega_0^2} & \frac{C}{1+PC} &= \frac{K(s+1/T_i)(s+1/T)}{s^2 + 2\zeta\omega_0 s + \omega_0^2} \\ \frac{P}{1+PC} &= \frac{K_p s/T}{s^2 + 2\zeta\omega_0 s + \omega_0^2} & \frac{1}{1+PC} &= \frac{s(s+1/T)}{s^2 + 2\zeta\omega_0 s + \omega_0^2} \\ \frac{PC_{ff}}{1+PC} &= \frac{b(2\zeta\omega_0 - 1/T)s + \omega_0^2}{s^2 + 2\zeta\omega_0 s + \omega_0^2} & \frac{C_{ff}}{1+PC} &= \frac{K(bs+1/T_i)(s+1/T)}{s^2 + 2\zeta\omega_0 s + \omega_0^2}. \end{aligned} \tag{6.7}$$

The largest value of the transfer function from a load disturbance at the process input to the process output is

$$\max_{\omega} |G_{xd}(i\omega)| = \max_{\omega} \left| \frac{P(i\omega)}{1 + P(i\omega)C(i\omega)} \right| = \frac{K_p}{\omega_0 T \min(1, \zeta)}.$$

To have good rejection of load disturbances it is thus desirable to choose ω_0 as large as possible. The largest value of ω_0 is limited by the magnitude of the control signals and the validity of the process model. The transfer function from measurement noise to the control signal has the magnitude K for high frequencies. If $K_{\max}$ is the largest permissible value of the controller gain it follows from (6.6) that

$$\omega_0 T < \frac{1 + K_p K_{\max}}{2\zeta}.$$

Let T_e be the sum of neglected time constants or time delays and using the rule of thumb that the phase error should be less than $\pm 15^\circ$ we find that ω_0 must be chosen so that $\omega_0 T_e < 0.25$. Compare with Section 2.8.

The frequency ω_0 chosen should not be too small. An indication of this is given by Equation 6.6, which shows that the proportional gain is negative if $2\zeta\omega_0 T < 1$. Further evidence is given in Figure 6.10, which shows Bode plots

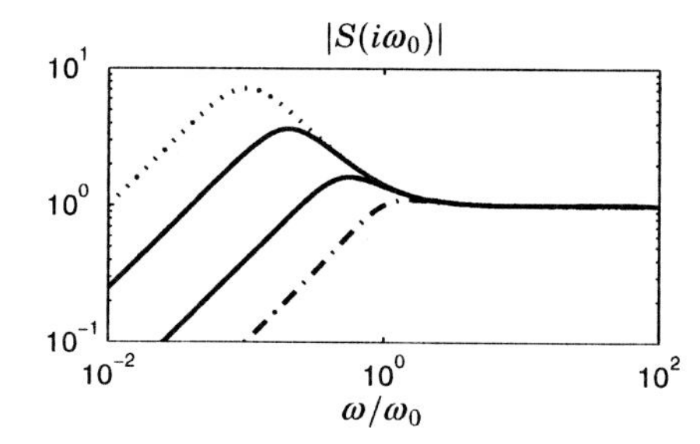

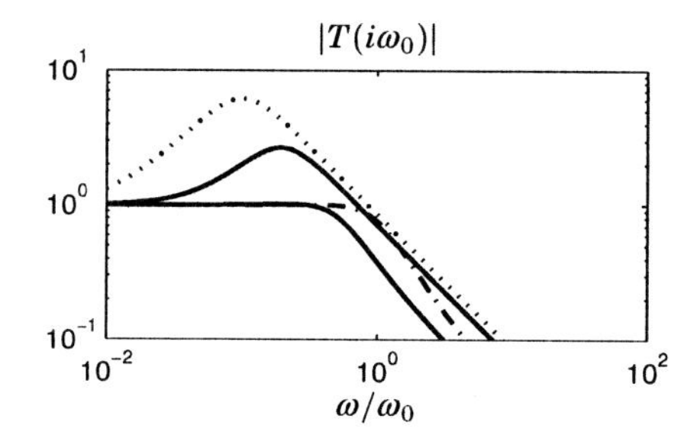

Figure 6.10 Gain curves of the sensitivity functions for $\zeta = 0.7$ and $\omega_0 T = 0.1, 0.2, 0.5,$ and 1. The dotted curve corresponds to $\omega_0 L = 0.1$ and the dash-dotted curve to $\omega_0 L = 1$.

of the gain curves of the sensitivity functions for different values of $\omega_0 T$. The figure shows that the sensitivities are large when $\omega_0 T$ is small. The maximum of the sensitivity function is approximately $M_s = 1/(2\zeta\omega_0 T)$. A reasonable choice of the parameter ω_0 is thus

$$\frac{1}{2\zeta} \leq \omega_0 T < \min\Big(\frac{0.25}{T_e}, \frac{1 + K_p K_{max}}{2\zeta}\Big). \tag{6.8}$$

The lower limit corresponds to pure integral control; see (6.6).

It follows from (6.7) that the transfer function from set point to process output has a zero at $s = -1/(bT_i)$. To avoid excessive overshoot in the set-point response, parameter b should be chosen so that the zero is to the left of the dominant closed-loop poles. A reasonable value is $b = 1/(\omega_0 T_i)$, which places the zero at $s = -\omega_0$. This gives

$$b = \frac{1}{2\zeta - 1/(\omega_0 T)}.$$

It is particularly important to use a small value of b when $\omega_0 T$ is small and for unstable systems where T is negative. A response to set-point changes that does not have an overshoot is obtained by choosing $b = 0$ and $\zeta \geq 1$.

The reason why the sensitivities are large for small values of $\omega_0 T$ is that the characteristic polynomial (6.5) is a poor choice for designs where the closed-loop system is slower than the open-loop system. In such cases it is better to make a design that cancels the process pole and gives a closed-loop system with a time constant T_0. Such a controller has the parameters

$$\begin{aligned} K &= \frac{T}{K_p T_0} \\ T_i &= T, \end{aligned} \tag{6.9}$$

and it gives a closed-loop system with $M_s = M_t = 1$. The controller is not suitable when $\omega_0 T_0 > 1$ because it follows from (6.7) that the transfer function from load disturbances to process output is

$$\frac{P}{1 + PC} = \frac{sK_p T_0}{(1 + sT)(1 + sT_0)}.$$

The attenuation of load disturbances is thus poor for large values of T_0/T. □

EXAMPLE 6.7—PI CONTROL OF PROCESS WITH TWO REAL POLES
Assume that the process is characterized by the second-order model

$$P(s) = \frac{K_p}{(1+sT_1)(1+sT_2)},$$

and that a PI controller is used. The loop transfer function becomes

$$G_\ell(s) = P(s)C(s) = \frac{K_pK(1+sT_i)}{sT_i(1+sT_1)(1+sT_2)} = \frac{K_pK(s+1/T_i)}{T_1T_2(s+1/T_1)(s+1/T_2)},$$

and the characteristic polynomial becomes

$$s^3 + \left(\frac{1}{T_1} + \frac{1}{T_2}\right)s^2 + \frac{1+K_pK}{T_1T_2}s + \frac{K_pK}{T_1T_2T_i}. \qquad (6.10)$$

The zeros of this third-order polynomial cannot be assigned arbitrary values since the controller only has two parameters. In particular, we find that the coefficient of s^2 is given by the time constants of the process. However, if we also consider the frequency ω_0 as a parameter it is possible to match the polynomial (6.10) to

$$(s+\alpha\omega_0)(s^2+2\zeta\omega_0 s+\omega_0^3).$$

Matching coefficients of equal powers of s we get

$$\begin{aligned}
\omega_0 &= \frac{T_1+T_2}{(\alpha+2\zeta)T_1T_2} \\
K &= \frac{(1+2\alpha\zeta)\omega_0^2T_1T_2-1}{K_p} \\
T_i &= \frac{K_pK}{\alpha\omega_0^3T_1T_2} \\
k_i &= \frac{\alpha\omega_0^3T_1T_2}{K_p}.
\end{aligned}$$

It is thus possible to obtain a design that gives a prescribed configuration of the poles with PI control, i.e., specified α and ζ. The parameter ω_0 is a scale factor that is determined by the process dynamics. □

EXAMPLE 6.8—PID CONTROL OF PROCESS WITH TWO REAL POLES
Suppose that the process is characterized by the second-order model

$$P(s) = \frac{K_p}{(1+sT_1)(1+sT_2)}.$$

This model has three parameters. By using a PID controller, which also has three parameters, it is possible to arbitrarily place the three poles of the closed-loop system. The transfer function of the PID controller can be written as

$$C(s) = \frac{K(1 + sT_i + s^2 T_i T_d)}{sT_i}.$$

The characteristic polynomial of the closed-loop system is

$$s^3 + s^2\left(\frac{1}{T_i} + \frac{1}{T_2} + \frac{K_p K T_d}{T_1 T_2}\right) + s\left(\frac{1}{T_1 T_2} + \frac{K_p K}{T_1 T_2}\right) + \frac{K_p K}{T_1 T_2 T_i}. \qquad (6.11)$$

A suitable closed-loop characteristic polynomial for a third-order system is

$$(s + \alpha\omega_0)(s^2 + 2\zeta\omega_0 s + \omega_0^2), \qquad (6.12)$$

which contains two dominant poles with relative damping ζ and frequency ω_0, and a real pole located in $-\alpha\omega_0$. Identifying the coefficients of equal powers of s in Equations 6.11 and 6.12 gives

$$\begin{aligned}\frac{1}{T_i} + \frac{1}{T_2} + \frac{K_p K T_d}{T_1 T_2} &= \omega_0(\alpha + 2\zeta)\\ \frac{1}{T_1 T_2} + \frac{K_p K}{T_1 T_2} &= \omega_0^2(1 + 2\zeta\omega_0)\\ \frac{K_p K}{T_1 T_2 T_i} &= \alpha\omega_0^3.\end{aligned}$$

Solving these equations gives the following controller parameters:

$$\begin{aligned}K &= \frac{T_1 T_2 \omega_0^2(1 + 2\alpha\zeta) - 1}{K_p}\\ T_i &= \frac{T_1 T_2 \omega_0^2(1 + 2\alpha\zeta) - 1}{T_1 T_2 \alpha\omega_0^3}\\ T_d &= \frac{T_1 T_2 \omega_0(\alpha + 2\zeta) - T_1 - T_2}{T_1 T_2 \omega_0^2(1 + 2\alpha\zeta) - 1}\\ k_i &= \frac{\alpha\omega_0^3 T_1 T_2}{K_p}.\end{aligned}$$

Provided that $c = 0$, the transfer function from set point to process output has one zero at $s = -1/(bT_i)$. To avoid excessive overshoot in the set-point response, parameter b can be chosen so that this zero cancels the pole at $s = -\alpha\omega_0$. This gives

$$b = \frac{1}{\alpha\omega_0 T_i} = \frac{\omega_0^2 T_1 T_2}{\omega_0^2 T_1 T_2(1 + 2\alpha\zeta) - 1}.$$

Also, notice that pure PI control is obtained for

$$\omega_0 = \omega_c = \frac{T_1 + T_2}{(\alpha + 2\zeta)T_1 T_2}.$$

The choice of ω_0 may be critical. The derivative time is negative for $\omega_0 < \omega_c$. Thus, frequency ω_c gives a lower bound to the bandwidth. The gain increases rapidly with ω_0. The upper bound to the bandwidth is given by the validity of the model. □

The General Case

Since there is a relation between the complexity of the model and the complexity of the controller it is natural to ask what is the most general model that will give PI and PID controllers. A PI controller has two parameters which are sufficient to characterize a second-order equation; this permits a process model of first order. The system in Example 6.6 is thus the most general system where pole placement will give a PI controller.

Since a PID controller has three parameters, it is possible to determine all parameters of a third-order equation. With PID control it is thus possible to use pole placement for a second-order system. The most general second-order system is not the one in Example 6.8, but the one in the next example.

If only a pattern of the pole is specified a PI controller suffices for a second-order system and a PID controller for a third-order system.

EXAMPLE 6.9—GENERAL SECOND-ORDER SYSTEM

Suppose that the process is characterized by the second-order model

$$P(s) = \frac{b_1 s + b_2}{s^2 + a_1 s + a_2}. \tag{6.13}$$

This model has four parameters. It has two poles that may be real or complex, and it has one zero. This model captures many processes, oscillatory systems, and systems with right half-plane zeros. The right half-plane zero can also be used as an approximation of a time delay. We assume that the process is controlled by a PID controller parameterized as

$$\begin{aligned} C(s) &= k + \frac{k_i}{s} + k_d s \\ C_{ff}(s) &= bk + \frac{k_i}{s} + ck_d s. \end{aligned}$$

The closed-loop system is of third order, and the characteristic polynomial is

$$s(s^2 + a_1 s + a_2) + (b_1 s + b_2)(k_d s^2 + ks + k_i).$$

A suitable closed-loop characteristic equation of a third-order system is

$$(s + \alpha\omega_0)(s^2 + 2\zeta\omega_0 s + \omega_0^2).$$

Equating coefficients of equal power in s in these equations gives the following equations:

$$\begin{aligned} a_1 + b_2 k_d + b_1 k &= (\alpha\omega_0 + 2\zeta\omega_0)(1 + b_1 k_d) \\ a_2 + b_2 k + b_1 k_i &= (1 + 2\alpha\zeta)\omega_0^2(1 + b_1 k_d) \\ b_2 k_i &= \alpha\omega_0^3(1 + b_1 k_d). \end{aligned}$$

This is a set of linear equations in the controller parameters. The solution is straightforward but tedious and is given by

$$\begin{aligned}
k &= \frac{a_2 b_2^2 - a_2 b_1 b_2(\alpha + 2\zeta)\omega_0 - (b_2 - a_1 b_1)(b_2(1 + 2\alpha\zeta)\omega_0^2 + \alpha b_1 \omega_0^3)}{b_2^3 - b_1 b_2^2(\alpha + 2\zeta)\omega_0 + b_1^2 b_2(1 + 2\alpha\zeta)\omega_0^2 - \alpha b_1^3 \omega_0^3} \\
k_i &= \frac{(-a_1 b_1 b_2 + a_2 b_1^2 + b_2^2)\alpha\omega_0^3}{b_2^3 - b_1 b_2^2(\alpha + 2\zeta)\omega_0 + b_1^2 b_2(1 + 2\alpha\zeta)\omega_0^2 - \alpha b_1^3 \omega_0^3} \\
k_d &= \frac{-a_1 b_2^2 + a_2 b_1 b_2 + b_2^2(\alpha + 2\zeta)\omega_0 - b_1 b_2 \omega_0^2(1 + 2\alpha\zeta) + b_1^2 \alpha \omega_0^3}{b_2^3 - b_1 b_2^2(\alpha + 2\zeta)\omega_0 + b_1^2 b_2(1 + 2\alpha\zeta)\omega_0^2 - \alpha b_1^3 \omega_0^3}.
\end{aligned} \tag{6.14}$$

These formulas are quite useful because many processes can be approximately described by the transfer function given by (6.13).

The transfer function from set point to process output is

$$G_{yy_{sp}}(s) = \frac{(b_1 s + b_2)(c k_d s^2 + b k s + k_i)}{(s + \alpha\omega_0)(s^2 + 2\zeta\omega_0 s + \omega_0^2)}.$$

The parameters b and c have a strong influence on the response of this transfer function. □

The formulas given in Example 6.9 are particularly useful in cases when we are "stretching" the PID controller to extreme situations. The standard tuning rules will typically not work in these cases. Typical examples are systems with zeros in the right half-plane and systems with poorly damped oscillatory modes. To illustrate this we will consider an example.

EXAMPLE 6.10—OSCILLATORY SYSTEM WITH RHP ZERO

Consider a system with the transfer function

$$P(s) = \frac{1 - s}{s^2 + 1}.$$

This system has one right half-plane zero and two undamped complex poles. The process is difficult to control. To provide damping for the undamped poles at $s = \pm i$ it is necessary to have a reasonable control gain at $\omega = 1$. This is difficult because the right-half plane zero at $s = 1$ implies that the gain crossover frequencies should be less than 0.5 in order to have a reasonably robust closed-loop system. None of the standard methods for tuning PID controllers work well for this system. To apply the pole placement method we specify that the closed-loop system has the characteristic polynomial

$$s^3 + 2s^2 + 2s + 1.$$

The formulas in Example 6.9 give a controller with the parameters $k = 0$, $k_i = 1/3$, and $k_d = 2/3$. This can also be verified with a simple calculation. Notice that the proportional gain is zero and that the controller has two complex zeros at $\pm i\sqrt{2}$. Such a controller can only be implemented with a PID controller having the non-interacting form. Compare with Section 3.2. □

Using Approximate Models

Since pole placement will only give PID controllers if the process model is of second order or less it is necessary to develop approximate models in order to use pole placement. Different approximation methods were discussed in Section 2.8. In this section we will illustrate the method with a few examples.

Consider a process described by the transfer function

$$P(s) = \frac{1}{(1+s)(1+0.2s)(1+0.05s)(1+0.01s)}. \tag{6.15}$$

This process has four lags with time constants 1, 0.2, 0.05, and 0.01. The approximations can be done in several different ways.

EXAMPLE 6.11—APPROXIMATION WITH A FIRST-ORDER SYSTEM

If the control requirements are not too severe, we can attempt to approximate the transfer function by

$$P(s) = \frac{1}{1+1.26s},$$

where the time constant is the average residence time of the system. As discussed in Section 2.8, this approximation is good at low frequencies. The sum of the neglected time constants is $T_e = 0.26$. The phase error is less than 15° for frequencies below 1 rad/s. Designing a PI controller with the pole placement method with $\zeta = 0.5$, the following controller parameters are obtained,

$$K = 1.26\omega_0 - 1$$
$$T_i = \frac{1.26\omega_0 - 1}{1.26\omega_0^2}$$
$$b = \frac{1.26\omega_0}{1.26\omega_0 - 1}.$$

where b is chosen so that the zero becomes $s = -\omega_0$. If the process model would be correct, the phase margin with $\zeta = 0.5$ would be 50°. Because of the approximations made, the phase margin will be less. It will decrease with ω_0. For $\omega_0 = 1$ the phase margin is $\varphi_m = 42°$. The closed-loop poles for the system are $-100, -20, -4.99, -0.46 \pm 1.02i$. The closed-loop poles obtained when the controller is applied to the simplified model are $-0.5 \pm 0.87i$. Because of the approximation the dominant poles differ from the design values. The difference increases with increasing ω_0. The system becomes unstable for $\omega_0 = 3.8255$.

Figure 6.11 shows the sensitivity functions for the approximate and the exact system. The maximum sensitivities are $M_t = 1.35$ and $M_s = 1.66$, respectively. This indicates that the closed-loop poles must be chosen with care when using pole placement. □

The next example shows what happens when the system is approximated with a second-order model.

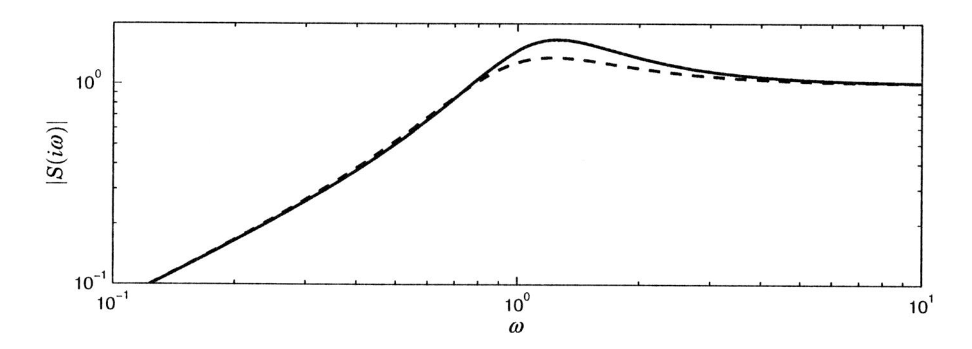

Figure 6.11 Sensitivity functions for the approximate system (dashed) and the true system in Example 6.11.

EXAMPLE 6.12—APPROXIMATION WITH A SECOND-ORDER MODEL

Consider the system given by (6.15). Approximate the transfer function by

$$P(s) = \frac{1}{(1+s)(1+0.26s)}.$$

It is obtained by keeping the longest time constant and approximating the three shorter time constants with their sum. The sum of the neglected time constants is $T_e = 0.06$. The phase error is less than 15° for frequencies below 4.4 rad/s. By making an approximation of the process model that is valid for higher frequencies than in the previous example, we can thus design a faster controller. If $\zeta = 0.5$ and $\alpha = 1$ are chosen in (6.16), the design calculations in Example 6.12 give the following PID parameters:

$$\begin{aligned} K &= 0.52\omega_0^2 - 1 \\ T_i &= \frac{0.52\omega_0^2 - 1}{0.26\omega_0^3} \\ T_d &= \frac{0.52\omega_0 - 1.26}{0.52\omega_0^2 - 1} \\ b &= \frac{0.26\omega_0^2}{0.52\omega_0^2 - 1}. \end{aligned} \tag{6.16}$$

In this case, pure PI control is obtained for $\omega_0 = 2.4$. The derivative gain becomes negative for lower bandwidths. The approximation neglects the time constant 0.05. If the neglected dynamics are required to give a phase error of, at most, 0.3 rad (17 deg) at the bandwidth, $\omega_0 < 6$ rad/s can be obtained. In Figure 6.12, the behavior of the control is demonstrated for $\omega_0 = 4, 5$, and 6.

The specification of the desired closed-loop bandwidth is crucial, since the controller gain increases rapidly with the specified bandwidth. It is also crucial to know the frequency range where the model is valid. Alternatively, an upper bound to the controller gain can be used to limit the bandwidth. Notice the effect of changing the design frequency ω_0. The system with $\omega_0 = 6$ responds

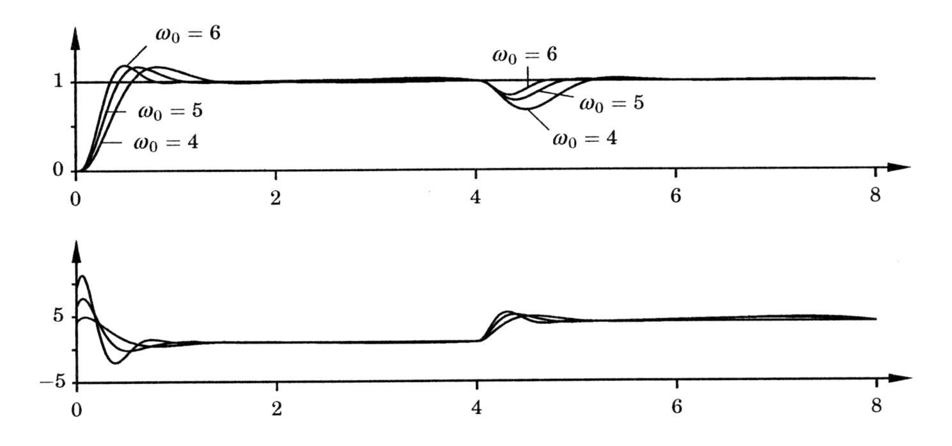

Figure 6.12 Set-point and load disturbance responses of the process with two poles controlled by a PID controller tuned according to Example 6.12. The responses for $\omega_0 = 4, 5$, and 6 are shown. The upper diagram shows set point $y_{sp} = 1$ and process output y, and the lower diagram shows control signal u.

faster and has a smaller error when subjected to load disturbances. The design will not work well when ω_0 is increased above 8. □

Dominant Pole Design

In pole placement design it is attempted to assign all closed-loop poles. One drawback with the method is that it is difficult to specify many closed-loop poles. In Section 4.5 it was mentioned that the behavior of a system can often be characterized with a few dominant poles. It can therefore be attempted to place a few dominant poles. We will illustrate this with a few examples.

EXAMPLE 6.13—AN INTEGRATING CONTROLLER

Consider a process with the transfer function $P(s)$ and an integrating controller

$$C(s) = \frac{k_i}{s}.$$

The closed-loop poles are given by

$$1 + k_i \frac{P(s)}{s} = 0.$$

Since the controller has one adjustable parameter, it is possible to assign one pole. To obtain a pole at $s = -a$ the controller parameter should be chosen as

$$k_i = \frac{a}{P(-a)}. \tag{6.17}$$

To obtain a good attenuation of load disturbances we will choose the closed-loop pole so that the integral gain k_i is as large as possible. For example, if

$$P(s) = \frac{1}{(s+1)^2},$$

we get

$$k_i = a(-a+1)^2 = a^3 - 2a^2 + a,$$

which has its maximum 4/27 for $a = 1/3$. □

EXAMPLE 6.14—PI CONTROL

A PI controller has two parameters. Consequently, it is necessary to assign two poles. Consider a process with transfer function $P(s)$, and let the controller be parameterized as

$$C(s) = k + \frac{k_i}{s}.$$

The closed-loop characteristic equation is

$$1 + \left(k + \frac{k_i}{s}\right)P(s) = 0.$$

Require that this equation has roots at

$$p_{1,2} = \omega_0\left(-\zeta_0 \pm i\sqrt{1-\zeta_0^2}\right) = \omega_0 e^{i(\pi\pm\gamma)} = \omega_0(-\cos\gamma \pm i\sin\gamma),$$

where $\gamma = \arccos\zeta_0$. The condition that the closed-loop system has a pole p_1 is thus

$$1 + \left(k + \frac{k_i}{p_1}\right)P(p_1) = 0. \tag{6.18}$$

This is a linear equation in complex variables with two unknown variables. To solve it we introduce $a(\omega_0)$ and $\phi(\omega_0)$, defined as

$$P\left(\omega_0 e^{i(\pi-\gamma)}\right) = a(\omega_0)e^{i\phi(\omega_0)}.$$

Notice that $P\left(\omega_0 e^{i(\pi-\gamma)}\right)$ represents the values of the transfer function on the ray $e^{i(\pi-\gamma)}$. When $\gamma = \pi/2$, then $P\left(\omega_0 e^{i(\pi-\gamma)}\right) = P(i\omega_0)$, which is the normal frequency response.

Equation 6.18 can be written as

$$1 + \left(k + \frac{k_i}{\omega_0 e^{i(\pi-\gamma)}}\right)a(\omega_0)e^{i\phi(\omega_0)} = 0.$$

This equation, which is linear in k and k_i, has the solution

$$\begin{aligned} k &= -\frac{\sin(\phi(\omega_0)+\gamma)}{a(\omega_0)\sin\gamma} \\ k_i &= -\frac{\omega_0\sin\phi(\omega_0)}{a(\omega_0)\sin\gamma}. \end{aligned} \tag{6.19}$$

Notice that $\phi(\omega_0)$ is zero for $\omega_0 = 0$ and typically negative as ω_0 increases. This implies that the proportional gain is negative and the integral gain positive but small for small ω_0. When ω_0 increases both k and k_i will increase initially. For larger values of ω_0 both parameters will decrease. Requiring that both parameters are positive, we find that ω_0 must be selected so that

$$\gamma < -\phi(\omega_0) < \pi.$$

The integral time of the controller is

$$T_i = \frac{k}{k_i} = \frac{\sin(\phi(\omega_0) + \gamma)}{\omega_0 \sin \phi(\omega_0)}.$$

Notice that T_i is independent of $a(\omega_0)$. □

EXAMPLE 6.15—A PURE DEAD-TIME PROCESS
Consider a process with the transfer function

$$P(s) = e^{-sL}.$$

Using pure integral control, it follows from Equation 6.17 that $k_i = ae^{-aL}$. The gain has its largest value $k_i = e^{-1}/L$ for $a = 1/L$. The loop transfer function for the system is then

$$G_l(s) = P(s)C(s) = \frac{1}{esL}\, e^{-sL}.$$

The sensitivity of the system is $M_s = 1.39$, which is a reasonable value.

With PI control it follows from Equation 6.19 that

$$\begin{aligned} k &= \frac{\sin(\omega_0 L \sin\gamma - \gamma)}{\sin\gamma}\, e^{-\omega_0 L \cos\gamma} \\ k_i &= \omega_0 \frac{\sin(\omega_0 L \sin\gamma)}{\sin\gamma}\, e^{-\omega_0 L \cos\gamma}. \end{aligned}$$

To minimize IE, we determine the value of ω_0 that maximizes k_i. The results are given in Table 6.6. This table also gives the M_s values and the IAE. The IAE has minimum at $\zeta \approx 0.6$. Notice that there are significant variations in the gain but that the values of integration time are fairly constant for all values of the design parameter ζ. The value of IAE should be small to give good rejection of load disturbances, and M_s should be small to give good robustness. The table illustrates the trade-offs between these goals. To obtain a reasonable robustness of $M_s < 2$, the relative damping should be greater than 0.5.

Notice that for $\zeta = 1$ we get $k = e^{-2}$ and $k_i = 4e^{-2}/L$. This can be compared with $k_i = e^{-1}L$ for pure I control. With PI control the integral gain can thus be increased by a factor of 1.5 compared with an I controller. For a well-damped system ($\zeta = 0.707$) the gain is about 0.2 and the integral time is $T_i = 0.28L$. This can be compared with the values 0.45 and 2L obtained with the Ziegler-Nichols frequency response method. The dominant pole design thus gives a controller with much stronger integral action than the Ziegler-Nichols method. In Example 6.4 we found that this was highly desirable.

Table 6.6 Controller parameters for dominant pole design of a PI controller for a pure time delay process.

ζ	k	k_iL	T_i/L	$\omega_0 L$	M_s	IAE/L
0.1	0.388	1.50	0.258	1.97	6.34	4.03
0.2	0.343	1.27	0.270	1.93	3.60	2.42
0.5	0.244	0.847	0.288	1.86	1.99	1.56
0.707	0.195	0.688	0.284	1.88	1.69	1.54
1.0	0.135	0.541	0.250	2.00	1.49	1.85

In summary, we find that a process with a pure delay dynamics can be controlled quite well with a PI controller. □

Dominant pole design is a special case of pole placement where it is only attempted to place a few dominant poles. For pure P, I, or D controllers one pole can be placed. For PI and PD controllers there are two dominant poles, which can conveniently be parameterized with the relative damping ζ. The method becomes more complicated for PID control. After the design it is necessary to check that the closed-loop poles obtained are actually dominating. It is also necessary to evaluate the robustness of the closed-loop system.

Commentary

Pole placement is a standard method for control system design. The specifications are given in terms of all poles of the closed-loop system or possibly only the pole pattern. Good judgment is required to choose the poles properly. When using pole placement the complexity of the controller is determined by the complexity of the process model. To obtain a PID controller it is required that the model is of low order or that the model is approximated by a low-order model. Time delay are often approximated when using pole placement. There is no natural way to introduce a robustness constraint in pole placement. The resulting closed-loop system must be analyzed to ensure that it is sufficiently robust.

6.5 Lambda Tuning

Lambda tuning is a special case of pole placement that is commonly used in the process industry. The process is modeled by the FOTD model

$$P(s) = \frac{K_p}{1 + sT} e^{-sL}.$$

Different approximations of time delay L result in both PI and PID controllers.

PI Control

If a PI controller with the transfer function

$$C(s) = K\frac{1+sT_i}{sT_i}$$

is used with integral time T_i chosen equal to the time constant T of the process, the loop transfer function becomes

$$G_l(s) = P(s)C(s) = \frac{K_pK}{sT}e^{-sL} \approx \frac{K_pK(1-sL)}{sT},$$

where the exponential function has been approximated using a Taylor series expansion. The characteristic equation of the closed-loop system is

$$s(T - K_pKL) + K_pK = 0.$$

Requiring that the closed-loop pole is $s = -1/T_{cl}$, where T_{cl} is the desired closed-loop time constant, we find

$$K_pK = \frac{T}{L+T_{cl}},$$

which gives the following simple tuning rule

$$\begin{aligned} K &= \frac{1}{K_p}\frac{T}{L+T_{cl}} \\ T_i &= T. \end{aligned} \qquad (6.20)$$

The closed loop response time T_{cl} is the design parameter. In the original work by Dahlin [Dahlin, 1968] it was denoted as $T_{cl} = \lambda$, which explains the name lambda tuning.

The choice of T_{cl} is critical. A common rule of thumb is to choose $T_{cl} = 3T$ for a robust controller and $T_{cl} = T$ for aggressive tuning when the process parameters are well determined. Both choices lead to controllers with zero gain and zero integral time for pure time delay systems. For delay-dominated processes it is therefore sometimes recommended to choose T_i as the largest of the values T and $3L$.

A drawback with lambda tuning is that the process pole is canceled. This is not serious if for delay dominated processes. The integral gain is

$$k_i = \frac{K}{T_i} = \frac{1}{K_p(L+T_{cl})}.$$

When T_{cl} is proportional to T integral gain is thus small for large T. The response to load disturbances is thus very poor for lag-dominated processes.

For lag-dominated processes it is therefore useful to make a design that does not cancel the process pole. When the FOTD process is controlled with a PI controller the loop transfer function is

$$G_l(s) = P(s)C(s) = \frac{K_pK(1+sT_i)e^{-sL}}{sT_i(1+sT)} \approx \frac{K_pK(1+sT_i)(1-sL)}{sT_i(1+sT)},$$

where the exponential function has been approximated by a Taylor series expansion. The characteristic equation is of second order:

$$s^2\Big(\frac{T_iT}{K_pK} - T_iL\Big) + s\Big(T_i + \frac{T_i}{K_pK} - L\Big) + 1 = 0.$$

Comparing this with the desired characteristic equation,

$$s^2T_{cl}^2 + 2\zeta T_{cl}s + 1 = 0,$$

gives the controller parameters

$$\begin{aligned} K &= \frac{L + 2\zeta T_{cl}}{T_{cl}^2 + T_{cl}^2/(K_pK) + 2\zeta T_{cl}L + L^2} \\ T_i &= \frac{K_pK(L + 2\zeta T_{cl})}{1 + K_pK}. \end{aligned} \tag{6.21}$$

These tuning rules can also be applied to integrating process provided that T_{cl} is chosen properly. For lag-dominated processes it is reasonable to choose T_{cl} proportional to L.

PID Control

For the derivation of the PID design, the interacting form of the PID controller (3.8) is used:

$$C'(s) = K'\frac{(1 + sT_i')(1 + sT_d')}{sT_i'}.$$

The time delay is approximated using (2.59), which gives the process transfer function

$$P(s) = \frac{K_p}{1 + sT}e^{-sL} \approx \frac{K_p(1 - sL/2)}{(1 + sT)(1 + sL/2)}.$$

The integral time is chosen to $T_i' = T$ and the derivative time to $T_d' = L/2$. The zeros of the controller will then cancel the poles of the process, and the loop transfer function becomes

$$G_l(s) = P(s)C'(s) \approx \frac{K_pK'(1 - sL/2)}{sT}.$$

The characteristic equation is

$$s(T - K_pK'L/2) + K_pK' = 0.$$

Requiring that the closed-loop pole is $s = -1/T_{cl}$ we find

$$K_pK' = \frac{T}{L/2 + T_{cl}},$$

which gives the following simple tuning rules:

$$
\begin{aligned}
K' &= \frac{1}{K_p} \frac{T}{L/2 + T_{cl}} \\
T_i' &= T \\
T_d' &= \frac{L}{2}.
\end{aligned}
$$

Using (3.9), the corresponding parameters for the noninteracting PID controller becomes

$$
\begin{aligned}
K &= \frac{1}{K_p} \frac{L/2 + T}{L/2 + T_{cl}} \\
T_i &= T + L/2 \\
T_d &= \frac{TL}{L + 2T}.
\end{aligned}
\tag{6.22}
$$

Notice that there is no derivative action for pure delay processes ($T = 0$).

Commentary

Lambda tuning is a special case of pole placement. It is a simple method that can give good results in certain circumstances provided that the design parameter is chosen properly. The basic method cancels a process pole which will lead to poor response to load disturbances for lag-dominated processes. Various ad hoc fixes can be made, but this requires insight.

6.6 Algebraic Design

There are several algebraic tuning methods where the controller transfer function is obtained from the specifications by a direct algebraic calculation. The methods are closely related to pole placement.

Standard Forms

A fundamental question is to determine transfer functions that give suitable responses to set-point changes. This can be done by starting with a transfer function of a given form and determining the parameters so that some error criterion such as IAE, ISE, or ITAE is minimized.

Typical examples are

$$
\begin{aligned}
G_1 &= \frac{\omega_0^2}{s^2 + 2\zeta\omega_0 + \omega_0^2} \\
G_2 &= \frac{\alpha\omega_0^3}{(s^2 + 2\zeta\omega_0 s + \omega_0^2)(s + \alpha\omega_0)} \\
G_3 &= \frac{\omega_0(s + \beta\omega_0)}{\beta(s^2 + 2\zeta\omega_0 s + \omega_0^2)} \\
G_4 &= \frac{\alpha\omega_0^2(s + \beta\omega_0)}{\beta(s^2 + 2\zeta\omega_0 s + \omega_0^2)(s + \alpha\omega_0)}.
\end{aligned}
\tag{6.23}
$$

The parameter ω_0 is a scale factor that determines the response speed. Parameters α, β, and ζ determine the shape of the transfer functions. Relative damping ζ is typically in the range of 0.5 to 1. The parameters α and β have a significant influence if they are less than one. Decreasing α makes the response slower and reduces the overshoot. Decreasing β makes the response faster and increases the overshoot. There have been many efforts to find parameters that optimize various criteria. Consider a system where the process has transfer function $P(s)$ and the controller transfer functions are

$$C(s) = K\Big(1 + \frac{1}{sT_i} + sT_d\Big)$$
$$C_{ff}(s) = K\Big(b + \frac{1}{sT_i} + scT_d\Big).$$

The closed-loop transfer function from set point to process output is then

$$G_{yy_{sp}} = \frac{PC_{ff}}{1+PC}.$$

The controller parameters K, T_i, and T_d are first chosen to match the denominator of the specified transfer function, and the set-point weights b and c are then chosen to match the numerator of the specified transfer function. Since the simple controllers only have a few parameters it is necessary that the chosen transfer functions be sufficiently simple.

For systems with error feedback where $C(s) = C_{ff}(s)$ it is possible to give an explicit expression for the controller transfer function:

$$C = \frac{1}{P} \cdot \frac{G_{yy_{sp}}}{1 - G_{yy_{sp}}}. \tag{6.24}$$

To make sure that the controller obtained is a PID controller it is necessary to make approximations or cancellations as was discussed in Section 2.8.

It follows from (6.24) that all process poles and zeros are canceled by the controller unless $G_{yy_{sp}}$ has corresponding poles and zeros. This means that error feedback cannot be applied when the process has poorly damped poles and zeros. The method will also give a poor load disturbance response when slow process poles are canceled.

There are many different versions of algebraic design methods. Let it suffice to present a few cases.

Haalman's Method

For systems with a time delay L, Haalman has suggested choosing the loop transfer function

$$G_l(s) = P(s)C(s) = \frac{2}{3Ls}\, e^{-sL}.$$

The value $2/3$ was found by minimizing the mean square error for a step change in the set point. This choice gives a sensitivity $M_s = 1.9$, which is a reasonable value. Notice that it is only the dead time of the process that influences the loop transfer function. All other process poles and zeros are canceled, which may lead to difficulties.

Applying Haalman's method to a process with the transfer function

$$P(s) = \frac{K_p}{1 + sT} e^{-sL}$$

gives the controller

$$C(s) = \frac{2(1 + sT)}{3K_p Ls} = \frac{2T}{3K_p L}\left(1 + \frac{1}{sT}\right),$$

which is a PI controller with $K = 2T/3K_pL$ and $T_i = T$. These parameters can be compared with the values $K = 0.9T/L$ and $T_i = 3L$ obtained by the Ziegler-Nichols step response method.

Comparing Haalman's method with lambda tuning we find that the integral times are the same and that the gains are the same if we choose $T_{cl} = L/2$. Since lambda tuning is based on approximations of the time delay it appears more reasonable to use Haalman's method when the time delay L is large.

Applying Haalman's method to a process with the transfer function

$$P(s) = \frac{K_p}{(1 + sT_1)(1 + sT_2)} e^{-sL}$$

gives a PID controller with parameters $K = 2(T_1+T_2)/3K_pL$, $T_i = T_1+T_2$, and $T_d = T_1T_2/(T_1+T_2)$. For more complex processes it is necessary to approximate the processes to obtain a transfer function of the desired form as was discussed in Section 2.8.

Figure 6.13 shows a simulation of Haalman's method for a system with normalized dead time $\tau = 0.5$. The figure shows that the responses are good.

Dangers of Cancellation of Slow Process Poles

A key feature of Haalman's method is that process poles and zeros are canceled by poles and zeros in the controller. When poles and zeros are canceled, there will be uncontrollable modes in the closed-loop system. This may lead to poor performance if the modes are excited. The problem is particularly severe if the canceled modes are slow or unstable. We use an example to illustrate what may happen.

EXAMPLE 6.16—LOSS OF CONTROLLABILITY DUE TO CANCELLATION

Consider a closed-loop system where a process with the transfer function

$$P(s) = \frac{1}{1 + sT} e^{-sL}$$

is controlled with a PI controller whose parameters are chosen so that the process pole is canceled. The transfer function of the controller is then

$$C(s) = K\left(1 + \frac{1}{sT}\right) = K\,\frac{1 + sT}{sT}.$$

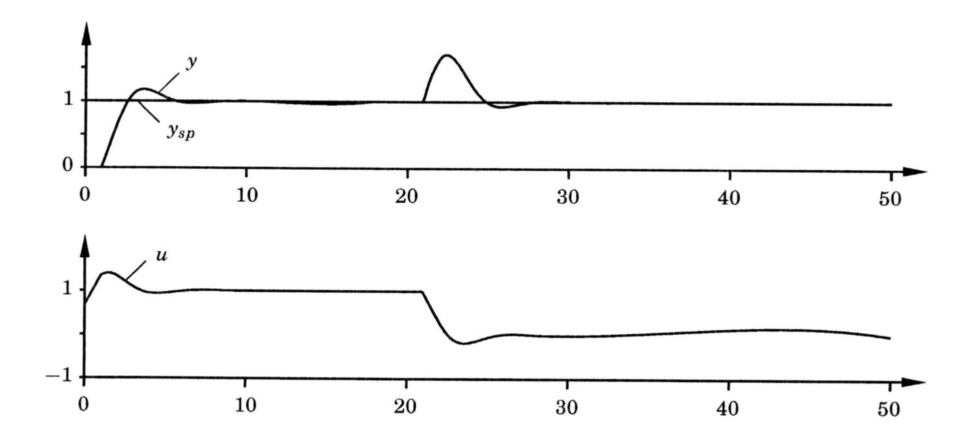

Figure 6.13 Simulation of a closed-loop system obtained by Haalman's method. The plant transfer function is $P(s) = e^{-s}/(s+1)$. The diagrams show set point y_{sp}, process output y, and control signal u.

The process can be represented by the equation

$$\frac{dy(t)}{dt} = \frac{1}{T}\left(u(t-L) - y(t)\right), \tag{6.25}$$

and the controller can be described by

$$\frac{du(t)}{dt} = -K\left(\frac{dy(t)}{dt} + \frac{y(t)}{T}\right). \tag{6.26}$$

Consider the behavior of the closed-loop system when the initial conditions are chosen as $y(0) = 1$ and $u(t) = 0$ for $-L < t < 0$. Without feedback the output is given by

$$y_{ol}(t) = e^{-t/T}.$$

To compute the output for the closed-loop system we first eliminate $y(t)$ between (6.25) and (6.26). This gives

$$\frac{du(t)}{dt} = -\frac{K}{T}u(t-L).$$

It thus follows that $u(t) = 0$, and (6.25) then implies that

$$y_{cl}(t) = e^{-t/T} = y_{ol}(t).$$

The trajectories of the closed-loop system and the open-loop system thus are the same. The control signal is zero, which means that the controller does not attempt to reduce the control error. □

The example clearly indicates that there are drawbacks with cancellation of process poles. Another illustration of the phenomenon is given in Figure 6.14,

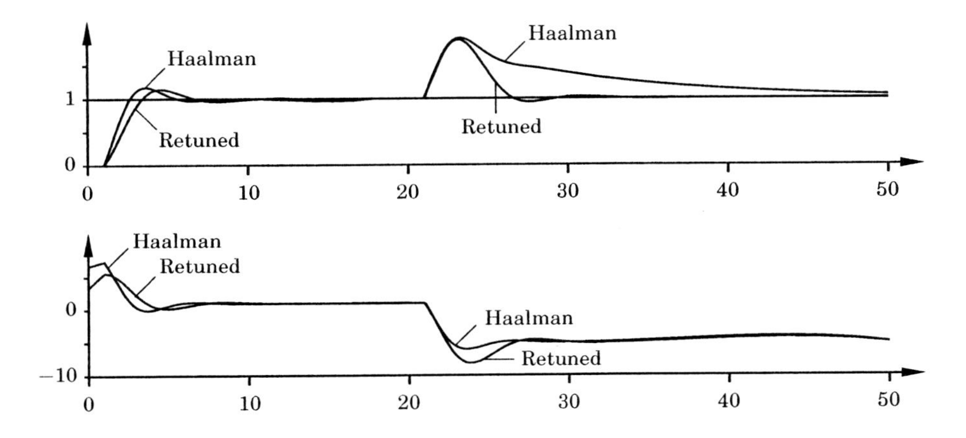

Figure 6.14 Simulation of a closed-loop system obtained by Haalman's method. The process transfer function is $P(s) = e^{-s}/(10s+1)$, and the controller parameters are $K = 6.67$ and $T_i = 10$. The upper diagram shows set point $y_{sp} = 1$ and process output y, and the lower diagram shows control signal u. The figure also shows the responses to a re-tuned controller with $K = 6.67$, $T_i = 3$, and $b = 0.5$.

which is a simulation of a closed-loop system where the controller is designed by Haalman's method. This simulation is identical to the simulation in Figure 6.13, but the process time constant is now 10 instead of 1 for the simulation in Figure 6.14.

In this case we find that the set-point response is excellent but that the response to load disturbances is very poor. The reason for this is that the controller cancels the pole $s = -0.1$ by having a controller zero at $s = -0.1$. Notice that the process output after a load disturbance decays with the time constant $T = 10$ but that the control signal is practically constant due to the cancellation. The attenuation of load disturbances is improved considerably by reducing the integral time of the controller as shown in Figure 6.14.

We have thus shown that cancellation of process poles may give systems with poor rejection of load disturbances. Notice that this does not show up in simulations unless the process is excited. For example, it will not be noticed in a simulation of a step change in the set point. We may also ask why there is such a big difference in the simulations in Figures 6.13 and 6.14. The reason is that the canceled pole in Figure 6.14 is slow in comparison with the closed-loop poles, but it is of the same magnitude as the closed-loop poles in Figure 6.13.

We can thus conclude that pole cancellation can be done for systems that are dead-time dominated but not for systems that are lag dominated.

Internal Model Control (IMC)

The internal model principle is a general method for design of control systems that can be applied to PID control. A block diagram of such a system is shown in Figure 6.15. In the diagram it is assumed that all disturbances acting on the process are reduced to an equivalent disturbance d at the process output. In the figure $\hat{P}$ denotes a model of the process, $\hat{P}^{\dagger}$ is an approximate inverse

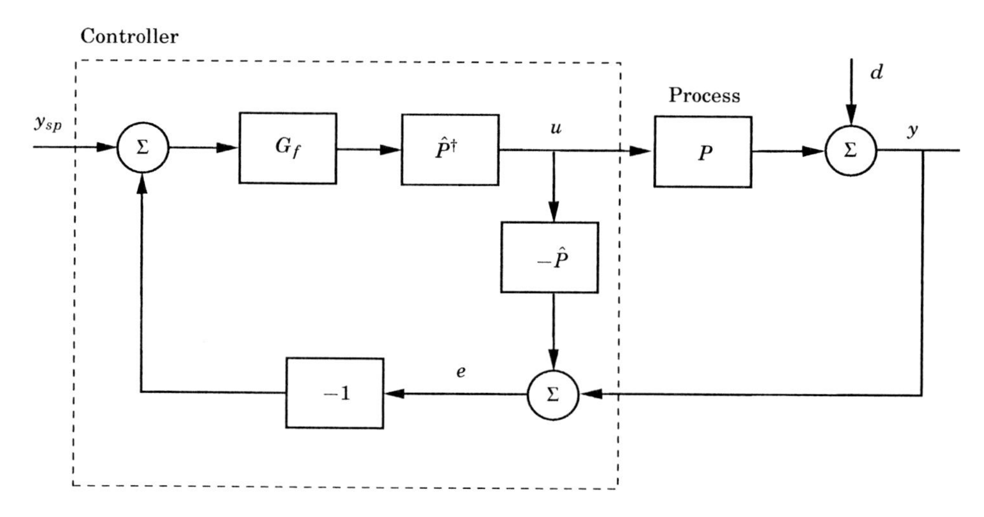

Figure 6.15 Block diagram of a closed-loop system with a controller based on the internal model principle.

of $\hat{P}$, and G_f is a low-pass filter. The name *internal model controller* derives from the fact that the controller contains a model of the process internally. This model is connected in parallel with the process.

If the model matches the process, i.e., $\hat{P} = P$, the signal e is equal to the disturbance d for all control signals u. If $G_f = 1$ and $\hat{P}^\dagger$ is an exact inverse of the process, then the disturbance d will be canceled perfectly. The filter G_f is introduced to obtain a system that is less sensitive to modeling errors. A common choice is $G_f(s) = 1/(1 + sT_f)$, where T_f is a design parameter.

The controller obtained by the internal model principle can be represented as an ordinary series controller with the transfer function

$$C = \frac{G_f \hat{P}^\dagger}{1 - G_f \hat{P}^\dagger \hat{P}}. \tag{6.27}$$

From this expression it follows that controllers of this type cancel process poles and zeros.

The internal model principle will typically give controllers of high order. By making special assumptions it is, however, possible to obtain PI or PID controllers from the principle. To see this consider a process with the transfer function

$$P(s) = \frac{K_p}{1 + sT} e^{-sL}. \tag{6.28}$$

An approximate inverse is given by

$$\hat{P}^\dagger(s) = \frac{1 + sT}{K_p}.$$

Notice that it is not attempted to find an inverse of the time delay. Choosing the filter

$$G_f(s) = \frac{1}{1 + sT_f},$$

and approximating the time delay by

$$e^{-sL} \approx 1 - sL,$$

Equation 6.27 now gives

$$C(s) = \frac{1 + sT}{K_p s(L + T_f)},$$

which is a PI controller. Notice that this controller is identical to the one obtained by lambda tuning if $T_f = T_{cl}$; see Equation 6.20.

If the time delay is approximated instead by a first-order Padé approximation,

$$e^{-sL} \approx \frac{1 - sL/2}{1 + sL/2},$$

Equation 6.27 gives instead the PID controller

$$C(s) = \frac{(1 + sL/2)(1 + sT)}{K_p s(L + T_f + sT_f L/2)} \approx \frac{(1 + sL/2)(1 + sT)}{K_p s(L + T_f)}.$$

For processes described by Equation 6.28, we thus find that the internal model principle will give PI or PID controllers. Approximations like the ones discussed in Section 2.8 can be used in the usual manner to obtain PI and PID controllers for more complex processes.

An interesting feature of the internal model controller is that robustness is considered explicitly in the design. Robustness can be adjusted by selecting the filter G_f properly. A trade-off between performance and robustness can be made by using the filter constant as a design parameter. The IMC can be designed to give excellent response to set-point changes. Since the design method inherently implies that poles and zeros of the plant are canceled, the response to load disturbances may be poor if the canceled poles are slow in comparison with the dominant poles. Compare with the responses in Figure 6.14. The IMC controller can also be viewed as an extension of the Smith predictor; see Chapter 8.

Skogestad's Internal Model Controller (SIMC)

Skogestad has developed a version of internal model control tuning method for PID control that avoids some of the drawbacks mentioned above. The method starts with a FOTD model for PI control or a SOTD model for PID control. It is required that the closed-loop system should have the transfer function

$$G_{yysp} = \frac{1}{1 + sT_{cl}} e^{-sL}$$

For an FOTD system it then follows from Equation 6.24 that the controller transfer function is

$$C(s) = \frac{1 + sT}{K_p(1 + sT_{cl} - e^{-sL})} \approx \frac{1 + sT}{sK_p(T_{cl} + L)}$$

where the exponential function is approximated using a Taylor series expansion. In contrast with the recommendation for IMC the closed-loop time constant is proportional to the time delay L. The choice $T_{cl} = L$ is recommended. The integral term is also modified for lag-dominated processes. The tuning rule for PI control is

$$K = \frac{T}{2K_pL} \qquad T_i = \min(T, 8L) \tag{6.29}$$

The same parameters are used for a PID controller in series form, and the derivative time is chosen as the shortest time constant.

Commentary

The analytical design methods are very closely related to pole placement. The main difference is that the complete transfer function is specified instead of just the closed-loop poles. A nice feature is that the calculations required are very simple. A drawback is that process poles are canceled. This is particularly serious for lag-dominated systems.

6.7 Optimization Methods

Optimization is a powerful tool for design of controllers. The method is conceptually simple. A controller structure with a few parameters is specified. Specifications are expressed as inequalities of functions of the parameters. The specification that is most important is chosen as the function to optimize. The method is well suited for PID controllers where the controller structure and the parameterization are given. There are several pitfalls when using optimization. Care must be exercised when formulating criteria and constraints; otherwise, a criterion will indeed be optimal, but the controller may still be unsuitable because of a neglected constraint. Another difficulty is that the loss function may have many local minima. A third is that the computations required may easily be excessive. Numerical problems may also arise. Nevertheless, optimization is a good tool that has successfully been used to design PID controllers. In this section we discuss some of these methods.

A Warning

Optimization algorithms are very powerful. They will solve whatever criterion is formulated. It is therefore very important to formulate the problems correctly and to introduce all relevant constraints. For PID control it is particularly important to introduce robustness constraints. This has frequently been disregarded in much work the on use of optimization for PID control. The following example illustrates what can happen.

EXAMPLE 6.17—A PI CONTROLLER OPTIMIZED FOR IAE
Consider a process with the transfer function

$$P(s) = \frac{1}{s+1}\, e^{-sL}.$$

Table 6.7 Controller parameters obtained from minimization of integrated absolute error, IAE. K_{hf} is the high-frequency gain of the controller.

L	IAE	M_s	K_{hf}	aK	T_i/L
0.0	0		∞		
0.2	0.14	3.3	4.7	0.94	2.9
0.5	0.60	3.0	2.0	1.0	2.2
1.0	1.5	2.4	1.0	1.0	1.4
2.0	3.2	2.1	0.60	1.2	1.0
5.0	7.7	2.0	0.42	2.1	0.6
10.0	15	1.9	0.37	3.7	0.53

Table 6.7 gives controller parameters that minimize IAE for load disturbances. Some of the other criteria are also given in the table. The table shows that the integrated absolute error increases with L, as can be expected. The table shows that the maximum sensitivity is large for practically all systems, particularly those with small L. The table also shows that the high-frequency gain of the controller is high for small values of L. For example, if we require $M_s < 1.8$, which is a fairly modest robustness requirement, none of the systems is acceptable. □

The example illustrates the necessity of considering all aspects of a problem when formulating the problem. Unfortunately, this was not observed in much of the early work on controller tuning.

Tuning Formulas Based on Optimization

Many studies have been devoted to development of tuning rules based on optimization. Very often a process described by

$$P = \frac{K_p}{1+sT} e^{-sL}$$

has been considered. The loss functions obtained for unit step changes in set point and process input have been computed and formulas of the type

$$p = a\left(\frac{L}{T}\right)^b,$$

where p is a controller parameter and a and b are constants, have been fitted to the numerical values obtained. In many cases, the criterion is IAE for load disturbances, which often gives systems with low damping and poor sensitivity. The formulas given often only hold for a small range of normalized dead times, e.g., $0.2 < \tau < 0.6$. It should also be observed that criteria based on set-point changes can often be misleading because it is often not observed that the set-point changes are drastically influenced by different set-point weightings.

Modulus and Symmetrical Optimum

Modulus Optimum (BO) and Symmetrical Optimum (SO) are two methods for selecting and tuning controllers that also can be viewed as analytical designs where the desired transfer functions given by Equations 6.23 are obtained by optimization. The acronyms BO and SO are derived from the German words Betrags Optimum and Symmetrische Optimum. The methods were developed for motor drives where the response to set-point changes is particularly important. The basic idea is to find a controller that makes the frequency response from set point to plant output as close to one as possible for low frequencies. If $G(s)$ is the transfer function from the set point to the output, the controller is determined in such a way that $G(0) = 1$ and that $d^n|G(i\omega)|/d\omega^n = 0$ at $\omega = 0$ for as many n as possible. An interesting property is that the design method takes account of unmodeled dynamics explicitly. We illustrate the idea with a few examples.

EXAMPLE 6.18—SECOND-ORDER SYSTEM
Consider the transfer function

$$G(s) = \frac{a_2}{s^2 + a_1 s + a_2},$$

which has been chosen so that $G(0) = 1$. Let us first consider how the parameters should be chosen in order to get a maximally flat frequency response. We have

$$|G(i\omega)|^2 = \frac{a_2^2}{a_1^2\omega^2 + (a_2 - \omega^2)^2} = \frac{a_2^2}{a_2^2 + \omega^2(a_1^2 - 2a_2) + \omega^4}.$$

By choosing $a_1 = \sqrt{2a_2}$ we find

$$|G(i\omega)|^2 = \frac{a_2^2}{a_2^2 + \omega^4}.$$

The first three derivatives of $|G(i\omega)|$ will vanish at the origin. The transfer function then has the form

$$G(s) = \frac{\omega_0^2}{s^2 + \omega_0 s\sqrt{2} + \omega_0^2}.$$

The step response of a system with this transfer function has an overshoot $o = 4\%$. The settling time to 2% of the steady-state value is $T_s = 6/\omega_0$. □

If the transfer function G in the example is obtained by error feedback of a system with the loop transfer function G_{BO}, the loop transfer function is

$$G_{\mathrm{BO}}(s) = \frac{G(s)}{1 - G(s)} = \frac{\omega_0^2}{s(s + \sqrt{2}\omega_0)}, \tag{6.30}$$

which is the desired loop transfer function for the method called modulus optimum.

The calculation in Example 6.18 can be performed for higher-order systems with more effort. We illustrate by another example.

EXAMPLE 6.19—THIRD-ORDER SYSTEM WITH NO ZEROS
Consider the transfer function

$$G(s) = \frac{a_3}{s^3 + a_1 s^2 + a_2 s + a_3}.$$

After some calculations we get

$$|G(i\omega)| = \frac{a_3}{\sqrt{a_3^2 + (a_2^2 - 2a_1a_3)\omega^2 + (a_1^2 - 2a_2)\omega^4 + \omega^6}}.$$

Five derivatives of $|G(i\omega)|$ will vanish at $\omega = 0$, if the parameters are such that $a_1^2 = 2a_2$ and $a_2^2 = 2a_1a_3$. The transfer function then becomes

$$G(s) = \frac{\omega_0^3}{s^3 + 2\omega_0 s^2 + 2\omega_0^2 s + \omega_0^3} = \frac{\omega_0^3}{(s + \omega_0)(s^2 + \omega_0 s + \omega_0^2)}. \tag{6.31}$$

The step response of a system with this transfer function has an overshoot $o = 8.1\%$. The settling time to 2% of the steady state value is $9.4/\omega_0$. A system with this closed-loop transfer function can be obtained with a system having error feedback and the loop transfer function

$$G_l(s) = P(s)C(s) = \frac{\omega_0^3}{s(s^2 + 2\omega_0 s + 2\omega_0^2)}.$$

The closed-loop transfer function (6.31) can also be obtained from other loop transfer functions if the controller has set-point weighting. For example, if a process with the transfer function

$$P(s) = \frac{\omega_0^2}{s(s + 2\omega_0)}$$

is controlled by a PI controller having parameters $K = 2$, $T_i = 2/\omega_0$, and $b = 0$, the loop transfer function becomes

$$G_{SO} = \frac{\omega_0^2(2s + \omega_0)}{s^2(s + 2\omega_0)}. \tag{6.32}$$

The symmetric optimum aims at obtaining the loop transfer function given by Equation 6.32. Notice that the Bode plot of this transfer function is symmetrical around the frequency $\omega = \omega_0$. This is the motivation for the name symmetrical optimum.

If a PI controller with $b = 1$ is used, the transfer function from set point to process output becomes

$$G(s) = \frac{G_{SO}(s)}{1 + G_{SO}(s)} = \frac{(2s + \omega_0)\omega_0^2}{(s + \omega_0)(s^2 + \omega_0 s + \omega_0^2)}.$$

This transfer function is not maximally flat because of the zero in the numerator. This zero will also give a set-point response with a large overshoot, about 43 percent. □

The methods BO and SO can be called loop-shaping methods since both methods try to obtain a specific loop transfer function. The design methods can be described as follows. It is first established which of the transfer functions, G_{BO} or G_{SO}, is most appropriate. The transfer function of the controller $C(s)$ is then chosen so that the loop transfer $G_l(s) = P(s)C(s)$ meet specifications. We illustrate the methods with the following examples.

EXAMPLE 6.20—BO CONTROL
Consider a process with the transfer function

$$P(s) = \frac{K_p}{s(1+sT)}. \tag{6.33}$$

With a proportional controller the loop transfer function becomes

$$G_\ell(s) = \frac{KK_p}{s(1+sT)}.$$

To make this transfer function equal to G_{BO} given by Equation 6.30 it must be required that

$$\omega_0 = \frac{\sqrt{2}}{2T}.$$

The controller gain should be chosen as

$$K = \frac{\omega_0\sqrt{2}}{2K_p} = \frac{1}{2K_pT}.$$

□

EXAMPLE 6.21—SO CONTROL
Consider a process with the same transfer function as in the previous example (Equation 6.33). With a PI controller having the transfer function

$$C(s) = \frac{K(1+sT_i)}{sT_i},$$

we obtain the loop transfer function

$$G_l(s) = P(s)C(s) = \frac{K_pK(1+sT_i)}{s^2T_i(1+sT)}.$$

This is identical to G_{SO} if we choose

$$K = \frac{1}{2K_pT}$$
$$T_i = 4T.$$

To obtain the transfer function given by Equation 6.31 the set-point weight b should be zero. □

A Design Procedure A systematic design procedure can be based on the methods BO and SO. The design method consists of two steps. In the first step the process transfer function is simplified to one of the following forms:

$$
\begin{aligned}
P_1(s) &= \frac{K_p}{1+sT} \\
P_2(s) &= \frac{K_p}{(1+sT_1)(1+sT_2)}, \quad T_1 > T_2 \\
P_3(s) &= \frac{K_p}{(1+sT_1)(1+sT_2)(1+sT_3)}, \quad T_1 > T_2 > T_3 \\
P_4(s) &= \frac{K_p}{s(1+sT)} \\
P_5(s) &= \frac{K_p}{s(1+sT_1)(1+sT_2)}, \quad T_1 > T_2.
\end{aligned}
$$

Process poles may be canceled by controller zeros to obtain the desired loop transfer function. A slow pole may be approximated by an integrator; fast poles may be lumped together as discussed in Section 2.8. The rule of thumb given in the original papers on the method is that time constants T such that $\omega_0 T < 0.25$ can be regarded as integrators.

The controller is derived in the same way as in Examples 6.20 and 6.21 by choosing parameters so that the loop transfer function matches either G_{BO} or G_{SO}. By doing this we obtain the results summarized in Table 6.8. Notice, for example, that Examples 6.20 and 6.21 correspond to the entries Process G_4 in the table. It is natural to view the smallest time constant as an approximation of neglected dynamics in the process. It is interesting to observe that it is this time constant that determines the bandwidth of the closed-loop system.

The set-point response for the BO method is excellent. Notice that it is necessary to use a controller with a two-degree-of-freedom structure or a pre-filter to avoid a high overshoot for the SO method. Notice also that process poles are canceled in the cases marked C1 or C2 in Table 6.8. The response to load disturbances will be poor if the canceled pole is slow compared to the closed-loop dynamics, which is characterized by ω_0 in Table 6.8.

These design principles can be extended to processes other than those listed in the table.

EXAMPLE 6.22—APPLICATION OF BO AND SO

Consider a process with the transfer function

$$P(s) = \frac{1}{(1+s)(1+0.2s)(1+0.05s)(1+0.01s)}. \tag{6.34}$$

Since this transfer function is of fourth order, the design procedure cannot be applied directly. We show how different controllers are obtained depending on the approximations made. The performance of the closed-loop system depends on the approximation. We use parameter ω_0 as a crude measure of performance.

If a controller with low performance is acceptable, the process (6.34) can be approximated with

$$P(s) = \frac{1}{1+1.26s}. \tag{6.35}$$

Table 6.8 Controller parameters obtained with the BO and SO methods. Entry P gives the process transfer function, entry C gives the controller structure, and entry M tells whether the BO or SO method is used. In the entry Remark, A1 means that $1 + sT_1$ is approximated by sT_1, and Ci means that the time constant T_i is canceled.

P	C	M	Remark	KK_p	T_i	T_d	ω_0	b	c
P_1	I	BO		0.5	T		$\frac{0.7}{T}$		
P_2	P	BO	A1	$\frac{T_1}{2T_2}$			$\frac{0.7}{T_2}$	1	
P_2	PI	BO	C1	$\frac{T_1}{2T_2}$	T_1		$\frac{0.7}{T_2}$	1	
P_2	PI	SO	A1	$\frac{T_1}{2T_2}$	$4T_2$		$\frac{0.5}{T_2}$	0	
P_3	PD	BO	A1, C2	$\frac{T_1}{2T_3}$		T_2	$\frac{0.7}{T_3}$	1	1
P_3	PID	BO	C1, C2	$\frac{T_1+T_2}{2T_3}$	T_1+T_2	$\frac{T_1T_2}{T_1+T_2}$	$\frac{0.7}{T_3}$	1	1
P_3	PID	SO	A1, C2	$\frac{T_1(T_2+4T_3)}{8T_3^2}$	T_2+4T_3	$\frac{4T_2T_3}{T_2+4T_3}$	$\frac{0.5}{T_3}$	$\frac{T_2}{T_2+4T_3}$	0
P_4	P	BO		$\frac{1}{2T}$			$\frac{0.7}{T}$	1	
P_4	PI	SO		$\frac{1}{2T}$	4T		$\frac{0.5}{T}$	0	
P_5	PD	BO	C1	$\frac{1}{2T_2}$		T_1	$\frac{0.7}{T_2}$	1	1
P_5	PD	SO	A1	$\frac{T_1}{8T_2^2}$		$4T_2$	$\frac{0.5}{T_2}$	1	0
P_5	PID	SO	C1	$\frac{T_1+4T_2}{8T_2^2}$	T_1+4T_2	$\frac{4T_1T_2}{T_1+4T_2}$	$\frac{0.5}{T_2}$	$\frac{T_1}{T_1+4T_2}$	0

Table 6.9 Results obtained with different controllers designed by the BO and SO methods in Example 6.22. The frequency ω_m defines the upper limit when the phase error is less than 10%.

Controller	K	T_i	T_d	k_i	b	c	ω_0	ω_m	IAE
1				0.4			0.55	1.12	2.7
2	1.92	1		0.52	1		2.7	5.15	0.52
3	10	1.2	0.17	8.3	1	1	11.7	26.6	0.12
4	15.3	0.44	0.11	35	0.45	0	8.3	26.6	0.029

The approximation has a phase error less than 10° for $\omega \leq 1.12$. It follows from Table 6.8 that the system (6.35) can be controlled with an integrating controller with

$$k_i = \frac{K}{T_i} = \frac{0.5}{1.26} = 0.4.$$

This gives a closed-loop system with $\omega_0 = 0.55$.

A closed-loop system with better performance is obtained if the transfer function (6.34) is approximated with

$$P(s) = \frac{1}{(1+s)(1+0.26s)}. \tag{6.36}$$

The slowest time constant is thus kept, and the remaining time constants are approximated by lumping them together. The approximation has a phase error less than 10° for $\omega \leq 5.15$. A PI controller can be designed using the BO method. The parameters $K = 1.92$ and $T_i = 1$ are obtained from Table 6.8. The closed-loop system has $\omega_0 = 2.7$.

If the transfer function is approximated as

$$P(s) = \frac{1}{(1+s)(1+0.2s)(1+0.06s)}, \tag{6.37}$$

the approximation has a phase error less than 10° for $\omega \leq 26.6$. The BO method can be used also in this case. Table 6.8 gives the controller parameters $K = 10$, $T_i = 1.2$, and $T_d = 0.17$. The controller structure is defined by the parameters $b = 1$ and $c = 1$. This controller gives a closed-loop system with $\omega_0 = 11.7$.

The method SO can also be applied to the system (6.37). Table 6.8 gives the controller parameters $K = 15.3$, $T_i = 0.44$, $T_d = 0.11$, and $b = 0.45$. For these parameters we get $\omega_0 = 8.3$.

Controllers with different properties can be obtained by approximating the transfer function in different ways. A summary of the properties of the closed-loop systems obtained is given in Table 6.9, where IAE refers to the load disturbance response. Notice that Controller 2 cancels a process pole with time constant 1 s and that Controller 3 cancels process poles with time constants 1 s and 0.25 s. This explains why the IAE drops drastically for Controller 4, which

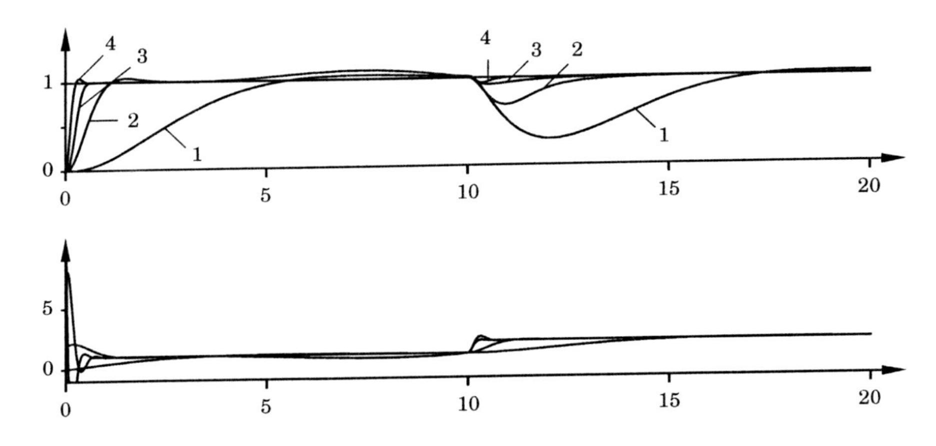

Figure 6.16 Simulation of the closed-loop system obtained with different controllers designed by the BO and SO methods given in Table 6.9. The upper diagram shows set point y_{sp} and process output y, and the lower diagram shows control signal u.

does not cancel any process poles. Controller 4 actually has a lower bandwidth ω_0 than Controller 3.

A simulation of the different controllers is shown in Figure 6.16. The simulation and the data shown in Table 6.9 clearly illustrate the benefits of improved modeling and more complicated controllers. □

Design for Disturbance Rejection

The design methods discussed so far have been based on a characterization of process dynamics. The properties of the disturbances have only influenced the design indirectly. A load disturbance in the form of a step was used, and in some cases a loss function based on the error due to a load disturbance was minimized. Measurement noise was also incorporated by limiting the high-frequency gain of the controller.

In this section, we briefly discuss design methods that directly attempt to make a trade-off between attenuation of load disturbances and amplification of measurement noise due to feedback.

Consider the system shown in Figure 6.17. Notice that the measurement signal is filtered before it is fed to the controller. Let D and N be the Laplace transforms of the load disturbance and the measurement noise, respectively. The process output and the control signal are then given by

$$
\begin{aligned}
X &= \frac{P}{1+G_\ell} D - \frac{G_\ell}{1+G_\ell} N \\
U &= -\frac{G_\ell}{1+G_\ell} D - \frac{CG_f}{1+G_\ell} N,
\end{aligned}
\tag{6.38}
$$

where $G_\ell = PCG_f$ is the loop transfer function. Different assumptions about the disturbances and different design criteria can now be given. We illustrate by an example.

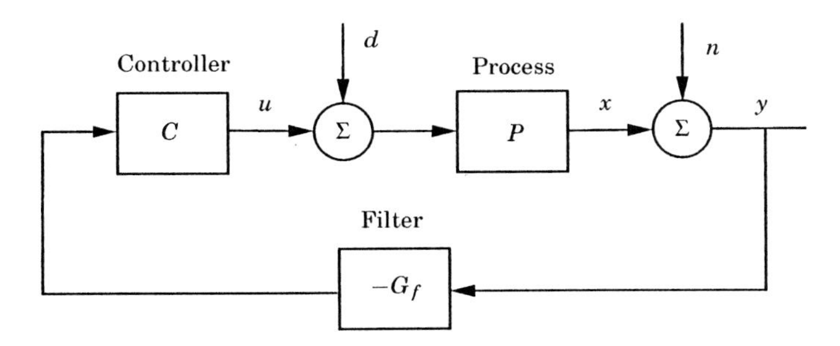

Figure 6.17 Block diagram of a closed-loop system.

EXAMPLE 6.23—DESIGN FOR DISTURBANCE REJECTION

Assume that the transfer functions in Figure 6.17 are given by

$$P = \frac{1}{s}, \qquad G_f = 1, \qquad C = k + \frac{k_i}{s}.$$

Furthermore, assume that n is stationary noise with spectral density ϕ_n and that d is obtained by sending stationary noise with the spectrum ϕ_d through an integrator. This is one way to model the situation that the load disturbance is drifting and the measurement noise has high frequency.

With the given assumptions, Equation 6.38 then becomes

$$X = \frac{s}{s^2 + ks + k_i}\,\frac{1}{s}\,D_1 - \frac{sk + k_i}{s^2 + ks + k_i}\,N$$

$$U = -\frac{sk + k_i}{s^2 + ks + k_i}\,\frac{1}{s}\,D_1 - \frac{s^2k + k_is}{s^2 + ks + k_i}\,N,$$

where we have assumed

$$D(s) = \frac{1}{s}\,D_1(s).$$

If n and d_1 are white noises, it follows that the variance of x is given by

$$J = Ex^2 = \frac{1}{2kk_i}\,\phi_d + \frac{1}{2}\left(k + \frac{k_i}{k}\right)\phi_n.$$

This equation clearly indicates the compromise in designing the controller. The first term of the right-hand side is the contribution to the variance due to the load disturbance. The second term represents the contribution due to the measurement noise. Notice that the attenuation of the load disturbances increases with increasing k and k_i, but that large values of k and k_i also increase the contribution of measurement noise.

We can attempt to find values of k and k_i that minimize J. A straightforward calculation gives

$$k = \sqrt{2}\left(\frac{\phi_d}{\phi_n}\right)^{1/4}$$

$$k_i = \sqrt{\frac{\phi_d}{\phi_n}}.$$

This means that the controller parameters are uniquely given by the ratio of the intensities of the process noise and the measurement noise. Also notice that with these parameters the closed-loop characteristic polynomial becomes

$$s^2 + \omega_0 s\sqrt{2} + \omega_0^2,$$

with $\omega_0 = \sqrt{\phi_v/\phi_e}$. The optimal system thus has a relative damping $\zeta = 0.707$ and a bandwidth that is given by the ratio of the intensities of load disturbance and measurement noise. □

Commentary

Optimization techniques are very powerful. When using them it is essential to include all relevant aspects of the problem in the formulation; otherwise, the so-called optimal controller may have very bad properties. In this section we have covered a few optimization methods that have been used for PID control.

The methods BO and SO are widely used for drive systems. The optimization is to find a transfer function from set point to process output that is maximally flat. The methods are primarily intended for systems without dead time. Small dead times can be dealt with by approximation.

An interesting feature of the procedure is the use of approximations; fast poles and slow time constants are neglected, and slow dynamics are approximated by integrators. Model uncertainty also appears explicitly in the design because the achievable bandwidth is determined by slowest neglected time constants.

The methods can be interpreted as pole placement where the desired closed-loop characteristic polynomial is

$$A_{\mathrm{BO}}(s) = s^2 + \omega_0 s\sqrt{2} + \omega_0^2$$

for the modulus optimum and

$$A_{\mathrm{SO}}(s) = (s + \omega_0)(s^2 + \omega_0 s + \omega_0^2)$$

for the symmetrical optimum. There are possibilities for combining the approaches. A drawback with all design methods of this type is that process poles may be canceled. This may lead to poor attenuation of load disturbances if the canceled poles are excited by disturbances and if they are slow compared to the dominant closed-loop poles.

6.8 Robust Loop Shaping

The design methods discussed so far all have the property that the robustness to process variations has to be checked after a design. One of the major advances in control theory in the end of the last century was the emergence of design methods with guaranteed robustness (the so called $\mathcal{H}^\infty$ theory). In this section we will present a method for design of PID controllers in the same spirit. In Section 4.6 it was shown that robustness conditions can be expressed in terms of circular discs that are forbidden regions for the Nyquist curve of the loop transfer function. For PID control these conditions give a set of admissible values of the controller parameters, called the robustness region. Attenuation of low-frequency load disturbances is inversely proportional to integral gain k_i. Measurement noise injection is captured by controller gain k for P and PI control or derivative gain k_d for PD and PID control. The design method is to maximize integral gain k_i subject to constraints on robustness and noise injection. Good set-point response is then obtained by set-point weighting or feedforward as discussed in Section 5.3. This design method brings design of PID controllers into the mainstream of control system design.

The Robustness Region

In Section 4.6 it was shown that robustness to process variations can be expressed by the maximum sensitivity M_s, the maximum complementary sensitivity M_t, or with the joint sensitivity M. All these conditions say that the Nyquist curve of the loop transfer function should avoid circles enclosing the critical point. For PID control of a process with given transfer function the robustness constraint translates into constraints on the controller parameters, called the robustness region. To determine the robustness region we consider a process with the transfer function $P(s)$ and an ideal PID controller with the transfer function $C(s)$. The loop transfer function is $G_l(s)$, and the square of the distance from a point on the Nyquist curve of the loop transfer function to the point $-c$ is

$$f(k, k_i, k_d, \omega) = |c + G_l(i\omega)|^2 = |c + (k + i(k_d\omega - k_i/\omega))P(i\omega)|^2;$$

and the robustness constraint becomes

$$f(k, k_i, k_d, \omega) \geq r^2. \tag{6.39}$$

Introduce

$$P(i\omega) = \alpha(\omega) + i\beta(\omega) = \rho(\omega)e^{i\varphi(\omega)}, \tag{6.40}$$

where

$$\begin{aligned}\alpha(\omega) &= \rho(\omega)\cos\varphi(\omega),\\ \beta(\omega) &= \rho(\omega)\sin\varphi(\omega).\end{aligned}$$

The following straightforward but tedious calculation shows that the function f can be written as

$$\begin{aligned}f(k, k_i, k_d, \omega) &= \Big|c + (k + i(k_d\omega - k_i/\omega))(\alpha(\omega) + i\beta(\omega))\Big|^2\\ &= \big|c + \alpha k + \beta(k_d\omega - k_i/\omega) + i(\beta k + \alpha(k_d\omega - k_i/\omega))\big|^2\\ &= c^2 + \rho^2k^2 + 2c\alpha k + \rho^2(k_d\omega - k_i/\omega)^2 - 2\beta c(k_d\omega - k_i/\omega)\\ &= \rho^2\Big(k + \frac{\alpha c}{\rho^2}\Big)^2 + \frac{\rho^2}{\omega^2}\Big(k_i + \frac{\omega\beta c}{\rho^2} - k_d\omega^2\Big)^2 \geq r^2,\end{aligned}$$

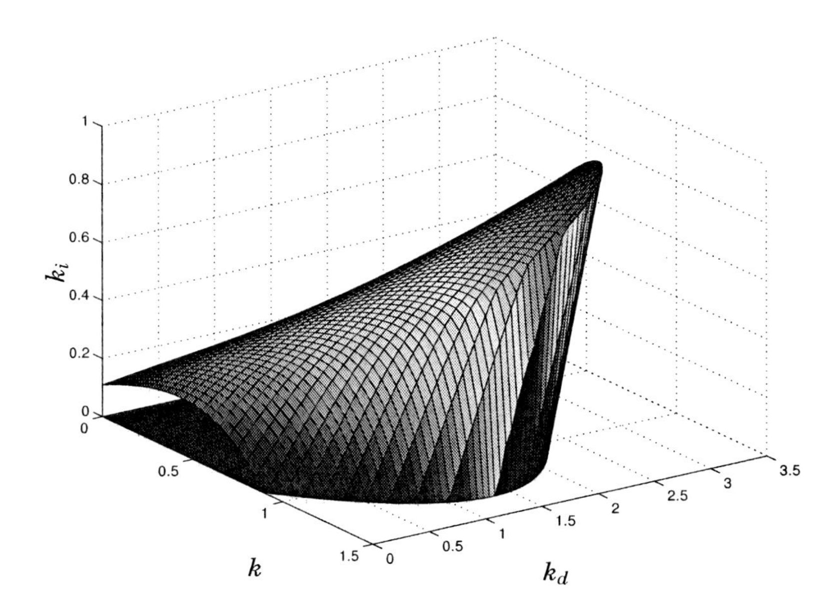

Figure 6.18 Robustness region for a process with the transfer function $P(s) = 1/(s+1)^4$ and the robustness criterion $M_s \leq 1.4$.

where the argument ω in the functions α and β have been dropped to simplify the writing. Inserting the arguments the robustness condition can be written as

$$\left(\frac{\rho(\omega)}{r}\right)^2\left(k + \frac{\alpha(\omega)c}{\rho(\omega)^2}\right)^2 + \left(\frac{\rho(\omega)}{\omega r}\right)^2\left(k_i + \frac{\omega\beta(\omega)c}{\rho(\omega)^2} - \omega^2 k_d\right)^2 \leq 1. \qquad (6.41)$$

To have a stable closed-loop system there is also an encirclement condition required by Nyquist's stability theorem. The robustness constraint thus implies that the controller parameters must belong to a region called the *robustness region*; see Figure 6.18. Design of PID controllers can thus be formulated as the following semi-infinite programming problem: maximize k_i subject to the robustness constraint (6.41) and constraints on k and k_d.

Figure 6.18 gives good insight into the design problem. The PI controller, which maximizes integral gain, can be found from the intersection of the robustness region with the plane $k_d = 0$. The best PI controller has $k = 0.4$ and $k_i = 0.2$. Five times larger values of the integral gain can be obtained by using derivative action.

The optimization problem is not straightforward since the constraint (6.41) must be satisfied for all ω, and the set of parameters that satisfy the constraint is not necessarily convex. Before solving the optimization problem we will therefore investigate the constraint set.

A Geometric Interpretation

The robustness constraint (6.41) has a nice interpretation. For fixed ω and k_d it represents the exterior of an ellipse in the k-k_i plane; see Figure 6.19.

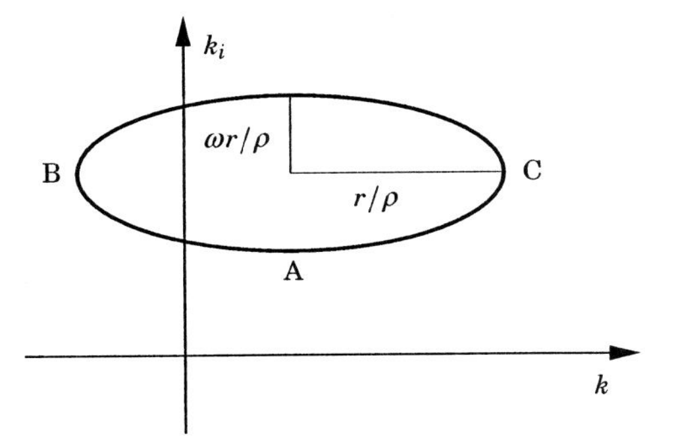

Figure 6.19 Graphical illustration of the sensitivity constraint (6.41).

The ellipse has its center in $k = \alpha c/\rho^2$ and $k_i = \omega\beta c/\rho^2$, and its axes are parallel to the coordinate axes. The horizontal half axis has length r/ρ, and the vertical half axis has length $\omega r/\rho$. The center of the ellipse lies on the stability boundary.

When ω ranges from 0 to ∞ the ellipses have an envelope

$$\begin{aligned} f(k, k_i, k_d, \omega) &= r^2, \\ \frac{\partial f}{\partial \omega}(k, k_i, k_d, \omega) &= 0, \end{aligned} \tag{6.42}$$

which defines one boundary of the sensitivity constraint. Assuming that the process has positive gain the other boundary is given by the $k - k_d$ plane. Since the function f is quadratic in k_i the envelope has two branches; only one branch corresponds to stable closed-loop systems.

Having understood the nature of the constraints it is conceptually easy to solve the optimization problem by finding the largest value of k_i on the envelope. There may be local minima and the envelope may have edges. This is illustrated in Figure 6.20, which shows the envelopes and the locus of the lowest vertex of the ellipse in two cases. The figure on the left has a smooth envelope, and the locus of the lowest vertex coincides with the envelope at the maximum. The figure on the right has an envelope with an edge at the maximum value of k_i. Since it is quite time-consuming to generate the envelope, it is desirable to find algorithms that can give a more effective solution. It is also of interest to characterize the cases where there is only one local minimum.

Smooth Envelope

We will first consider the case where the envelope is smooth and does not have corners near the maximum. The largest value of k_i for fixed k_d then occurs at a tangency with the lower vertex of the ellipse; see Figure 6.19. The locus of the lower vertical vertex is given by

$$\begin{aligned} k(\omega) &= -\frac{\alpha c}{\rho^2} = -\frac{c}{\rho(\omega)}\cos\varphi(\omega), \\ k_i(\omega) &= -\frac{\omega\beta c}{\rho^2} - \frac{\omega r}{\rho} + \omega^2 k_d = -\frac{\omega}{\rho(\omega)}(r + c\sin\varphi(\omega)) + \omega^2 k_d. \end{aligned} \tag{6.43}$$

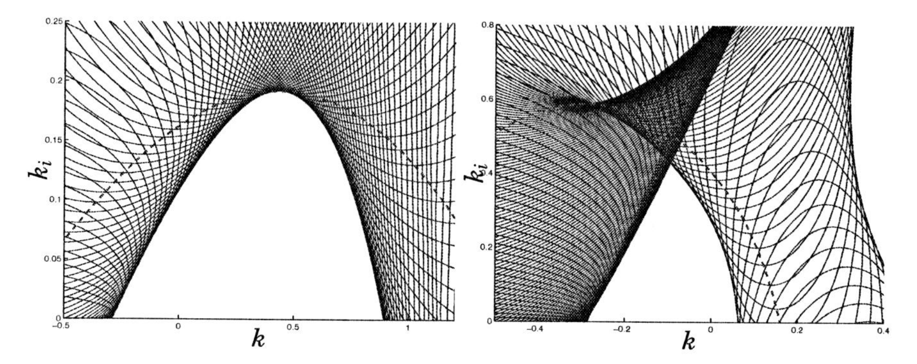

Figure 6.20 Geometrical illustration of the ellipses generated by the sensitivity constraint (6.39) and the envelope generated by it. The curves on the left are generated for a system with the transfer function $P(s) = (s+1)^{-4}$ with the constraint $M_s = 1.4$. The curves on the right are generated for a system with the transfer function $P(s) = 1/(s+1)(s^2+0.2s+9)$ with the constraint $M_s = 1.4$.

It is shown as a dashed line in the Figure 6.20. The largest value of k_i can thus be found by maximizing k_i on the locus of the lowest vertex. Differentiating the expression for k_i in (6.43) gives

$$\begin{aligned}\frac{dk_i}{d\omega} &= -\frac{d}{d\omega}\Big(\frac{\omega(r + c\sin\varphi)}{\rho}\Big) + 2\omega k_d \\ &= (r + c\sin\varphi)\Big(\frac{\omega\rho'}{\rho^2} - \frac{1}{\rho}\Big) - \frac{\omega\varphi' c\cos\varphi}{\rho} + 2\omega k_d = 0.\end{aligned}$$

To simplify the writing we have dropped the argument ω of the functions α, β, and φ. Dividing the above equation with ω and multiplying it with ρ, the condition for extremum becomes

$$h_{PID}(\omega) = (r + c\sin\varphi)\Big(\frac{\rho'}{\rho} - \frac{1}{\omega}\Big) - c\varphi'\cos\varphi + 2\rho k_d = 0. \tag{6.44}$$

To find the optimum we thus have to find the solution ω^*_{PID} of this equation; the controller parameters are then obtained from Equation 6.43.

Equation 6.44 is satisfied for a minimum, a maximum, or saddle point. To ensure that there is a maximum it must be required that

$$\frac{d^2 f}{d\omega^2}(\omega^*) > 0. \tag{6.45}$$

To guarantee that the constraint (6.41) is satisfied globally we have to evaluate it for all ω. This can be done by the Nyquist plot of the loop transfer function.

Equation 6.44 can be solved iteratively by bisection or with the Newton-Raphson method, both methods converge very fast, but they require appropriate initial conditions. Notice, however, that in general, there may be several solutions that can be found by starting the iteration from different initial conditions.

For special classes of systems it is possible to give good initial conditions. Consider systems where the transfer function $P(s)$ has positive low-frequency gain and

$$\begin{aligned}\frac{d \arg P(i\omega)}{d\omega} &< 0,\\ \frac{d \log_{10} |P(i\omega)|}{d \log_{10} \omega} &< 1.\end{aligned} \tag{6.46}$$

These conditions imply that the quantity $\rho'/\rho - 1/\omega$ is negative. For PI control, when $k_d = 0$, it follows from (6.44) and (6.46) that $h_{PI}(\omega_{90}) > 0$ and that $h_{PI}(\omega_{180-\arcsin r/c}) < 0$. Equation 6.44 then has a root in the interval

$$\omega_{90} < \omega_{PI}^* \leq \omega_{180-\arcsin(r/c)}. \tag{6.47}$$

The monotonicity condition (6.46) thus only has to be valid in the interval (6.47). If condition (6.46) holds it follows from Equation 6.43 and 6.47 that both k and k_i are positive. Many processes satisfy this condition.

PD Control

For PD control it is natural to maximize proportional gain subject to the robustness constraint. Working out the details for the case of smooth envelopes we find that the problem can be solved as follows: Find a value of ω such that

$$h_{PD}(\omega) = (r + c\cos\varphi)\frac{\rho'}{\rho} + c\varphi' \sin\varphi = 0. \tag{6.48}$$

Then compute the controller gains from the equations

$$\begin{aligned}k(\omega) &= -\frac{\alpha c}{\rho^2} - \frac{r}{\rho} = -\frac{r + c\cos\varphi}{\rho},\\ k_d(\omega) &= \frac{\beta c}{\omega\rho^2} = \frac{c\sin\varphi}{\omega\rho}.\end{aligned} \tag{6.49}$$

If ρ'/ρ is negative (6.48) always has a solution ω_{PD}^* in the interval

$$\omega_{180} < \omega_{PD}^* < \omega_{270-\arcsin(r/c)} = \omega_{180+\arccos(r/c)}. \tag{6.50}$$

The formula and the code for design of PD controllers can also be used simply by making the observation that designing a PD controller for the system $P(s)$ is the same as designing a PI controller for the system $sP(s)$.

A Design Algorithm

We obtained the following algorithm for solving the design problem in the case of smooth envelopes.

ALGORITHM 6.1—CONTROLLER DESIGN FOR SMOOTH ENVELOPE

1. Design a PD controller by solving (6.48) by bisection starting with the interval $(\omega_{180}, \omega_{180+\arccos(r/c)})$. The solution gives the frequency ω^*_{PD}.
2. Design a PI controller by solving (6.44) with $k_d = 0$ by bisection starting with the interval $(\omega_{90}, \omega_{180-\arcsin(r/c)})$. The solution gives the frequency ω^*_{PI}.
3. Design a PID controller for fixed k_d by solving (6.44) by bisection starting with the interval $(\omega^*_{PI}, \omega^*_{PD})$. Increase k_d to the largest value for which the robustness constraint is satisfied.
4. Verify that there is a smooth envelope by computing (6.45) or by the Nyquist plot of the loop transfer function.

□

If the envelope is not smooth the solution obtained by iteration corresponds to a local maximum of the distance from the critical point to the Nyquist curve. The Nyquist curve then enters the constraint region for points around the maximum.

Envelope with Corners

The largest value of k_i may also occur at a point where the envelope has an edge. This is illustrated in Figure 6.21. The vertices B and C of the ellipse in Figure 6.19 are given by

$$\begin{aligned} k(\omega) &= -\frac{\alpha c}{\rho^2} \pm \frac{r}{\rho} = -\frac{\alpha(\omega) c \cos\phi(\omega)}{\rho(\omega)} \pm \frac{r}{\rho(\omega)}, \\ k_i(\omega) &= -\frac{\omega\beta c}{\rho^2} + \omega^2 k_d = -\frac{\omega c \sin\phi(\omega)}{\rho(\omega)} + \omega^2 k_d, \end{aligned} \tag{6.51}$$

where the left vertex corresponds to a minus sign and the right vertex to a plus sign. The loci of these vertices are shown in thin dotted lines, and the loci of the center of the ellipses are shown in thin dashed lines. The envelope is shown as a thick solid line, and the locus of the lowest vertex of the ellipse by thick dashed lines. Notice that the maximum occurs at the intersection of ellipses corresponding to two different frequencies, ω_1 and ω_2; see Figure 6.21. The envelope condition (6.42) is then satisfied for both frequencies. This gives the condition

$$\begin{aligned} f(k, k_i, k_d, \omega_1) &= R^2, \\ \frac{\partial f}{\partial \omega}(k, k_i, k_d, \omega_1) &= 0, \\ f(k, k_i, k_d, \omega_2) &= R^2, \\ \frac{\partial f}{\partial \omega}(k, k_i, k_d, \omega_2) &= 0. \end{aligned} \tag{6.52}$$

In the Nyquist plot this corresponds to the case when the loop transfer function is tangent to the M circle at two points.

It is thus possible to characterize the point where k_i has its largest value by algebraic equations. This means that the design problem is reduced to solving

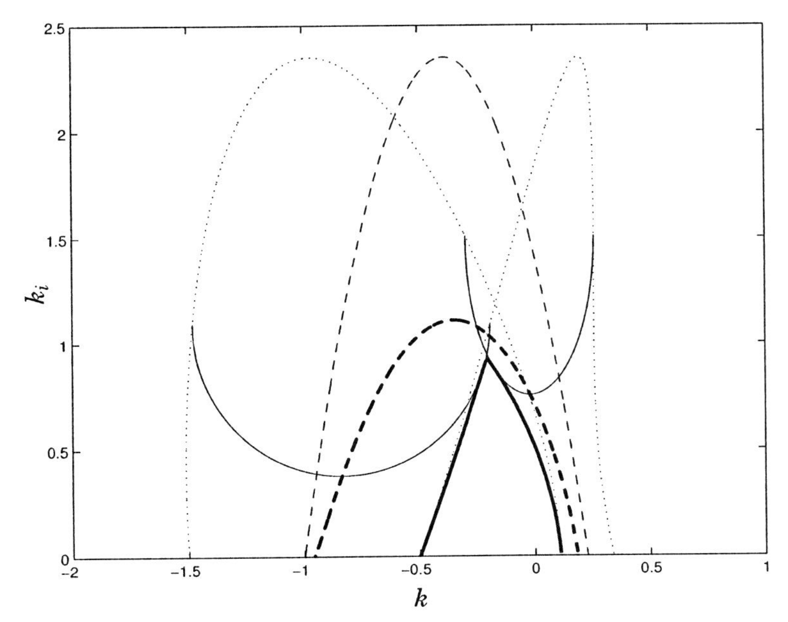

Figure 6.21 Geometrical illustration of the sensitivity constraint (6.39) and the envelope generated by it. The envelope is shown by the thick solid line; the locus of the lower vertex by the thick dashed line. One half ellipse is shown as a thin solid line. The locus of the center of the ellipses is shown as a thin dashed line, and the loci of the vertical vertices by dotted lines. The curves are generated for a system with the transfer function $P(s) = 1/(s+1)(s^2+0.2s+9)$ with the constraint $M_s = 2.0$.

algebraic equations, (6.52), and that elaborate search procedures are avoided. The equation can be solved using the Newton-Raphson method.

Good initial values essential for the Newton-Raphson iteration can be obtained by approximating the envelope by the loci of the right horizontal locus and the locus of the lowest vertex of the ellipse; see Figure 6.21. We illustrate the case of envelopes with corners with an example.

EXAMPLE 6.24—AN OSCILLATORY SYSTEM
Consider the process with the transfer function

$$P(s) = \frac{9}{(s+1)(s^2+as+9)}.$$

This is an interesting process from two points of view. First, the system has two oscillatory poles with relative damping $\zeta = a/6$. When parameter a is decreased it becomes more and more difficult to control the process. Second, depending on the value of parameter a the envelope may have a continuous derivative, $a \geq 1.0653$, or a corner, $a < 1.0653$.

For the case when the envelope has a corner, a PI controller was designed for $M_s = 2.0$. In Figure 6.22 the Nyquist curves and the time responses are shown for the cases $a = 0.2$, 0.5, and 1.0. The controller behaves reasonably well in spite of the poorly damped poles. In Table 6.10 the controller parameters and

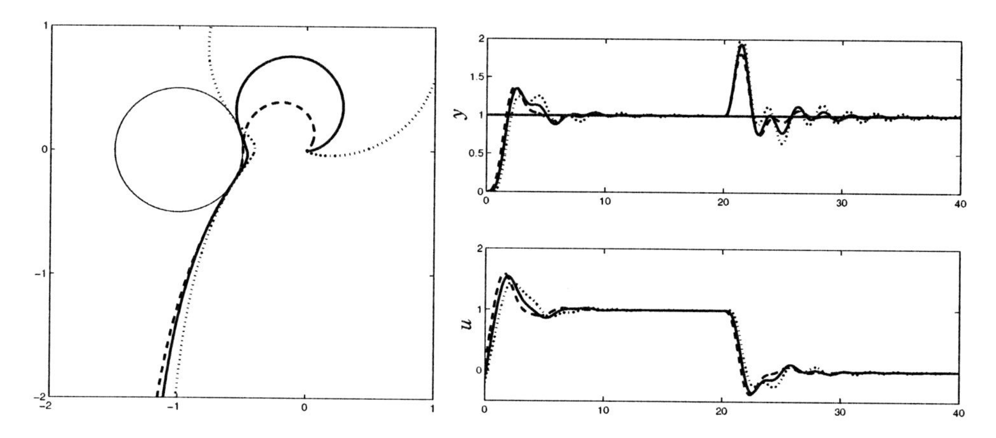

Figure 6.22 Nyquist curves of the loop transfer function and time responses for Example 6.24 with $a = 0.2$ (dotted), 0.5, and 1.0 (dashed), when designing for $M_s = 2.0$.

Table 6.10 Interesting parameters when designing a controller for $M_s = 2.0$ and different values of a in Example 6.24.

a	K	k_i	ω_1	ω_2
0.0	−0.29	0.68	0.97	2.75
0.1	−0.25	0.82	1.08	2.71
0.2	−0.20	0.93	1.16	2.67
0.5	−0.09	1.17	1.37	2.55
1.0	0.09	1.38	1.65	2.30
2.0	0.48	1.54	2.79	2.79

the frequencies at which the loop transfer function is tangent to the M_s-circle are shown. Notice in Table 6.10 how the proportional gain is negative for small values of a. This is the only way to increase the damping of the oscillatory poles with a PI controller.

Finally, we illustrate how our design method will provide a reasonable PI controller for the extreme case $a = 0$. With the design parameter $M_s = 1.4$ we obtain the controller parameters $K = -0.183$, $k_i = 0.251$, and $b = 0$. The time responses are shown in Figure 6.23. We observe that the set-point response is quite reasonable, even if there is a trace of poorly damped modes. The load disturbance will, however, excite the oscillatory modes. The fact that the PI controller is unable to provide damping of these modes is clearly noticeable in the figure. □

The Derivative Cliff

Smooth envelopes are frequently encountered for PI control of systems with essentially monotone frequency responses, and for PID control with moderate

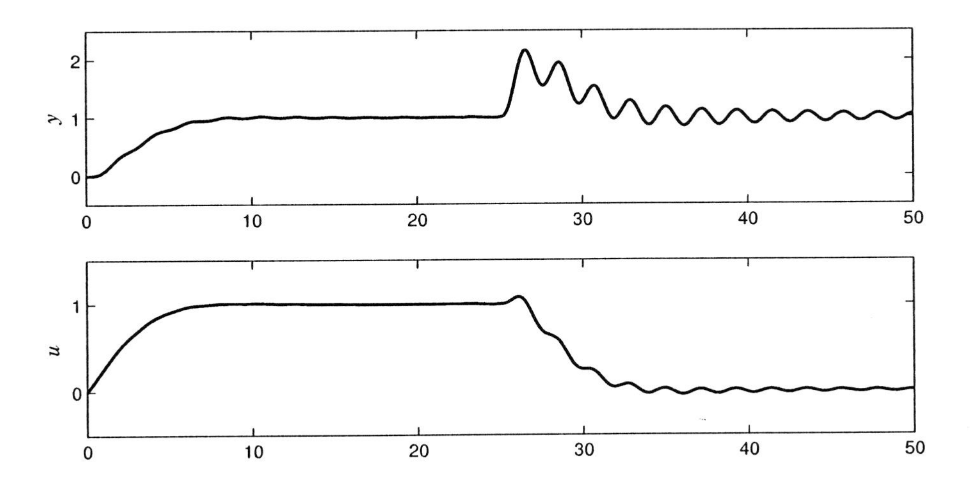

Figure 6.23 Time response of the closed-loop system of Example 6.24 obtained for $a = 0$, when designing the PI controller for $M_s = 1.4$.

values of k_d. However, optimization of k_i over the robustness region often gives controllers with undesirable properties. This can be understood from the plot of the robustness region in Figure 6.18, which shows that the largest value of k_i occurs at an edge. Such a solution is undesirable because small changes in controller parameters give drastic changes in k_i. This is also illustrated in Figure 6.24, which shows intersections of the robustness surface for fixed values of k_d. The figure shows that for PI control ($k_d = 0$) the envelope is smooth at the maximum $k_i = 0.2$ which occurs for $k = 0.4$. Integral gain k_i can be increased substantially by introducing derivative action. With higher values of k_d the maximum of k_i does, however, occur at an edge. Integral gain has its maximum $k_i = 0.9$ for $k = 0.925$ and $k_d = 2.86$. The performance is very sensitive to variations in the controller parameters at the maximum. Figure 6.24 shows that a marginal increase of proportional gain makes the system unstable. The controller that maximizes k_i also has other drawbacks, which are illustrated by the following example.

EXAMPLE 6.25—THE DERIVATIVE CLIFF

Consider a process with the transfer function

$$P(s) = \frac{1}{(s+1)^4}.$$

Maximizing integral gain k_i subject to the robustness constraint $M_s \leq 1.4$, gives the controller parameters $k = 0.925$, $k_i = 0.9$, and $k_d = 2.86$. The Nyquist plot of the loop transfer function is shown in Figure 6.25. Notice that the Nyquist curve has a loop. This will always occur when the maximum occurs where the envelope has an edge. The controller obtained has excessive phase lead, which is obtained by having a PID controller with complex zeros, $T_i < 4T_d$. In the particular case we have $T_i = 0.33T_d$. Time plots showing the response of

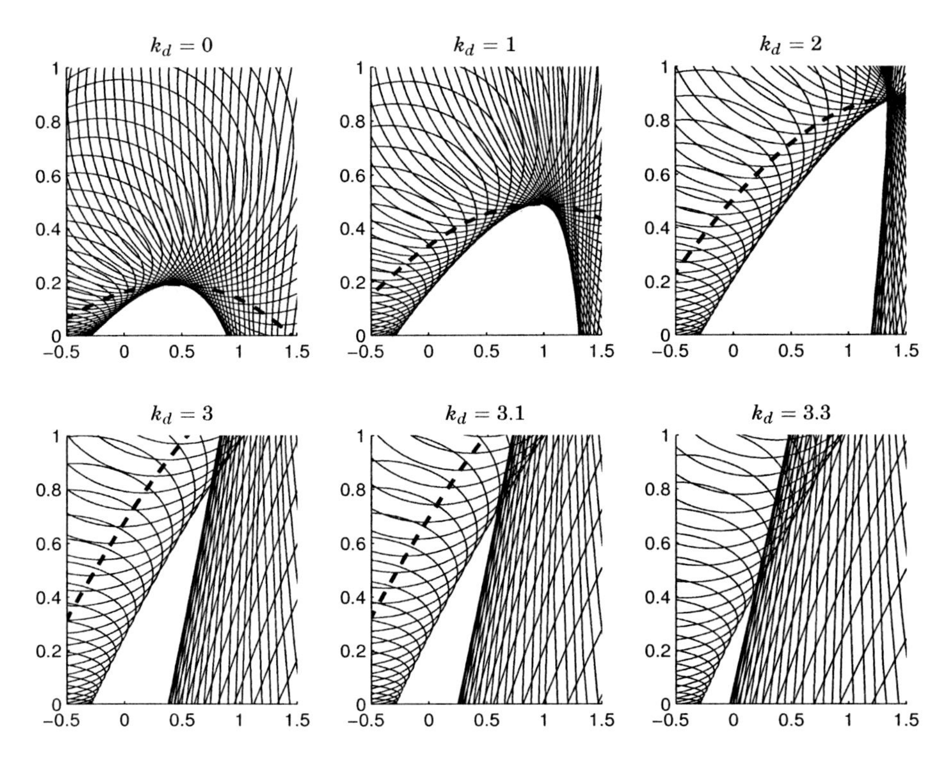

Figure 6.24 Cuts of the robustness region for constant derivative gain k_d. The curves are computed for PID control of the process $P(s) = 1/(s+1)^4$. Notice the sharp corners of the region for large k_d (the derivative cliff).

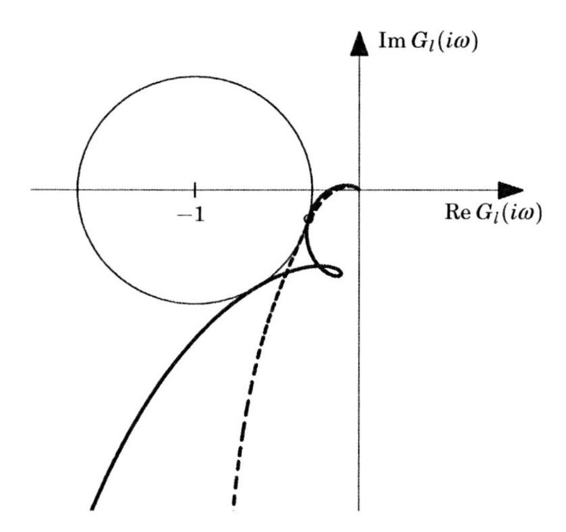

Figure 6.25 Nyquist curve of the loop transfer function for PID control of the process $P(s) = 1/(s+1)^4$, with a controller having parameters $k = 0.925$, $k_i = 0.9$, and $k_d = 2.86$.

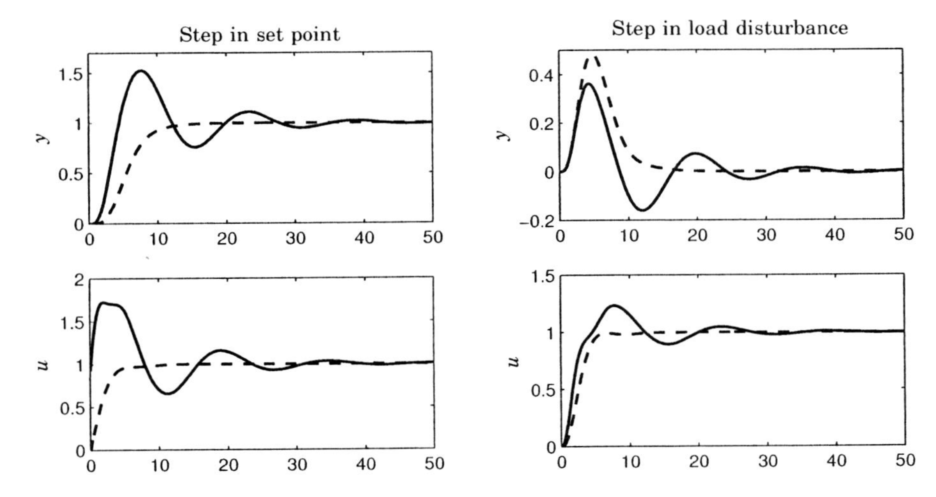

Figure 6.26 Time responses for PID control of the process $P(s) = 1/(s+1)^4$, with controller having parameters $k = 0.925$, $k_i = 0.9$, and $k_d = 2.86$ (solid lines) and $k = 1.1$, $k_i = 0.36$, and $k_d = 0.9$ (dashed lines).

the system to step changes in set point and load disturbances are shown in Figure 6.26. The responses are oscillatory.

For comparison we have shown Nyquist plots and time plots for a PID controller where $T_i = 4T_d$. The controller parameters are $k = 1.1$, $k_i = 0.36$, and $k_d = 0.9$. The responses of this controller are better, even if the peak in the response to load disturbances is larger. □

Avoiding the Derivative Cliff

There are several ways to modify the design problem to avoid the difficulties associated with the derivative cliff. One way is to introduce conditions that do not allow the Nyquist curve to have loops. Another alternative is to require that $T_i > \alpha T_d$. It has also been attempted to fix derivative gain to the value obtained by a PD controller. This does not eliminate the loops on the Nyquist curve in all cases. Maximization of k_i can also be replaced by maximizing the absolute integral error due to load disturbances.

The MIGO Method

After many attempts it has been found that a simple solution is to restrict derivative gain so that the maximum occurs at a point where $\partial k_i/\partial k = 0$. This avoids having a maximum at a ridge. The algorithm is straightforward.

ALGORITHM 6.2—MIGO DESIGN OF PID CONTROLLER

1. Fix derivative gain k_d. Find controller parameters by solving (6.44); then compute controller gains from (6.43).
2. Compute the value of M for a range of frequencies around ω^*, and test the robustness constraint $M \geq M_{crit}$.

3. Increase k_d until the largest value that satisfies the robustness constraint is obtained.

□

A good initial value of integral gain is the value obtained for a PD controller. This particular design method is called MIGO (M constrained Integral Gain Optimization).

An Algorithm for a Controller in Series Form

It frequently happens that the MIGO design method gives controller parameters such that $T_i < 4T_d$. In Section 3.2 it was shown that such controllers cannot be implemented in series form. It is therefore of interest to have controllers where the parameters are constrained to $T_i \geq 4T_d$. When the ratio $n = T_i/T_d \geq 4$, the controller can be written as

$$C(s) = k\left(1 + \frac{1}{sT_i} + sT_d\right) = k'\frac{(T_i's+1)(T_d's+1)}{T_i's}, \tag{6.53}$$

where

$$\begin{aligned} k &= k'\frac{T_i' + T_d'}{T_i'} \\ T_i &= T_i' + T_d' \\ T_d &= \frac{T_i'T_d'}{T_i' + T_d'}. \end{aligned} \tag{6.54}$$

Introducing $n' = T_i'/T_d'$, it also follows that

$$n = \frac{(1+n')^2}{n'}. \tag{6.55}$$

Notice that $n' = 1$ corresponds to $n = 4$.

It follows from Equation 6.54 that $T_i = nT_d$ gives the following relation between the controller parameters

$$k_i = \frac{k^2}{nk_d}.$$

A simple algorithm for maximizing the integral gain of a PID controller with $T_i = nT_d$ subject to a robustness constraint will now be developed. We first make the observation that PID control of the process $P(s)$ gives the loop transfer function

$$G_l(s) = P(s)C(s) = k'\frac{(1+sT_i')(1+sT_d')}{sT_i'}P(s) = k'\frac{(1+sT_i')}{sT_i'}\Big((1+sT_d')P(s)\Big).$$

This is identical to the loop transfer function for PI control of the process

$$P'(s) = (1+sT_d')P(s).$$

Since there are efficient algorithms for PI control we obtain the following iterative algorithm.

Table 6.11 Controller parameters obtained by loop-shaping design with $M_s = 1.4$ for a process with the transfer function $P(s) = (s+1)^{-4}$.

Controller	K	k_i	k_d	b	T_i	T_d	IAE
PD	1.333	0	1.333	1	0	1	∞
PI	0.433	0.192	0	0.14	2.25	0	5.20
PID MIGO	1.305	0.758	1.705	0*	1.72	1.31	2.25
PID $T_i = 4T_d$	1.132	0.356	0.900	0.9	3.18	0.80	2.51

ALGORITHM 6.3—DESIGN OF PID CONTROLLER WITH $T_i = 4T_d$

1. Start by designing a PI controller for the process $P(s)$. This gives a controller with the integral time $T_i = k/k_i$. Set $T_1' = T_i/2$ and $j = 1$.
2. Design a PI controller for the process $P'(s) = (1 + sT_1')P(s)$. Let the integral time of the controller be T_i'. Set $T_{j+1}' = (T_j' + T_i')/2$ and repeat until T_j' converges to T'. Let the controller gain be k'.
3. The controller parameters are $k = 2k'$, $T_i = 2T'$ and $T_d = T'/2$.

□

Examples

The design method will be illustrated by two examples.

EXAMPLE 6.26—FOUR EQUAL LAGS
Consider a system with the transfer function

$$P(s) = \frac{1}{(s+1)^4}.$$

Table 6.11 summarizes properties of PD, PI, and PID controllers designed for $M_s = 1.4$. The PD controller was designed by maximizing proportional gain; the PI and PID controllers by maximizing integral gain. A PID controller with the additional constraint $T_i = 4T_d$ was also designed. Responses to set-point changes and load disturbances for are shown in Figure 6.27.

The PID controllers have better performance than the PI controller. Integral gain is 2 to 3 times larger and IAE a factor of 2 smaller. The controller PID MIGO has $T_i = 1.3T_d$. The table shows that performance is decreased when controller parameters are constrained to $T_i = 4T_d$. Notice that many commercial PID controllers have the constraint $T_i \geq 4T_d$ built in, because they are based on the series form; see (3.10).

The parameter b is calculated as described in Section 5.3. The calculation for the PID MIGO controller shows that the overshoot cannot be reduced sufficiently by using zero set-point weight, which is indicated by the entry 0* in the table. In this case it is recommended to use a proper feedforward design for a system with two degrees of freedom. Such a design can improve set-point response significantly. □

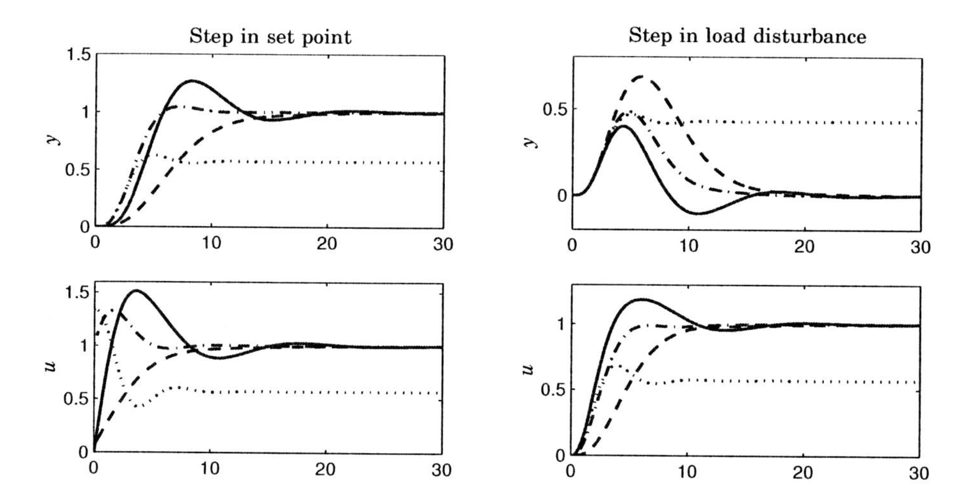

Figure 6.27 Responses for the system $P(s) = (s+1)^{-4}$ with the controllers in Table 6.11 to unit step changes in set point (left) and load disturbances (right). The dotted lines show responses with the PD controller, dashed with the PI controller, dash-dotted with the PID controller with parameters constrained to $T_i = 4T_d$, and solid lines with the PID controller designed using the MIGO method.

EXAMPLE 6.27—FOUR WIDELY DISTRIBUTED LAGS
Consider a system with the transfer function

$$P(s) = \frac{1}{(s+1)(0.1s+1)(0.01s+1)(0.001s+1)}.$$

Table 6.12 summarizes properties of PD, PI and PID controllers. All controllers were designed with the constraint that the maximum sensitivity is not larger than $M_s = 1.4$. The PD controller was designed by maximizing proportional gain, and the PI and PID controllers by maximizing integral gain. A PID with the additional constraint $T_i = 4T_d$ was also designed. Responses to set-point changes and load disturbances for the different controllers are shown in Figure 6.28.

Table 6.12 and Figure 6.28 show that derivative action improves performance drastically. The proportional gains of the controllers with derivative

Table 6.12 Controller parameters obtained by loop-shaping design with $M_s = 1.4$ for a process with the transfer function $P(s) = 1/(s+1)(0.1s+1)(0.01s+1)(0.001s+1)$.

Controller	K	k_i	k_d	b	T_i	T_d	IAE
PD	91.7	0	4.4	1	0	0.048	∞
PI	4.21	8.53	0	1	0.494	0	0.1044
PID MIGO	85.5	1488	3.87	0	0.057	0.045	0.00143
PID $T_i = 4T_d$	86.7	518	3.63	0.6	0.168	0.042	0.00143

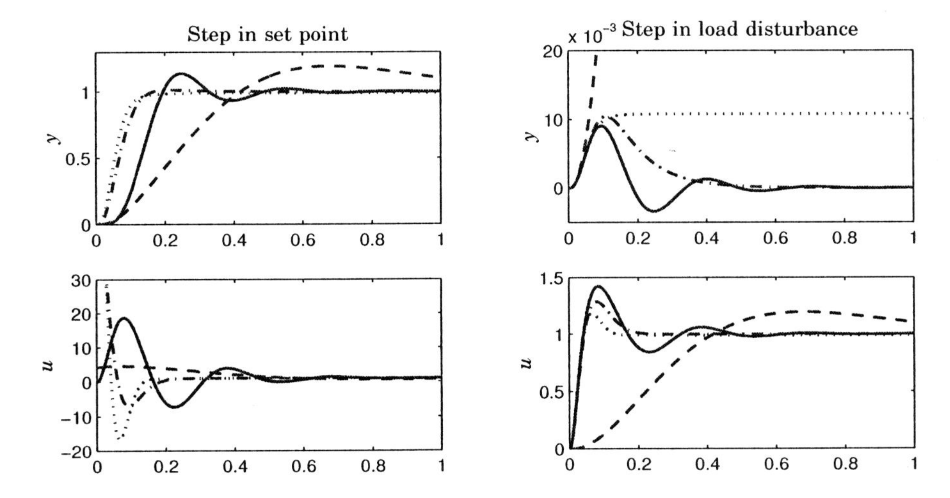

Figure 6.28 Responses for the system $P(s) = 1/(s+1))(0.1s+1)(0.01s+1)(0.001s+1)$ with the controllers in Table 6.12. The dotted lines show responses with the PD controller, dashed with the PI controller, dash-dotted with the PID controller with parameters constrained to $T_i = 4T_d$, and solid lines with the PID controller designed using the MIGO method. The load-disturbance response for the PI controller is out of scale.

action are around 90, while the PI controller has the gain 4.2. It follows from (4.40) that the largest peak of the load-disturbance response is around 0.01 for controllers with derivative action and about 20 times larger for PI control. The peak is so large that the load disturbance response for PI control is way outside the graph. The response time is also drastically increased when derivative action is used. The integral gains of controllers with derivative action are also much larger than for PI control. □

It is interesting to compare Examples 6.26 and 6.27. For the system with four equal lags in Example 6.26 the integral gain can be increased by a factor of 3 by introducing derivative action, while it can be increased by a factor of 200 for the system with distributed lags in Example 6.27. The main differences between the system is that the system in Example 6.27 is lag dominated; it has a normalized time delay $\tau = 0.07$. The system in Example 6.26 has $\tau = 0.33$. The normalized time delay is a good indicator for the benefits of derivative action. In Chapter 7 it will be shown that the large performance improvement with derivative action is possible for processes with small normalized time delays (lag dominated processes).

6.9 Summary

A number of techniques for designing PID controllers have been presented in this chapter, starting with methods of the Ziegler-Nichols type, where process dynamics were characterized by a few features that could be obtained from simple experiments. These methods have been very influential and have been used extensively by vendors. In spite of their popularity there are two drawbacks with the Ziegler-Nichols method, the fundamental assumption of quarter amplitude damping, which results in systems with very bad robustness, and the limited process knowledge used. Methods which avoids both difficulties will be developed in Chapter 7.

Standard methods for control system design can also be adapted to design of PID controllers. When using analytical techniques there is a correspondence between model and controller complexity, and it is necessary to approximate process dynamics by first- and second-order systems. Model reduction techniques are therefore necessary to apply pole placement to PID control. In particularly it is necessary to approximate the time delay. The unmodeled dynamics limit the performance that can be achieved, and the closed-loop poles that can be chosen.

Another way to use pole placement is to fix a pole configuration and to determine both the controller parameters and the magnitude of the poles. In this way it is possible to use second-order models to design PI controllers and third-order models to design PID controllers. Another way to apply pole placement to PID control is to place the dominant poles only. The advantage of this approach is that it can be applied to models of arbitrary order. A particular pole placement technique called lambda tuning, which has been used extensively in the process industry, is given particular attention.

A number of so-called algebraic design methods have also been discussed. In these methods the closed-loop transfer function is given and the controller parameters are obtained by algebraic calculations. The controller parameters can also be determined by optimization methods, where it is attempted to optimize criteria that specify performance subject to various constraints. There are many methods reflecting the richness of the control problem. Two methods, BO and SO, which are commonly used in motion control, have been given particular attention.

A novel design method developed by the authors and their coworkers is also presented. In this method it is attempted to optimize disturbance attenuation subject to constraints on robustness. The method gives a simple way to balance attenuation of load disturbances with the injection of measurement noise that is inevitable when feedback is used. Combining this method with set-point weighting, or more elaborate feedforward, gives a nice way to also achieve good response to set-point changes. The method can be viewed as an adaptation of robust design to PID control.

6.10 Notes and References

There is a very large literature on tuning of PID controllers. Good general sources are the books [Smith, 1972; Deshpande and Ash, 1981; Shinskey, 1988; McMillan, 1983; Corripio, 1990; Suda *et al.*, 1992; Oquinnaike and Ray, 1994; Marlin, 2000; Wang and Cluett, 2000; Wang *et al.*, 2000; Quevedo and Escobet, 2000; Cominos and Munro, 2002; Seborg *et al.*, 2004; O'Dwyer, 2003; Choi and Chung, 2004; Michael and Moradi, 2005]. The books clearly show the need for a variety of techniques, simple tuning rules, as well as more elaborate procedures that are based on process modeling, formulation of specifications, and control design. Even if simple heuristic rules are used, it is important to realize that they are not a substitute for insight and understanding. Successful controller tuning cannot be done without knowledge about process modeling and control theory. It is also necessary to be aware that there are many different types of control problems and consequently many different design methods. To only use one method is as dangerous as to only believe in empirical tuning rules. Control problems can be specified in many different ways. A good review of different ways to specify requirements on a control system is given in [Truxal, 1955; Maciejowski, 1989; Boyd and Barratt, 1991]. To formulate specifications it is necessary to be aware of the factors that fundamentally limit the performance of a control system.

The seminal papers [Ziegler and Nichols, 1942; Ziegler and Nichols, 1943] are the first attempts to develop systematic methods for tuning PID controllers. An interesting perspective on these paper is given in an interview with Ziegler; see [Blickley, 1990]. The CHR-method, described in [Chien *et al.*, 1952], is a modification of the Ziegler-Nichols method. This is one of the first papers where it is mentioned that different tuning methods are required for set-point response and for load disturbance response. Good response to load disturbances is often the relevant criterion in process control applications. Notice that the responses can be tuned independently by having a controller that admits a two-degree-of-freedom structure. The usefulness of a design parameter is also mentioned in the CHR-paper. In spite of its shortcomings, the Ziegler-Nichols method has been the foundation for many tuning methods; see [Tan and Weber, 1985; Mantz and Tacconi, 1989; Hang *et al.*, 1991]. Tuning charts were presented in [Wills, 1962b; Wills, 1962a; Fox, 1979].

The loop-shaping methods were inspired by classical control design methods based on frequency response; see [Truxal, 1955]. Applications to PID control are found in [Pessen, 1954; Habel, 1980; Chen, 1989; Yuwana and Seborg, 1982].

The idea of algebraic design was presented in [Truxal, 1955] and [Newton *et al.*, 1957] as a systematic method of design to given specifications; a more recent presentation is found in [Boyd and Barratt, 1991]. Algebraic design was applied to process control in [Smith, 1957; Atherton, 1999; Hansen, 2000]. The original papers on the λ-tuning method are [Dahlin, 1968] and [Higham, 1968]. The method is sometimes called the Dahlin method; see [Deshpande and Ash, 1981]. The method is very popular in the pulp and paper industry where it has been used to develop standardized tuning procedures; see [Sell, 1995] and [Anonymous, 1997]. Lambda tuning is closely related to the Smith predictor and the internal model controller; see [Smith, 1957; Chien, 1988; Chien and Fruehauf, 1990; Rivera *et al.*, 1986]. The tuning techniques developed in [Smith and Murrill, 1966; Pemberton, 1972a; Pemberton, 1972b; Smith *et al.*, 1975; Hwang and Chang, 1987] are other examples of the analytical approach to design. In [Rivera *et al.*, 1986] it was shown that internal model control reduces to PI and PID control when proper approximations of the time delay are done. A novel algebraic design method, described in [Hansen, 2000; Hansen, 2003] is used in a PID controller developed by Foxboro. An interesting feature is that the desired response is given as a high-order system.

The analytical tuning method gives controllers that cancel poles and zeros in the transfer function of the process. This leads to lack of observability or controllability. There are severe drawbacks in this as has been pointed out many times, e.g., in [Chien and Fruehauf, 1990; Shinskey, 1991b; Morari and Lee, 1991]. The response to load disturbances will be very sluggish for processes with lag dominated dynamics. A modification that does not cancel the process pole is given in [Chien and Fruehauf, 1990]. Skogestad's internal model controller is presented in [Skogestad, 2003]. This controller avoids the cancellation by an ad hoc modification of integral time for lag dominated dynamics.

Many methods for control design are based on optimization techniques. This approach has the advantage that it captures many different aspects of the design problem. There is also powerful software that can be used. A general discussion of the use of optimization for control design is found in [Boyd and Barratt, 1991]. The papers [Rovira *et al.*, 1969; Lopez *et al.*, 1969] give controllers that are optimized with respect to the criteria ISE, IAE, and ITAE. Other applications to PID control are given in [Hazebroek and van der Waerden, 1950; Wolfe, 1951; Oldenburg and Sartorius, 1954; van der Grinten, 1963; Lopez *et al.*, 1967; Marsili-Libelli, 1981; Yuwana and Seborg, 1982; Patwardhan *et al.*, 1987; Wong and Seborg, 1988; Polonoyi, 1989; Zhuang and Atherton, 1991]. The methods BO and SO were introduced in [Kessler, 1958a; Kessler, 1958b]. A discussion of these methods with many examples is found in [Fröhr, 1967; Fröhr and Orttenburger, 1982].

Pole placement is a straightforward algebraic design method much used in control engineering; see [Truxal, 1955]. It has the advantage that the closed-loop poles are specified directly. Many other design methods can also be interpreted as pole placement. The papers [Elgerd and Stephens, 1959; Graham and Lathrop, 1953] show how many properties of the closed-loop system can be deduced from the closed-loop poles. This gives good guidance for choosing the suitable closed-loop poles. An early example of pole placement is [Cohen and Coon, 1953; Coon, 1956a; Coon, 1956b]. It may be difficult to choose desired closed-loop poles for high-order systems. This is avoided by specifying only a few poles, as in the dominant pole design method described in [Persson, 1992; Persson and Åström, 1992; Persson and Åström, 1993].

The development of robust control was a major advance of control theory which made it possible to explicitly account for robustness in control design; see [Doyle *et al.*, 1992; Horowitz, 1993; Green and Limebeer, 1995], [Skogestad and Postlethwaite, 1996; Zhou *et al.*, 1996; Vinnicombe, 2000]. These ideas were applied to PID control in the papers [Panagopoulos *et al.*, 1997; Åström *et al.*, 1998; Panagopoulos and Åström, 2000; Panagopoulos, 2000; Kristensson, 2003]. The method discussed in Section 6.8 are based on these papers.

There are comparatively few papers on PID controllers that consider the random nature of disturbances. The papers [van der Grinten, 1963; Goff, 1966a; Fertik, 1975] are exceptions.

There are many papers on comparisons of control algorithms and tuning methods. The paper [McMillan, 1986] gives much sound advice; other useful papers are [Miller *et al.*, 1967; Gerry, 1987; Gerry, 1999].

7

A Ziegler-Nichols Replacement

7.1 Introduction

Since PID controllers are so common it is useful to have simple tuning rules that can be applied to a wide range of processes. This is testified to by the longevity of the Ziegler-Nichols rules. They have been used for more than half a century even though they have severe drawbacks. In this chapter we present new tuning methods in the spirit of Ziegler and Nichols.

Control system design is a rich problem, as was discussed in Section 4.2. Any design problem should take account of load disturbances, measurement noise, robustness, and set-point following. When developing the simple rules we will follow the main ideas used by Ziegler and Nichols. We will thus focus on load disturbances by maximizing integral gain, but we will depart from Ziegler and Nichols by also adding a robustness constraint. In this chapter we have chosen to require that the joint sensitivity is larger than $M = 1.4$. Measurement noise is handled by detuning the controllers if the gains are too large, and set-point following is dealt with by set-point weighting.

The procedure we use is essentially the same as the one employed by Ziegler and Nichols. We will select a large batch of representative processes. This includes a wide variety of systems with essentially monotone step responses, which are typically encountered in process control. Controllers for each process in the batch are then obtained by applying the MIGO design described in Section 6.8, which is based on the criteria given above. Having obtained the controller parameters we will then try to find correlations with normalized process parameters. The simple tuning rules obtained are called AMIGO, which stands for Approximate MIGO design.

The procedure shows that it is indeed possible to obtain simple tuning formulas. A major result is that it is necessary to use more process information than used by Ziegler and Nichols. Tuning based on step responses can be based on FOTD models. It is necessary to use all *three* process parameters K_p, L, and T and not just two parameters $a = K_pL/T$ and L, as was suggested by Ziegler and Nichols. For PI control it is possible to obtain tuning rules that are close to the optimal rules for the whole test batch. For PID control rules that are close to optimal can be obtained for balanced and delay-dominated processes. For lag-dominated processes it is necessary to have better process information. It is, however, possible to obtain efficient rules for balanced and delay dominant processes.

For the frequency response method where Ziegler and Nichols characterized the process by two parameters, K_{180} and T_{180}, we have shown that it is necessary to add a third parameter, e.g., the static gain K_p. Even with these parameters it is not possible to obtain rules that are close to optimal for all processes in the test batch. It is, however, possible to obtain conservative tuning rules both for PI and PID controllers.

The design method used can give high controller gains for processes that are lag dominated. This may result in large variations in the control signal due to noise. In some cases, it may therefore be necessary to make a trade-off between attenuation of load disturbances and injection of measurement noise. This can be accomplished by detuning the controllers. Methods for doing this are also included in this chapter.

Analysis of all the controllers in the test batch has also given much insight into PI and PID control. It is shown that derivative action only gives moderate improvement for balanced and delay-dominated processes but that very large improvements can be obtained for lag-dominated processes. It is also shown that there is a wide range of processes where it is advantageous to have $T_i < 4T_d$. Notice that controllers implemented on series form do not permit this. Nice formulas that give the ratio of the average residence time for open- and closed-loop systems are also given. This makes it possible to estimate the closed-loop response times that can be expected.

In the next section, the test batch is presented. Using this batch, AMIGO tuning rules based on step response experiments are derived for PI controllers in Section 7.3 and PID controllers in Section 7.4. AMIGO tuning rules based on frequency response experiments are presented in Section 7.5. More efficient tuning rules for PID control of lag-dominant processes can be obtained if a second order model is used. This is discussed in Section 7.6. In Section 7.7, the MIGO and AMIGO rules are compared for three different processes, one lag-dominant, one balanced, and one delay dominant, respectively. Section 7.8 and Section 7.9 treat noise filtering and high-frequency gain reduction of the controllers by detuning.

7.2 The Test Batch

PID control is not suitable for all processes. In [Hägglund and Åström, 2002] it is suggested that the class of processes where PID is suitable can be characterized as having essentially monotone step responses. One way to characterize such processes is to introduce the monotonicity index:

$$\alpha = \frac{\int_0^\infty g(t)dt}{\int_0^\infty |g(t)|dt}, \tag{7.1}$$

where g is the impulse response of the system. Systems with $\alpha = 1$ have monotone step responses, and systems with $\alpha > 0.8$ are considered essentially monotone. The tuning rules presented in this paper are derived using a test batch of essentially monotone processes.

The following 134 processes were used to derive the tuning rules:

$$
\begin{aligned}
P_1(s) &= \frac{e^{-s}}{1+sT}, \\
T &= 0.02, 0.05, 0.1, 0.2, 0.3, 0.5, 0.7, 1, \\
&\quad 1.3, 1.5, 2, 4, 6, 8, 10, 20, 50, 100, 200, 500, 1000 \\
P_2(s) &= \frac{e^{-s}}{(1+sT)^2}, \\
T &= 0.01, 0.02, 0.05, 0.1, 0.2, 0.3, 0.5, 0.7, 1, \\
&\quad 1.3, 1.5, 2, 4, 6, 8, 10, 20, 50, 100, 200, 500 \\
P_3(s) &= \frac{1}{(s+1)(1+sT)^2}, \\
T &= 0.005, 0.01, 0.02, 0.05, 0.1, 0.2, 0.5, 2, 5, 10 \\
P_4(s) &= \frac{1}{(s+1)^n}, \\
n &= 3,\ 4,\ 5,\ 6,\ 7,\ 8 \\
P_5(s) &= \frac{1}{(1+s)(1+\alpha s)(1+\alpha^2 s)(1+\alpha^3 s)}, \\
\alpha &= 0.1, 0.2, 0.3, 0, 4, 0.5, 0.6, 0.7, 0.8, 0.9 \\
P_6(s) &= \frac{1}{s(1+sT_1)} e^{-sL_1}, \\
L_1 &= 0.01, 0.02, 0.05, 0.1, 0.3, 0.5, 0.7, 0.9, 1.0, \quad T_1 + L_1 = 1 \\
P_7(s) &= \frac{T}{(1+sT)(1+sT_1)} e^{-sL_1}, \qquad T_1 + L_1 = 1, \\
T &= 1, 2, 5, 10 \qquad L_1 = 0.01, 0.02, 0.05, 0.1, 0.3, 0.5, 0.7, 0.9, 1.0 \\
P_8(s) &= \frac{1-\alpha s}{(s+1)^3}, \\
\alpha &= 0.1, 0.2, 0.3, 0.4, 0.5, 0.6, 0.7, 0.8, 0.9, 1.0, 1.1 \\
P_9(s) &= \frac{1}{(s+1)((sT)^2 + 1.4sT + 1)}, \\
T &= 0.1, 0.2, 0.3, 0.4, 0.5, 0.6, 0.7, 0.8, 0.9, 1.0.
\end{aligned}
\tag{7.2}
$$

The processes are representative for many of the processes encountered in process control. The test batch includes both delay-dominated, lag-dominated, and integrating processes. All processes have monotone step responses except P_8 and P_9. The parameters range for processes P_8 and P_9 are chosen so that the systems are essentially monotone with $\alpha \geq 0.8$. The normalized time delay ranges from 0 to 1 for the process P_1 but only from 0.14 to 1 for P_2. Process P_6 is integrating, and therefore $\tau = 0$. The rest of the processes have values of τ in the range $0 < \tau < 0.5$.

7.3 PI Control

The processes in test batch (7.2) are first approximated by the simple FOTD model

$$P(s) = \frac{K_p}{1+sT}e^{-sL}, \tag{7.3}$$

where K_p is the static gain, T the time constant (also called lag), and L the time delay. Processes with integration are approximated by the model

$$P(s) = \frac{K_v}{s}e^{-sL}, \tag{7.4}$$

where K_v is the velocity gain and L the time delay. The model (7.4) can be regarded as the limit of (7.3) as K_p and T go to infinity in such a way that $K_p/T = K_v$ is constant. The parameters of (7.3) and (7.4) can be determined from a step response experiments using the methods presented in Section 2.7.

The tuning rules were obtained in the following way. The MIGO design method (see Section 6.8) with $M = 1.4$ was applied to all processes in the test batch (7.2). This gave the PI controller parameters K and T_i. The AMIGO rules were then obtained by finding relations between the controller parameters and the process parameters.

Figure 7.1 illustrates the relations between the controller parameters and the process parameters for all processes in the test batch. The controller gain is normalized by multiplying it either with the static process gain K_p or with the parameter $a = K_pL/T = K_vL$. The integral time is normalized by dividing it by T or by L. The parameters for the integrating processes P_6 are only normalized with a and L since K_p and T are infinite for these processes. The controller parameters in Figure 7.1 are plotted versus the normalized dead time $\tau = L/(L+T)$.

The figure shows that there is a good correlation between the normalized controller parameters and normalized time delay. This indicates that it is possible to develop good tuning rules based on the FOTD model. Notice, however, that there are significant variations in the parameters with the normalized time delay τ.

Ziegler and Nichols tried to find rules that do not depend on τ. Figure 7.1 shows that the normalized parameters KK_p, aK, T_i/T, and T_i/L vary as much as two orders of magnitude with τ. It is thus not possible to find efficient rules that do not depend on τ.

The solid lines in Figure 7.1 correspond to the AMIGO tuning formula,

$$\begin{aligned} K &= \frac{0.15}{K_p} + \left(0.35 - \frac{LT}{(L+T)^2}\right)\frac{T}{K_pL} \\ T_i &= 0.35L + \frac{13LT^2}{T^2+12LT+7L^2}, \end{aligned} \tag{7.5}$$

and the dotted lines show the limits for 15 percent variations in the controller parameters. Almost all processes included in the test batch fall within these limits.

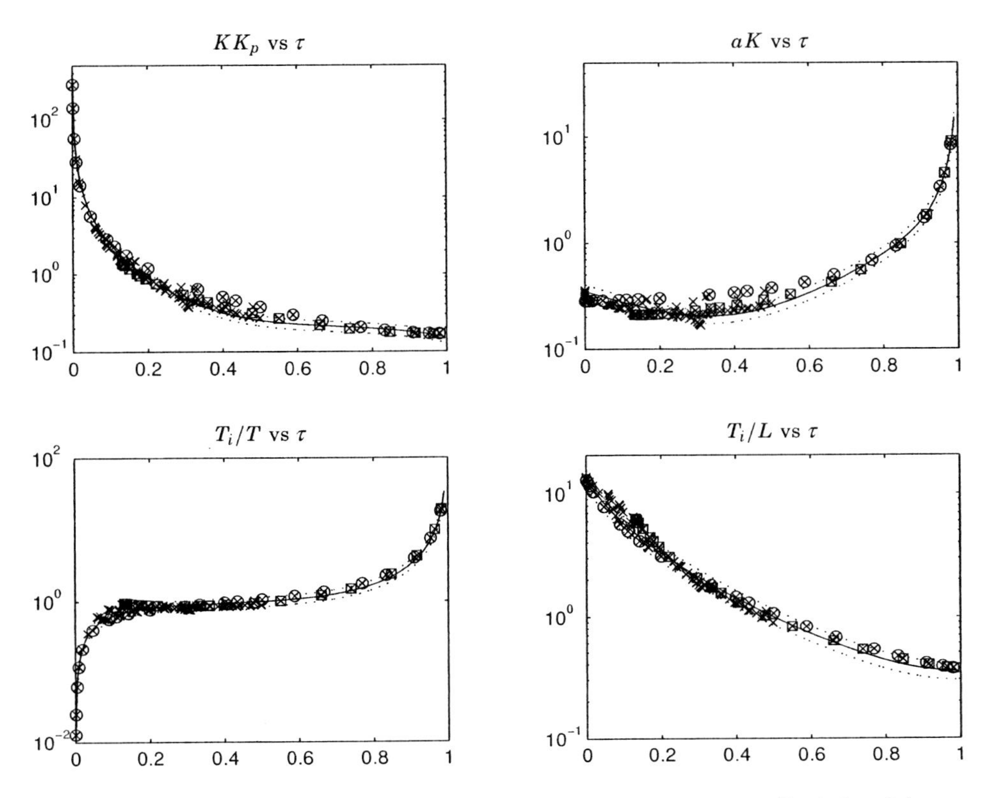

Figure 7.1 Normalized PI controller parameters plotted versus normalized time delay τ. The solid lines correspond to the AMIGO design rule (7.5), and the dotted lines indicate 15 percent parameter variations. The circles mark parameters obtained from process P_1, and squares parameters obtained from process P_2.

For integrating processes, K_p and T go to infinity and $K_p/T = K_v$. Therefore, the AMIGO tuning rules (7.5) can be simplified to

$$\begin{aligned} K &= \frac{0.35}{K_v L} \\ T_i &= 13.4L. \end{aligned} \tag{7.6}$$

for integrating processes.

The tuning rule (7.5) can be seen as a replacement for the Ziegler-Nichols' step response method for PI control. Notice that the rule was designed for the sensitivity $M = 1.4$. Similar rules can be found for other values of the design parameter.

Set-Point Weighting

The MIGO design method also gives suitable values of b. It is determined so that the resonance peak of the transfer function between set point and process output becomes close to one, as discussed in Section 5.3. Figure 7.2 shows the values of the b-parameter for the test batch (7.2).

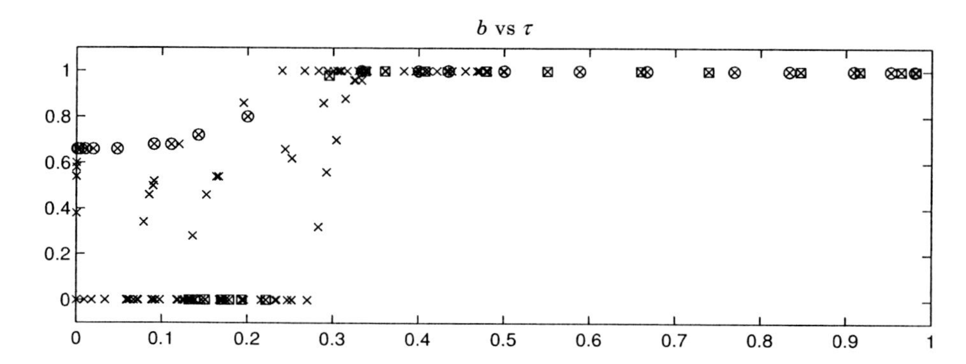

Figure 7.2 Set-point weighting as a function of τ for the test batch (7.2). The circles mark parameters obtained from the process P_1, and the squares mark parameters obtained from the process P_2.

Figure 7.2 shows that the correlation with parameter τ is not as good as for the feedback parameters and that there is a larger difference between the pure FOTD model P_1 and the other processes. The set-point weight should be $b = 1$ for processes with $\tau > 0.3$.

7.4 PID Control

Figure 7.3 illustrates the relations between the PID controller parameters obtained from the MIGO design and the process parameters for all processes in the test batch. The robustness criterion $M = 1.4$ is used together with the additional constraint $\partial k_i / \partial k = 0$; see Section 6.8. The normalized controller parameters in Figure 7.3 are plotted versus the normalized dead time τ.

The figure indicates that the variations of the normalized controller parameters are several orders of magnitude. We can thus conclude that it is not possible to find good universal tuning rules that do not depend on the normalized time delay τ. Recall that Ziegler and Nichols suggested the rules $aK = 1.2$, $T_i = 2L$, and $T_d = 0.5L$. Figure 7.3 shows that these parameters are only suitable for very few processes in the test batch.

The controller parameters for processes P_1 are marked with circles and those for P_2 are marked with squares in Figure 7.3. For $\tau < 0.5$, the gains for P_1 are typically smaller than for the other processes, and the integral time is larger. This is the opposite to what happened for PI control; see Figure 7.1. Process P_2 has a gain that is larger and an integral time that is shorter than for the other processes.

For PI control, it was possible to obtain simple tuning rules, where the controller parameters obtained from the AMIGO rules differed less than 15 percent from those obtained from the MIGO rules for most processes in the test batch. Figure 7.3 indicates that universal tuning rules for PID control can be obtained only for $\tau \geq 0.5$.

For $\tau < 0.5$ there is a significant spread of the normalized parameters. This

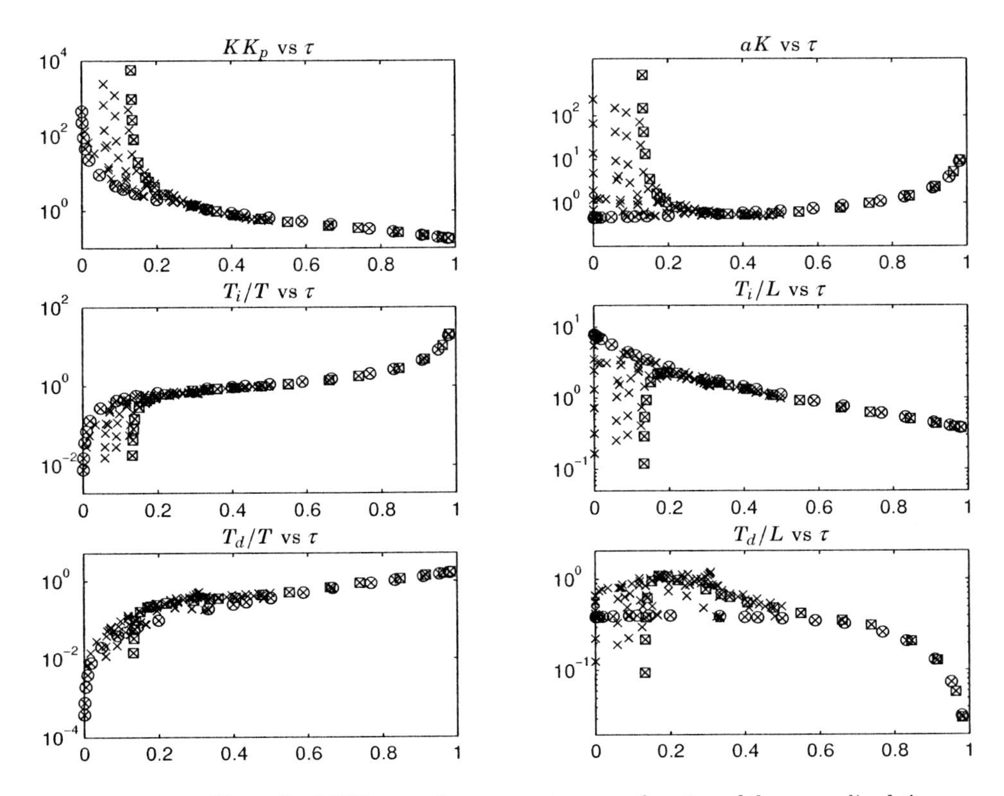

Figure 7.3 Normalized PID controller parameters as a function of the normalized time delay τ. The controllers for the process P_1 are marked with ○ and controllers for P_2 with □.

implies that it is not possible to find universal tuning rules for lag-dominated processes. Notice that the gain and the integral time are well defined for $0.3 < \tau < 0.5$ but that there is a considerable variation of the normalized derivative time in that interval.

Because of the large spread in parameter values for $\tau < 0.5$ it is worthwhile to model the process more accurately to obtain good tuning of PID controllers. The process models (7.3) and (7.4) model stable processes with three parameters and integrating processes with two parameters. In practice, it is very difficult to obtain more process parameters from the simple step response experiment. A step response experiment is thus not sufficient to tune PID controllers with $\tau < 0.5$ accurately.

However, it is possible to find *conservative* tuning rules for $\tau < 0.5$ by choosing controllers with parameters that correspond to the lowest gains and the largest integral times of Figure 7.3. Before developing such rules, we will discuss the reason why universal tuning rules for lag-dominant processes cannot be found.

Problems with FODT Structure

The criterion used is to maximize integral gain k_i. The fundamental limitations are given by the *true* time delay of the process, which we denote L_0. The integral gain is proportional to the gain crossover frequency ω_{gc} of the closed-loop system. The gain crossover frequency ω_{gc} is typically limited to

$$\omega_{gc} L_0 < 0.5.$$

When a process is approximated by the FOTD model the apparent time delay L is longer than the true time delay L_0 because lags are approximated by additional time delays. This implies that the integral gain obtained for the FOTD model will be lower than for a design based on the true model. The situation is particularly pronounced for systems with small τ.

Consider PI control of first-order systems, i.e., processes with the transfer functions

$$P(s) = \frac{K_p}{1 + sT} \quad \text{or} \quad P(s) = \frac{K_v}{s}.$$

Since these systems do not have time delays there is no dynamics limitation, and arbitrarily high integral gain can be obtained. Since these processes can be matched perfectly by the models (7.3) and (7.4), the design rule reflects this property. The process parameters are $L = 0$, $a = 0$, and $\tau = 0$, and both the design method MIGO and the approximate AMIGO rule (7.5) give infinite integral gains.

Consider PID control of second-order systems with the transfer functions

$$P(s) = \frac{K_v}{s(1 + sT_1)} \quad \text{and} \quad P(s) = \frac{K_p}{(1 + sT_1)(1 + sT_2)}.$$

Since the systems does not have time delays it is possible to have controllers with arbitrarily high integral gains. The first transfer function has $\tau = 0$. The second process has values of τ in the range $0 \le \tau < 0.13$, where $\tau = 0.13$ corresponds to $T_1 = T_2$. When these transfer functions are approximated with a FOTD model one of the time constants will be approximated with a time delay. Since the approximating model has a time delay there will be limitations in the integral gain.

We can thus conclude that for $\tau < 0.13$ there are processes in the test batch that permit infinitely large integral gains. This explains the wide spread of controller parameters for small τ. The spread is infinitely large for $\tau < 0.13$, and it decreases for larger τ. Therefore, for small τ improved modeling gives a significant benefit.

One way to avoid the difficulty is to use a more complicated model, such as

$$P(s) = \frac{b_1 s + b_2 s}{s^2 + a_1 s + a_2} e^{-sL}.$$

It is, however, very difficult to estimate the parameters of this model accurately from a simple step response experiment. Design rules for models having five parameters may also be cumbersome. Since the problem occurs for small values of τ it may be possible to approximate the process with

$$P(s) = \frac{K_v}{s(1+sT)}e^{-sL},$$

which only has three parameters. Instead of developing tuning rules for more complicated models it may be better to simply compute the controller parameters based on the estimated model.

Conservative Tuning Rules (AMIGO)

Figure 7.3 shows that it is not possible to find optimal tuning rules for PID controllers that are based on the simple process models (7.3) or (7.4). It is, however, possible to find conservative robust tuning rules with lower performance. The rules are close to the MIGO design for the process P_1, i.e., the process that gives the lowest controller gain and the longest integral time; see Figure 7.3.

The suggested AMIGO tuning rules for PID controllers are

$$\begin{aligned} K &= \frac{1}{K_p}\left(0.2 + 0.45\frac{T}{L}\right) \\ T_i &= \frac{0.4L + 0.8T}{L + 0.1T}L \\ T_d &= \frac{0.5LT}{0.3L + T}. \end{aligned} \tag{7.7}$$

For integrating processes, Equation 7.7 can be written as

$$\begin{aligned} K &= 0.45/K_v \\ T_i &= 8L \\ T_d &= 0.5L. \end{aligned} \tag{7.8}$$

Figure 7.4 compares the tuning rule (7.7) with the controller parameters given in Figure 7.3. The tuning rule (7.7) describes the controller gain K well for a process with $\tau > 0.3$. For small τ, the controller gain is well fitted to processes P_1, but the AMIGO rule underestimates the gain for other processes.

The integral time T_i is well described by the tuning rule (7.7) for $\tau > 0.2$. For small τ, the integral time is well fitted to processes P_1, but the AMIGO rule overestimates it for other processes.

The tuning rule (7.7) describes the derivative time T_d well for process with $\tau > 0.5$. In the range $0.3 < \tau < 0.5$ the derivative time can be up to a factor of 2 larger than the value given by the AMIGO rule. If the values of the derivative time for the AMIGO rule are used in this range the robustness is decreased; the value of M may be reduced by about 15 percent. For $\tau < 0.3$, the AMIGO tuning rule gives a derivative time that sometimes is shorter and sometimes longer than the one obtained by MIGO. Despite this, it appears that AMIGO gives a conservative tuning for all processes in the test batch, mainly because of the decreased controller gain and increased integral time.

The tuning rule (7.7) has the same structure as the Cohen-Coon method, but the parameters differ significantly.

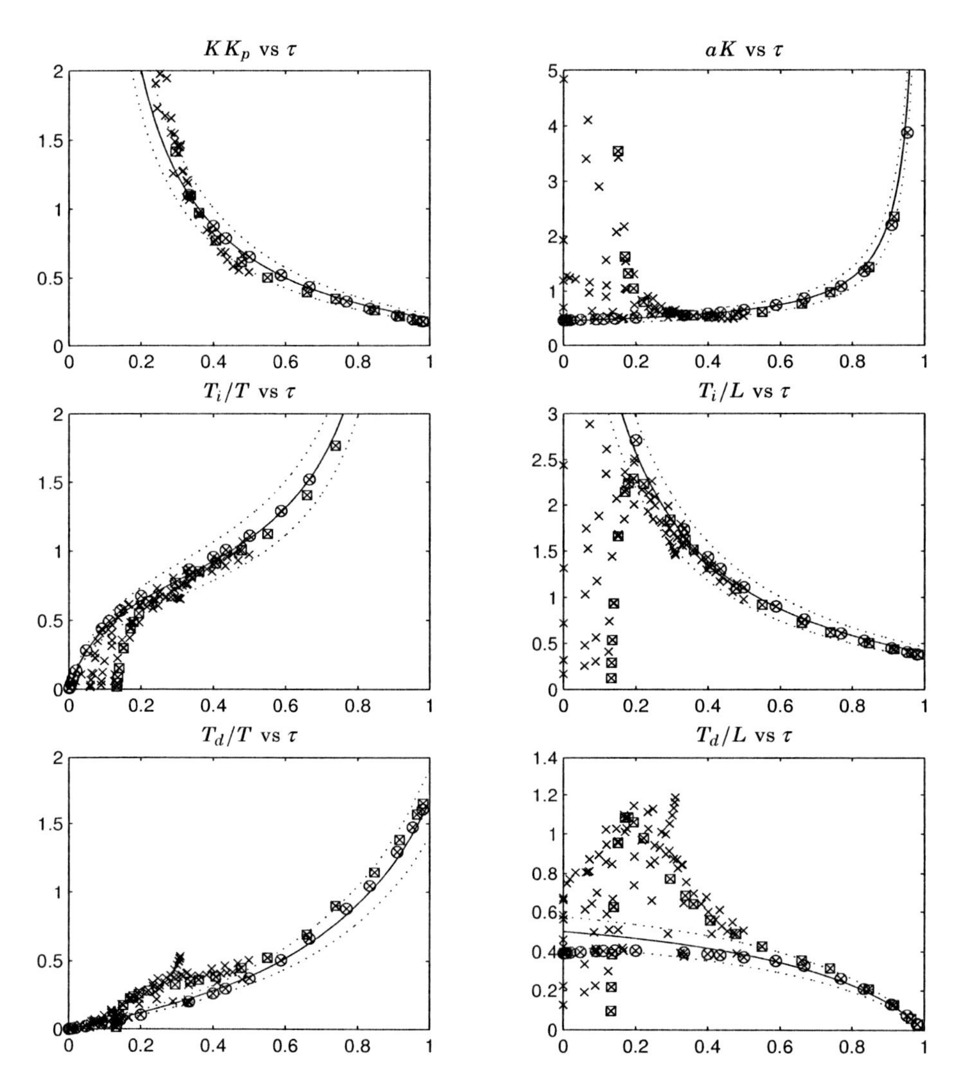

Figure 7.4 Normalized controller parameters as a function of normalized time delay τ. The solid line corresponds to the tuning rule (7.7), and the dotted lines indicate 15 percent parameter variations. The circles mark parameters obtained from the process P_1, and the squares mark parameters obtained from the process P_2.

Robustness

Figure 7.5 shows the Nyquist curves of the loop transfer functions obtained when the processes in the test batch (7.2) are controlled with the PID controllers tuned with the conservative AMIGO rule (7.7). When using MIGO all Nyquist curves are outside the M-circle. With AMIGO there are some processes where the Nyquist curves are inside the circle. An investigation shows that the derivative action is too small in these cases; compare with the curves of T_d/L vs τ in Figure 7.4. The increase of M is at most about 15 percent

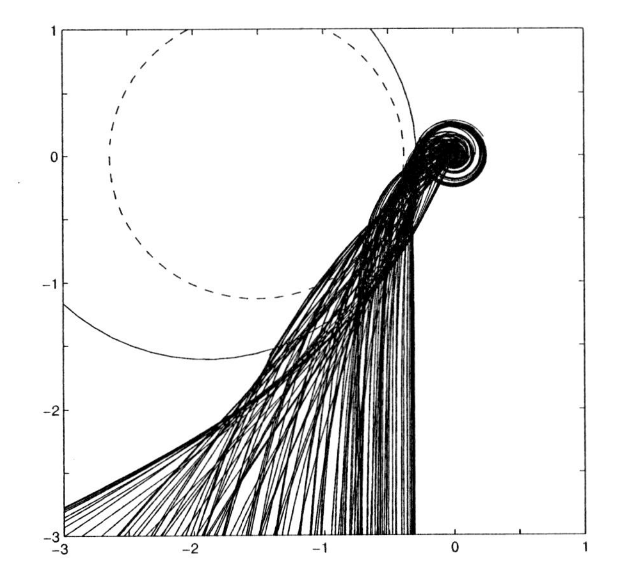

Figure 7.5 Nyquist curves of loop transfer functions obtained when PID controllers tuned according to (7.7) are applied to the test batch (7.2). The solid circle corresponds $M = 1.4$, and the dashed to a circle where M is increased by 15 percent.

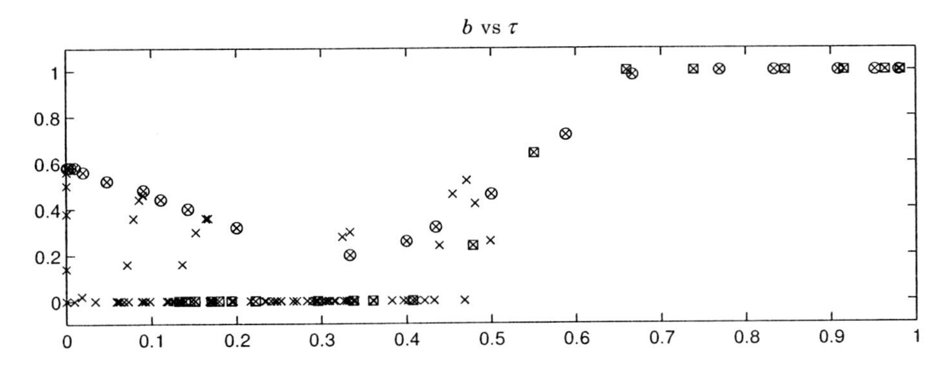

Figure 7.6 Set-point weighting as a function of τ for the test batch (7.2). The circles mark parameters obtained from the process P_1, and the squares mark parameters obtained from the process P_2.

with the AMIGO rule. If this increase is not acceptable derivative action can be increased or the gain can be decreased with about 15 percent.

Set-Point Weighting

Figure 7.6 shows the values of the b-parameter for the test batch (7.2).

The correlation between b and τ is not so good, but a conservative and simple rule is to choose b as

$$b = \begin{cases} 0, & \text{for } \tau \leq 0.5 \\ 1, & \text{for } \tau > 0.5. \end{cases} \tag{7.9}$$

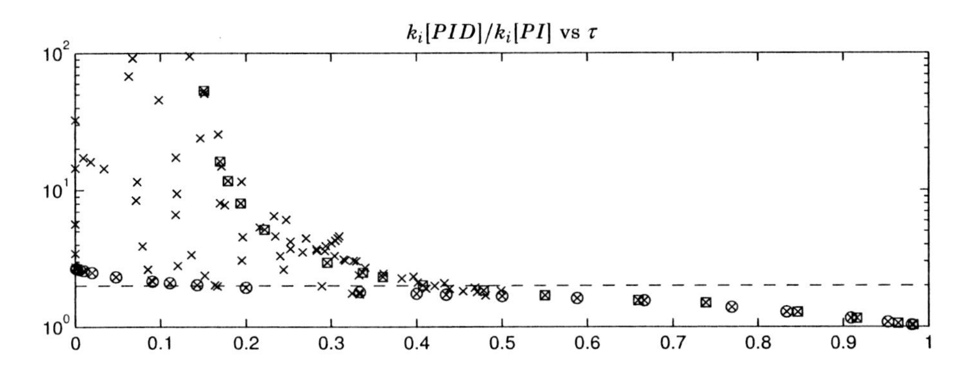

Figure 7.7 The ratio of integral gain with PID and PI control as a function of normalized time delay τ. The dashed line corresponds to the ratio $k_i[PID]/k_i[PI] = 2$. The controllers for the process P_1 are marked with circles and controllers for P_2 with squares.

The Benefits of Derivative Action

Since maximization of integral gain was chosen as design criterion we can judge the benefits of derivative action by the ratio of integral gain for PID and PI control. Figure 7.7 shows this ratio for the test batch, except for a few processes with a high ratio at small values of τ.

The figure shows that the benefits of derivative action are marginal for delay-dominated processes but that the benefits increase with decreasing τ. For $\tau = 0.5$ the integral gain can be doubled, and for values of $\tau < 0.15$ integral gain can be increased arbitrarily for some processes.

The Ratio T_i/T_d

The ratio T_i/T_d is of interest for several reasons. It is a measure of the relative importance of derivative and integral action. Many PID controllers are implemented in series form, which requires that the ratio be larger than 4. Many classical tuning rules therefore fix the ratio to 4. Figure 7.8 shows the ratio for the full test batch. The figure shows that there is a significant variation in the ratio T_i/T_d, particularly for small τ. The ratio is close to 2 for $0.5 < \tau < 0.9$, and it increases to infinity as τ approaches 1 because the derivative action is zero for processes with pure time delay.

Figure 7.8 also shows the ratio obtained by the AMIGO tuning rule (7.7). The ratio is less than four for processes with $0.3 < \tau < 0.9$, which means that the tuning rule cannot be used for controllers in series form for these processes. However, it appears that the changes of performance and robustness are marginal if the tuning rule (7.7) is modified so that $T_d = T_i/4$ for these processes. Figure 7.9 shows the Nyquist curves of the loop transfer functions obtained when the processes in the test batch with $0.3 < \tau < 0.9$ are tuned such that gain K and integral time T_i are obtained from (7.7), and the derivative time is obtained as $T_d = T_i/4$. The figure shows that the robustness is about the same as for (7.7); compare with Figure 7.5.

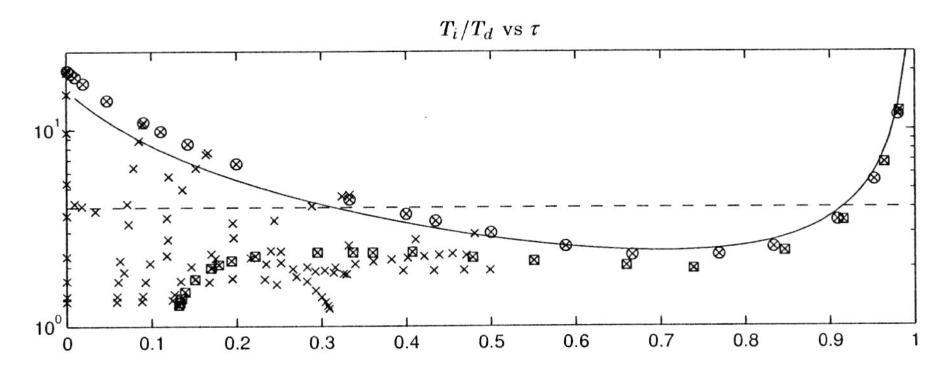

Figure 7.8 The ratio between T_i and T_d as a function of normalized time delay τ. Process P_1 is marked with circles and process P_2 with squares. The dashed line corresponds to the ratio $T_i/T_d = 4$, and the solid line to the ratio given by the AMIGO tuning rule (7.7).

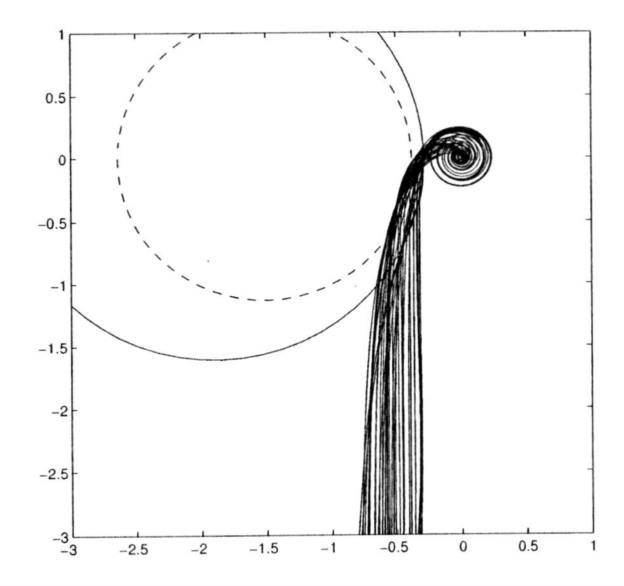

Figure 7.9 Nyquist curves of the loop transfer functions obtained from the processes in the test batch with $0.3 < \tau < 0.9$ when the controller is tuned with $T_d = T_i/4$.

The Average Residence Time

The parameter T_{63}, which is the time when the step response has reached 63 percent, a factor of $(1-1/e)$, of its steady-state value, is a reasonable measure of the response time for stable systems. It is easy to determine the parameter by simulation, but not by analytical calculations. For the FOTD process we have $T_{ar} = T_{63}$. The average residence time T_{ar} is in fact a good estimate of T_{63} for systems with essentially monotone step response. For all stable processes in the test batch we have $0.99 < T_{63}/T_{ar} < 1.08$.

The average residence time is easy to compute analytically. Consider the closed-loop system obtained when a process with transfer function $P(s)$ is con-

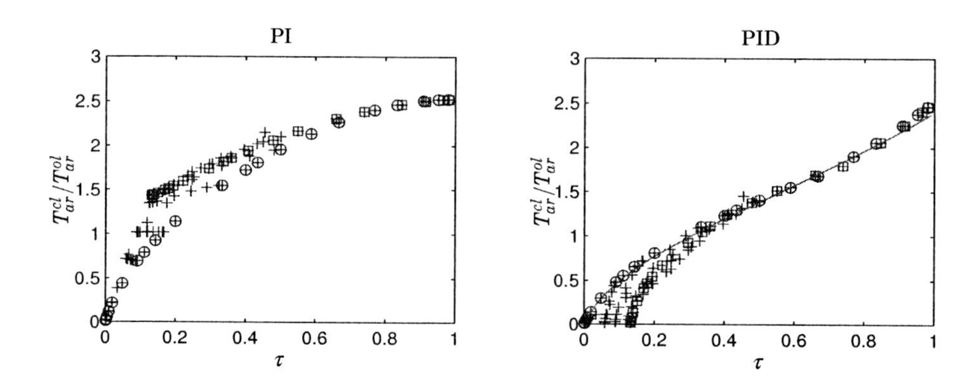

Figure 7.10 The ratio of the average residence time of the closed loop system and the open loop system for PI control left and PID control right.

trolled with a PID controller with set-point weighting. The closed-loop transfer function from set point to output is

$$G_{yy_{sp}}(s) = \frac{P(s)C_{ff}(s)}{1 + P(s)C(s)},$$

where

$$C_{ff}(s) = bK + \frac{k_i}{s}.$$

Straightforward but tedious calculations give

$$T_{ar} = -\frac{G'_{yy_{sp}}(0)}{G_{yy_{sp}}(0)} = T_i\Big(1 - b + \frac{1}{KK_p}\Big). \tag{7.10}$$

Figure 7.10 shows the average residence times of the closed-loop system divided with the average residence time of the open-loop system. Figure 7.10 shows that for PID control the closed-loop system is faster than the open-loop system when $\tau < 0.3$ and slower for $\tau > 0.3$.

7.5 Frequency Response Methods

In this section we will investigate if it is possible to obtain simple tuning rules similar to the Ziegler-Nichols frequency response method.

Parameterization

Ziegler-Nichols characterized the processes by two parameters K_{180} and T_{180} when they developed their frequency response method for controller tuning. The Ziegler-Nichols tuning rules do not use sufficient information, and they give too aggressive tuning, which does not give robust closed-loop systems.

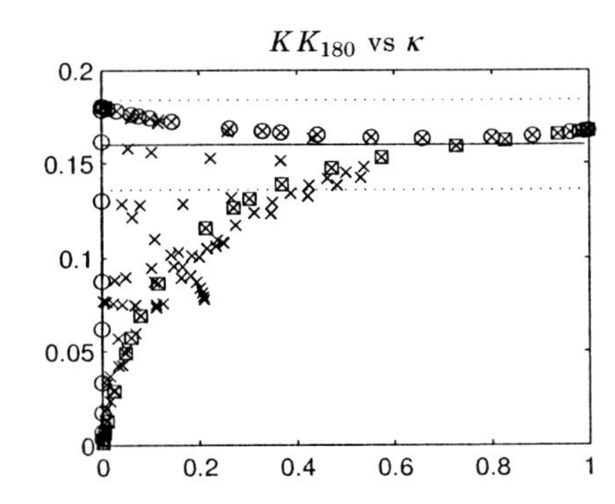

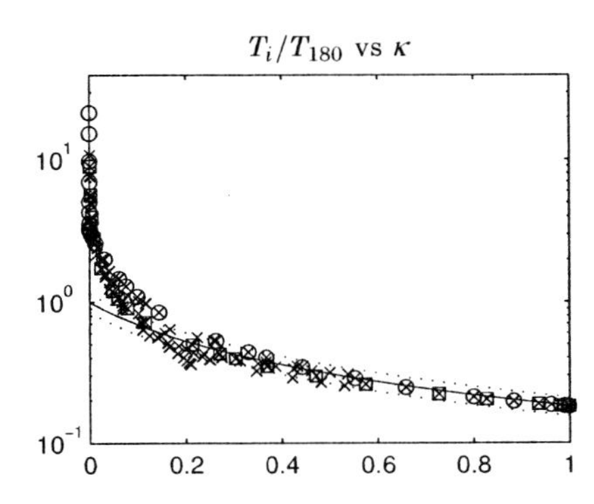

Figure 7.11 Normalized controller parameters plotted versus gain ratio κ for stable processes for $M = 1.4$. The solid lines correspond to the tuning rule (7.11), and the dotted lines indicate 15 percent variations from the rule. The circles mark data from process P_1, and squares data from P_2.

When investigating the step response method it was found that significant improvement could be obtained by including an additional third process parameter, the static process gain. In this section it will be investigated if similar improvements can be obtained for the frequency domain method.

For the step response method we used the normalized time delay τ as a parameter to characterize the process. The corresponding frequency domain parameter is the gain ratio $\kappa = K_{180}/K_p$.

PI Tuning Rules for Balanced and Delay-Dominated Processes

The MIGO design method has been applied to all processes in the test batch (7.2). Figure 7.11 shows the controller parameters obtained for $M = 1.4$. The figure shows that there is a significant spread of controller parameters for lag-dominant processes.

The Ziegler-Nichols tuning rules have constant values $KK_{180} = 0.4$ and $T_i/T_{180} = 0.8$, for all values of κ. Figure 7.11 shows that it may be reasonable to have a constant value KK_{180} for $\kappa > 0.5$, but not for smaller values of κ. The gain $KK_{180} = 0.4$ suggested by Ziegler and Nichols is clearly too high, which explains the poor robustness of their method. The integral time suggested by Ziegler and Nichols, $T_i = 0.8T_{180}$, is too high except for processes with very small values of κ.

Figure 7.11 shows that it is not possible to capture all data by one tuning rule. It may, however, be possible to obtain a rule for balanced and delay-dominated processes. Figure 7.11 shows the graphs corresponding to the following tuning rule.

$$
\begin{aligned}
KK_{180} &= 0.16 \\
\frac{T_i}{T_{180}} &= \frac{1}{1 + 4.5\kappa}.
\end{aligned}
\tag{7.11}
$$

The tuning rule (7.11) is not appropriate for lag-dominant processes, but it gives controller parameters that are fairly close to the optimal for processes with $\kappa > 0.2$. Notice in particular that the ratio T_i/T_{180} is reduced by a factor

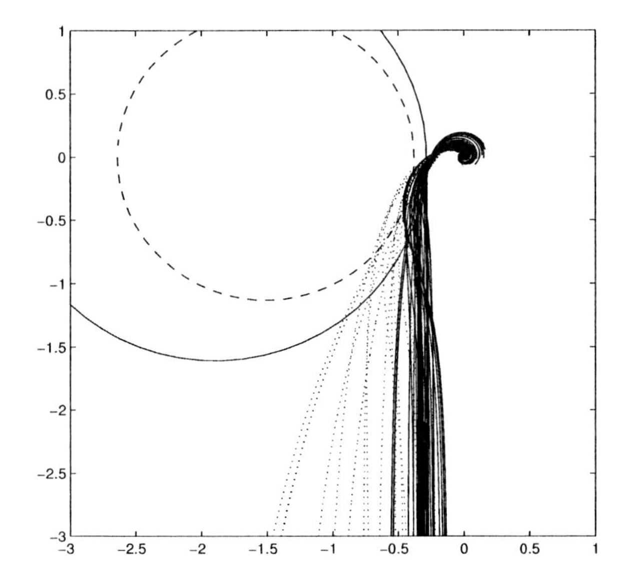

Figure 7.12 Nyquist curves of loop transfer functions obtained when PI controllers tuned according to (7.11) are applied to processes in the test batch with $\kappa > 0.1$. Transfer functions corresponding to processes with $0.1 < \kappa < 0.2$ are shown with dotted lines. The solid circle corresponds to $M = 1.4$ and the dashed to a circle where M is increased by 15 percent.

of three when κ increases from 0.2 to 1.

Figure 7.12 shows the Nyquist curves obtained for all processes in the test batch with $\kappa > 0.1$ when the tuning rule (7.11) is used. The figure shows that all loop transfer functions are close to the M-circle.

PID Tuning Rules for Balanced and Delay-Dominated Processes

Parameters of PID controllers for all the processes in the test batch (7.2) were computed using the MIGO design with the constraints described in the previous section. The design parameter was chosen to $M = 1.4$.

Figure 7.13 illustrates the relations between the controller parameters obtained from the MIGO design and the process parameters for all processes in the test batch.

The controller parameters for processes P_1 are marked with circles, and those for P_2 are marked with squares in Figure 7.13. For $\kappa < 0.3$, the gain for P_1 is typically smaller than for the other processes, and the integral time is larger. This is opposite to what happened for PI control. Process P_2 has a gain that is larger and an integral time that is shorter than for most other processes.

The figure indicates that the variations of the normalized controller parameters are more than an order of magnitude. Therefore, it is not possible to find good universal tuning rules that do not depend on the gain ratio κ. Ziegler and Nichols suggested the rule $KK_{180} = 0.6$, $T_i/T_{180} = 0.5$, and $T_d/T_{180} = 0.125$. The rule is indicated by the dashed lines in Figure 7.13. The Ziegler-Nichols

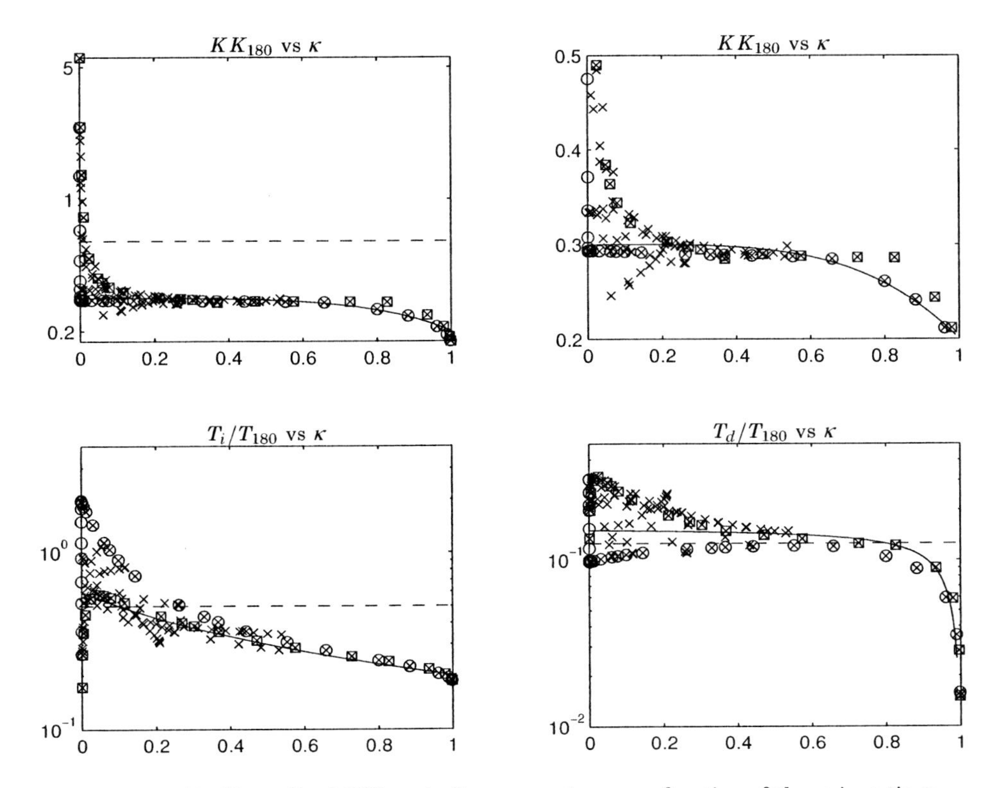

Figure 7.13 Normalized PID controller parameters as a function of the gain ratio κ. Parameters obtained for process P_1 are marked with circles, and parameters obtained for process P_2 with squares. The dashed lines indicate Ziegler-Nichols' tuning rule and the solid lines corresponds to the rule (7.12).

rule is only suitable for very few processes in the test batch. The controller gain is too high except for some processes with very small values of κ.

Even if Figure 7.13 indicates that it is not possible capture all data by one tuning rule it is clear that a good tuning rule can be found for balanced and delay-dominated processes. Figure 7.13 shows the graphs corresponding to the following tuning rule as solid lines.

$$\begin{aligned} K &= (0.3 - 0.1\kappa^4)/K_{180} \\ T_i &= \frac{0.6}{1+2\kappa} T_{180} \\ T_d &= \frac{0.15(1-\kappa)}{1-0.95\kappa} T_{180}. \end{aligned} \tag{7.12}$$

The tuning rule (7.12) is not appropriate for lag-dominant processes, but it gives controller parameters that are fairly close to the optimal for processes with $\kappa > 0.2$. Figure 7.14 shows the Nyquist curves obtained for all processes in the test batch with $\kappa > 0.1$ when the tuning rule (7.12) is used. The figure shows that all loop transfer functions remain fairly close to the M-circle.

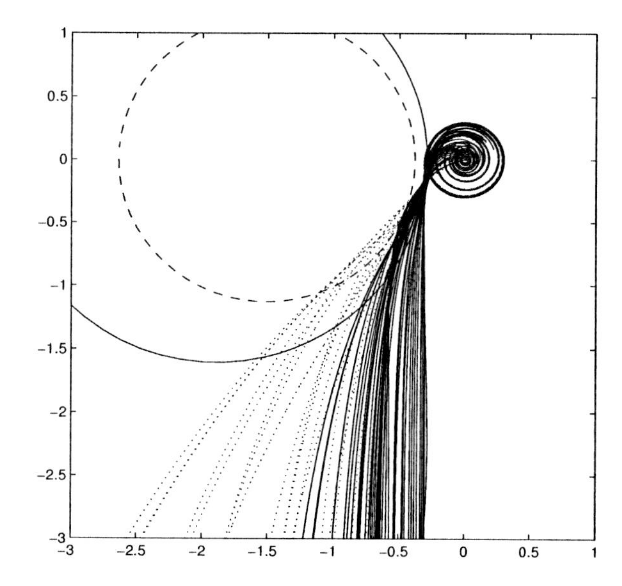

Figure 7.14 Nyquist curves of loop transfer functions obtained when PID controllers tuned according to (7.12) are applied to processes in the test batch with $\kappa > 0.1$. Transfer functions corresponding to processes with $0.1 < \kappa < 0.2$ are shown with dotted lines. The solid circle corresponds to $M = 1.4$, and the dashed to a circle where M is increased by 15 percent.

7.6 PID Control Based on Second-Order Model

In this section, tuning rules based on the SOTD model (2.47) are presented. The SOTD model may be obtained using the combined step and frequency response method presented in Section 2.7.

Figure 7.15 shows controller parameters K, $k_i = K/T_i$, and $k_d = KT_d$ for all processes in the test batch except the integrating process, plotted against the normalized time delay $\tau_1 = L_1/T_{63}$. Figures 7.16 shows the controller parameters T_i and T_d with different normalizations. Notice that the scales are also different. A comparison with Figure 7.3, where the simpler FOTD model is used, shows a significant improvement, particularly for small normalized time delays. This is not surprising because the achievable performance is primarily given by the time delay, and the improvement is mainly due to improved estimates of the true time delay.

The figures show that there is a considerable span of the parameter values. In Figure 7.15 the parameters KL_1/T_{63}, k_iT_{63}, and k_iL_1 range over two decades. The range of variation is larger for other normalizations; for example, the parameter KK_p ranges over five decades. Also notice that there is a spread in the values, particularly in k_d and T_d. This means that we cannot expect to find nice formulas where the normalized parameters are functions only of τ_1.

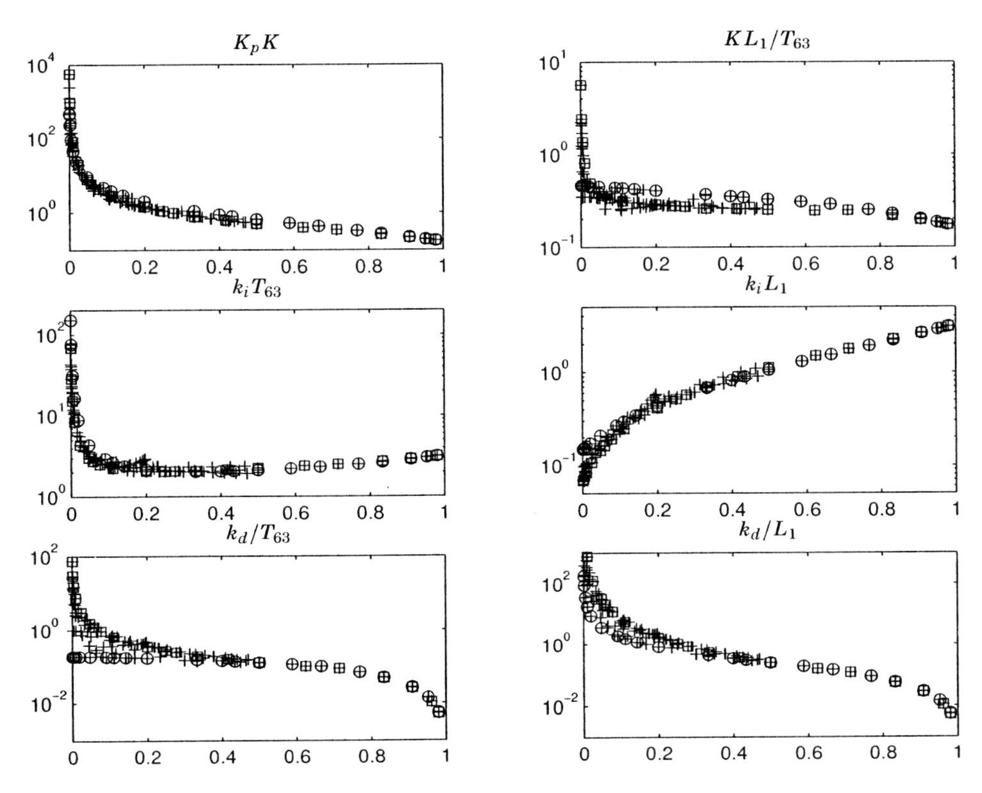

Figure 7.15 Normalized controller parameters K, k_i, and k_d for the processes in the test batch plotted versus τ_1. Data for the processes P_1 are marked with circles, and those for P_2 with squares.

Structure of Tuning Formulas

To get insight into suitable parameterizations we will consider some special systems.

For delay-dominated processes where $L_1 \gg T_1 > T_2$ the model (2.47) can be approximated by

$$P(s) = K_p e^{-sL_1}.$$

Derivative action cannot be used for this process. Designing a PI controller for the process we find

$$C(s) = K + \frac{k_i}{s} = \frac{0.1677}{K_p} + \frac{0.4618}{sL_1K_p},$$

where the numerical values are given for design with $M = 1.4$. Neglecting the time delay and using the numerical values of the controller parameters we find that the closed-loop system is of first order with the pole $sL_1 = -0.4$.

If the process dynamics is a time delay with a small lag,

$$P(s) = \frac{K_p}{1+sT}e^{-sL},$$

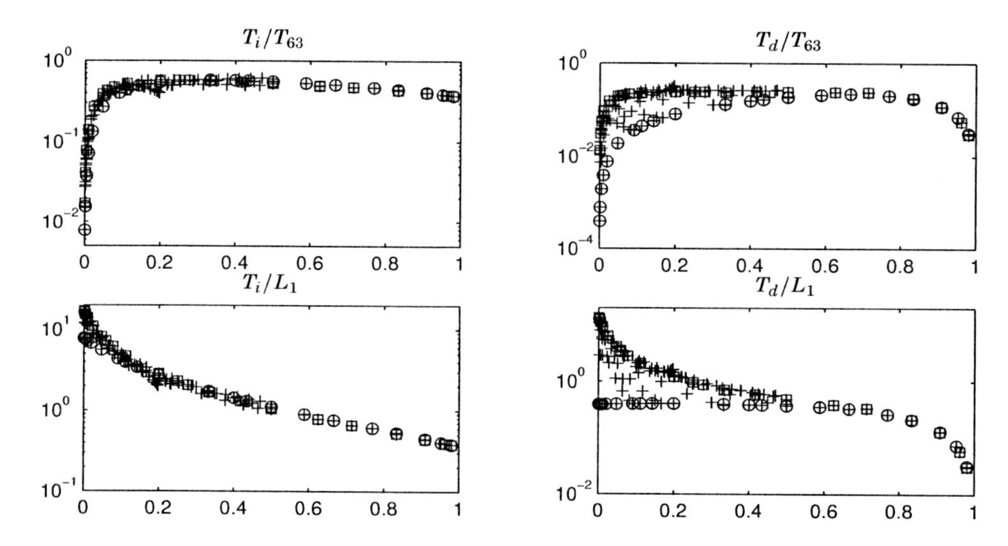

Figure 7.16 Normalized controller parameters T_i and T_d for the processes in the test batch plotted versus τ_1. Data for the processes P_1 is marked with circles, and those for P_2 with squares.

we find that the loop transfer functions under PID control with derivative gain k_d approaches

$$G_l(s) \approx \frac{k_d K_p}{T} e^{-sL}.$$

The Nyquist curve of this transfer function is a circle around the origin with radius $k_d K_p/T$. The design criterion using the combined sensitivity requires that the radius is less than $(M-1)/M$. The largest permissible derivative gain is thus

$$k_d = \frac{T}{K_p}\frac{M-1}{M}.$$

For delay-dominated processes the derivative time is thus proportional to the lag.

The PID controller for the process

$$P(s) = \frac{K_p}{sT} e^{-sL_1}$$

is

$$C(s) = K + \frac{k_i}{s} + k_d s = \frac{0.4603T}{K_p L_1} + \frac{0.05841T}{sK_p L_1^2} + \frac{0.1796sT}{K_p},$$

where the numerical values are given for design with $M = 1.4$. Neglecting the time delay and using the numerical values of the controller parameters we find that the closed loop system is of second order with the poles $sL_1 = -0.2 \pm 0.11i$; the dominant pole is thus $\omega_d = 0.2$.

The PID controller for the process

$$P(s) = \frac{K_p}{s^2 T_1 T_2} e^{-sL_1}$$

is

$$C(s) = K + \frac{k_i}{s} + k_d s = \frac{0.02140 T_1 T_2}{K_p L_1^2} + \frac{0.001218 T_1 T_2}{K_p L_1^3 s} + \frac{0.3 T_1 T_2 s}{K_p L_1},$$

where the numerical values are given for design with $M = 1.4$. Neglecting the time delay and using the numerical values of the controller parameters we find that the closed-loop system is of third order with the poles $sL_1 = -0.23$ and $sL_1 = 0.035 \pm 0.064i$; the dominant pole is thus $\omega_d = 0.07$.

Parameterization

Based on the special cases given above it is reasonable to try tuning formulas having the form

$$\begin{aligned} K_p K &= \alpha_1 + \alpha_2 \frac{T_1}{L_1} + \alpha_3 \frac{T_2}{L_1} + \alpha_4 \frac{T_1 T_2}{L_1^2} \\ K_p k_i &= \beta_1 \frac{1}{L_1} + \beta_2 \frac{T_1}{L_1^2} + \beta_3 \frac{T_2}{L_1^2} + \beta_4 \frac{T_1 T_2}{L_1^3} \\ K_p k_d &= \left(\gamma_1 L_1 + \gamma_2 T_1 + \gamma_3 T_2 + \gamma_4 \frac{T_1 T_2}{L_1} \right) \frac{T_1 + T_2}{T_1 + T_2 + L_1}. \end{aligned} \tag{7.13}$$

This will match the controllers for the special cases. The coefficients of proportional and integral gain are simply obtained by adding the coefficients for the prototype processes. Because of the structure of the formula this will automatically give an interpolation between processes with pure delay and double integrator with delay. This procedure will not work for the derivative gain. In this case we have simply taken the weighted average with weights L_1 and $T_1 + T_2$.

Making a least squares fit of the parameters in (7.13) using the parameters of the test batch gives the results in Table 7.1.

Final Parameters

It seems reasonable to make the following approximations.

$$\begin{array}{llll} \alpha_1 = 0.19 & \alpha_2 = 0.37 & \alpha_3 = 0.18 & \alpha_4 = 0.02 \\ \beta_1 = 0.48 & \beta_2 = 0.03 & \beta_3 = -0.0007 & \beta_4 = 0.0012 \\ \gamma_1 = 0.29 & \gamma_2 = 0.16 & \gamma_3 = 0.20 & \gamma_4 = 0.28. \end{array} \tag{7.14}$$

The parameters β_3 and β_4 are quite small. This means that integral gain is essentially determined by parameters K_p, T_1, and L_1. This explains why there is a good correlation in the data for integral gain in Figure 7.15. The correlation for proportional gain in Figure 7.15 is good but not as good as for k_i, because the parameters α_3 and α_4 are larger. The correlation is poor for k_d because parameters γ_3 and γ_4 are large.

Integrating Processes

To investigate that the formulas also work for integrating processes we will investigate the process P_6. A model for integrating processes can be obtained

Table 7.1 Parameters fitted to the tuning formula for different data sets; $P^{\dagger}$ denotes all processes except the integrating process P_6.

Par	P_1	P_2	P_1, P_2	$P^{\dagger}$	e^{-s}	e^{-s}/s	e^{-s}/s^2
α_1	0.1755	0.1815	0.1823	0.1903	0.1677	-	-
α_2	0.4649	−0.0215	0.4607	0.3698	-	0.4603	-
α_3	0	0.6816	0.0930	0.1777	-	-	-
α_4	0	0.0210	0.0211	0.0196	-	-	0.02140
β_1	0.5062	0.4613	0.4800	0.4767	0.4618	-	-
β_2	0.0587	−0.2028	0.0596	0.0310	-	0.05841	-
β_3	0	0.2877	−0.0367	0.0017	-	-	-
β_4	0	0.0013	0.0013	0.0012	-	-	0.001218
γ_1	0.3026	0.2864	0.2971	0.2918	-	-	-
γ_2	0.1805	0.0590	0.1814	0.1654	-	0.1796	-
γ_3	0	0.2464	0.0814	0.2033	-	-	-
γ_4	0	0.3090	0.3096	0.2772	-	-	0.3

by taking the limit of

$$P(s) = \frac{K_p}{(1 + sT_1)(1 + sT_2)} e^{-sL_1}$$

as K_p and T_1 goes to infinity in such a way that $K_p/T_1 = K_v$. The model then becomes

$$P(s) = \frac{K_v}{s(1 + sT_2)} e^{-sL_1}.$$

The tuning formula (7.13) becomes

$$\begin{aligned} K_v K &= \alpha_2 \frac{1}{L_1} + \alpha_4 \frac{T_2}{L_1^2} \\ K_v k_i &= \beta_2 \frac{1}{L_1^2} + \beta_4 \frac{T_2}{L_1^3} \\ K_v k_d &= \gamma_2 + \gamma_4 \frac{T_2}{L_1}. \end{aligned} \tag{7.15}$$

Validation

Figure 7.17 shows the Nyquist curves of the loop transfer functions obtained when the processes in the test batch (7.2) are controlled with the PID controllers tuned with the rules (7.13), (7.15), and (7.14). When using MIGO all Nyquist curves are outside the M-circle in the figure. With the approximative rule there are some processes where the Nyquist curves are inside the circle. The increase of M is, however, less than 15 percent for all processes in the test batch.

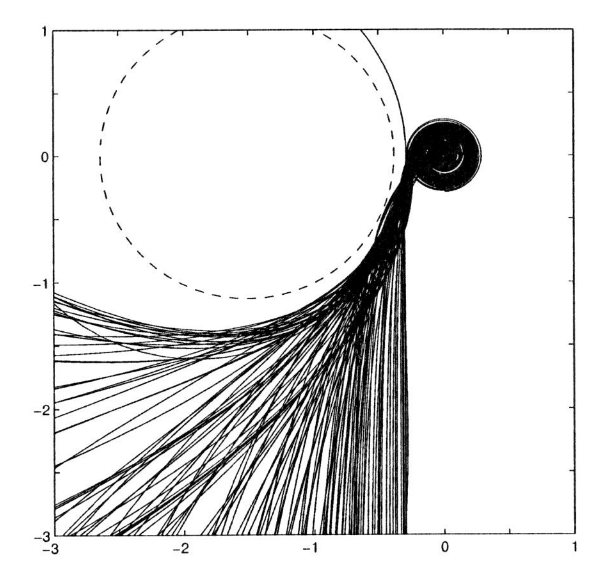

Figure 7.17 Nyquist curves of loop transfer functions obtained when PID controllers tuned according to (7.13), (7.15), and (7.14) are applied to the test batch (7.2). The solid circle corresponds $M = 1.4$, and the dashed to a circle where M is increased by 15 percent.

7.7 Comparison of the Methods

This section presents a few examples that illustrate the AMIGO method and compares it with the MIGO designs for PI and PID controllers. Three examples are given, one with a lag-dominant process, one with a delay-dominant process, and one with a process with balanced lag and delay.

EXAMPLE 7.1—LAG-DOMINATED DYNAMICS
Consider a process with the transfer function

$$P(s) = \frac{1}{(1+s)(1+0.1s)(1+0.01s)(1+0.001s)}.$$

Fitting the model (7.3) to the process we find that the apparent time delay and time constants are $L = 0.075$ and $T = 1.04$, which gives $\tau = 0.067$. The dynamics are thus lag dominated. The corresponding frequency response data needed for the AMIGO design are $K_{180} = 0.0091$ and $T_{180} = 0.199$. Since the static gain is $K_p = 1$, the gain ratio becomes $\kappa = K_{180}/K_p = 0.0091$. Since the process is lag dominant with $\kappa < 0.1$, the AMIGO rules based on frequency response data cannot be used for this process. Fitting the second-order model (2.47) gives the parameters $T_1 = 0.980$, $T_2 = 0.108$, and $L_1 = 0.010$.

The controller parameters obtained from the MIGO and AMIGO tuning rules are presented in Table 7.2.

Figure 7.18 shows the responses of the system to changes in set point and load disturbances when the controllers are tuned with the MIGO and AMIGO design. The figure shows that the AMIGO rule gives responses that are close

Table 7.2 Controller parameters obtained from the MIGO and AMIGO tuning rules for the lag-dominant process in Example 7.1.

Controller	Design	K	T_i	T_d	b	k_i
PI	MIGO	3.56	0.660		0	5.39
	AMIGO–step	4.13	0.539		0	7.66
PID	MIGO	56.9	0.115	0.0605	0	495
	AMIGO–step	6.44	0.361	0.0367	0	17.8
	AMIGO–step+frequency	59.6	0.127	0.0523	0	468

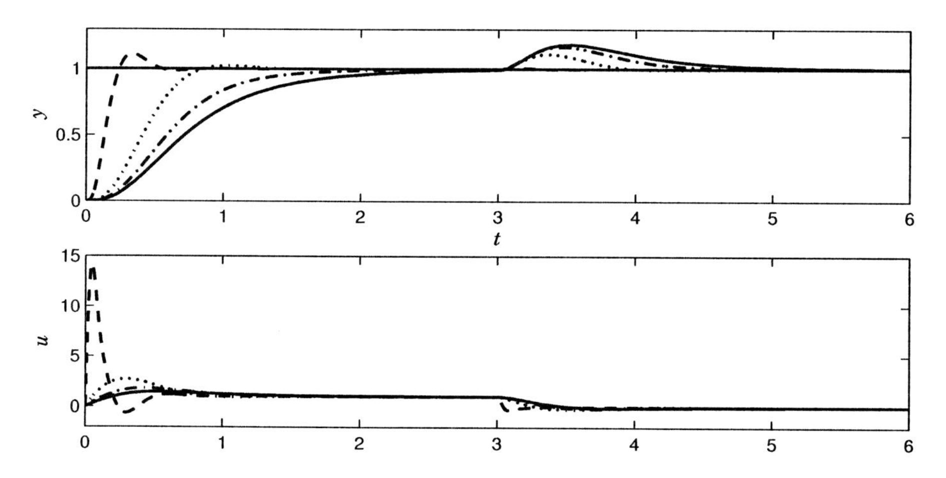

Figure 7.18 Responses to a unit step change at time 0 in set point and a unit load step at time 3 for PI controllers designed by MIGO (solid line) and AMIGO (dash-dotted line), and PID controllers designed by MIGO (dashed line) and AMIGO-step (dotted line) for the lag-dominant process in Example 7.1.

to the MIGO rule for PI control. However, since this is a lag-dominant process, the AMIGO tuning rule for PID control is conservative compared to the MIGO rule. This is obvious in the figure.

The responses obtained using the SOTD model are not presented in the figure, but Table 7.2 shows that the controller parameters are close to the MIGO design.

Notice that the magnitudes of the control signals are about the same at load disturbances, but that there is a major difference in the response time. The differences in the responses clearly illustrate the importance of reacting quickly.

The example shows that derivative action can give drastic improvements in performance for lag-dominated processes. It also demonstrates that the control performance can be increased considerably by obtaining better process models than (7.3). □

Table 7.3 Controller parameters obtained from the MIGO and AMIGO tuning rules for the process with balanced lag and delay in Example 7.2.

Controller	Design	K	T_i	T_d	b	k_i
PI	MIGO	0.432	2.43		1	0.178
	AMIGO–step	0.414	2.66		0	0.156
	AMIGO–frequency	0.640	2.96			0.216
PID	MIGO	1.19	2.22	1.21	0	0.536
	AMIGO–step	1.12	2.40	0.619	0	0.467
	AMIGO–frequency	1.20	2.51	0.927		0.478
	AMIGO–step+frequency	1.15	2.17	1.32	0	0.506

Next we will consider a process where the lag and the delay are balanced.

EXAMPLE 7.2—BALANCED LAG AND DELAY
Consider a process with the transfer function

$$P(s) = \frac{1}{(s+1)^4}.$$

Fitting the model (7.3) to the process we find that the apparent time delay and time constants are $L = 1.42$ and $T = 2.90$. Hence, $L/T = 0.5$ and $\tau = 0.33$. The frequency response data needed for the AMIGO design are $K_{180} = 0.250$ and $T_{180} = 6.28$. The gain ratio becomes $\kappa = K_{180}/K_p = 0.25$. Fitting the second-order model (2.47) gives the parameters $T_1 = 1.73$, $T_2 = 1.73$, and $L_1 = 1.05$.

The controller parameters obtained from the MIGO and AMIGO tuning rules are presented in Table 7.3.

Figure 7.19 shows the responses of the system to changes in set point and load disturbances when the MIGO and AMIGO-step designs are used. The figure shows that the load disturbance responses obtained by MIGO and AMIGO are quite similar, which can be expected because of the similarity of the controller parameters. The difference in set-point response between the MIGO and AMIGO design is caused by the different set-point weightings b of the two designs.

The integral gain k_i is about three times higher for PID control than for PI control. This is in accordance with Figure 7.7. □

Finally, we will consider an example where the dynamics are dominated by the time delay.

EXAMPLE 7.3—DELAY-DOMINATED DYNAMICS
Consider a process with the transfer function

$$P(s) = \frac{1}{(1+0.05s)^2}e^{-s}.$$

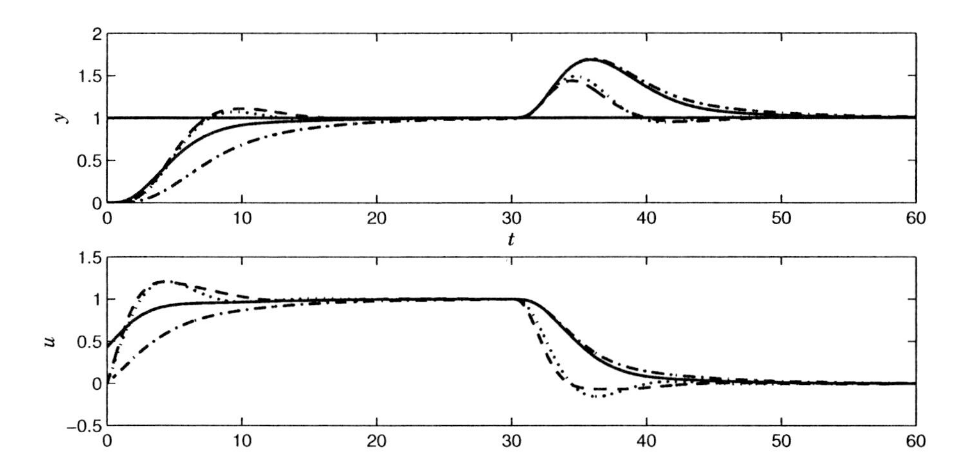

Figure 7.19 Responses to a unit step change at time 0 in set point and a unit load step at time 30 for PI controllers designed by MIGO (solid line) and AMIGO-step (dash-dotted line), and PID controllers designed by MIGO (dashed line) and AMIGO-step (dotted line) for the process with balanced lag and delay in Example 7.2.

Table 7.4 Controller parameters obtained from the MIGO and AMIGO tuning rules for the delay-dominant process in Example 7.3.

Controller	Design	K	T_i	T_d	b	k_i
PI	MIGO	0.170	0.404		1	0.421
	AMIGO–step	0.175	0.360		1	0.486
	AMIGO–frequency	0.163	0.407			0.400
PID	MIGO	0.216	0.444	0.129	1	0.486
	AMIGO–step	0.242	0.474	0.119	1	0.511
	AMIGO–frequency	0.212	0.446	0.0957		0.475
	AMIGO–step+frequency	0.218	0.453	0.129	1	0.481

Approximating the process with the model (7.3) gives the process parameters $L = 1.01$, $T = 0.0932$, and $\tau = 0.92$. The large value of τ shows that the process is delay dominated. The frequency response data needed for the AMIGO design is $K_{180} = 0.980$ and $T_{180} = 2.20$. The gain ratio becomes $\kappa = K_{180}/K_p = 0.98$. The process has the same structure as (2.47) so the parameters of this model become $T_1 = T_2 = 0.05$, and $L_1 = 1$.

The controller parameters obtained from the MIGO and AMIGO tuning rules are presented in Table 7.4.

Figure 7.20 shows the responses of the system to changes in set point and load disturbances. The responses obtained from the MIGO and AMIGO designs are similar. It also shows that there are small differences between PI and PID control, which was expected since the process is delay dominant. □

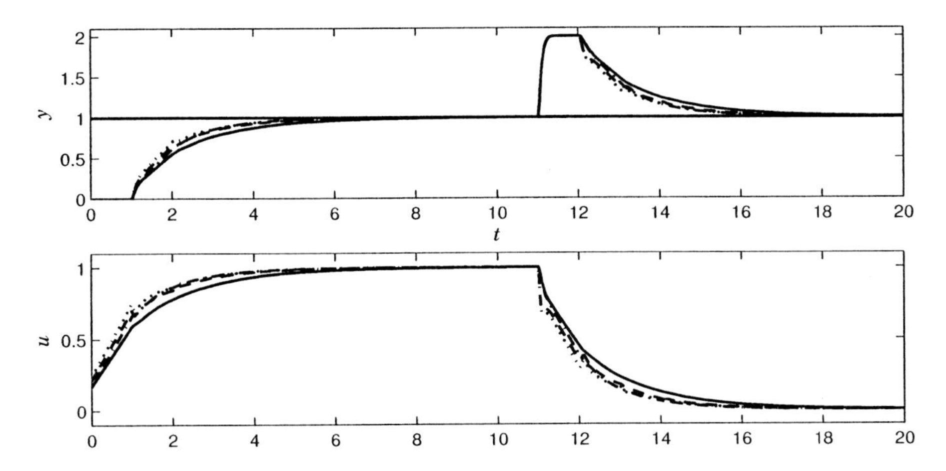

Figure 7.20 Responses to a unit step change at time 0 in set point and a unit load step at time 10 for PI controllers designed by MIGO (solid line) and AMIGO-step (dash-dotted line), and PID controllers designed by MIGO (dashed line) and AMIGO-step (dotted line) for the delay-dominant process in Example 7.3.

7.8 Measurement Noise and Filtering

So far we have focused on attenuation of load disturbances and robustness to process variations. In many cases it is also necessary to consider measurement noise. This is particularly the case for lag-dominated processes where maximization of integral gain gives controllers with high gain. Measurement noise can then create large control actions. In extreme cases the control signals can be so large that the actuator is saturated. The effect of measurement noise can be estimated from the transfer function from measurement noise to control signal:

$$G_{un} = -\frac{C}{1+PC}. \tag{7.16}$$

Since measurement noise typically has high frequencies, the high-frequency properties of the transfer function are particularly important.

The effect of measurement noise can be alleviated by filter the measurement signal as is shown in Figure 4.3. The transfer function from measurement noise to controller output is then

$$G_{un} = -\frac{CG_f}{1+PCG_f}. \tag{7.17}$$

A typical filter transfer function is given by

$$G_f(s) = \frac{1}{1+sT_f+(sT_f)^2/2}; \tag{7.18}$$

see (3.16). Adding a filter will reduce the robustness of the controller. It is easy to recover robustness by redesigning a controller for a process with the

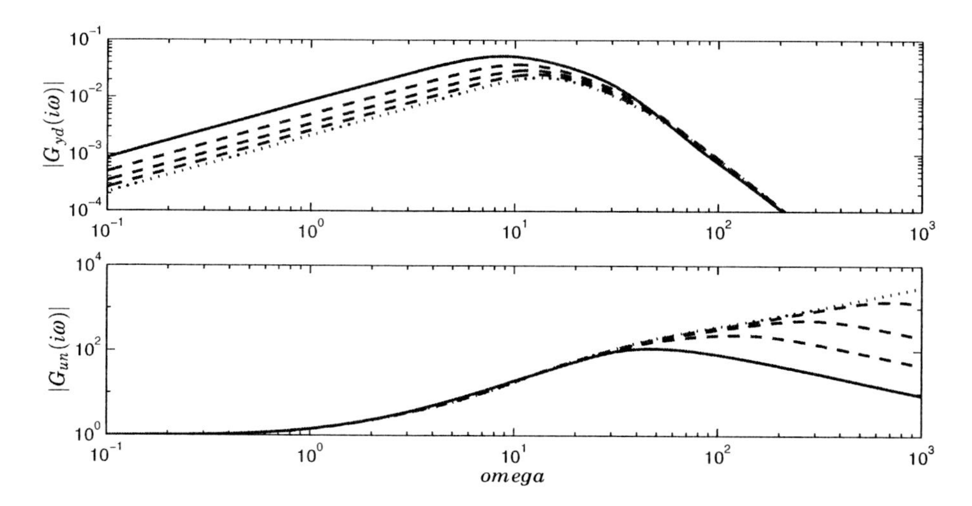

Figure 7.21 Gain curves of the transfer functions from load disturbance to process output (upper) and from measurement noise to controller output (lower). Curves for ideal PID control are shown in dotted lines, and for PID control with filtering with time constants $T_f = 0.002$, 0.005, and 0.010 in dashed lines, and for $T_f = 0.02$ in solid lines.

transfer function $P(s)G_f(s)$. The design procedure starts by designing an ideal PID controller for the process $P(s)$. The design gives guidance for choosing the filter time constant T_f; typically a fraction of the integral time for PI control or the derivative time for PID control. An ideal PID controller is then designed for the process $P(s)G_f(s)$, and the controller for the process $P(s)$ is then $C(s)G_f(s)$. If necessary, the procedure can be iterated a few times. Adding a filter improves attenuation of measurement noise at the cost of poorer load disturbance attenuation. The final design choice is thus a compromise. The procedure is illustrated by an example.

EXAMPLE 7.4—EFFECT OF FILTERING.
Consider the lag-dominated system in Example 7.1. Table 7.2 shows that the MIGO design gives a controller with high gain, $k = 56.9$,, which gives good attenuation of load disturbances with integral gain $k_i = 495$. The transfer function from measurement noise to controller output has high gain at high frequencies, as is shown by the Bode plot in Figure 7.21. The derivative time is $T_f = 0.06$ and reasonable filter time constants are in the range $T_f = 0.002$–0.020.

To design controllers for the system $P(s)G_f(s)$ we approximate the transfer function using Skogestad's half-rule. Starting with the SOTD model used in Example 7.1 we account for filtering by adding $T_f/2$ to the time constant T_2 and to the time delay L_1. The combination or process $P(s)$ and filter $G_f(s)$ is then represented by the SOTD model (2.47) with $T_1 = 0.980$, $T_2 = 0.108 + T_f/2$, and $L_1 = 0.010 + T_f/2$. Equation (7.13) then gives the controller parameters shown in Table 7.5. Controller gain decreases by a factor of 2 with increasing values of the filter constant, integral time increases by a factor of 2 and the derivative time increases by about 40 percent. Integral gain k_i decreases with

Table 7.5 Controller parameters obtained in Example 7.4. Compare with Example 7.1.

T_f	K	T_i	T_d	k_i	k_d/T_f	M_{un}
0.000	59.6	0.127	0.0523	468	∞	∞
0.002	52.6	0.138	0.0546	382	1436	1436.
0.005	44.7	0.153	0.0578	293	516	520
0.010	35.6	0.176	0.0624	203	222	234
0.020	25.1	0.220	0.0705	115	88.6	112

a factor of 4 and the largest high-frequency gain of G_{un} decreases with several orders of magnitude.

Table 7.5 also shows the largest gain, M_{un}, of the transfer function $G_{un}(s)$ and its estimate k_d/T_f given by (4.44). The simple estimate is remarkable accurate for small filter-time constants.

The properties of the different controllers are also illustrated in Figure 7.22 which shows the responses of the system to load disturbances and measurement noise for different controllers designed with different values of the filter-time constant T_f.

Notice that there are large variations in the control signal for $T_f = 0.002$ even if the noise in the process output is not too large. The reason for this is that the controller gain is quite large.

Figures 7.21 and 7.22 give a good illustration of the trade-off between attenuation of load disturbances and injection of measurement noise. The final trade-off is always subjective, but a moderate amount of filtering is always useful because the effect of measurement noise can be decreased significantly with only moderate increase of integral gain. In the particular case a value of T_f around 0.01 is a reasonable choice. □

7.9 Detuning

The AMIGO tuning rule lends itself naturally to detuning. For PI control, load disturbance rejection can be characterized by integral gain $k_i = K/T_i$. Amplification of measurement noise can be characterized by controller gain K. Since measurement noise typically has high frequencies the variation of the control signal generated by measurement noise is approximately $Kn(t)$ where $n(t)$ is the measurement noise.

Figure 7.23 shows the robustness domain of a PI controller for typical first-order processes. All gains in the white area satisfy the robustness condition that the combined sensitivities are less than $M = 1.4$. Any combination of controller parameter in that range is thus admissible from the point of view of robustness. Load disturbance attenuation is captured by the integral gain k_i. Assuming that load disturbances enter at the process input the transfer

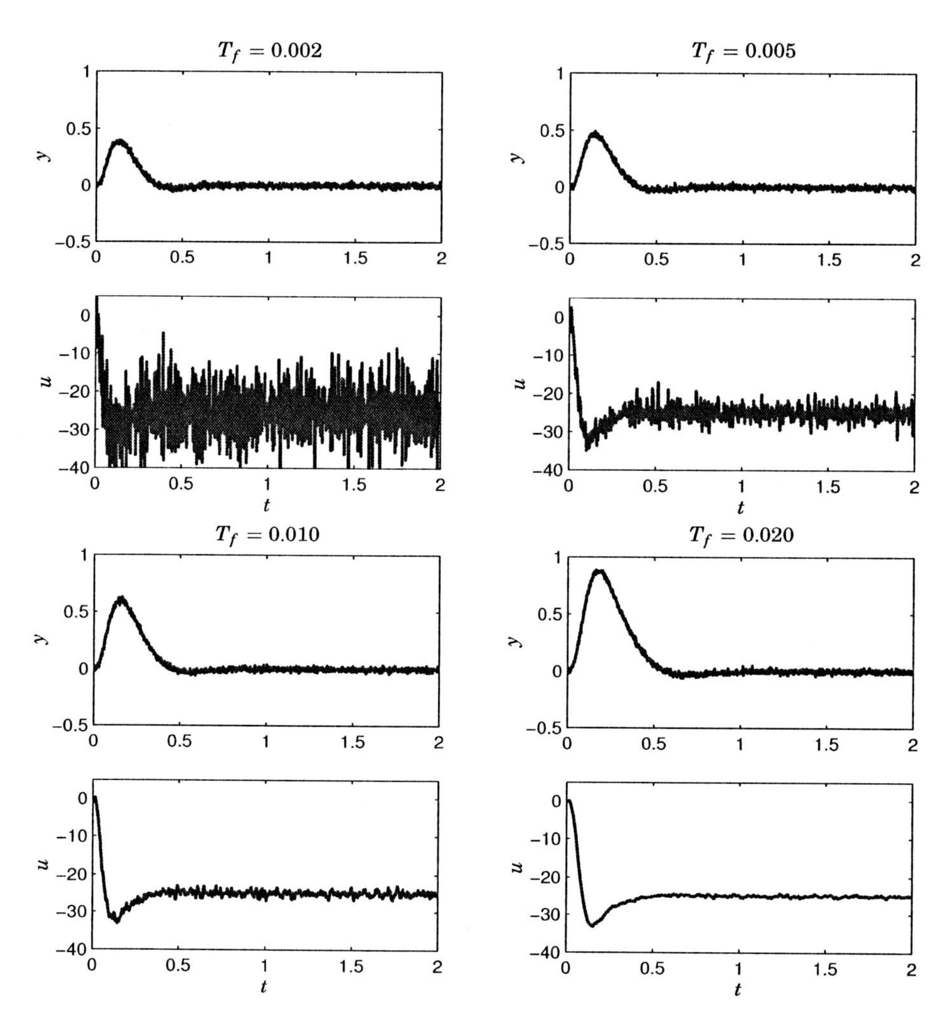

Figure 7.22 Simulation of PID control of the system in Example 7.4. The measurements are filtered with the second order filter (7.18) where time constants are $T_f = 0.002$, 0.005, 0.010, and 0.020. For each filter constant the controller parameters are chosen to maximize integral gain subject to the robustness constraint $M = 1.4$. A load disturbance of 25 is applied at time 0 and measurement noise is acting on the system.

function from load disturbances to process output is approximately given by

$$G(s) = \frac{s}{k_i}.$$

Load disturbance attenuation is thus inversely proportional to k_i. Measurement noise typically has high frequencies. For high frequencies the transfer function from measurement noise to the control signal is approximately given by

$$G(s) = K.$$

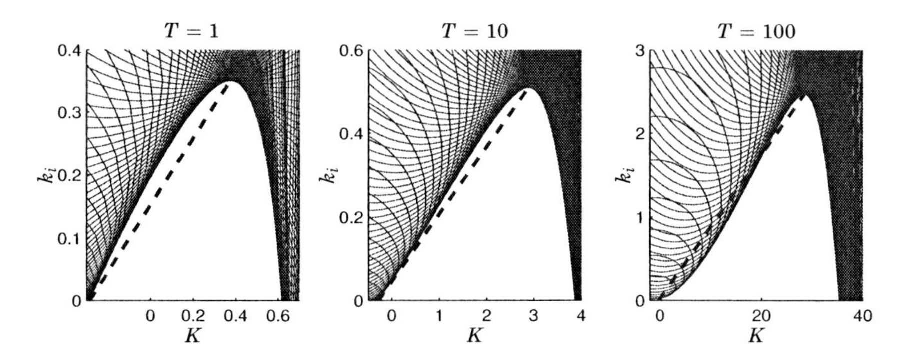

Figure 7.23 The sensitivity constraint for a system with $M = 1.4$ and the transfer function $P(s) = e^{-s}/(1 + sT)$, with $T = 1$, 10 and 100.

Injection of measurement noise is thus proportional to controller gain K. Since all values of K and k_i that satisfy the robustness requirement are given in Figure 7.23, it is straightforward to make a trade-off between load disturbance attenuation and injection of measurement noise.

Figure 7.23 indicates that variations of the control variable due to measurement noise can be reduced simply by reducing proportional gain. The penalty for this is poorer attenuation of load disturbances. A proper quantitative trade-off is easily done based on Figure 7.23. Instead of choosing the largest value of integral gain we should simply choose a combination of proportional and integral gain on the left border of the robustness region in the figure. Since the figure is not available when the simple tuning formulas are used we will develop an approximate formula for the left boundary of the robustness region.

A First Attempt

One possibility is to reduce the gains as indicated by the straight line in Figure 7.23. This line goes through the peak with parameters K^0 and k_i^0 obtained by the nominal design. When integral gain is zero the robustness boundary goes through the point

$$\begin{aligned} KK_p &= -1 + \frac{1}{M_s} = -\frac{M_s - 1}{M_s} = -\alpha \\ k_i &= 0. \end{aligned} \tag{7.19}$$

A line through this point and the extremum is

$$k_i = k_i^0 \frac{KK_p + \alpha}{K^0 K_p + \alpha}. \tag{7.20}$$

Notice that it is not useful to reduce proportional gain below the value $K = 0$, when the controller is reduced to a pure integral controller. Figure 7.23 shows that the formula (7.20) is conservative for $T = 1$ and $T = 10$ but not for $T = 100$ since the line will be partially outside the robustness boundary for this process. We will use the detuning formula (7.20) for processes with $\tau > 0.1$. For lag-dominant processes with $\tau < 0.1$, better approximations of the robustness boundary are required.

Figure 7.23 indicates that for $T = 100$ the lower left-hand side of the robustness boundary has the shape of a parabola. To obtain a better approximation of the left-hand boundary of the robustness region we will first consider a simple example where the robustness boundary can be computed explicitly. An integrator with delay is the extreme case of a lag-dominated process, but we will start by determining the robustness bound for an even simpler case.

A Pure Integrator

Consider a pure integrator

$$P(s) = \frac{1}{s}.$$

The loop transfer function with PI control is

$$G_l(s) = \frac{Ks + k_i}{s^2} = \frac{k_i}{s^2} + \frac{K}{s}.$$

Requiring that the loop transfer function is outside a circle with radius r and center at $-c$ gives

$$\left| c + G_l(i\omega) \right|^2 \geq r^2. \tag{7.21}$$

But

$$\begin{aligned} \left| c + G_l(i\omega) \right|^2 &= \left| c - \frac{k_i}{\omega^2} - i\frac{K}{\omega} \right|^2 = \left(c - \frac{k_i}{\omega^2} \right)^2 + \left(\frac{K}{\omega} \right)^2 \\ &= \frac{k_i^2}{\omega^4} + \frac{K^2 - 2ck_i}{\omega^2} + c^2 = \left(\frac{k_i}{\omega^2} + \frac{K^2 - 2ck_i}{2k_i} \right)^2 + c^2 - \left(\frac{K^2 - 2ck_i}{2k_i} \right)^2. \end{aligned}$$

The robustness condition can thus be written as

$$\left| c + G_l(i\omega) \right|^2 = \left(\frac{k_i}{\omega^2} + \frac{K^2 - 2ck_i}{2k_i} \right)^2 + c^2 - \left(\frac{K^2 - 2ck_i}{2k_i} \right)^2 \geq r^2.$$

The left-hand side has its smallest value for

$$\omega^2 = \frac{2k_i^2}{2ck_i - K^2},$$

where we require that $2ck_i \geq K^2$. The robustness condition thus imposes the following constraint between integral and proportional gain:

$$\left(\frac{2ck_i - K^2}{2k_i} \right)^2 \leq c^2 - r^2.$$

Equality is achieved for

$$\frac{2ck_i - K^2}{2k_i} = \sqrt{c^2 - r^2},$$

or

$$k_i = \frac{K^2}{2\left(c - \sqrt{c^2 - r^2}\right)} = \frac{K^2 (c + \sqrt{c^2 - r^2})}{2r^2}, \tag{7.22}$$

which is a parabola in the K, k_i plane.

Non-normalized Variables

So far we have used scaled variables. If we consider a process with the transfer function

$$P(s) = \frac{K_v}{s} = \frac{K_p}{sT},$$

the equation becomes

$$k_i = \frac{K_p K^2 (c + \sqrt{c^2 - r^2})}{2Tr^2}.$$

For a design based on a constraint on M_s we have $c = 1$ and $r = 1/M_s$; hence

$$\frac{c + \sqrt{c^2 - r^2}}{2r^2} = \frac{M_s\left(M_s + \sqrt{M_s^2 - 1}\right)}{2}.$$

For a design with equal constraints on both sensitivity and complementary sensitivity we have

$$r = \frac{2M - 1}{2M(M-1)}$$
$$c = \frac{2M^2 - 2M + 1}{2M(M-1)}.$$

This implies $c^2 - r^2 = 1$, and we get

$$\frac{c + \sqrt{c^2 - r^2}}{2r^2} = \frac{c+1}{2r^2} = M(M-1).$$

Summarizing, we find that the robustness constraint for a pure integrator becomes

$$k_i = \beta \frac{K_p K^2}{T}, \tag{7.23}$$

where

$$\beta = \begin{cases} M_s\left(M_s + \sqrt{M_s^2 - 1}\right)/2 & \text{for design based on } M_s \\ M(M-1) & \text{for design based on } M. \end{cases} \tag{7.24}$$

Equation 7.23 implies that integral gain is reduced by the factor n^2 when gain is reduced by the factor n. Since $T_i = K/k_i$ we find that the integral time increases with the factor n.

The detuning rule (7.23) is derived for an integrator without time delay. To deal with the process (7.3) we first observe from Figure 7.23 and Equation 7.19 that the parabola passes through the point $KK_p = -\alpha$ for $k_i = 0$. For the process (7.3) the detuning rule (7.23) should therefore be replaced by

$$k_i = \beta \frac{(\alpha + KK_p)^2}{K_p(L+T)}, \tag{7.25}$$

where the time constant T has been replaced by the effective time constant $T + L$.

Combining the Results

We have obtained two formulas for detuning. The formula (7.20) based on linear extrapolation gives good results for processes with $\tau > 0.1$, and processes with $\tau < 0.1$ as long as the gain reduction is moderate. The formula (7.25) gives good results for strongly lag-dominated processes with large gain reduction. It is then natural to combine the formulas. This will give a good match to the left part of the robustness constraint in Figure 7.23.

The formulas (7.20) and (7.25) give the same result for

$$k_i^0 \frac{\alpha + KK_p}{\alpha + K^0 K_p} = \beta \frac{(\alpha + KK_p)^2}{K_p(L+T)}$$

or

$$KK_p = \frac{k_i^0 K_p (L+T)}{\beta(\alpha + K^0 K_p)} - \alpha. \tag{7.26}$$

Summarizing we obtain the following formula for detuning the PI controller. First choose a gain $K < K_0$. Then determine the integral gain in the following way. For process with $\tau > 0.1$, determine k_i from (7.20). For processes with $\tau < 0.1$, compute integral gain from

$$k_i = \begin{cases} k_i^0 \dfrac{\alpha + KK_p}{\alpha + K^0 K_p} & \text{for } KK_p \geq \dfrac{k_i^0 K_p (L+T)}{\beta(\alpha + K^0 K_p)} - \alpha \\ \beta \dfrac{(\alpha + KK_p)^2}{K_p(L+T)} & \text{for } KK_p < \dfrac{k_i^0 K_p (L+T)}{\beta(\alpha + K^0 K_p)} - \alpha. \end{cases} \tag{7.27}$$

Notice that this equation is an approximation of the left-hand side of the robustness constraint in Figure 7.23.

Examples

The detuning rule (7.27) will be illustrated by some examples. First, we treat one single process with the structure (7.3).

EXAMPLE 7.5—DETUNING
A PI controller designed for the process

$$P(s) = \frac{1}{1 + 1000s} e^{-s} \tag{7.28}$$

using the AMIGO design (7.5) with the robustness constraint $M = 1.4$ has the controller parameters $K = 349$ and $T_i = 13.2$, which gives the integral gain $k_i = 26.4$. The process is almost an integrator with delay, $P(s) \approx 0.001e^{-s}/s$, with a normalized time delay $\tau \approx 0.001$. This explains the high gain in the controller. Figure 7.24 shows Nyquist curves of the loop transfer function, as well as the curves obtained when the gain is reduced by the factors 0.5, 0.1, 0.05, 0.01, and 0.005, respectively, using the detuning rule (7.27). The figure shows that the loop transfer functions of the detuned systems remain close to the robustness region.

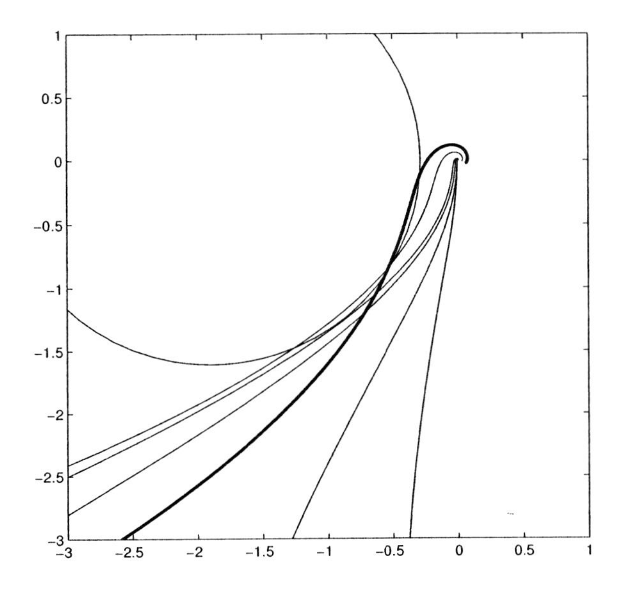

Figure 7.24 Nyquist curves for loop transfer functions for PI control of the process (7.28). The thick line corresponds to the optimal controller, and the thin lines to controllers where the gain is reduced by the factors 0.5, 0.1, 0.05, 0.01, and 0.005, respectively. The circle shows the robustness constraint $M = 1.4$.

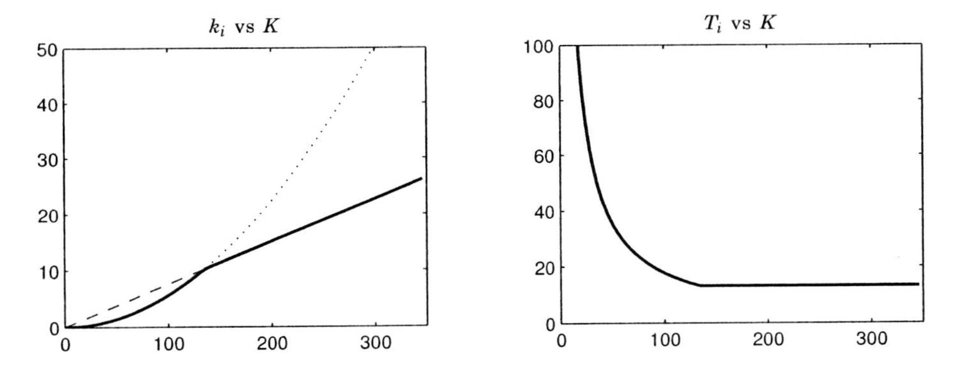

Figure 7.25 Relations between the reduced gain K and the integral gain k_i (left) and the integral time T_i (right). The dashed line corresponds to the detuning rule (7.20) and the dotted line to the rule (7.25)

Figure 7.25 shows how integral gain k_i and integral time T_i are changed when the gain is reduced. Notice that the integral time remains almost constant as long as the gain reduction is made according to the linear part of (7.27). The linear reduction is replaced by the quadratic reduction when the gain is lower than $K \approx 135$. The gain reduction at this point is $K/K^0 \approx 135/349 \approx 0.4$.
□

In the next example, the detuning rule (7.27) is applied to a large test batch of processes.

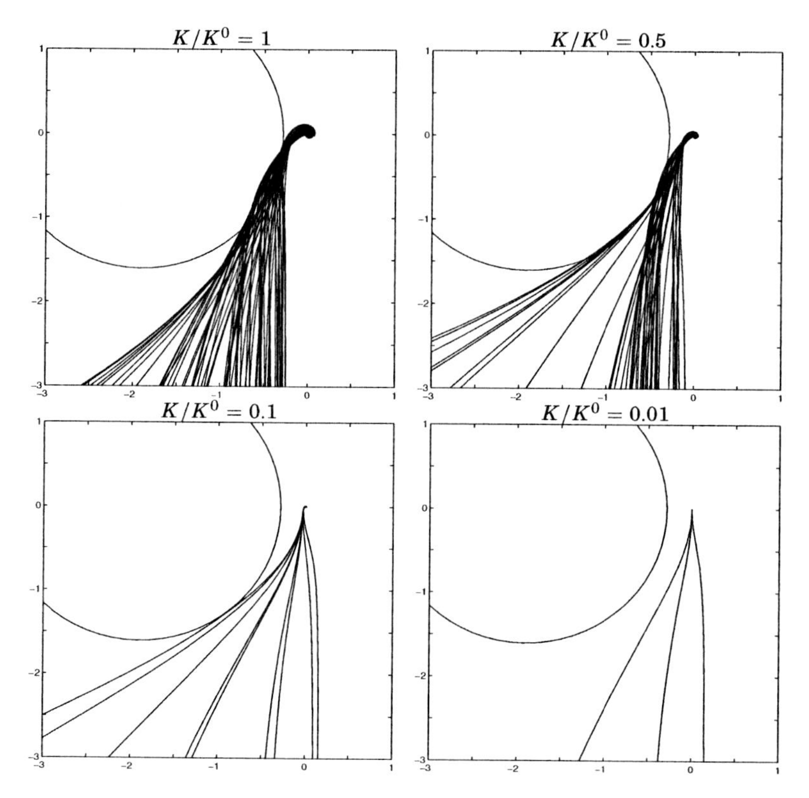

Figure 7.26 Nyquist curves of loop transfer functions where the controller is detuned using the rule (7.27). The circles show the robustness margin $M = 1.4$.

EXAMPLE 7.6—DETUNING APPLIED TO THE TEST BATCH
The detuning rule (7.27) has been applied to all processes in the test batch (7.2). Figure 7.26 shows the Nyquist plots of the loop transfer functions obtained when the PI controllers are detuned using (7.27). The figure shows four cases; the original loop and loops where the controller gain is reduced by factors 0.5, 0.1, and 0.01. Only those systems where the controller gain is larger than 0.5 are shown. This is the reason why only three cases are left when the controller gain is reduced by a factor of 0.01.

The example shows that the loop transfer functions remain close to the robustness region and that the detuning rule works well for the processes in the test batch. □

A Pole Placement Interpretation

There are situations when the response time and the bandwidth are of great importance. In this case the detuning problem can be solved using a simple pole placement approach. Neglecting the the time delay the loop transfer function

Table 7.6 Controller parameters, frequency and damping.

T	K^0	k_i^0	ω	ζ	ω^e	ζ^e
5	1.21	0.296	0.343	0.910	0.222	0.830
10	2.82	0.513	0.226	0.845	0.216	0.804
20	6.24	0.99	0.222	0.815	0.217	0.794
100	34	4.94	0.222	0.788	0.211	0.784

obtained when a PI controller is combined with the process model (7.3) becomes

$$G_l(s) = \frac{K_p K s + K_p k_i}{s(1+sT)}.$$

The characteristic polynomial is

$$s^2 + s\frac{1+K_pK}{T} + \frac{k_i K_p}{T}.$$

Comparing this with the standard polynomial $s^2 + 2\zeta\omega s + \omega^2$ we find

$$\begin{aligned} 1 + K_pK &= 2\zeta\omega T \\ K_p k_i &= \omega^2 T. \end{aligned} \tag{7.29}$$

Using the numerical values for the process (7.3) with $K_p = 1$ and $L = 1$ we get the values in Table 7.6 for different values of time constant T. The optimal controller parameters K^0 and k_i^0 are determined from the MIGO design with $M = 1.4$. The last two columns are the frequency and the damping when the time constant T is replaced by the effective time constant $T_e = T + L$. Notice that the frequency and the damping are practically constant for the whole range of parameters.

Another way to detune the controller is to use Equation 7.29 and reduce the natural frequency. This gives

$$\omega = \frac{1+KK_p}{2\zeta T_e}$$

and

$$k_i = \frac{(1+KK_p)^2}{4\zeta^2 K_p T_e}. \tag{7.30}$$

This is similar but of somewhat different form than the parabolic expression in (7.27).

PID Control

For PID control, it is natural to start a high-frequency gain reduction by reducing the derivative gain. One way to do this is the following. Let K^{PID}, k_i^{PID}, and k_d^{PID} denote the gains for PID controllers obtained by the AMIGO tuning formula, and let K^{PI} and k_i^{PI} be the corresponding controller gains for the PI controller. Following the ideas used in PI control we will obtain detuned controller by linear interpolation. This gives

$$\begin{aligned} K &= K^{PI} + \frac{k_d}{k_d^{PID}}(K^{PID} - K^{PI}) \\ k_i &= k_i^{PI} + \frac{k_d}{k_d^{PID}}(k_i^{PID} - k_i^{PI}). \end{aligned} \tag{7.31}$$

This gives a natural way to detune the PID controller until it becomes a PI controller. If further gain reductions are required we can proceed as for PI controllers.

EXAMPLE 7.7—DETUNING APPLIED TO THE TEST BATCH
The detuning rule (7.31) has been applied to the test batch (7.2). Figure 7.27 shows the Nyquist plots of the loop transfer functions obtained when the PID controllers are detuned using (7.31). The figure shows four cases, the original loop tuned with the AMIGO tuning rules (7.7) and loops where the derivative gain is reduced by factors 0.1, 0.01, and 0. The last case gives a pure PI controller.

The example shows that the loop transfer functions remain close to the robustness region and that the detuning rule works well for the processes in the test batch. □

7.10 Summary

In this section it has been attempted to develop simple tuning rules in the spirit of the work done by Ziegler and Nichols in the 1940s. The goal has been to make rules that can be used both for manual tuning and in auto-tuners for a wide range of processes. The methods were developed by applying the techniques for robust loop shaping presented in Section 6.8 to a large test batch of representative processes. The controller parameters obtained were then correlated with simple features of process dynamics.

One interesting observations was that there are significant differences between processes with delay-dominated and lag-dominated dynamics. To capture this difference, process dynamics must be characterized by at least three parameters. Notice that Ziegler and Nichols used only two parameters. One possible choice is: process gain K_p, apparent time constant T, and apparent time delay L. These parameters can be obtained from a step response experiment. Section 2.7. The relative time delay $\tau = L/(L+T)$, which ranges from 0 to 1, is used for a crude characterization of dynamics. Processes with small τ are called lag dominated, processes with τ close to one are called delay dominated and processes with τ around 0.5 are called balanced.

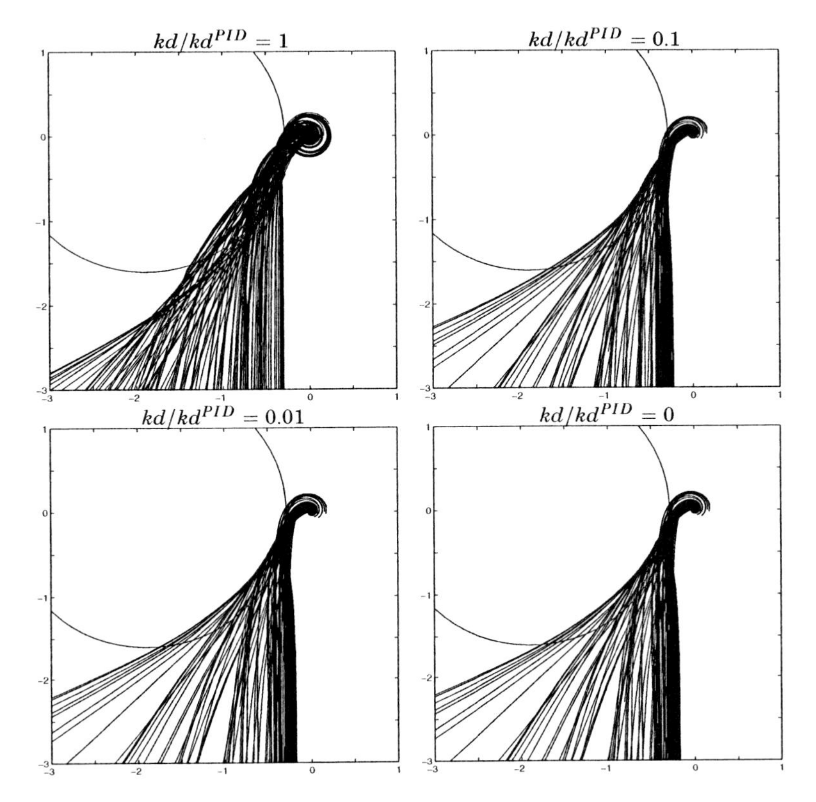

Figure 7.27 Nyquist curves of loop transfer functions where the controller is detuned using the rule (7.31). The circles show the robustness constraint $M = 1.4$.

Very satisfactory results were obtained for PI control, where the parameters from MIGO tuning can be matched with

$$
\begin{aligned}
K &= \frac{0.15}{K_p} + \left(0.35 - \frac{LT}{(L+T)^2}\right)\frac{T}{K_p L} \\
T_i &= 0.35L + \frac{13LT^2}{T^2 + 12LT + 7L^2},
\end{aligned}
\tag{7.32}
$$

for the full test batch. The tuning rule, which we called AMIGO (Approximate MIGO), gave good results for all processes in the test batch ranging from process with integration to processes with pure time delay.

The numerical values in (7.32) are based on a combined sensitivity $M = 1.4$. The form of the tuning rules are the same for other values of M but the numerical values of the coefficients are different.

For PID control of processes with $\tau > 0.3$ it was also possible to find the simple tuning rule

$$
\begin{aligned}
K &= \frac{1}{K_p}\left(0.2 + 0.45\frac{T}{L}\right)\\
T_i &= \frac{0.4L + 0.8T}{L + 0.1T}L\\
T_d &= \frac{0.5LT}{0.3L + T}.
\end{aligned}
\tag{7.33}
$$

This tuning rule also gave a conservative tuning rule for lag-dominant processes. It can thus be used for the full range of processes provided that a conservative tuning is acceptable. Derivative action can give substantial benefits for lag-dominated processes. A quantitative estimate can be obtained by comparing the integral gains k_i of of (7.32) and (7.33).

For some lag-dominated processes it is possible to give tuning rules with much better performance than (7.33). When process dynamics is characterized by the parameters K_p, L, and T both time delay and small time constants are captured in L. For lag-dominated processes improved performance can be obtained if the time constant and the delay are separated by better modeling. For a model characterized by an SOTD model with four parameters the AMIGO tuning rule is

$$
\begin{aligned}
K_pK &= \alpha_1 + \alpha_2\frac{T_1}{L_1} + \alpha_3\frac{T_2}{L_1} + \alpha_4\frac{T_1T_2}{L_1^2}\\
K_pk_i &= \beta_1\frac{1}{L_1} + \beta_2\frac{T_1}{L_1^2} + \beta_3\frac{T_2}{L_1^2} + \beta_4\frac{T_1T_2}{L_1^3}\\
K_pk_d &= \left(\gamma_1L_1 + \gamma_2T_1 + \gamma_3T_2 + \gamma_4\frac{T_1T_2}{L_1}\right)\frac{T_1+T_2}{T_1+T_2+L_1},
\end{aligned}
\tag{7.34}
$$

where the parameters are given by

$$
\begin{array}{llll}
\alpha_1 = 0.19 & \alpha_2 = 0.37 & \alpha_3 = 0.18 & \alpha_4 = 0.02\\
\beta_1 = 0.48 & \beta_2 = 0.03 & \beta_3 = -0.0007 & \beta_4 = 0.0012\\
\gamma_1 = 0.29 & \gamma_2 = 0.16 & \gamma_3 = 0.20 & \gamma_4 = 0.28.
\end{array}
\tag{7.35}
$$

This tuning rule is similar to (7.33) for processes with balanced and delay-dominated dynamics, but it typically gives higher gain for lag-dominated processes. This tuning rule requires improved process models. It is difficult to obtain two time constants from a step response experiment. System identification or the combined frequency and step response methods described in Section 2.7 can be used.

Tuning rules based on frequency response data have also been developed. In this case the parameters were chosen as: process gain K_p, ultimate gain K_{180}, and ultimate period T_{180}. The parameter $\kappa = K_{180}/K_p$ was used to classify the processes. This choice matches what is used in auto-tuners based on relay feedback. The AMIGO tuning rule for PI controllers based on frequency response data is

$$
\begin{aligned}
KK_{180} &= 0.16\\
\frac{T_i}{T_{180}} &= \frac{1}{1+4.5\kappa}
\end{aligned}
\tag{7.36}
$$

and the tuning rules for PID controllers are

$$K = (0.3 - 0.1\kappa^4)/K_{180}$$
$$T_i = \frac{0.6}{1+2\kappa}T_{180} \qquad (7.37)$$
$$T_d = \frac{0.15(1-\kappa)}{1-0.95\kappa}T_{180}.$$

These tuning rules give good tuning for balanced and delay-dominant processes with $\kappa > 0.2$, but are not appropriate for lag-dominant processes.

The AMIGO tuning rules optimize load disturbance attenuation with a specified robustness. Measurement noise can be dealt with by filtering the process output. There are significant advantages of using a second-order filter. The dynamics of the filter can be accounted for in a simple way by applying the AMIGO rules with $T_f/2$ added to T and L for the FOTD model and with $T_f/2$ added to T_2 and L for the SOTD model. In this way it is possible to make the trade-off between attenuation of load disturbances and injection of measurement noise.

A systematic method of detuning the controllers to give a specific controller gain has also been developed.

7.11 Notes and References

This chapter is based on work by the authors and their students. The motivation was to gain improved understanding in the information required to develop good tuning rules and to find tuning rules that can be used for manual and automatic tuning. The basis for the work is the robust design method (MIGO) which is developed in [Åström *et al.*, 1998] for PI control and in [Åström and Hägglund, 2001; Panagopoulos *et al.*, 2002] for PID control. In certain circumstances it is advantageous to have $T_i < 4T_d$ as has been noted by [Kristiansson and Lennartsson, 2002]. Tuning rules for that case are given by [Wallén *et al.*, 2002]. The MIGO method requires knowledge of the transfer function of the process. The AMIGO tuning rules for PI and PID presented in [Hägglund and Åström, 2002; Hägglund and Åström, 2004b; Hägglund and Åström, 2004a] can be applied when only features of step and frequency responses are known. Much of the material in the chapter have not been published before.

8

Predictive Control

8.1 Introduction

A PI controller only considers present and past data, and a PID controller also predicts the future process behavior by linear extrapolation. There have been many attempts to find other ways of predicting future process behavior and to take this into account when making the control actions. Good predictions can improve controller performance, particularly when the process has time delays, which are common in process control. Time delays can arise from a pure delay mechanism caused by transport or time for computation and communication. Delays may also be caused by measurements obtained by off-line analysis. They also appear when a high-order system or a partial differential equation is approximated with a low-order model as in heat conduction. Time delays appear in many of the models discussed in this book. A new controller that could deal with processes having long time delays was proposed by Smith in 1957. The controller is now commonly known as the Smith predictor. It can be viewed as a new type of controller but it can also be interpreted as an augmentation of a PID controller. There are also many other controllers that have predictive abilities. The model predictive controller is a large class of controller that is becoming increasingly popular.

In this chapter we start by presenting the Smith predictor in Section 8.2. This controller can give significant improvements in the response to set-point changes, but the Smith predictor can also be very sensitive to model uncertainties. This is shown in Section 8.3 where we analyze the closed-loop system when a Smith predictor is used. The analysis also shows that the concepts of gain and phase margin are not sufficient to characterize the robustness of the system. The reason for this is that the Nyquist curve of the loop transfer function can have large loops at frequencies larger than the gain crossover frequency. The robustness is well captured by the properties of the *Gang of Four*, and there is also another classical robustness measure, the delay margin, that gives good insight. A special type of the Smith predictor called the PPI controller is discussed in Section 8.4. This controller is simpler and more robust. Model predictive control, a more general form of prediction that is gaining in popularity, is discussed in Section 8.6.

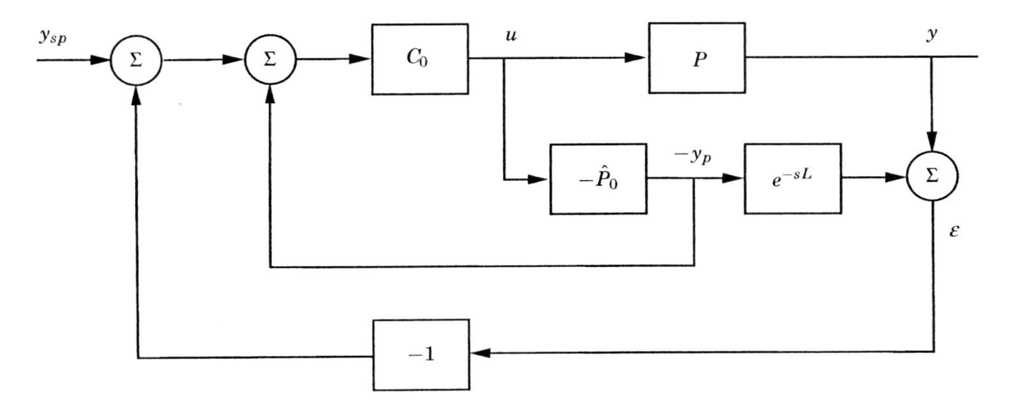

Figure 8.1 Block diagram of a system with a Smith predictor.

8.2 The Smith Predictor

To describe the idea of a Smith predictor we consider a process with a time delay L, and we factor the process transfer function as

$$P(s) = P_0(s)e^{-sL}, \tag{8.1}$$

where the transfer function P_0 does not have any time delays. Figure 8.1 shows a block diagram of a closed-loop system with a Smith predictor. The controller consists of an ordinary PI or PID controller C_0 and a model of the process $\hat{P}$, factored in the same way as the process, connected in parallel with the process. If the model is identical with the process the signal y_p represents the output without the delay or, equivalently, a prediction of what the output would be if there were no delays. By using the model it is thus possible to generate a prediction of the output. The signal y_p is fed back to the controller, and there is also an additional feedback from the process output y to cope with load disturbances. If the model $\hat{P}$ is identical to the process P and if there are no disturbances acting on the process the signal ε is zero. This means that the outer feedback loop gives no contribution, and the input-output relation of the system is given by

$$G_{yy_{sp}} = \frac{PC_0}{1 + P_0C_0} = \frac{P_0C_0}{1 + P_0C_0}e^{-sL}. \tag{8.2}$$

The controller C_0 can thus be designed as if the process has no time delay, and the response of the closed-loop system will simply have an additional time delay.

The system shown in Figure 8.1 can also be represented by the block diagram in Figure 8.2, which is an ordinary feedback loop with a process P and a controller C, where the controller has the transfer function

$$C = \frac{C_0}{1 + C_0(\hat{P}_0 - \hat{P})} = \frac{C_0}{1 + C_0\hat{P}_0(1 - e^{-sL})}. \tag{8.3}$$

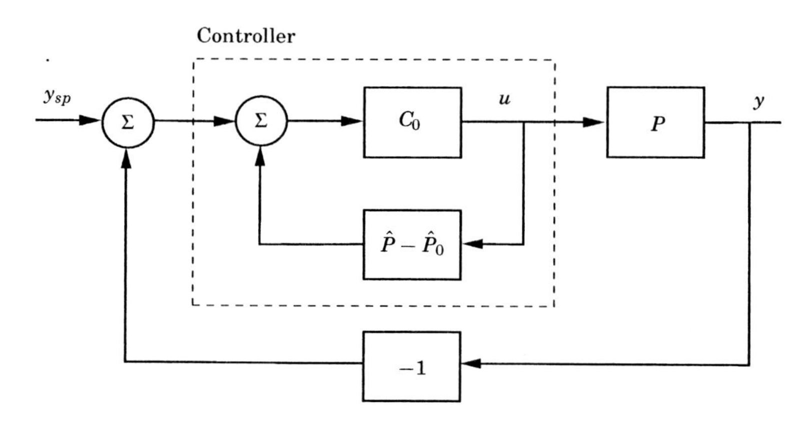

Figure 8.2 Another representation of a system with a Smith predictor.

The transfer function $\hat{P}_0 e^{-sL}$ is the transfer function of the process model used to design the controller. The controller C is thus obtained by wrapping a feedback around the controller C_0. The input-output relation of the controller C can be written as

$$U(s) = C_0(s)\big(E(s) - \hat{P}_0(s)(1 - e^{-sL})U(s)\big), \tag{8.4}$$

where $U(s)$ and $E(s)$ are the Laplace transforms of the control signal and the error. The term $\hat{P}_0(s)(1 - e^{-sL})U(s)$ can be interpreted physically as the predicted effect on the output of control signals in the interval $(t - L, t)$. The Smith predictor can thus be interpreted as an ordinary PI controller where the effects of past control actions are subtracted from the error. The controller can be compared with a PID controller, which predicts by extrapolating the current process output linearly, as is illustrated in Figure 3.5. This type of prediction is less effective for systems with time delays because future process outputs are strongly influenced by past control control actions rather than current inputs.

The properties of the Smith predictor will be illustrated by an example.

EXAMPLE 8.1—FIRST-ORDER SYSTEM WITH TIME DELAY
Consider a process with transfer function

$$P(s) = \frac{K_p}{1 + sT} e^{-sL}. \tag{8.5}$$

A PI controller that gives the characteristic polynomial

$$s^2 + 2\zeta\omega_0 s + \omega_0^2$$

for the process without delay is designed as described in Section 6.4. The controller is

$$C_0(s) = K\Big(1 + \frac{1}{sT_i}\Big),$$

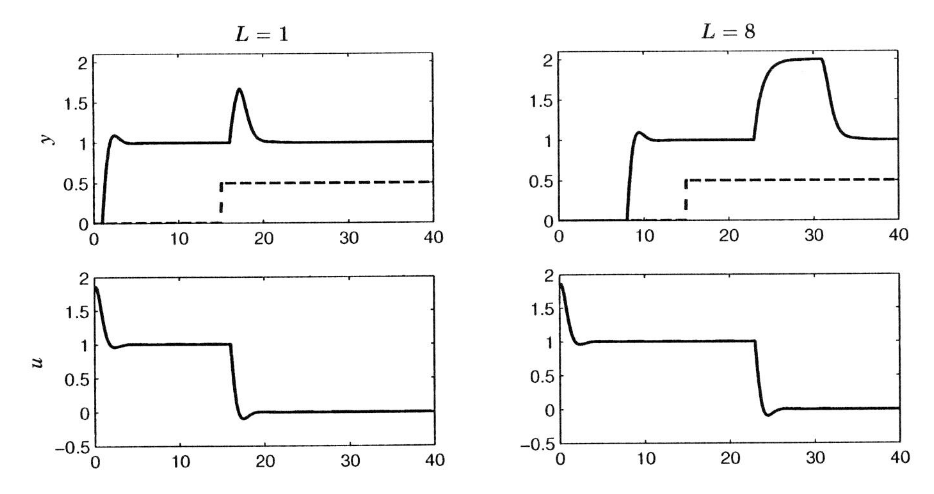

Figure 8.3 Responses of a closed-loop system with Smith predictor. The process has the transfer function $P(s) = e^{-sL}/(s+1)$, and the figure shows response for $L = 1$ and 8. The dashed line is the load disturbance.

where

$$\begin{aligned} K &= \frac{2\zeta\omega_0 T - 1}{K_p} \\ T_i &= \frac{K_p K}{\omega_0^2 T}. \end{aligned} \tag{8.6}$$

Figure 8.3 shows the responses of the system to a unit step change in the set point and a load disturbance in the form of a unit step in the process input. The load disturbance is applied at time $t = 15$ in all cases. The time constant is equal to one in all cases, and the time delay L is changed. The PI controller is designed to give a closed-loop system with $\omega_0 = 2$ and $\zeta = 0.7$ for the process without delays. The figure shows that the responses to set point have the same shape but with a delay that changes with the process delay. The shape is the same as for a system without the time delay. This property of the system is quite remarkable.

The shapes of the responses to load disturbances change with the time delay L. With increasing time delay it will take a longer time for the system to react. The initial part of the responses are similar but with different delays. Because of the varying delay the time to recover from the disturbance varies with the time delay. □

Analyzing the results it may appear remarkable that it is possible to obtain such good responses even when the time delay is as long as $L = 8$. In the following we will analyze the systems obtained when using the Smith's predictor to better understand its behavior.

The Predictor

It follows from (8.3) that the Smith predictor can be viewed as the cascade connection of an ordinary controller C_0 and a block with the transfer function

$$C_{pred} = \frac{1}{1 + C_0(\hat{P}_0 - \hat{P})} = \frac{1}{1 + C_0\hat{P}_0(1 - e^{-sL})}. \tag{8.7}$$

To obtain the responses shown in Figure 8.3 the transfer function C_{pred} compensates for the time delay of the process. Intuitively this can be understood in the following way. Assume $C_0\hat{P}_0 \approx -1$; it then follows from (8.7) that

$$C_{pred} \approx e^{sL}.$$

This means that the transfer function $C_{pred}(s)$ acts like an ideal predictor. We can therefore expect that the transfer function C_{pred} behaves like an ideal predictor for frequencies where $C_0(i\omega)\hat{P}_0(i\omega)$ is close to -1. Notice that it is not possible to have $C_0(i\omega)\hat{P}_0(i\omega) = -1$ for any frequency because the transfer function (8.2) is then unstable. The properties of the transfer function (8.7) will be illustrated by an example.

EXAMPLE 8.2—PREDICTOR FOR FIRST-ORDER SYSTEM WITH TIME DELAY
Consider the same system as in Example 8.1. Assuming that there are no modeling errors it follows that $\hat{P} = P = P_0e^{-sL}$. Combined with a PI controller the predictor becomes

$$C_{pred} = \frac{1}{1 + C_0(P_0 - P)} = \frac{1}{1 + \dfrac{K_pK(1 + sT_i)}{sT_i(1 + sT)}(1 - e^{-sL})}. \tag{8.8}$$

It follows from (8.8) that $C_{pred}(i\omega) = 1$ for $\omega L = 2\pi, 4\pi, 6\pi, \ldots$ and that $C_{pred}(s)$ goes to 1 for large s. The transfer function C_{pred} has the series expansion

$$C_{pred}(s) = \frac{T_i}{T_i + K_pKL}\Big(1 + \frac{K_pKL}{T_i + K_pKL}(T + \frac{L}{2} - T_i)s + \ldots\Big).$$

The static gain of C_{pred} decreases with increasing L and is always less than one. Figure 8.4 shows the Bode plot for the transfer function for $L = 8$. The figure shows that the transfer function gives a very large phase advance, more than 800°. A comparison with the phase curve of an ideal predictor shows that the system does approximate an ideal predictor well for certain frequencies. The solid and dashed curves are very close for those frequencies where the gain curve has peaks. Notice, however, that the gain curves are different. The ideal predictor has constant gain, but the gain of the transfer function C_{pred} changes with several orders of magnitude.

We will now investigate how the large phase advance is created. Figure 8.5 shows Nyquist curves of the transfer function C_{pred} for $K_p = 1$, $T = 1$, $K = 1.8$, $T_i = 0.45$, and $L = 1$, 2.5, 4, and 8. For $L = 1$ the largest phase advance is close to 90°. The phase advance increases with increasing L, as is indicated in the curve for $L = 2.5$ where the circular part of the Nyquist curve increases. The Nyquist curve goes to infinity for $L = 2.99$, which indicates that the

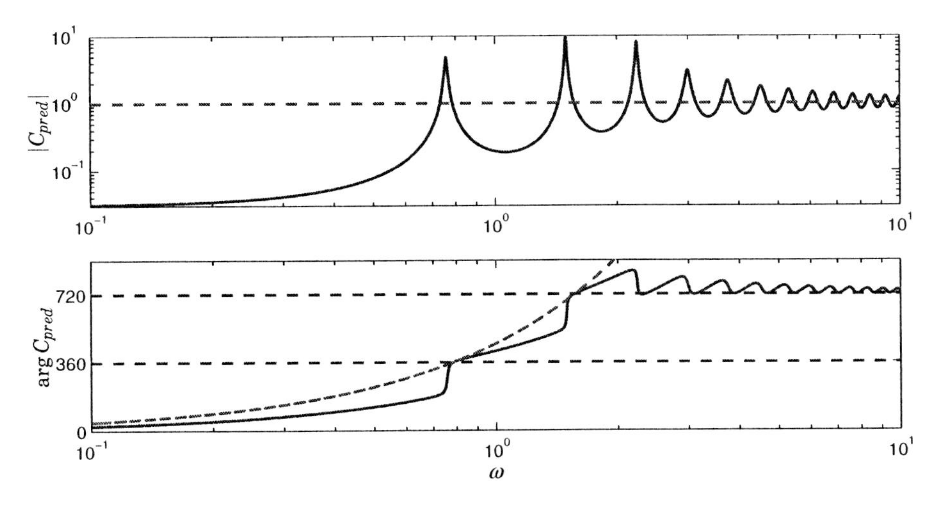

Figure 8.4 Bode plot of the loop transfer functions $C_{pred}(s)$ given by (8.8) for $L = 8$ (solid) and for the ideal predictor e^{sL} (dashed).

transfer function has poles on the imaginary axis. For larger L the Nyquist curve encircles the origin, which means that the phase advance is more than 360°. The curve for $L = 4$ shows that the largest phase advance is more than 450°. As L is increased further the Nyquist curve again goes to infinity for $L = 6.40$, and for larger L there are two encirclements of the origin, indicating that the phase advance is more than 720°. The curve for $L = 8$ shows that the largest phase advance is more than 800°.

To deform the curve for $L = 2.5$ continuously to the curve for $L = 4$ in Figure 8.5 the curve must go to infinity for some intermediate value of L. In the particular case the Nyquist curve of C_{pred} goes to infinity for $L = 2.99$, 6.40, 9.80, 13.40, 17,00, 20.6, ... This means that the transfer function C_{pred} is unstable for some values of L. It has two poles in the right-half plane for $2.99 < L < 6.40$, four poles in the right half plane for $6.40 < L < 9.80$, etc. For the simulation with $L = 10$ in Figure 8.3 the predictor transfer function has six poles in the right half plane. The predictor (8.3) thus achieves very large phase advances through poles in the right half plane. □

There are severe drawbacks with unstable controllers. It follows from Bode's integral (4.28) that poles in the right half plane increase the sensitivity. The remarkable response to set-point changes shown in Figure 8.3 thus comes at a price. Some of these issues will be discussed in the next section.

8.3 Analysis of Smith Predictor Control

The closed-loop system obtained when a process is controlled using a Smith predictor will now be investigated. Let the process transfer function be P, the

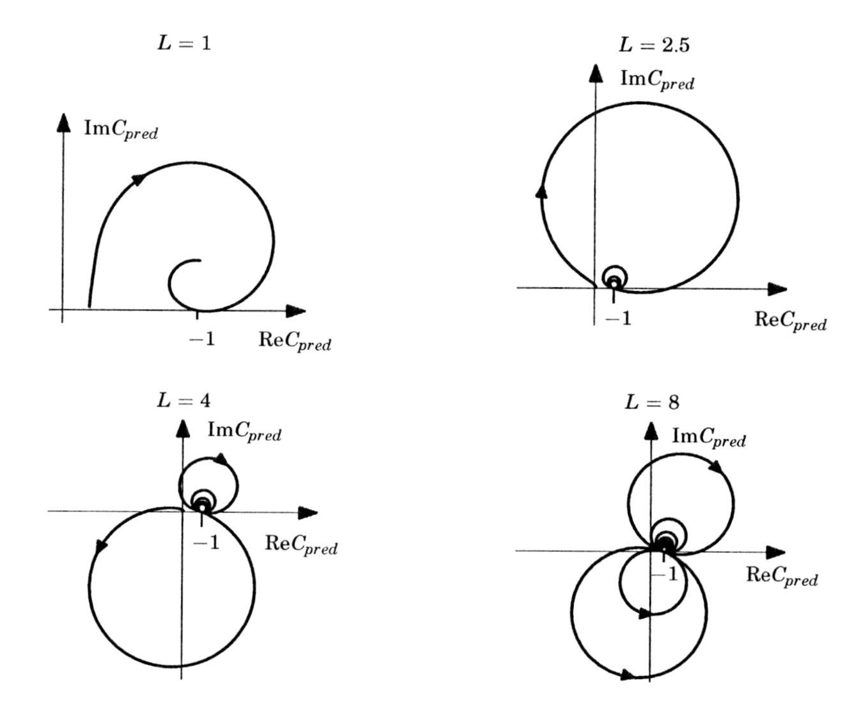

Figure 8.5 Nyquist plots of the transfer function C_{pred} for the system in Example 8.2 with $L = 1$, 2.5, 4, and 8. The plot for $L = 1$ and 2.5 all circles clockwise, the plot for $L = 4$ first makes one counterclockwise loop before making the clockwise loops, the plot for $L = 8$ first makes tow counterclockwise the remaining loops are clockwise.

transfer function of the Smith predictor (8.3), and we find

$$
\begin{aligned}
G_{yy_{sp}} &= \frac{PC}{1+PC} = \frac{PC_0}{1+\hat{P}_0C_0+(P-\hat{P})C_0} = \frac{P_0C_0}{1+P_0C_0}e^{-sL}\\
G_{yd} &= \frac{P}{1+PC} = \frac{P(1+(\hat{P}_0-\hat{P})C_0)}{1+\hat{P}_0C_0+(P-\hat{P})C_0} = P\Big(1-\frac{P_0C_0}{1+P_0C_0}e^{-sL}\Big)\\
-G_{un} &= \frac{C}{1+PC} = \frac{C_0}{1+\hat{P}_0C_0+(P-\hat{P})C_0} = \frac{C_0}{1+P_0C_0}\\
-G_{yn} &= \frac{1}{1+PC} = \frac{1+(\hat{P}_0-P)C_0}{1+\hat{P}_0C_0+(P-\hat{P})C_0} = 1-\frac{P_0C_0}{1+P_0C_0}e^{-sL},
\end{aligned}
\tag{8.9}
$$

where the last equality is obtained by assuming that the model is perfect, i.e., $\hat{P} = P$. The form of the transfer function from set point to process output $G_{yy_{sp}}$ shows that apart from the time delay the set-point responses are the same as for the system without time delays. The transfer G_{un} from measurement noise to the control signal is the same as the transfer function from set point to controller output. This transfer function is the same as for a system without a delay.

Stability

It follows from (8.9) that the closed-loop system has poles at the open-loop process poles and at the zeros of of the function

$$1 + \hat{P}_0 C_0 + (P - \hat{P}) C_0 \approx 1 + \hat{P}_0 C_0,$$

where the approximation is valid when $\hat{P} \approx P$. The zeros of this function can be chosen to be stable by a proper controller C_0. To have a stable closed-loop system it must also be required that the process be stable. This means that the Smith predictor does not work for processes with unstable open-loop dynamics. Modifications to eliminate this difficulty will be given in Section 8.5.

Response to Load Disturbances

When modeling errors are neglected the response to a load disturbance at the process input is given by the transfer function

$$G_{yd} = P\left(1 - \frac{P_0 C_0}{1 + P_0 C_0} e^{-sL}\right);$$

see (8.9). The second term has a time delay L. If a disturbance occurs at time 0 it follows that the response in the interval $0 \leq t < L$ is the same as the response of the open-loop system. A typical illustration is given in Figure 8.3.

Assume that the process P is stable with static gain K_p and that controller C_0 has integral action with integral gain k_i. A series expansion of G_{yd} for small s gives

$$G_{yd}(s) \approx K_p \left(1 - \frac{K_p k_i}{s + K_p k_i}(1 - Ls)\right) = K_p \frac{s + K_p k_i L s}{s + K_p k_i} \approx \left(K_p L + \frac{1}{k_i}\right) s. \tag{8.10}$$

Since $G_{yd}(0) = 0$ there is no steady-state error for a step change in the load disturbance. Furthermore, the integrated error for a load disturbance in the form of a unit step is

$$IE = K_p L + \frac{1}{k_i}. \tag{8.11}$$

Notice that the first term $K_p L$ only depends on the process and that the second term $1/k_i$ only depends on the controller.

The transfer function P has a pole at the origin for processes that have integral action. For such processes and a controller with integral action we have $P(s) \approx K_v/s$ and $C(s) \approx k_i/s$ for small values of s. This implies that

$$G_{yd}(s) \approx \frac{K_v}{s}\left(1 - \frac{K_v k_i}{s^2 + K_v k_i}(1 - Ls)\right) = \frac{K_v}{s} \frac{s^2 + K_v k_i L s}{s^2 + K_v k_i} \approx K_v L. \tag{8.12}$$

This means that there will be a steady-state error for processes with integration even if the controller has integral action. The recovery from load disturbances will therefore be very slow for processes with slow dynamics. Notice that the closed-loop system is stable even though P contains an integrator. The reason is that the integrator of P is canceled with a zero of the transfer function $1 - PC_0/(1 + P_0 C_0)$. Several modifications of the Smith predictor have been proposed for processes with integration. This will be discussed in Section 8.5.

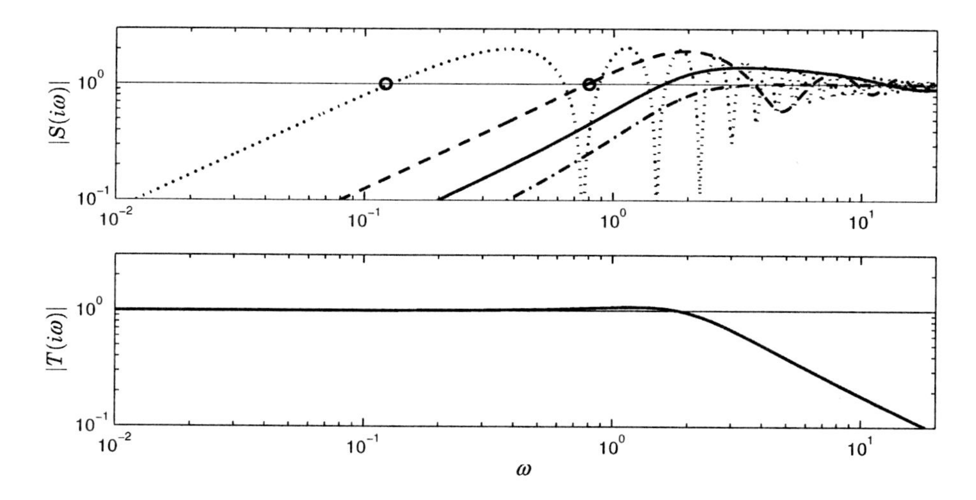

Figure 8.6 Gain curves for the sensitivity functions for the system in Example 8.3 with $L = 0$ (dash-dotted), 0.25 (solid), 1 (dashed), and 8 (dotted).

The Sensitivity Functions

In the ideal case $\hat{P} = P$, it follows from (8.9) that the sensitivity and the complementary sensitivity functions are

$$\begin{aligned} S &= 1 - \frac{PC_0}{1 + P_0C_0} = 1 - \frac{P_0C_0}{1 + P_0C_0}e^{-sL} = 1 - T_0e^{-sL} \\ T &= \frac{PC_0}{1 + P_0C_0} = \frac{P_0C_0}{1 + P_0C_0}e^{-sL} = T_0e^{-sL}, \end{aligned} \tag{8.13}$$

where T_0 is the complementary sensitivity function for the nominal system without delay. Notice that the gain curves of T and T_0 are identical. The gain curve of the complementary sensitivity function is independent of L.

EXAMPLE 8.3—SYSTEM OF FIRST ORDER WITH TIME DELAY

For the first-order system in Example 8.1 where the controller C_0 was designed to give $\omega_0 = 2$. The sensitivity functions are

$$\begin{aligned} T(s) &= \frac{K_pK(1 + sT_i)}{sT_i(1 + sT) + K_pK(1 + sT_i)}e^{-sL} = \frac{s\omega_0^2T/(2\zeta\omega_0T - 1) + \omega_0^2}{s^2 + 2\zeta\omega_0 s + \omega_0^2}e^{-sL} \\ S(s) &= 1 - T(s). \end{aligned}$$

Figure 8.6 shows the gain curves of the sensitivity functions for $L = 0$, 0.25, 1, and 8, which corresponds to $\omega_0 L = 0$, 0.5, 2, and 16. The largest sensitivity increases rapidly with L; we have $M_s = 1.1$, 1.4 1.6, and 2 for $L = 0$, 0.24, 0.4, and 1.2, respectively. For $L > 1.2$ the maximum sensitivity remains close to $M_s = 2$. Also notice that the sensitivity for low frequencies increases rapidly with increasing L.

The differences between the low-frequency properties of the sensitivity functions in Figure 8.6 are easily explained from (8.10). The low-frequency asymptote of the gain curves of the sensitivity function intersects the unit magnitude line for $\omega = k_i/(1 + k_i K_p L)$. For the system in the figure we have $K_p = 1$ and $k_i = 4$, and the intersections are denoted by circles in Figure 8.6. □

The sensitivity functions shown in Figure 8.6 are typical for systems with Smith predictors. The complementary sensitivity function is close to one for frequencies up to the bandwidth ω_b of the nominal system without time delay. The sensitivity function has the typical oscillatory behavior shown in the figure. It intersects the line $|S| = 1$ several times. For large delays the sensitivity crossover frequency is approximately $\omega_{sc} = k_i/(1 + k_i K_p L)$, reflecting the fact that the attenuation of load disturbances is poor for large L. Also notice that the largest peaks of the sensitivity function are close to $M_s = 2$ in the frequency range where $|T(i\omega)| \approx 1$.

Robustness

For controllers with integral action we have $T(0) = 1$. Let ω_b be a frequency such that $|T(i\omega)|$ is close to 1 for $0 \leq \omega \leq \omega_b$. If $\omega_b L \geq \pi$ it then follows from (8.13) that the maximum sensitivity is around $M_s = 2$. In order to have smaller sensitivities it is therefore necessary to require that $\omega_b L$ is not too large. It follows from (4.32) that it is possible to have perturbations in the process such that

$$\frac{|\Delta P(i\omega)|}{|P(i\omega)|} < \frac{1}{|T(i\omega)|}$$

without making the system unstable. For frequencies less than ω_b the right-hand side is equal to one. The inequality then implies that the uncertainty region is a circle with center at $P(i\omega)$ that passes through the origin. If we only consider variations in the phase admissible variations are therefore 60° or $\pi/3$ rad. Since the phase change is ωL we find

$$|\omega_b \Delta L| < \frac{\pi}{3},$$

which gives the following estimate of permissible variations in the time delay

$$\frac{|\Delta L|}{L} < \frac{\pi}{3\omega_b L} \approx \frac{1}{\omega_b L}. \tag{8.14}$$

Controllers with large values of $\omega_b L$ thus require that the time delay be known accurately. Consider, for example, the system in Figure 8.3 with $L = 8$. In this case we have $\omega_b L = 16$, which implies that the permissible error in the time delay is at most 6 percent.

The Loop Transfer Function

Analysis of the sensitivity functions indicates that the robustness of a closed-loop system with a Smith predictor may be poor when $\omega_b L$ is large. An analysis of the loop transfer function gives additional insight.

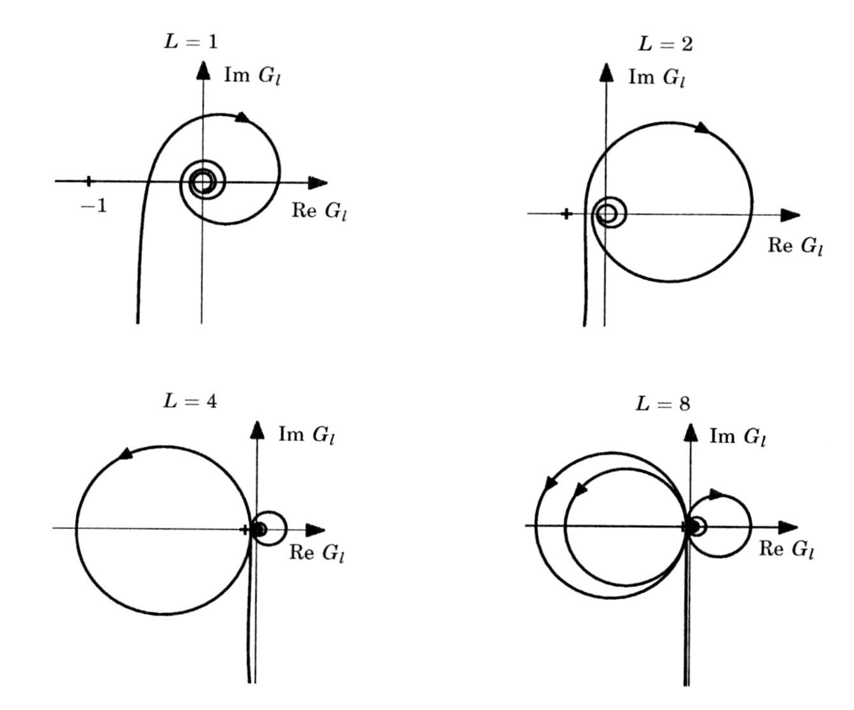

Figure 8.7 Nyquist plots of the loop transfer function for a FODT system, (8.5), with a Smith predictor controller. The critical point -1 is marked with a $+$.

When there are no modeling errors the loop transfer function obtained using a Smith predictor is

$$PC = \frac{PC_0}{1 + C_0(P_0 - P)} = \frac{P_0 C_0 e^{-sL}}{1 + P_0 C_0 (1 - e^{-sL})}. \tag{8.15}$$

Figure 8.7 shows the Nyquist plots of the loop transfer function for different values of L. For $L = 1$ the Nyquist plot has a loop of moderate size. The loop increases with increasing L, as is seen by comparing the cases $L = 1$ and $L = 2$ in Figure 8.7. The loop is almost circular for L larger than 2. For $L = 2.99$ the loop is infinitely large, and for $2.99 < L < 6.40$ the loop transfer function has two encirclements of the critical point, one for positive and another for negative ω. Notice that we have only shown the branch of the Nyquist plot corresponding to $0 \leq \omega < \infty$. The unstable poles are the poles of the predictor transfer function (8.7). The number of encirclements increases as L increases. For $L = 8$ there are four encirclements of the critical point.

Figure 8.8 shows the Bode plots of the loop transfer function for the cases $L = 1$ and $L = 8$. The loop transfer functions change drastically with L. The gain crossover frequency is 0.82 for $L = 1$ and decreases to about 0.13 for $L = 8$. These values agree quite well with the performance limit $\omega_{gc} L \approx 1$ given by (4.57). Notice that the gain curve for $L = 8$ has several crossings at higher frequencies. The gain crossover frequency is smaller for $L = 8$ even if the rise

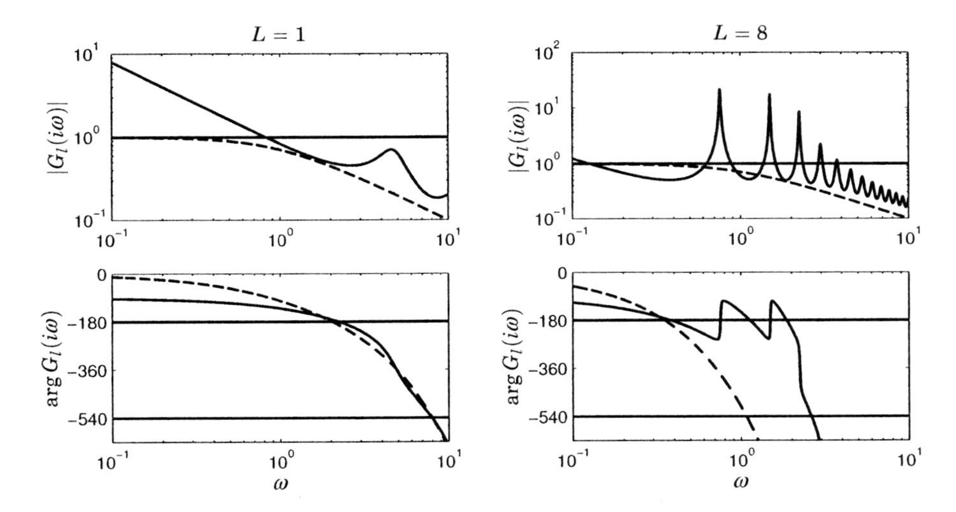

Figure 8.8 Bode plots of the loop transfer function (solid) and the process transfer function (dashed) for a FOTD system (8.5) with a Smith predictor. The curves on the left are for $L = 1$, and those on the right for $L = 8$.

time for set-point changes are the same for both systems. The high peaks of the gain curve correspond to the loops in the Nyquist plot in Figure 8.7.

The Bode plot of the open-loop system is shown in dashed lines in Figure 8.8. Notice that the controller gives a large phase advance at the frequencies corresponding to the first two peaks, which represent the unstable poles of the controller.

The Delay Margin

The classical robustness measures, gain margin and phase margin, do not capture the properties of Nyquist curves of the type shown in Figure 8.7, where the Nyquist curve has large loops. This is illustrated in Figure 8.9, which shows Nyquist plots of the loop transfer function for the case $L = 2$ and for a system where the time delay of the process has been increased with 30 percent. The figure shows that the system becomes unstable when the time delay is increased by 30 percent. Notice that it is the large loop that crosses the critical point -1 and not the part of the Nyquist curve close to the gain margin. The robustness measure called *the delay margin* is introduced to capture this effect. The delay margin is defined as the change in the time delay required to make a system unstable. For the systems with $L = 2$ and $L = 8$ in Figure 8.7 the delay margins are 27 percent and 7 percent, respectively.

Notice that the sensitivity functions also capture the robustness in the cases of loop transfer functions like the ones shown in Figure 8.7. The sensitivity to variations in the time delay can be estimated by (8.14), which gives delay margins of 25 percent and 6 percent for the systems in Figure 8.7 with $L = 2$ and $L = 8$. These numbers are close to the numbers obtained by using the delay margin.

Another way to quantify robustness is to explore the sensitivity of the closed

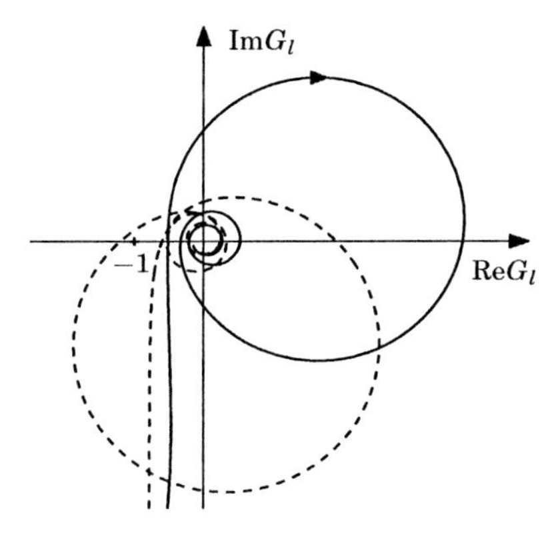

Figure 8.9 Nyquist plots of the loop transfer functions for the system in Example 8.1 with $L = 2$ in the nominal case (solid line) and when the time delay is increased by 30 percent (dashed line).

loop to variations in the process parameters. For the FOTD process we have

$$P(s) = \frac{K_p}{1+sT}e^{-sL}$$

Hence,

$$\log P = \log K_p - \log(1+sT) - sL$$

Differentiation this gives

$$\frac{dP}{P} = \frac{dK_p}{K_p} - \frac{sdT_p}{1+sT_p} - sdL = \frac{dK_p}{K_p} - \frac{sdT_p}{1+sT_p} - sL\frac{dL}{L}$$

For systems with large time delays the last term is dominating, which means that the sensitivity to time variations in the time delay is the critical constraint. Equation (4.32) then gives

$$\frac{|dL|}{L} < \frac{1}{\omega L|T(i\omega)|},$$

and we obtain the following estimate of the delay margin:

$$d_m = \max\frac{|dL|}{L} < \max\frac{1}{\omega L|T(i\omega)|}.$$

Summary

The Smith predictor makes it possible to obtain dramatic improvements of the set-point response as illustrated in Figure 8.3. The controller is obtained in a very simple way by first designing a controller C_0 for a nominal system P_0 that does not have the time delay. The Smith predictor is then obtained by cascading C_0 with a predictor C_{pred}, which effectively eliminates the time delay. An interesting feature of the Smith predictor is that it uses past control actions for prediction. It is in principle possible to compensate for any delay. The controller may, however, have unstable poles. The product $\omega_b L$, where ω_b is the bandwidth of the nominal closed-loop system $T_0 = P_0C_0/(1+P_0C_0)$ and L is the time delay L, are crucial parameters. The number of unstable controller poles grows with $\omega_b L$. Controllers with poles in the right half plane have poor robustness. Admissible variations in the time delay are inversely proportional to $\omega_b L$. To have a robust closed-loop system it is therefore necessary to restrict $\omega_b L$. In Example 8.3 we found, for example, that to have $M_s = 1.4$ it was necessary to have $\omega_b L < 0.5$.

8.4 The PPI Controller

In this section we will discuss special cases of the Smith predictor that give controllers of a particularly simple form. The Smith predictor discussed in Example 8.1 was based on the FOTD model. The design criterion was to find a controller that gives a second-order system with poles having relative damping ζ and frequency ω_0 for the system without delay. Another possible design is to choose a controller that cancels the process pole and makes the other closed-loop pole equal to $s = -1/T_{cl}$, where T_{cl} is the desired response time of the closed-loop system. This design method gives the following controller parameters;

$$K = \frac{T}{T_{cl}K_p}, \quad T_i = T.$$

The loop transfer function of the nominal system without delay is $P_0C_0 = 1/(sT_{cl})$, and the controller has the transfer function

$$C(s) = \frac{1+sT}{K_p s T_{cl}} \frac{1}{1+\frac{1}{sT_{cl}}(1-e^{-sL})}. \tag{8.16}$$

The loop transfer function is

$$P(s)C(s) = \frac{1}{sT_{cl}} \frac{1}{1+\frac{1}{sT_{cl}}(1-e^{-sL})}. \tag{8.17}$$

Since the process pole is canceled it should be required that the process pole is fast in comparison with the dominant closed-loop dynamics; see Section 6.6. There is one tuning parameter: the closed-loop response time T_{cl}.

The input-output relation of the controller (8.16) can be written as

$$\begin{aligned} U(s) &= \frac{1+sT}{K_p s T_{cl}} E(s) - \frac{1}{sT_{cl}}(1-e^{-sL})U(s) \\ &= \frac{1+sT}{K_p s T_{cl}} \left(E(s) - \frac{K_p}{1+sT}(1-e^{-sL})U(s) \right) = \frac{1+sT}{K_p s T_{cl}} E_p(s), \end{aligned} \tag{8.18}$$

where $E_p(s)$ is the Laplace transform of the predicted error

$$e_p(t) = y_{sp}(t) - y(t) - \tilde{y}(t),$$

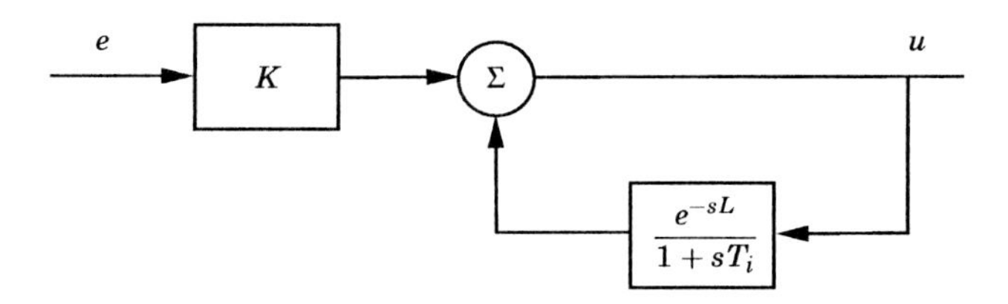

Figure 8.10 Block diagram of an implementation the PPI controller with $T_{cl} = T = T_i$.

and

$$\tilde{Y}(s) = \frac{K_p}{1+sT}(1-e^{-sL})U(s).$$

The term $\tilde{y}(t)$ represents the effect on the output of control actions taken in the interval $(t-L,t)$. The controller can thus be interpreted as a PI controller that acts on a predicted error, which is the actual error compensated for past control actions that have not yet appeared at the output. The controller is called the *predicting PI controller* or the *PPI controller.*

The controller is particularly simple if $T_{cl} = T$. The input-output relation of the controller then becomes

$$U(s) = KE(s) + \frac{e^{-sL}}{1+sT_i}U(s).$$

A block diagram describing this equation is given in Figure 8.10. Notice the strong similarity with the PI controller shown in Figure 3.3. There are also versions of this controller where the gain is replaced by a PD controller.

The Predictor

The PPI controller (8.16) is a cascade combination of a PI controller and a predictor with the transfer function

$$C_{pred}(s) = \frac{1}{1+\frac{1}{sT_{cl}}(1-e^{-sL})}. \tag{8.19}$$

Apart from frequency scaling the predictor is completely characterized by the ratio T_{cl}/L. It can be shown that the predictor does not have poles in the right half plane for any values of T_{cl}. The reason for this is that the loop transfer function of the nominal system without delay has constant phase.

A series expansion of the transfer function (8.19) for small s gives

$$\begin{aligned} C_{pred}(s) &\approx \frac{1}{1+L/T_{cl} - sT_{cl}(L/T_{cl})^2/2+\dots} \\ &\approx \frac{1}{1+L/T_{cl}}\left(1+\frac{1}{2}\frac{(L/T_{cl})^2}{1+L/T_{cl}}T_{cl}s+\dots\right). \end{aligned} \tag{8.20}$$

The static gain is $C_{pred}(0) = 1/(1+L/T_{cl})$, and it also follows that C_{pred} goes to 1 as s goes to infinity. Figure 8.11 shows the Bode plot of the predictor (8.19).

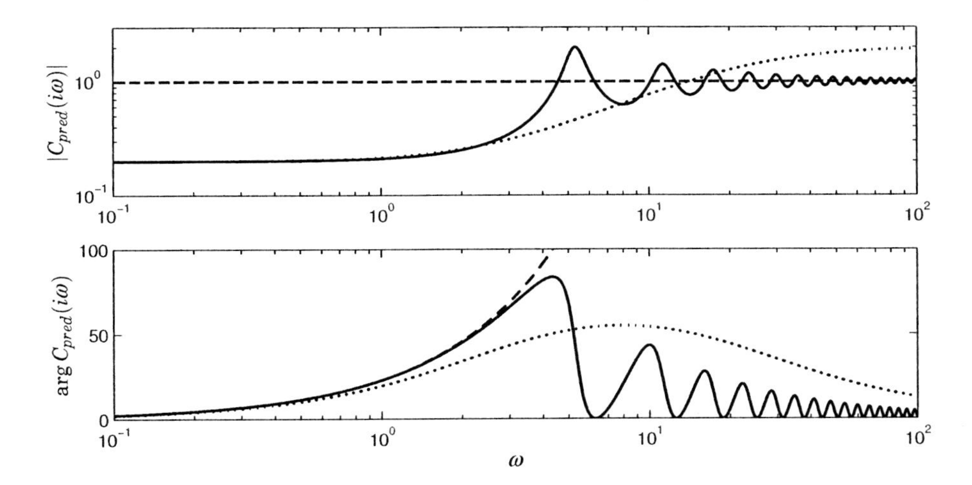

Figure 8.11 Bode plots for the predictor (8.19) (solid), a predictor based on differentiation (dotted), and the ideal predictor e^{sT} (dashed).

For comparison we have also given the Bode plots for an ideal predictor $e^{sT_{pred}}$, where

$$T_{pred} = \frac{1}{2}\frac{(L/T_{cl})^2}{1 + L/T_{cl}}, \tag{8.21}$$

and a predictor based on differentiation. The predictor based on differentiation has been adjusted to give the same maximum gain as the predictor (8.19). There are differences between the predictors. The ideal predictor has unit gain for all frequencies; the other predictors have higher gains at high frequencies and lower gains at lower frequencies. The predictor (8.19) provides larger phase advance than the predictor based on differentiation, but the phase advance falls off rapidly for higher frequencies.

Design Choices

The choice of the design parameter T_{cl} is a compromise between robustness and performance. The response time is directly given by T_{cl}; fast response time requires a small T_{cl}. Robustness is governed by the ratio T_{cl}/L. The sensitivity function is given by

$$S = 1 - \frac{e^{-sL}}{1 + sT_{cl}}.$$

Figure 8.12 shows the maximum sensitivity as a function of T_{cl}/L. Notice that the largest sensitivity has the property $M_s \leq 2$. To have $M_s \leq 1.6$ requires $T_{cl} > 0.66L$ and $M_s \leq 1.4$ requires $T_{cl} > 1.4L$. To have a reasonable robustness the desired response time cannot be chosen much shorter than L. It follows from (8.14) that the largest relative error in the time delay is given by

$$\frac{|\Delta L|}{L} \leq \frac{T_{cl}}{L}.$$

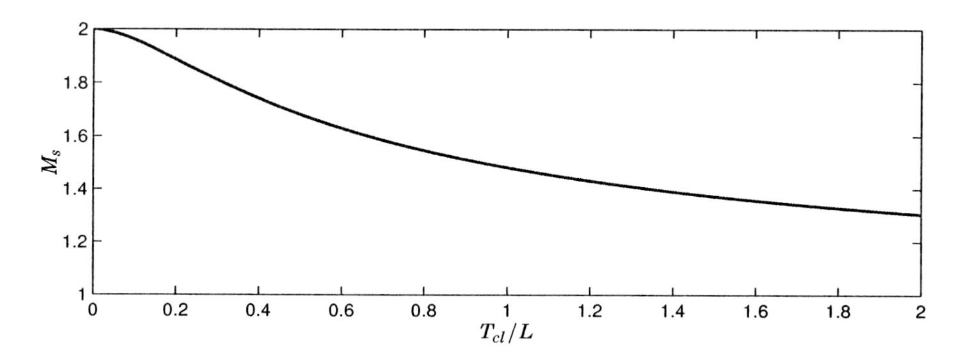

Figure 8.12 Maximum sensitivity M_s for the closed-loop system with the PPI controller (8.16) as a function of T_{cl}/L.

If maximum sensitivities as high as $M_s = 2$ are allowed and if the time delay is known precisely it is possible to allow smaller ratios T_{cl}/L.

For well-damped systems the integrated error IE is a performance measure that is easy to compute. From (8.18), the PPI controller in time domain is

$$u(t) = \frac{T}{K_p T_{cl}} e(t) + \frac{1}{K_p T_{cl}} \int_0^t e(t)dt - \frac{1}{T_{cl}} \int_0^t (u(t) - u(t-L))dt. \tag{8.22}$$

To compute the integral error for the PPI controller it will be assumed that the system is initially at rest and that a load disturbance in the form of a unit step is applied to the process input. Since the controller has integral action, we have $u(\infty) = 1$. Therefore,

$$\int_0^\infty (u(t) - u(t-L))dt = L.$$

After a unit load disturbance, it follows from (8.22) that

$$u(\infty) - u(0) = 1 = \frac{1}{K_p T_{cl}} \int_0^\infty e(t)dt - \frac{L}{T_{cl}}.$$

The integral error thus becomes

$$IE_{\text{PPI}} = K_p(L + T_{cl}).$$

The integrated error consists of two terms. The first term, K_pL, is due to the time delay and cannot be influenced by the controller. The second term, K_pT_{cl}, may be made small by specifying a short closed-loop time constant T_{cl}. A small value of T_{cl} will, however, result in poor robustness.

It is interesting to compare the performance of the PPI controller with the performance of PID controller. In Section 4.9, it was shown that the integral error for a PID controller is

$$IE_{\text{PID}} = \frac{T_i}{K} = \frac{1}{k_i}.$$

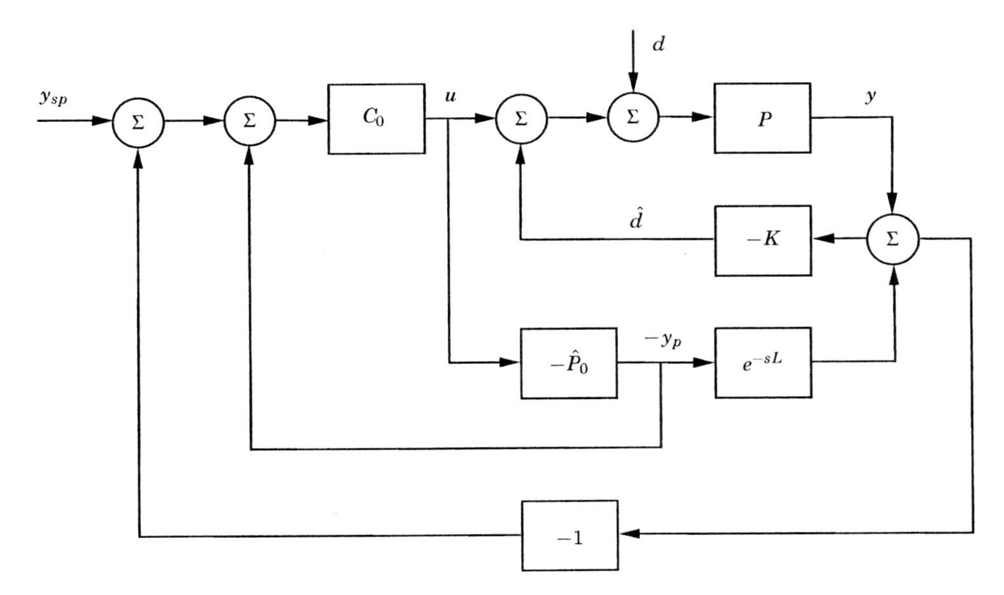

Figure 8.13 Modified Smith predictor for integrating processes.

It follows from (7.7) that a PID controller for delay-dominated processes tuned for $M_s = 1.4$ has $k_i K_p L = 0.5$. This gives $IE = 2K_p L$, which is close to the value $IE = 2.4K_p L$ obtained for the PPI controller. We thus obtain the conclusion that the PPI controller does not give significantly better performance at load disturbances than a PI controller if both controllers have the same robustness. The main advantage of the PPI controller is its ability to improve set-point responses; see Figure 8.3.

8.5 Predictors for Integrating Processes

The basic Smith predictor has useful properties, but it also has some severe drawbacks. It cannot be used for unstable systems, and it gives a steady-state error for load disturbances for processes with integration. Several modifications have therefore been proposed.

For processes with integration it has been suggested to modify the Smith predictor, as shown in Figure 8.13, in order to obtain zero steady-state error for a constant load disturbance. The reason for the modification can be understood from the principle of internal model control. The signal $\hat{d}$ that is fed back is an estimate of the load disturbance.

From Figure 8.13 the transfer functions from set point y_{sp} and load distur-

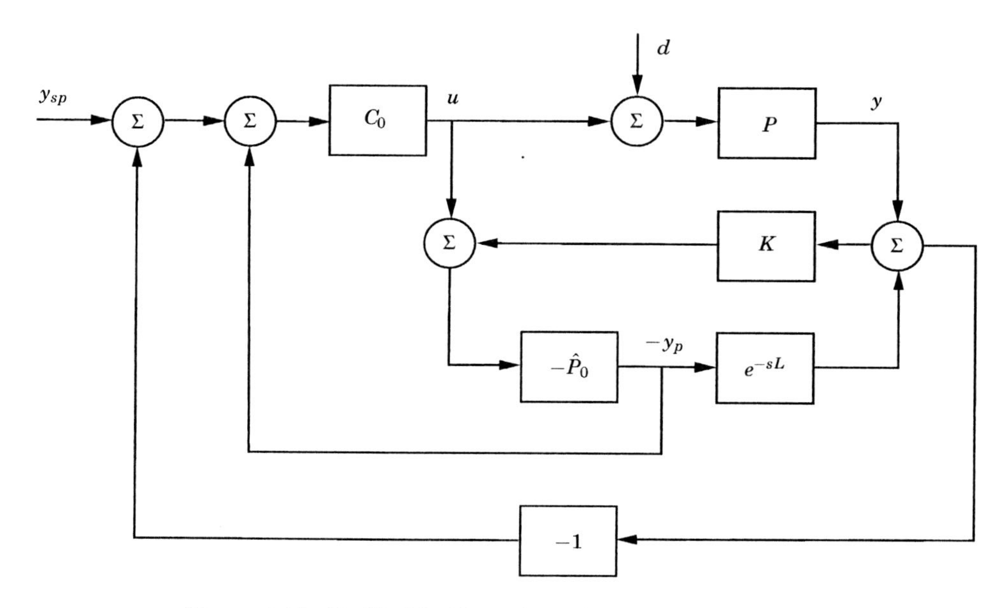

Figure 8.14 Modified Smith predictor for integrating processes.

bance d to output y are given by

$$
\begin{aligned}
Y =& \frac{PC_0(1 + K\hat{P})}{1 + C_0(\hat{P}_0 - \hat{P}) + P(K + KC_0\hat{P}_0 + C_0)} Y_{sp} \\
&+ \frac{P(1 + C_0(\hat{P}_0 - \hat{P}))}{1 + C_0(\hat{P}_0 - \hat{P}) + P(K + KC_0\hat{P}_0 + C_0)} D.
\end{aligned}
\tag{8.23}
$$

When $s \to 0$, the following approximations hold:

$$C_0 \approx \frac{k_i}{s} \qquad \hat{P}_0 - \hat{P} \approx \frac{K_v}{s}(1 - e^{-sL}) \approx K_v L.$$

If we also assume that $\hat{P} = P$, it can be shown that the transfer function between y_{sp} and y becomes one, and the transfer function between d and y becomes zero when $s \to 0$.

Another modification for integrating processes is given in Figure 8.14. The variable y_p is an estimate of the undelayed measurement signal

$$Y = P_0(U + D).$$

The estimation is given by

$$Y_p = \hat{P}_0(U + K(Y - \hat{Y})).$$

When $\hat{P}_0$ is stable, the value $K = 0$ can be used, corresponding to the original Smith predictor. For integrating processes, it is, however, necessary to have $K \neq 0$.

From Figure 8.14 the transfer functions from set point y_{sp} and load disturbance d to output y are given by

$$\begin{aligned}Y =& \frac{PC_0(1+KP)}{1+K\hat{P}+\hat{P}_0-\hat{P}+PC_0(1+K\hat{P}_0)}Y_{sp}\\ &+\frac{P(1+K\hat{P}+\hat{P}_0-\hat{P})}{1+K\hat{P}+\hat{P}_0-\hat{P}+PC_0(1+K\hat{P}_0)}D.\end{aligned} \tag{8.24}$$

Under the assumption that $\hat{P}=P$, it can be shown that the transfer function between y_{sp} and y becomes one, and the transfer function between d and y becomes zero when $s \to 0$.

8.6 Model Predictive Control

Model predictive control is based on the prediction of future process behavior based on a process model and optimization of the process behavior over a finite time horizon. Feedback is obtained by applying the initial part of the control signal and repeating the process over a shifted time horizon. This procedure is called *receding horizon control* or *moving horizon control.* Referring to Figure 8.15 the algorithm can be described as follows:

1: Develop a process model.

2: Consider the situation at time t. Past process inputs u and past process outputs y are observed; see Figure 8.15. The future behavior of the process is predicted under the assumption that the process model and the future control signals $u_f = u(\tau)$, $t \le \tau < t+t_h$ are known.

3: The control signal u_f is determined to give the desired future behavior.

4: The initial part of control signal u_f is applied over the interval $[t, t+h]$.

5: Change time to $t+h$, and repeat the procedure from Step 2.

The steps can be performed in many different ways, and there are a large number of algorithms. Different process models can be used; physical models, input-output models, and state models. The method can be applied both to single-input single-output systems and to systems with many inputs and many outputs.

The desired behavior can be specified in many ways. A common procedure is to specify the desired future behavior by a mathematical model, for example, one that tells how to approach the set point. The deviation from the desired behavior can be formulated as an optimization problem to minimize the deviation between actual and desired behavior, possibly with a penalty on control actions. Step 2 is an open-loop optimization problem where optimization is carried out over a finite time horizon. Feedback is obtained by only applying the initial part of the control signal. The horizon is then shifted forward, and the optimization is then repeated.

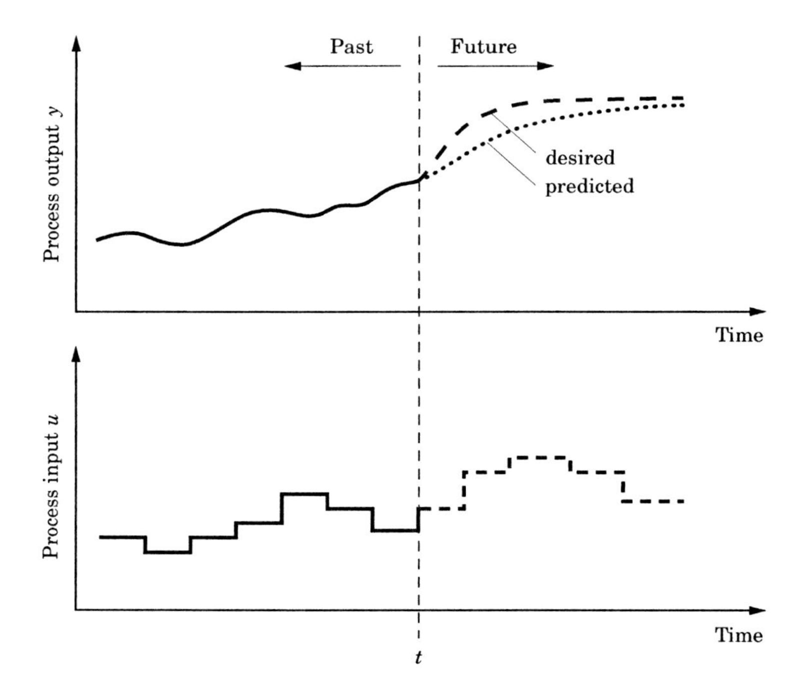

Figure 8.15 Illustration of the model predictive control.

Model predictive control is particularly simple for sampled systems where the control signal is constant over the sampling intervals. Parameter h can then be chosen as the sampling interval, and the prediction horizon t_h is typically chosen as a small number of sampling intervals. Most predictive controllers are also developed for sampled systems.

A very useful property of model predictive control is that constraints on the control signal and the process output can be taken into account. A common choice is to formulate the problem so that efficient algorithms for quadratic programming can be used. A key difficulty with model predictive control is to ensure stability when the prediction horizon is finite. Much research has been devoted to this problem.

A Simple Example

To illustrate the ideas we will give details in a simple case. Consider the sampled process model

$$y(t)+a_1y(t-h)+\cdots+a_ny(t-nh) = b_1u(t-h)+b_2u(t-2h)+\cdots+b_nu(t-nh), \tag{8.25}$$

where y is the process output and u the process input. Consider the situation at time t. The past behavior is completely characterized by

$$\mathcal{Y}_t = (y(t), y(t-h), \cdots, u(t-h), u(t-2h), \cdots). \tag{8.26}$$

Using the model it is straightforward to predict future values of process output as a function of current and future control signals:

$$\mathcal{U}_t = (u(t), u(t+h), \ldots, u(t+Nh)). \tag{8.27}$$

The desired future behavior can be characterized by specifying a reference trajectory for future process outputs, as indicated in Figure 8.15 and giving a loss function that penalizes deviations $e(t) = y(t) - y_d(t)$ from the desired output $y_d(t)$ and the increments of the control actions $\Delta u(t) = u(t) - u(t-h)$

$$J(u(t), u(t+h), \ldots, u(t+Nh)) = \sum_{k=1}^{t+N} e(t+kh)^2 + \rho(\Delta u(t+(k-1)h))^2. \tag{8.28}$$

There may also be constraints on process inputs and outputs and on the increment of the control signal.

Future control signals $\mathcal{U}_t$ are then computed by minimizing J subject to the constraints. The control signal $u(t)$ is then applied, and the whole procedure is repeated. The control signal is a function of past inputs and past outputs

$$u(t) = F(y(t), y(t-h), \ldots, y(t-nh), u(t-h), y(t-2h), \ldots, u(t-nh)),$$

where the function F is obtained implicitly by solving an optimization problem.

A particularly simple case is when the process model is of first order in the increments of process inputs and outputs, which we illustrate by an example.

EXAMPLE 8.4—MPC FOR FIRST-ORDER SYSTEM

Let the process model be

$$\Delta y(t+h) = -a\Delta y(t) + b\Delta u(t),$$

where $\Delta y(t) = y(t) - y(t-h)$ and $\Delta u(t) = u(t) - u(t-h)$. Let the desired trajectory be a signal $y_d(t)$ which starts at $y(t)$ and approaches the set point y_{sp} exponentially with time constant T_{cl}. The desired process output at time $t+h$ is then

$$y_d(t+h) = y(t) + \left(1 - e^{-h/T_{cl}}\right)(y_{sp} - y(t)).$$

Assuming that there are no penalties on the control actions the desired process output can then be achieved in the next sampling period. Equating $y(t+h)$ with $y_d(t+h)$ gives

$$y(t+h) = y(t) + \Delta y(t+h) = y(t) - a\Delta y(t) + b\Delta u(t) = y(t) + \left(1 - e^{-h/T_{cl}}\right)(y_{sp} - y(t)).$$

Solving this equation for $\Delta u(t)$ gives

$$\Delta u(t) = \frac{1 - e^{-h/T_{cl}}}{b}(y_{sp} - y(t)) + \frac{a}{b}\Delta y(t),$$

which is a PI controller with gains

$$k = \frac{a}{b}$$
$$k_i = \frac{1 - e^{-h/T_{cl}}}{b}.$$

Notice that the proportional gain only depends on the process model and that the integral gains depend on the desired response rate T_{cl}. □

It is straightforward to deal with systems having many inputs and outputs. It is also possible to include constraints. There are many special cases and variants of model predictive control. A few of them will be discussed briefly; for more details we refer to the references.

The Dahlin-Higham Algorithm

One of the earliest model predictive controllers was developed for control of paper machines. The algorithm is based on a process model in terms of the FOTD model

$$P(s) = \frac{K_p}{1+sT}e^{-sL},$$

and the desired response to set-point changes is given by

$$G_{yy_{sp}} = \frac{1}{1+sT_{cl}}e^{-sL}.$$

Assuming that the control signal is constant over sampling intervals of length $h = L/n$, where n is an integer, gives the sampled process model

$$y(t+h) = ay(t) + K_p(1-a)u(t-nh).$$

The desired response to set points is given by the difference equation

$$y_d(t+h) = a_d y_d(t) + (1-a_d)y_{sp}(t-nh).$$

Introducing the backward shift operator q^{-1} defined by

$$q^{-1}y(t) = y(t-h), \tag{8.29}$$

the process model can be written as

$$y(t) = \frac{K_p q^{-(n+1)}}{1-aq^{-1}}u(t) = P(q^{-1})u(t).$$

Let the controller be characterized by

$$u(t) = C(q^{-1})(y_{sp}(t) - y(t)).$$

The input-output relation for the closed-loop system is then

$$y(t) = \frac{P(q^{-1})C(q^{-1})}{1+P(q^{-1})C(q^{-1})}y_{sp}(t).$$

Using the backward shift operator the desired response is given by

$$y_d(t) = \frac{(1-a_d)q^{-(n+1)}}{1-a_d q^{-1}}y_{sp}(t) = G_d(q^{-1})y_{sp}(t),$$

where $a_d = e^{-h/T_{cl}}$. Equating this with the process output gives

$$\frac{P(q^{-1})C(q^{-1})}{1+P(q^{-1})C(q^{-1})} = G_d(q^{-1}) = \frac{(1-a_d)q^{-(n+1)}}{1-a_dq^{-1}}.$$

Solving this equation with respect to $C(q^{-1})$ gives

$$C(q^{-1}) = \frac{G_d(q^{-1})}{P(q^{-1})(1-G_d(q^{-1}))} = \frac{(1-a_d)(1-aq^{-1})}{K_p(1-a_dq^{-1}-(1-a_d)q^{-(n+1)})}.$$

The controller can then be described by

$$u(t) = \frac{1-a_d}{K_p}(e(t)-ae(t-h)) + a_du(t-h) + (1-a_d)u(t-(n+1)h).$$

This controller has integral action, and past inputs are used for prediction.

Dynamic Matrix Control (DMC)

In dynamic matrix control the process is modeled by the finite impulse response model

$$y(t) = b_1u(t-h) + b_2u(t-2h) + \cdots + b_nu(t-nh), \tag{8.30}$$

and the criterion is to minimize the loss function

$$J\big(u(t), u(t+h), \ldots, u(t+(n-1)h)\big) = \sum_{k=1}^{n} e^2(t+kh),$$

where

$$e(t+kh) = y_d(t+kh) - b_1u(t+kh-h) + b_2u(t+kh-2h) + \cdots + b_nu(t+kh-nh).$$

Since e is a linear function of future control variables and the loss function is quadratic the optimization is straightforward. Notice that the model (8.30) also holds if there are many inputs and outputs. The coefficients b_i are then matrices. They were called *dynamic matrices* since they reflect the dynamics of the response in the original paper, which motivated the name DMC. In standard control terminology the parameters are simply the coefficients of the impulse response. In the early use of dynamic matrix control it was common practice to determine the matrices b_i from a simple impulse or step response measurement.

A drawback with DMC is that a large number of parameters may be required if the process dynamics are slow. The DMC algorithm was later generalized to QDMC (Quadratic Dynamic Matrix Control), which also can handle constraints on the control signal.

Minimum Variance Control

The minimum variance controller is a predictive controller for systems with random disturbances where the criterion is to minimize the variance of the fluctuations in process output. The algorithm was originally developed for control of paper machines where the stochastic nature of the disturbances is as important as the process dynamics. We start by a simple example.

EXAMPLE 8.5—MINIMUM VARIANCE CONTROL
Consider a model

$$y(t+h) = -ay(t) + bu(t) + e(t+h) + ce(t),$$

where u is the control variable, y the process output, and e a sequence of independent random variables with zero mean value and standard deviation σ. The sampling period is h.

Consider the situation at time t. The process output $y(t)$ is known and the output at time $t+h$ can be given arbitrary values by choosing the control signal $u(t)$. The random signal $e(t+h)$ is independent of past inputs and outputs $\mathcal{Y}_t$ given by (8.26). Furthermore, $e(t)$ can be computed from past inputs and outputs $\mathcal{Y}_t$. The control law that minimizes the deviation from the set point y_{sp} is given by

$$u(t) = \frac{ay(t) - ce(t)}{b},$$

If this control law is used we find that $y(t) = e(t)$, which means that the output is white noise. The computation of $e(t)$ from past inputs is thus trivial, and the control law becomes

$$u(t) = \frac{a-c}{b} y(t).$$

□

In the general case, the process model is

$$a(q^{-1})y(t) = b(q^{-1})u(t) + c(q^{-1})e(t). \tag{8.31}$$

where u is the process input, y the process output, and e is a sequence of independent Gaussian random variables with zero mean and variance σ. $a(q^{-1})$, $b(q^{-1})$, and $c(q^{-1})$ are polynomials in the backward shift operator

$$\begin{aligned} a(q^{-1}) &= 1 + a_1 q^{-1} + a_2 q^{-2} + \ldots + a_n q^{-n} \\ b(q^{-1}) &= \quad b_\ell q^{-\ell} + b_{\ell+1} q^{-\ell-1} + \ldots + b_n q^{-n} \\ c(q^{-1}) &= 1 + c_1 q^{-1} + c_2 q^{-2} + \ldots + c_n q^{-n}. \end{aligned}$$

For simplicity we have chosen to let all polynomials be of the same degree. This is no lack of generality because we can allow trailing coefficients to be zero. The coefficient b_ℓ is the first non-vanishing coefficient in the polynomial $b(q^{-1})$. The number ℓ is an important parameter called the input-output delay, and we also introduce the polynomial $b'(q^{-1}) = q^\ell b(q^{-1})$.

It is natural to assume that there are no factors common to all three polynomials $a(q^{-1})$, $b(q^{-1})$, and $c(q^{-1})$. The polynomial $c(q^{-1})$ is assumed to have all its zeros outside the unit disc. The model (8.31) captures the dynamics both of the process and its disturbances.

Minimum variance control is closely related to prediction, and we will therefore first determine a predictor for the process output when the input u is zero. The prediction of y ℓ steps ahead is given by

$$c(q^{-1})\hat{y}(t+\ell) = g(q^{-1})y(t),$$

where the polynomial $g(q^{-1})$ is given by

$$a(q^{-1})f(q^{-1}) + q^{-l}g(q^{-1}) = c(q^{-1}).$$

Notice that the dynamics of the predictor are given by the polynomial $c(q^{-1})$ in the model (8.31). The prediction error

$$\varepsilon(t) = f(q^{-1})e(t)$$

has the variance

$$E\varepsilon^2 = \sigma^2 \sum_0^{\ell-1} f_k^2. \tag{8.32}$$

The simple minimum variance control strategy is given by

$$u(t) = -\frac{s(q^{-1})}{r(q^{-1})}y(t) = -\frac{g(q^{-1})}{b'(q^{-1})f(q^{-1})}y(t), \tag{8.33}$$

and the control error is

$$y(t) = f(q^{-1})e(t). \tag{8.34}$$

The error under minimum variance control is thus equal to the error in predicting the output ℓ steps ahead. The control error is a moving average of order $\ell - 1$. It is thus easy to determine if a process is under minimum variance control simply by computing the correlation function of the output. Since the control error is a moving average of order $\ell - 1$ its covariance function is zero for all lags greater than ℓ.

The robustness of minimum variance control is strongly influenced by the choice of sampling interval. It is good practice to choose h larger than $L/2$.

8.7 Summary

The performance of a PI controller can be improved by adding predictive capability. Derivative action is one possibility, but there are many other alternatives. The Smith predictor and model predictive control are useful for systems with time delays when good models are available. Drastic improvements in the response to set-point changes can be obtained when good models are available. The predictive PI controller is a simple version of the Smith predictor. It has the advantage over a PID controller that the achievable phase advance is larger. A paradox is that the predictive controller only gives modest improvements compared to PI controllers for processes with delay-dominated dynamics but the performance improvements can be significant for lag-dominated processes. Model predictive controllers are more general than Smith predictors, and they can also deal with systems having many inputs and many outputs. Constraints can also be taken into account.

Since predictive controllers are based on mathematical models it is important that the models are accurate. It is particularly important to have a good estimate of the time delay. A fairly complete robustness analysis was given for the Smith predictor. Similar results are available for other predictive controllers. The key result is that sensitivity to modeling errors is closely related to the parameter $\omega_b L$, where ω_b is the closed-loop bandwidth or L/T_{cl} where T_{cl} is the desired closed-loop response time when time delay L is neglected. Robustness required that both parameters are not too small. A reasonable rule of thumb is that the parameters should be larger than 0.5.

8.8 Notes and References

A controller for systems having time delay was proposed by [Smith, 1957]; it is also treated in the book [Smith, 1958]. An explanation of the mechanism that generates the large phase advance is given in [Åström, 1977]. Many modifications of the Smith predictor have been presented; see [Åström *et al.*, 1994; Matausek and Micic, 1996; Matausek and Micic, 1999; Kaya and Atherton, 1999; Kristiansson and Lennartson, 1999]. The controller in [Haalman, 1965], the PPI controller in [Hägglund, 1996], and the PIDτ controller in [Shinskey, 2002] are all special cases of the Smith predictor. The papers [Ross, 1977; Meyer *et al.*, 1976; Ingimundarson and Hägglund, 2002] compare Smith predictors with PID controllers.

Minimum variance control was developed in the early phase of computer control of paper machines as an attempt to find a control strategy that minimizes fluctuations in quality variables. A key result is that the smallest variance that can be achieved is the variance of the error in predicting the output over the time delay of the process. Minimum variance control was first published by [Åström, 1967] and a perspective on its use is given in [Åström, 2001]. Minimum variance control requires a model of disturbances and process dynamics. A method to obtain this information directly from process experiments was developed in [Åström and Bohlin, 1965] and applied to modeling and control of paper machines [Åström, 1970]. The self-tuning controller [Åström and Wittenmark, 1973] can be viewed as automation of system identification and minimum variance control.

The controller presented in [Dahlin, 1968] and [Higham, 1968] can be viewed as a discrete-time version of the Smith predictor. Both the Smith Predictor and the Dahlin-Higham controller, which are early versions of model predictive control [Shinskey, 1991b], were first developed for process control applications. There are many versions of model predictive control; see [Richalet *et al.*, 1976], [Cutler and Ramaker, 1980] and [Garcia and Morshedi, 1986]. There are several recent books on model predictive control [Allgower and Zheng, 2000; Kouvaritakis and Cannon, 2001; Maciejowski, 2002]. The survey papers [Rawlings, 2000; Qin and Badgwell, 2003] contain many references. The papers [Kulhavy *et al.*, 2001], [Downs, 2001] and [Young *et al.*, 2001] and the book [Blevins *et al.*, 2003] give an industrial perspective. Although model predictive control was originally intended for multi-variable systems it has also been suggested to use it as a replacement for PID control; see [Lu, 2004].

9

Automatic Tuning and Adaptation

9.1 Introduction

Automatic tuning, or auto-tuning, is a method where the controller is tuned automatically on demand from a user. Typically, the user will either push a button or send a command to the controller. Automatic tuning of PID controllers can be accomplished by combining the methods for determining process dynamics, described in Chapter 2, with the methods for computing the parameters of a PID controller, described in Chapters 4, 6, and 7. An automatic tuning procedure consists of three steps:

- Generation of a process disturbance.
- Evaluation of the disturbance response.
- Calculation of controller parameters.

This is the same procedure that an experienced engineer uses when tuning a controller manually. The process must be disturbed in some way in order to determine the process dynamics. This can be done in many ways, e.g., by adding steps, pulses, or sinusoids to the process input. The evaluation of the disturbance response may include a determination of a process model or a simple characterization of the response.

Industrial experience has clearly shown that automatic tuning is a highly desirable and useful feature. Automatic tuning is sometimes called tuning on demand or one-shot tuning. Commercial PID controllers with automatic tuning facilities have been available since the beginning of the eighties.

Automatic tuning can be built into a controller. It can also be performed using external devices that are connected to the control loop only during the tuning phase. Controller parameters are displayed when the tuning experiment is finished. Since the tuning devices are supposed to work together with controllers from different manufacturers, they must be provided with a lot of information about the controller in order to give an appropriate parameter suggestion.

Even when automatic tuning devices are used, it is important to obtain a certain amount of process knowledge. This is discussed in the Section 9.2. Automatic tuning is only one way to use the adaptive technique. Section 9.3 gives an overview of several adaptive techniques, as well as a discussion about their use. The automatic tuning approaches can be divided into two categories, namely, model-based approaches and rule-based approaches. In the model-based approaches, a model of the process is obtained explicitly, and the tuning is based on this model. Section 9.4 treats approaches where the model is obtained from transient response experiments, frequency response experiments, and parameter estimation. In the rule-based approaches, no explicit process model is obtained. The tuning is instead based on rules similar to those rules that an experienced operator uses to tune the controller manually. The rule-based approach is treated in Section 9.5. Section 9.7 treats iterative feedback tuning, which is an iterative method to tune the controllers.

A few industrial products with adaptive facilities are presented in Section 9.8. This section illustrates how some of the ideas are used in products. It is not intended as an exhaustive presentation of products. The chapter ends with conclusions and references in Sections 9.9 and 9.10.

9.2 Process Knowledge

In this chapter we will discuss several methods for automatic tuning. Before going into details we must remark that poor behavior of a control loop can not always be corrected by tuning the controller. It is absolutely necessary to understand the reason for the poor behavior.

The process may be poorly designed so that there are long dead times, long time constants, nonlinearities, and inverse responses. Sensors and actuators may be poorly placed or badly mounted, and they may have bad dynamics. Typical examples are thermocouples with heavy casings that make their response slow or on-off valve motors with long travel time. Valves may be over-sized so that they only act over a small region. The sensor span may be too wide so that poor resolution is obtained, or it may also have excessive sensor noise.

There may also be failure and wear in the process equipment. Valves may have excessive stiction. There may be backlash due to wear. Sensors may drift and change their properties because of contamination.

If a control loop is behaving unsatisfactorily, it is essential that we first determine the reason for this before tuning is attempted. It would, of course, be highly desirable to have aids for the process engineer to do the diagnosis. Automatic tuning may actually do the wrong thing if it is not applied with care. For example, consider a control loop that oscillates because of friction in the actuator. Practically all tuning devices will attempt to stabilize the oscillation by reducing the controller gain. This will only increase the period of the oscillation! These important questions are treated in a separate chapter in this book, Chapter 10. Remember that no amount of so called “intelligence” in equipment can replace real process knowledge.

9.3 Adaptive Techniques

Techniques for automatic tuning grew out of research in adaptive control. Adaptation was originally developed to deal with processes with characteristics that were changing with time or with operating conditions. Practically all adaptive techniques can be used for automatic tuning. The adaptive controller is simply run until the parameters have converged, and the parameters are then kept constant. The drawback with this approach is that adaptive controllers may require prior information. There are many special techniques that can be used. Industrial experience has shown that automatic tuning is probably the most useful application of adaptive techniques. Gain scheduling is also a very effective technique to cope with processes that change their characteristics with operating conditions. An overview of these techniques will be given in this section. In this book the phrase *adaptive techniques* will include auto-tuning, gain scheduling, and adaptation.

Adaptive Control

An adaptive controller adjusts its parameters continuously to accommodate changes in process dynamics and disturbances. Adaptation can be applied both to feedback and feedforward control parameters. It has proved particularly useful for feedforward control. The reason for this is that model fidelity is crucial for feedforward control. Adaptive control is sometimes called continuous adaptation to emphasize that parameters are changed continuously.

There are two types of adaptive controllers based on direct and indirect methods. In a direct method, controller parameters are adjusted directly from data in closed-loop operation. In indirect methods, the controller parameters are obtained indirectly by first updating a process model on line, and then determining the controller parameters from some method for control design. The model reference system is a direct adaptive controller. The self-tuning regulator can be implemented both for direct and indirect control. There is a large number of methods available both for direct and indirect methods. They can conveniently be described in terms of the methods used for modeling and control design.

A block diagram of an indirect adaptive controller is shown in Figure 9.1. There is a parameter estimator that determines the parameters of the model based on observations of process inputs and outputs. There is also a design block that computes controller parameters from the model parameters. If the system is operated as a tuner, the process is excited by an input signal. The parameters can either be estimated recursively or in batch mode. Controller parameters are computed, and the controller is commissioned. If the system is operated as an adaptive controller, parameters are computed recursively, and controller parameters are updated when new parameter values are obtained.

Automatic Tuning

By automatic tuning (or auto-tuning) we mean a method where a controller is tuned automatically on demand from a user. Typically, the user will either push a button or send a command to the controller. Industrial experience has clearly indicated that this is a highly desirable and useful feature. Automatic

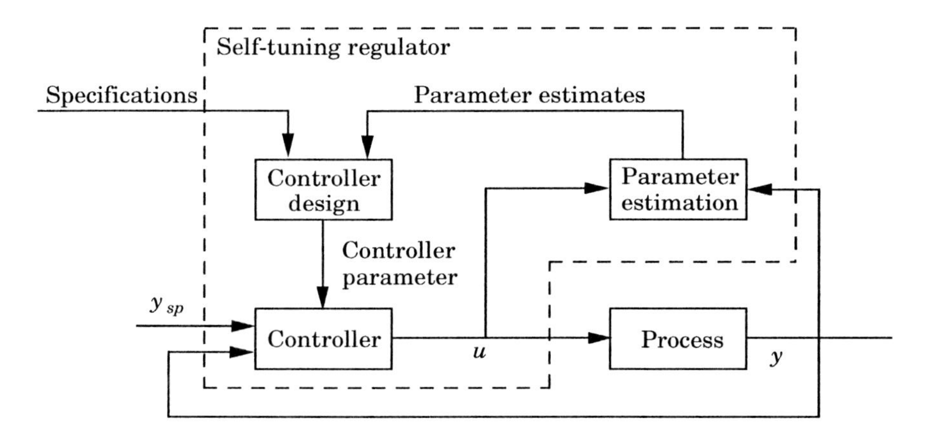

Figure 9.1 Block diagram of an adaptive controller.

tuning is sometimes called tuning on demand or one-shot tuning. Auto-tuning can be built into the controllers. Practically all controllers can benefit from tools for automatic tuning. This will drastically simplify the use of controllers. Single-loop controllers and distributed systems for process control are important application areas. Most of these controllers are of the PID type. Automatic tuning is currently widely used in PID controllers.

Auto-tuning can also be performed with external devices that are connected to a process. Since these systems have to work with controllers from different manufacturers, they must be provided with information about the controller structure in order to give an appropriate parameter suggestion. Such information includes controller structure (standard, series, or parallel form), sampling rate, filter time constants, and units of the different controller parameters (gain or proportional band, minutes or seconds, time or repeats/time).

Gain Scheduling

Gain scheduling is a technique that deals with nonlinear processes, processes with time variations, or situations where the requirements on the control change with the operating conditions. To use the technique it is necessary to find measurable variables, called scheduling variables, that correlate well with changes in process dynamics. The scheduling variable can be, for instance, the measured signal, the control signal, or an external signal. For historical reasons the phrase *gain scheduling* is used even if other parameters than the gain, e.g., derivative time or integral time, are changed. Gain scheduling is a very effective way of controlling systems whose dynamics change with the operating conditions. Gain scheduling has not been used much because of the effort required to implement it. When combined with auto-tuning, however, gain scheduling is very easy to use.

A block diagram of a system with gain scheduling is shown in Figure 9.2. The system can often be viewed as having two loops. There is an inner loop, composed of the process and the controller, and an outer loop, which adjusts the controller parameters based on the operating conditions. There are also

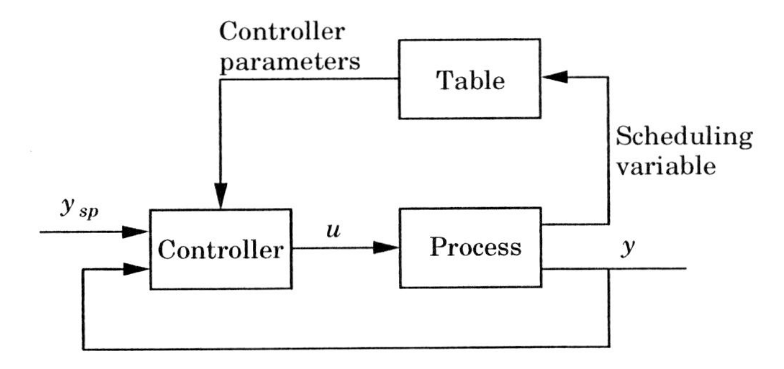

Figure 9.2 Block diagram of a system with gain scheduling.

situations when there is no outer loop, and the scheduling variable is unaffected by the controller output.

The notion of gain scheduling was originally used for flight control systems, but it is being used increasingly in process control. It is, in fact, a standard ingredient in some single-loop PID controllers. For process control applications significant improvements can be obtained by using just a few sets of controller parameters.

Gain scheduling is often an alternative to adaptation. It has the advantage that it can follow rapid changes in the operating conditions. The key problem is finding suitable scheduling variables. Possible choices are the control signal, the process variable, or an external signal. Production rate is often a good choice in process control applications, since time constants and time delays are often inversely proportional to production rate.

Development of a schedule may take a substantial engineering effort. The availability of automatic tuning can significantly reduce the effort because the schedules can then be determined experimentally. A scheduling variable is first determined. Its range is quantified into a number of discrete operating conditions. The controller parameters are then determined by automatic tuning when the system is running in one operating condition. The parameters are stored in a table. The procedure is repeated until all operating conditions are covered. In this way it is easy to install gain scheduling into a computer-controlled system by programming a table for storing and recalling controller parameters and appropriate commands to accomplish this.

Uses of Adaptive Techniques

We have described three techniques that are useful in dealing with processes that have properties changing with time or with operating conditions. In Figure 9.3 is a diagram to guide in choosing the different adaptive techniques.

Controller performance is the first thing to consider. If the requirements are modest, a controller with constant parameters and conservative tuning can be used. With higher demands on performance, other solutions should be considered. If the process dynamics are constant, a controller with constant parameters should be used. The parameters of the controller can be obtained using auto-tuning.

If the process dynamics or the nature of the disturbances are changing, it

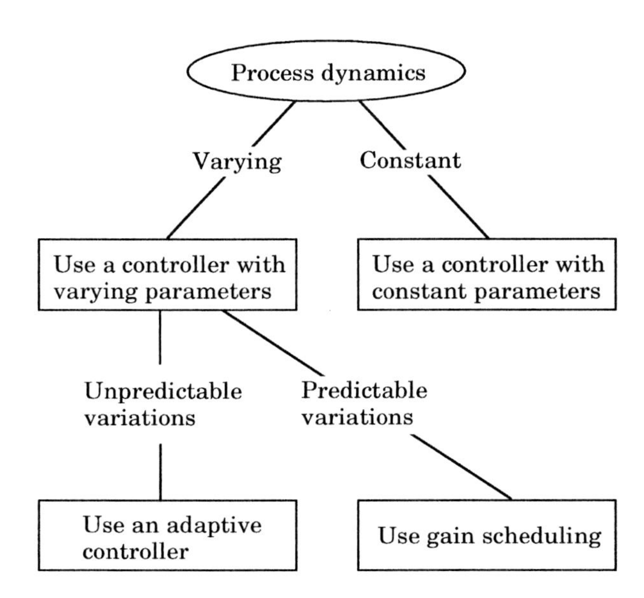

Figure 9.3 When to use different adaptive techniques.

is useful to compensate for these changes by changing the controller. If the variations can be predicted from measured signals, gain scheduling should be used because it is simpler and gives superior and more robust performance than the continuous adaptation. Typical examples are variations caused by nonlinearities in the control loop. Auto-tuning can be used to build up the gain schedules.

There are also cases where the variations in process dynamics are not predictable. Typical examples are changes due to unmeasurable variations in raw material, wear, fouling, etc. These variations cannot be handled by gain scheduling, since no scheduling variable is available, but must be dealt with by adaptation. An auto-tuning procedure is often used to initialize the adaptive controller. It is then sometimes called pre-tuning or initial tuning.

Feedforward control deserves special mentioning. It is a very powerful method for dealing with measurable disturbances. Use of feedforward control, however, requires good models of process dynamics. It is difficult to tune feedforward control loops automatically on demand, since the operator often cannot manipulate the disturbance used for the feedforward control. To tune the feedforward controller it is necessary to wait for an appropriate disturbance. Adaptation, therefore, is particularly useful for the feedforward controller.

9.4 Model-Based Methods

This section gives an overview of automatic tuning approaches that are based on an explicit derivation of a process model. Models can be obtained in many ways, as seen in Chapter 2. In this section we discuss approaches based on transient responses, frequency responses, and parameter estimation. The methods can also be characterized in terms of open and closed loop methods.

Transient Response Methods

Auto-tuners can be based on open-loop or closed-loop transient response analysis. Methods for determining the transient response were discussed in Section 2.7. The most common methods are based on step or pulse responses, but there are also methods that can use many other types of perturbations.

Open-Loop Tuning A simple process model can be obtained from an open-loop transient response experiment. A step or a pulse is injected at the process input, and the response is measured. To perform such an experiment, the process must be stable. If a pulse test is used, the process may include an integrator. It is important that the process be in equilibrium when the experiment is begun.

There are only one or two parameters that must be set *a priori*, namely, the amplitude and the signal duration. The amplitude should be chosen sufficiently large so that the response is easily visible above the noise level. On the other hand, it should be as small as possible in order not to disturb the process more than necessary and to keep the dynamics linear. The noise level can be determined automatically at the beginning of the tuning experiment. However, even if the noise level is known, we cannot decide a suitable magnitude of a step in the control signal without knowing the gain of the process. Therefore, it must be possible for the operator to decide the magnitude.

The duration of the experiment is the second parameter that normally is set *a priori*. If the process is unknown, it is very difficult to determine whether a step response has settled or not. An intuitive approach is to say that the measurement signal has reached its new steady state if its rate of change is sufficiently small. The rate of change is related, however, to the time constants of the process, which are unknown. If a pulse test is used, the duration of the pulse should also be related to the process time constants.

Many methods can be used to extract process characteristics from a transient response experiment. Most auto-tuners determine the static gain, the apparent time constant, and the apparent dead time. The static gain is easy to find accurately from a step-response experiment by comparing the stationary values of the control signal and the measurement signal before and after the step change. The time constant and the dead time can be obtained in several ways, see Section 2.7.

The transient response methods are often used in a pre-tuning mode in more complicated tuning devices. The main advantage of the methods, namely, that they require little prior knowledge, is then exploited. It is also easy to explain the methods to plant personnel. The main drawback with the transient response methods is that they are sensitive to disturbances. This drawback is less important if they are used only in the pre-tuning phase.

Closed-Loop Tuning Automatic tuning based on transient response identification can also be performed in closed loop. The steps or pulses are then added

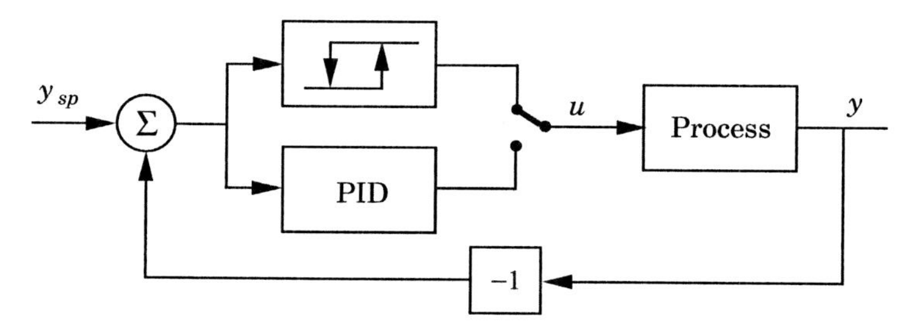

Figure 9.4 The relay auto-tuner. In the tuning mode the process is connected to relay feedback.

either to the set point or to the control signal. There are also auto-tuners that do not introduce any transient disturbances. Perturbations caused by set-point changes or load disturbances are used instead. In these cases it is necessary to detect that the perturbations are sufficiently large compared to the noise level.

Closed-loop tuning methods cannot be used on unknown processes. Some kind of pre-tuning must always be performed in order to close the loop in a satisfactory way. On the other hand, they do not usually require any additional *a priori* information. The magnitude of the step changes in set point are easily determined from the desired, or accepted, change in the measurement signal.

Since a proper closed-loop transient response is the goal for the design, it is appealing to base tuning on closed-loop responses. It is easy to give design specifications in terms of the closed-loop transient response, e.g., damping, overshoot, closed-loop time constants, etc. The drawback is that the relation between these specifications and the PID parameters is normally quite involved. Heuristics and logic are required therefore.

Frequency Response Methods

There are also auto-tuners that are based on frequency response methods. In Section 2.7, it was shown how frequency response techniques could be used to determine process dynamics.

Use of the Relay Method In traditional frequency response methods, the transfer function of a process is determined by measuring the steady-state responses to sinusoidal inputs. A difficulty with this approach is that appropriate frequencies of the input signal must be chosen *a priori*. A special method, where an appropriate frequency of the input signal is generated automatically, was described in Section 2.7. The idea was simply to introduce a nonlinear feedback of the relay type in order to generate a limit cycle oscillation. With an ideal relay the method gives an input signal to the process with a period close to the ultimate frequency of the open-loop system.

A block diagram of an auto-tuner based on the relay method is shown in Figure 9.4. Notice that there is a switch that selects either relay feedback or ordinary PID feedback. When it is desired to tune the system, the PID function is disconnected and the system is connected to relay feedback control. The system then starts to oscillate. The period and the amplitude of the oscillation is determined when steady-state oscillation is obtained. This gives the ultimate period and the ultimate gain. The parameters of a PID controller can then be determined from these values. The PID controller is then automatically switched in again, and the control is executed with the new PID parameters.

The initial amplitude of the relay must be specified in advance. A feedback loop from measurement of the amplitude of the oscillation to the relay amplitude can be used to ensure that the output is within reasonable bounds during the oscillation. It is also useful to introduce hysteresis in the relay. This reduces the effects of measurement noise and also increases the period of the oscillation. With hysteresis there is an additional parameter. This can be set automatically, however, based on a determination of the measurement noise level. Notice that there is no need to know time scales *a priori* since the ultimate frequency is determined automatically from the experiment.

In the relay method, an oscillation with suitable frequency is generated by a static nonlinearity. Even the order of magnitude of the time constant of the process can be unknown. Therefore, this method is not only suitable as a tuning device; it can also be used in pre-tuning. It is also suitable for the determination of sampling periods in digital controllers.

The relay tuning method also can be modified to identify several points on the Nyquist curve. This can be accomplished by making several experiments with different values of the amplitude and the hysteresis of the relay. A filter with known characteristics can also be introduced in the loop to identify other points on the Nyquist curve.

On-Line Methods Frequency response analysis can also be used for on-line tuning of PID controllers. By introducing bandpass filters, the signal content at different frequencies can be investigated. From this knowledge, a process model given in terms of points on the Nyquist curve can be identified and tracked on line. In this auto-tuner the choice of frequencies in the bandpass filters is crucial. This choice can be simplified by using the tuning procedure described above in a pre-tuning phase.

Parameter Estimation Methods

A common tuning procedure is to use recursive parameter estimation to determine a low-order discrete time model of the process. The parameters of the low-order model obtained are then used in a design scheme to calculate the controller parameters. An auto-tuner of this type can also be operated as an adaptive controller that changes the controller parameters continuously. Auto-tuners based on this idea, therefore, often have an option for continuous adaptation.

The main advantage of auto-tuners of this type is that they do not require any specific type of excitation signal. The control signal can be a sequence of manual changes of the control signal, for example, or the signals obtained during normal operation. A drawback with auto-tuners of this type is that they require significant prior information. A sampling period for the identification procedure must be specified; it should be related to the time constants of the closed-loop system. Since the identification is performed on line, a controller that at least manages to stabilize the system is required. Systems based on

Table 9.1 Rules of thumb for the effects of the controller parameters on speed and stability in the control loop.

	Speed	Stability
K increases	increases	reduces
T_i increases	reduces	increases
T_d increases	increases	increases

this identification procedure need a pre-tuning phase, which can be based on the methods presented earlier in this section.

9.5 Rule-Based Methods

This section treats automatic tuning methods that do not use an explicit model of the process. Tuning is based instead on the idea of mimicking manual tuning by an experienced process engineer.

Controller tuning is a compromise between performance and robustness. Table 9.1 shows how stability and speed change when the PID controller parameters are changed. Note that the table only contains rules of thumb. There are exceptions. For example, an increased gain often results in more stable control when the process contains an integrator. The same rules can also be illustrated in tuning maps. See, for example, the tuning map for PI control in Figure 6.7.

The rule-based automatic tuning procedures wait for transients, set-point changes, or load disturbances in the same way as the model-based methods. When such a disturbance occurs, the behavior of the controlled process is observed. If the control deviates from the specifications, the controller parameters are adjusted based on some rules.

Figures 9.5 and 9.6 show set-point changes of control loops with a poorly tuned PI controller. The response in Figure 9.5 is very sluggish. Here, a correct rule is to increase the gain and to decrease the integral time. Figure 9.6 also shows a sluggish response because of a too large integral time. The response is also oscillatory because of a too high gain. A correct rule, therefore, is to decrease both the gain and the integral time.

If graphs like those in Figures 9.5 and 9.6 are provided, it is easy for an experienced operator to apply correct rules for controller tuning. To obtain a rule-based automatic tuning procedure, the graphs must be replaced by quantities that characterize the responses. Commonly used quantities are overshoot and decay ratio to characterize the stability of the control loop and time constant and oscillation frequency to characterize the speed of the loop.

It is rather easy to obtain relevant rules that tell whether the different controller parameters should be decreased or increased. However, it is more difficult to determine *how much* they should be decreased or increased. The rule-based methods are, therefore, more suitable for continuous adaptation

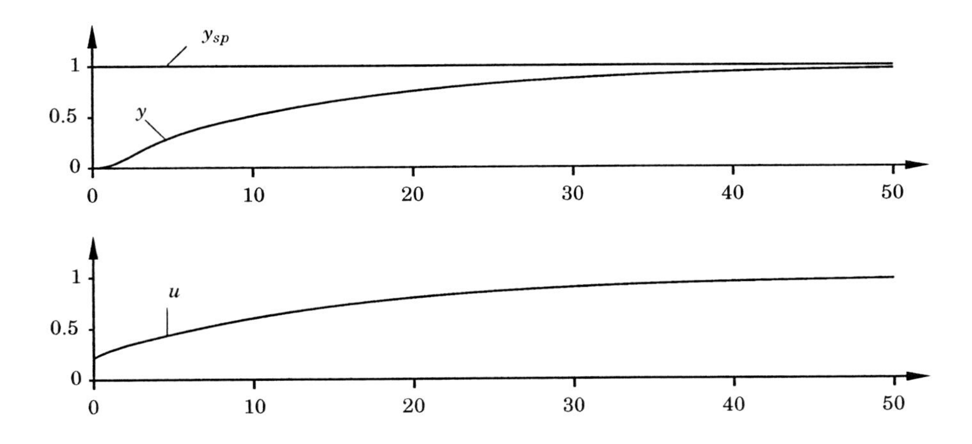

Figure 9.5 A set-point response where a correct rule is to increase the gain and decrease the integral time. The upper diagram shows set-point y_{sp} and process output y, and the lower diagram shows control signal u.

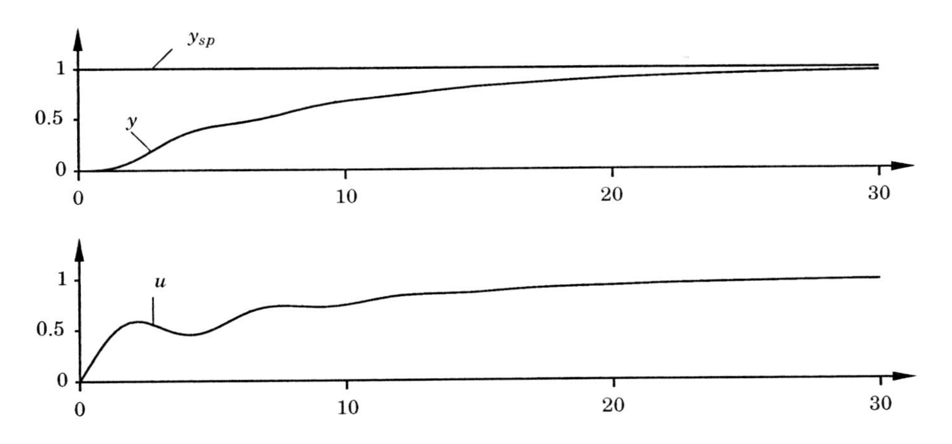

Figure 9.6 A set-point response where a correct rule is to decrease the gain and decrease the integral time. The upper diagram shows set point y_{sp} and process output y, and the lower diagram shows control signal u.

where rather small successive changes in the controller parameters are performed after each transient.

The rule-based methods have a great advantage compared to the model-based approaches when they are used for continuous adaptation, namely, that they handle load disturbances efficiently and in the same way as set-point changes. The model-based approaches are well suited for set-point changes. However, when a load disturbance occurs, the transient response is caused by an unknown input signal. To obtain an input-output process model under such circumstances is not so easy.

A drawback with the rule-base approaches is that they normally assume that the set-point changes or load disturbances are isolated steps or pulses.

Two set-point changes or load disturbances applied shortly after each other may result in a process output that invokes an erroneous controller tuning rule.

9.6 Supervision of Adaptive Controllers

Automatic tuning and gain scheduling have been well accepted by the process industry and are now common both in single-station controllers and distributed control systems. There are many well-engineered auto-tuners that are very easy to use. The industrial use of the "true" adaptive controller is, however, more limited. There are several reasons for this. One is that many controllers that have been tested industrially have not been sufficiently robust. This has tarnished the technique with a somewhat bad reputation. The adaptive algorithms must be provided with a supervisory shell that takes care of those operating conditions that the algorithm is not designed for.

The problem is not unique to adaptive controllers. *Every* controller needs a supervisory shell. The simple PID controller, e.g., has antiwindup functions to treat the situation when the control signal saturates, functions for bumpless transfer at mode switches between manual and automatic control, functions for bumpless transfer at parameter changes, and sometimes dead-zones and control signal rate limitations. This section discusses some supervisory functions for adaptive controllers.

Initialization

The first topic to consider is the initialization of the adaptive controller. Initialization should ensure that suitable controller parameters are used when the adaptation starts. An adaptive controller also requires additional parameters that should be obtained in the initialization phase. For example, the adaptive controllers, both the model based and the rule based, need to know the time scale of the process. It is used to set sampling periods and time constants.

In special-purpose adaptive controllers, the initialization can be performed manually by an experienced user. However, in multi-purpose adaptive controllers, this phase should not be left to unexperienced users. It should be performed automatically. Therefore, almost all industrial multi-purpose adaptive controllers have some kind of automatic tuning or pre-tuning function that initializes the adaptive controller. These procedures may be based on step response experiments, which provide the time scale of the process in terms of the apparent dead time and the apparent time constant. They can also be based on a relay feedback experiment. In this case the time scale of the process is obtained in terms of the ultimate frequency ω_u.

In the following, it is assumed that the time scale of the process has been obtained and is available in the adaptive controller. It is denoted by T_p. It is also assumed that the design calculations are performed in such a way that T_p also is proportional to the closed-loop time constant. The initialization procedure is not only invoked once when the adaptive controller is installed. Parts of the initialization procedure have to be used at mode transitions and parameter changes too.

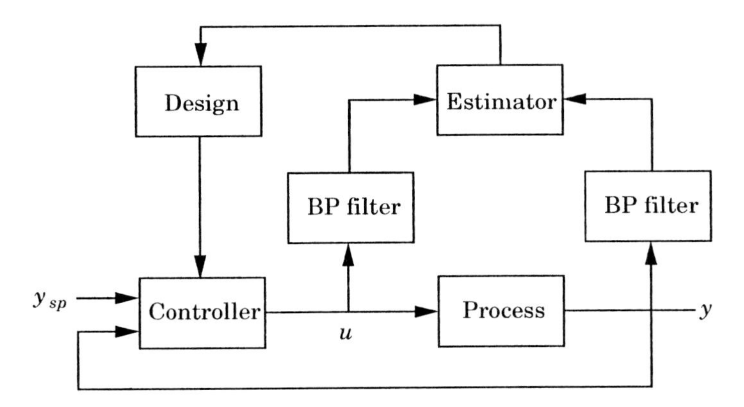

Figure 9.7 Filtering of control and measurement signals.

An important feature of the initialization procedure is to obtain suitable time constants of filters. Low-frequency components of the signals should be reduced in order to eliminate bias terms. High-frequency components are normally corrupted with measurement noise that disturbs the parameter estimator. Therefore, the control signal and the measurement signal should be band-pass filtered before entering the parameter estimator. See Figure 9.7.

It is important that parameter estimation is based on relevant data. If the model order is low, it is particularly important that the model be fitted to data in a frequency region that is suitable for controller design, namely, the frequency range around the ultimate frequency. This frequency range is determined by the choice of time constants in the band-pass filters. The frequency range can be made narrow or wide, depending on the control objective and the estimated model order. In the ECA600 controller, see Section 9.8, a narrow band-pass filter is used, and a process model consisting of only two parameters is identified. Models with more parameters require wider filters.

Excitation Detection

The parameter estimator is the central part of an adaptive controller. A recursive least squares estimator is normally used. This can be described by

$$
\begin{aligned}
\hat{\theta}(t) &= \hat{\theta}(t-1) + P(t)\varphi(t)\varepsilon(t) \\
\varepsilon(t) &= y(t) - \varphi(t)^T\hat{\theta}(t-1) \\
P(t) &= P(t-1) - \frac{P(t-1)\varphi(t)\varphi(t)^T P(t-1)}{1+\varphi(t)^T P(t-1)\varphi(t)},
\end{aligned}
\tag{9.1}
$$

where $\hat{\theta}$ is the parameter estimates, P is the covariance matrix, and φ is the regression vector, which normally contains delayed measurement and control signals. To be able to track variations in process dynamics, it is necessary to rely more on recent data than on older. This is often ensured by introducing a forgetting factor λ and modifying the covariance matrix according to

$$P(t) = \frac{1}{\lambda}\left(P(t-1) - \frac{P(t-1)\varphi(t)\varphi(t)^T P(t-1)}{\lambda + \varphi(t)^T P(t-1)\varphi(t)}\right). \tag{9.2}$$

A forgetting factor in the range $0 < \lambda < 1$ prevents the covariance matrix from converging to zero. The choice of λ is a compromise between adaptation rate and robustness. Decreasing λ will, e.g., result in an increased adaptation rate but also decreased robustness. The introduction of a forgetting factor may cause problems if the excitation is not good enough. Suppose, e.g., that φ is zero for a certain period. From Equation 9.2 it then follows that the covariance matrix will increase exponentially. There are ways to overcome this problem, e.g., by using a variable forgetting factor or by using directional forgetting. It has also been proposed to reinitialize the covariance matrix periodically to ensure that P stays within certain bounds. This will surely solve the numerical problem but in such a way that the estimation uncertainty is varying periodically, which is unsatisfactory.

The excitation problem is not only a numerical problem. The problem is also to ensure that the parameter estimator is provided with enough relevant data to produce a reliable process model. There are in principle two solutions to this problem:

1. Ensure that excitation always is present by adding excitation signals to the process input.

2. Ensure that estimation is performed only when there is enough natural excitation of the process.

The first approach might seem appealing. An excitation signal that is so small that it is hardly noticeable compared to the normal measurement noise will not do much harm. Unfortunately, such an excitation is not of much help for parameter estimation. The excitation signal must have a significant amplitude to be of any use. Friction or other nonlinearities may otherwise distort or even eliminate the response from the process output, and the excitation is lost. An excitation signal with a significant amplitude causes degradation of the control, and can therefore only be accepted during short periods such as during an automatic tuning experiment. For these reasons, the first approach is seldom used in industrial controllers. Instead, the second approach is used.

To ensure that estimation is only performed after significant changes in set point or load, when there is enough excitation, a procedure that measures the excitation is needed.

A convenient approach for excitation detection that is similar to the one used in the ECA600 controller will now be described. The basic idea is to make a high-pass filtering of the measurement signal. When the magnitude of the filtered variable exceeds a certain threshold, it is concluded that the excitation is high enough for adaptation. The high-pass filter is given by

$$Y_{hp} = \frac{s}{s + \omega_{hp}} Y, \tag{9.3}$$

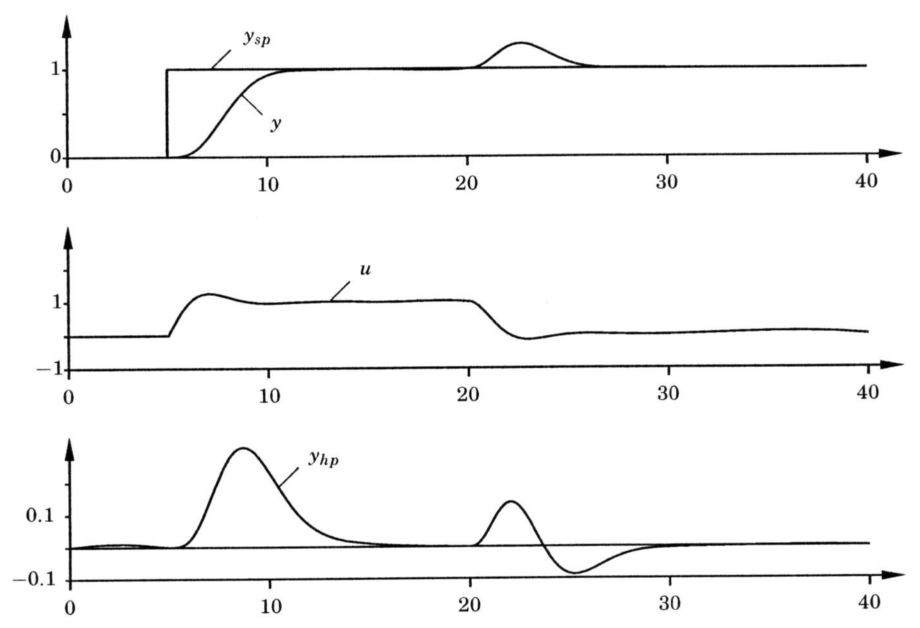

Figure 9.8 Excitation detection using high-pass filtering of the measurement signal. The figure shows responses to a set-point change at $t = 5$ and a load disturbance at $t = 20$.

where Y is the Laplace transform of the process output y, and Y_{hp} is the corresponding high-pass filtered signal. The filter has unit gain at high frequencies. The frequency ω_{hp} is chosen to be inversely proportional to the process time scale T_p.

Figure 9.8 shows a simulation where the measurement signal is passed through the high-pass filter (9.3). From the figure, it is obvious that the output from the high-pass filter is suitable for excitation detection. Excitation is high and adaptation can be initiated when the magnitude of $|y_{hp}|$ becomes large.

The next problem is to decide when the excitation is so low that the estimation should be interrupted again. One approach is to allow adaptation as long as $|y_{hp}|$ remains large. A drawback with this approach is that there are delays in the estimator. This means that even if $|y_{hp}|$ is small, there might still be excitation in the filtered signals in the parameter estimator. A solution to the problem is to simply allow adaptation for a fixed time after excitation has been detected.

Load Disturbance Detection

Model-based adaptive controllers have problems with load disturbances. To see this, consider the block diagram in Figure 9.9. The process output y is given by

$$Y(s) = P(s)\,(U(s) + D(s)) + N(s),$$

where $P(s)$ is the process transfer function, $U(s)$ is the Laplace transform of control signal u, $D(s)$ is the Laplace transform of load disturbance d, and $N(s)$ is the Laplace transform of measurement noise n. It is assumed that the

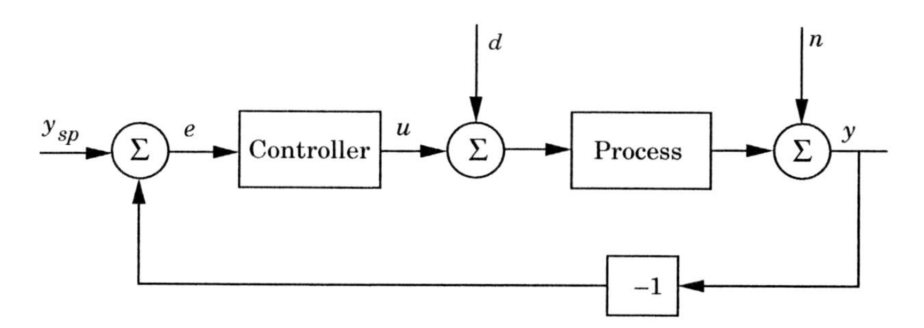

Figure 9.9 Block diagram of a simple feedback loop.

measurement noise only contains high frequencies and that these are filtered out by the filters in the controller. The noise term is therefore not considered in the sequel. The process output can be decomposed into two terms,

$$y(t) = y_u(t) + y_d(t), \tag{9.4}$$

where y_u is caused by the control signal and y_d is caused by the load disturbance.

In Equation 9.1, the prediction error in the least-squares estimator is given by

$$\varepsilon(t) = y(t) - \hat{y}(t) = y(t) - \varphi(t)^T\hat{\theta}(t-1). \tag{9.5}$$

The least-squares estimator tries to minimize $\varepsilon(t)$, i.e., to make the predicted process output $\hat{y}(t)$ equal to the true process output $y(t)$. It is implicitly assumed that

$$y(t) = y_u(t) = \varphi(t)^T\theta(t-1),$$

where $\theta(t)$ are the true process parameters. If this assumption is valid, i.e., if $y(t) = y_u(t)$, parameter estimates $\hat{\theta}(t)$ will converge to the true values $\theta(t)$, provided that the excitation is sufficient. However, if the process output is given by Equation 9.4, and if $y_d(t)$ has frequency components in the estimation region, the parameter estimates will not converge to their true values.

This is a very serious problem in process control applications. In process control, set-point changes are often performed only during production changes. (Exceptions are secondary controllers in cascade configurations.) This means that load disturbances often are the only excitation signals. For rule-based as well as model-based adaptive controllers there are possibilities to obtain useful information provided that the load disturbances come in the form of isolated transients. Such a solution will now be presented.

Figure 9.10 shows the different components of the process output after a step change in the load disturbance. Shortly after the load change, $y(t) \approx y_d(t)$, i.e., the changes in the process output are caused by load d only. After a while, the contribution from the control signal u is the dominating component.

A solution to the identification problem is to avoid adaptation during the first phase of the response, where y_d dominates over y_u. Adaptation should be initiated in the second phase where the major excitation in $y(t)$ is caused by the control signal.

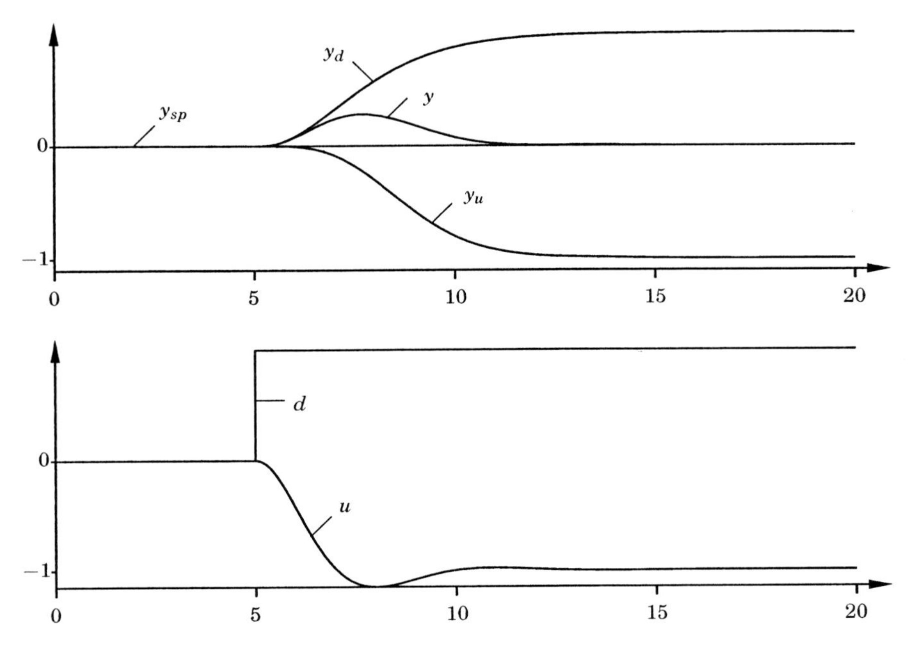

Figure 9.10 The upper diagram shows set point $y_{sp} = 0$, measurement signal y, its load component y_d, and its control signal component y_u. The lower diagram shows load disturbance d and control signal u.

To use this solution, a procedure that detects load disturbances is needed. This detection must be fast, so that adaptation is interrupted as quickly as possible. The detection can be done in the following way. First, the control signal is high-pass filtered in the same way as the measurement signal in Equation 9.3:

$$U_{hp} = \frac{s}{s + \omega_{hp}} U. \tag{9.6}$$

Figure 9.11 illustrates the same experiment as in Figure 9.8, but the high-pass filtered value of the control signal is also presented. In the following, it is assumed that the process has a positive static gain, i.e., $P(0) > 0$, and that all zeros are in the left-half plane. After a set-point change, both y_{hp} and u_{hp} then go in the same direction, whereas they go in opposite directions when a load disturbance occurs. This difference can be used to distinguish between set-point changes and load disturbances. In this way, it is possible to delay the adaptation and avoid adaptation during the first phase of the load disturbance response, and perform adaptation only during the second phase.

Another simpler way to avoid adaptation during the first phase of a load disturbance response can be obtained from the prediction error $\varepsilon(t)$; see Equation 9.5. The prediction error can be written as

$$\begin{aligned} \varepsilon(t) &= y(t) - \varphi(t)^T \hat{\theta}(t-1) \\ &= \left(y_u(t) - \varphi(t)^T \hat{\theta}(t-1)\right) + y_d(t). \end{aligned} \tag{9.7}$$

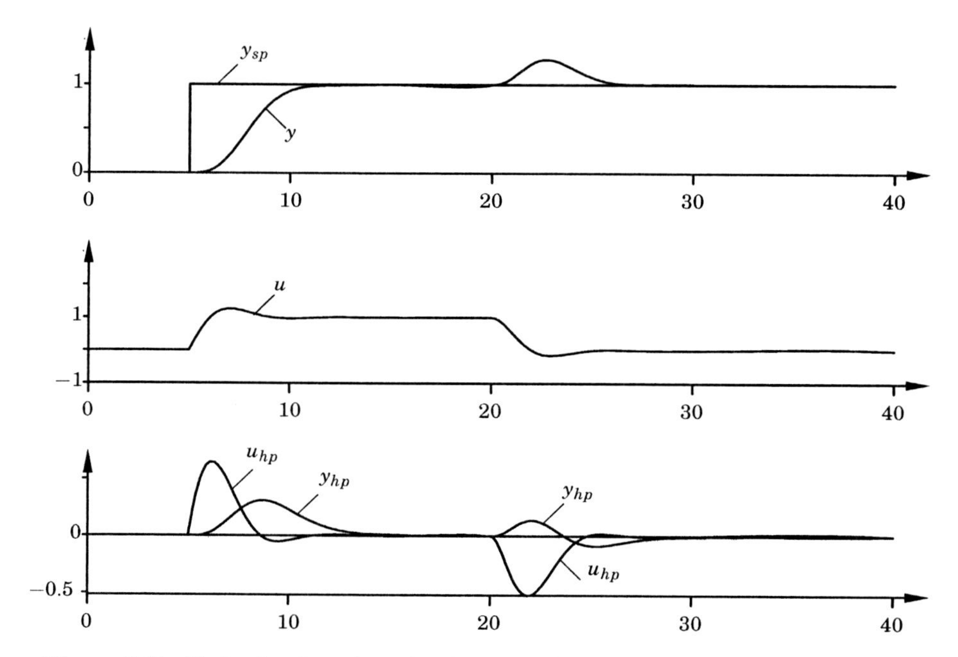

Figure 9.11 Excitation detection using high-pass filters. The figure shows responses to a set-point change at time t=5 and a load disturbance at time t=20.

Hence, $\varepsilon(t)$ consists of two terms, the true prediction error that we want to minimize and the load disturbance component of the process output. If we assume that process dynamics change slowly, the first term will remain bounded, and large load disturbances can be detected through the magnitude of $|\varepsilon(t)|$. If we restrict adaptation to those periods when $|\varepsilon(t)|$ is small, we will be able to track slow variations in the process, and we will also avoid adaptation when $|y_d(t)|$ becomes large.

Oscillation Detection

Oscillations with a high-frequency content near the ultimate frequency form an ideal excitation for adaptive control if they are caused by set-point variations or high controller gains resulting in small stability margins. In these cases, $y(t) = y_u(t)$.

Unfortunately, oscillations in control loops are normally generated by other sources. A common cause of oscillations is stick-slip motion because of valve friction. See Section 10.2. If no precautions are taken, the adaptive controller will interpret these oscillations as caused by a too high loop gain. This means that the controller will be detuned. This happens both for model-based and rule-based adaptive controllers. Stick-slip motion can be modeled as a load disturbance.

Another reason for oscillations in control loops might be that an external oscillating load disturbs the process. This disturbance may, e.g., be caused by a neighboring control loop with stick-slip motion.

In these cases, it is no longer true that $y(t) = y_u(t)$, but the load component

$y_d(t)$ dominates. This means that these disturbances will provide the process estimator with disinformation in the same way as the load disturbances discussed in the previous subsection. To avoid the problem, oscillations have to be detected in the same way as load disturbances, so that adaptation can be inhibited when these disturbances are present. Such detection procedures are presented in Section 10.4.

Signal Saturation

When the process output saturates, it is no longer true that $y(t) = y_u(t)$. Suppose that the process output becomes saturated at the limit y_{limit}. The process output can then described in the following way:

$$y(t) = y_u(t) + (y_{\text{limit}} - y_u(t)).$$

The second term on the right-hand side can be interpreted as a load disturbance component. Hence, we have the same problem as was discussed in the previous subsections. Therefore, the estimation should be interrupted when y saturates. Again, it is useful to have a timer connected to this interrupt, so that the estimation is kept off for a while after the saturation period. This is to avoid erroneous estimates during the transients.

It may also be desirable to interrupt adaptation when control signal u saturates. This might seem confusing, since $y(t) = y_u(t)$ in this case. However, if load disturbances are present it may no longer be true that $y_u(t)$ dominates over $y_d(t)$ in the second phase of a load disturbance response in this case.

Mode Transitions

A constant-parameter controller runs mainly in two modes:

- Manual mode
- Automatic mode

Bumpless transfer between the different operating modes is performed by ensuring that all states are assigned suitable values at the transitions. If this is not done properly, "bumps" will occur in the control signal at the mode transitions. See Section 13.5.

An adaptive controller has three modes of operation:

- Manual mode
- Automatic mode
- Adaptive mode

Here, it is important to ensure bumpless transfer also between the first two modes and the third adaptive mode. The parameter estimation is normally disconnected when the controller is in manual mode and often also when it is in automatic mode without adaptation. It is therefore important to initialize all the additional states that are given in the parameter estimator when the adaptive controller is started.

An erroneous initialization of the parameter estimator will result in "bumps" in the parameter estimates. These bumps are not always immediately visible as a "bump" in the control signal, but they may deteriorate the control in other ways since they provide an erroneous process model.

The most important states of the parameter estimator are given by Equations 9.1 and 9.2. The covariance matrix $P(t)$ should be assigned a large value when the parameter estimates are uncertain. However, when the controller parameters are initialized by an automatic tuning procedure or when they for other reasons are believed to be accurate, $P(t)$ should not be reinitialized to a large value but kept close to its stationary value.

The residual vector $\varphi(t)$ normally contains delayed control and measurement signals. This vector should be initialized by actual values of these signals.

It is also important that filters as well as the supervisory functions be provided with correct states. This can sometimes be accomplished by introducing a delay in the estimator. Suppose, e.g., that the controller is switched from manual to adaptive mode and that the process output is not close to the set point. This means that we immediately get a transient at the mode switch. If the excitation detection procedure is active, the adaptation mechanism may then start before the states have got their appropriate values. This problem can be avoided by delaying the excitation detection procedure at the mode transition.

A re-initialization of the adaptive controller must also be made if parameters related to the adaptation are changed. Suppose, e.g., that the sampling period is changed by the user or that an automatic tuning procedure is run, resulting in new values of the sampling period. This means that a total re-initialization of the adaptive controller must be made, with new filter-time constants, etc.

Another mode transition occurs if the adaptive controller is combined with gain scheduling. In this case, a re-initialization should be performed whenever there is a switch in the gain schedule.

It is also often possible to reset the adaptive controller, so that the parameter estimates $\hat{\theta}(t)$ are reinitialized to some pre-specified values, normally those obtained during the initialization phase.

Bounds on Parameter Estimates

There is a region in the parameter space where the information provided during the initialization phase is relevant. Inside this region, the a priori information about the process time T_p is correct, sampling periods and filter-time constants are suitable. If the process dynamics change so much that the parameter estimates tend to go outside this region, the behavior of the controller might be poor.

It is therefore advantageous to bound the parameter estimates to an allowable region. The adaptation may continue outside the region, but the algorithm should be reinitialized so that new parameters suitable for the new region are obtained. Using gain scheduling it is, e.g., possible to have several regions with different sampling periods and filter-time constants.

It may be difficult to find such regions if the estimated model is of high order. It is easier when the model order is lower, and it perhaps is possible to find physical interpretations of the parameters. In adaptive PID controllers, there are often bounds on the gain, integral time, and derivative time.

There is another reason for bounding the parameter estimates, which is related to the excitation needed for the parameter estimation. Suppose, e.g., that the parameter estimates change so much that a very low closed-loop bandwidth is obtained. The excitation in the interesting frequency band will then be low, and we will get a very slow adaptation.

It may also be advantageous to have bounds on the rate of estimate changes. This is done to decrease the effects of sudden outliers or other errors. This feature can be compared with the rate limiters that often are used in standard controllers.

9.7 Iterative Feedback Tuning

Iterative feedback tuning, IFT, is an iterative on-line method for adjusting controller parameters. The key idea is a clever way of computing the gradient of the controller error with respect to controller parameters.

Consider a standard system with error feedback. Assume that it is desired to minimize the loss function

$$J = \int_0^T f(y(t), u(t)) dt$$

for a PID controller with the parametrization

$$C(s) = k + \frac{k_i}{s} + k_d s.$$

To minimize the criterion it is useful to know the gradient of the loss function with respect to the controller parameters. The partial derivative of J with respect to controller gain k is given by

$$\frac{\partial J}{\partial k} = \int_0^T \Big(\frac{\partial f(y(t), u(t))}{\partial y} \frac{\partial y}{\partial k} + \frac{\partial f(y(t), u(t))}{\partial u} \frac{\partial u}{\partial k} \Big) dt. \tag{9.8}$$

To evaluate the right-hand side we need the partial derivatives

$$y_k = \frac{\partial y}{\partial k}, \quad u_k = \frac{\partial u}{\partial k}.$$

They can conveniently be computed from the Laplace transforms. We have

$$\begin{aligned} Y_k &= \frac{\partial Y}{\partial C}\frac{\partial C}{\partial k} = \frac{\partial Y}{\partial C} \\ Y_{k_i} &= \frac{\partial Y}{\partial C}\frac{\partial C}{\partial k_i} = \frac{1}{s}\frac{\partial Y}{\partial C} \\ Y_{k_d} &= \frac{\partial Y}{\partial C}\frac{\partial C}{\partial k_d} = s\frac{\partial Y}{\partial C}. \end{aligned} \tag{9.9}$$

The process output is given by

$$\begin{aligned} Y &= \frac{PC}{1+PC}Y_{sp} + \frac{P}{1+PC}D + \frac{1}{1+PC}N \\ &= \Big(1 - \frac{1}{1+PC}\Big)Y_{sp} + \frac{P}{1+PC}D + \frac{1}{1+PC}N, \end{aligned}$$

and the control error is given by

$$E = Y_{sp} - Y = \frac{1}{1+PC}Y_{sp} - \frac{P}{1+PC}D - \frac{1}{1+PC}N.$$

Using this expression for the error we find

$$\begin{aligned} \frac{\partial Y}{\partial C} &= \frac{P}{(1+PC)^2}Y_{sp} - \frac{P^2}{(1+PC)^2}D - \frac{P}{(1+PC)^2}N \\ &= \frac{P}{1+PC}\Big(\frac{1}{1+PC}Y_{sp} - \frac{P}{1+PC}D - \frac{1}{1+PC}N\Big). \end{aligned}$$

Hence,

$$\frac{\partial Y}{\partial C} = \frac{P}{1+PC}E = \frac{1}{C}\frac{PC}{1+PC}E. \tag{9.10}$$

The partial derivatives of the output with respect to the controller parameters can be computed in a similar way. We have

$$\begin{aligned} U_k &= \frac{\partial U}{\partial C}\frac{\partial C}{\partial k} = \frac{\partial U}{\partial C} \\ U_{k_i} &= \frac{\partial U}{\partial C}\frac{\partial C}{\partial k_i} = \frac{1}{s}\frac{\partial U}{\partial C} \\ U_{k_d} &= \frac{\partial U}{\partial C}\frac{\partial C}{\partial k_d} = s\frac{\partial U}{\partial C}. \end{aligned} \tag{9.11}$$

Straightforward calculations show that the sensitivity derivative of the output is given by

$$\frac{\partial U}{\partial C} = \frac{1}{1+PC}E. \tag{9.12}$$

Equations 9.10 and 9.12 can be used to compute the sensitivity derivatives needed for the optimization. The error E is known, but there is a difficulty because the process transfer function P is not known. This difficulty can be circumvented in the following way:

- Make an experiment, and store the output y_1 and the control error signal e_1.
- Make a second experiment of the same duration where the set point is chosen as the control error e_1 from the first experiment. Store the output y_2 and the control error e_2 of this experiment.

The output and the control error of the second experiment are given by

$$\begin{aligned}
Y_2 &= \frac{PC}{1+PC}E_1 + \frac{P}{1+PC}D_2 + \frac{1}{1+PC}N_2 \\
&= \frac{1}{C}\frac{\partial Y_1}{\partial C} + \frac{P}{1+PC}D_2 + \frac{1}{1+PC}N_2 \\
E_2 &= \frac{1}{1+PC}E_1 - \frac{P}{1+PC}D_2 - \frac{1}{1+PC}N_2 \\
&= \frac{\partial U_1}{\partial C} + \frac{P}{1+PC}D_2 + \frac{1}{1+PC}N_2.
\end{aligned}$$

The terms D_2 and N_2 are uncorrelated with E_1 if the experiments are well separated in time. Their effect can be made arbitrarily small by choosing long data sequences. Hence,

$$\begin{aligned}
\frac{\partial Y_1}{\partial C} &\approx CY_2 \\
\frac{\partial U_1}{\partial C} &\approx E_2.
\end{aligned} \tag{9.13}$$

The second experiment thus gives an estimate of the sensitivity derivatives of the input and the output with respect to the controller parameters. Combining this with the input and the output from the first experiment we can now compute the gradient of the loss function with respect to the controller parameter from (9.8). The controller parameters can then be adjusted recursively. Summarizing we obtain the following algorithm.

ALGORITHM 9.1—ITERATIVE FEEDBACK TUNING

1. Make an experiment of fixed duration, and store the output y_1 and the control error signal e_1.
2. Make a second experiment of the same duration where the set point is chosen as the control error e_1 from the first experiment. Store the output y_2 and the control error e_2 of this experiment.
3. Compute the gradient of the loss function from Equations 9.8, 9.9, 9.11, and 9.13.
4. Modify the controller parameters using the gradient.
5. Repeat from 1 until the gradient is sufficiently small.

□

The same idea can be applied to a controller with two degrees of freedom but a third experiment is then required. A nice property of iterative feedback tuning is that it can be used for many different controllers and criteria. It is particularly well suited to optimization with respect to stationary stochastic disturbances.

9.8 Commercial Products

To illustrate how adaptive techniques are used industrially we present some features of industrial controllers. Rather than to give an exhaustive presentation we have selected a few products to show the wide range of techniques, and we have chosen products that have a good track record. We have also selected products where reasonably detailed descriptions are published; more products are described in the book [Van Doren, 2003] and in reviews in trade journals.

Foxboro EXACT^{TM}(760/761)

Foxboro was one of the first companies to announce products using adaptive techniques. The single-loop controller Foxboro EXACT^{TM}(760/761) , which used adaptation based on pattern recognition , was released by Foxboro in October 1984. The controller was later augmented with more features and Foxboro has continued to expand their use of adaptation in a range of products including their DCS system Foxboro I/A^{TM}. The ideas are described in [Bristol *et al.*, 1970] and [Bristol, 1977] and details about the system are found in [Bristol and Kraus, 1984] and [Bristol, 1986]. Foxboro continued the development of adaptation, and auto-tuning and adaptation are now available in their distributed control system under the trade name Exact MV^{TM}. A presentation of the details of the system are found in [Hansen, 2003]. Three function blocks, PIDA, FBTUNE, and FFTUNE, are used to implement the controller. PIDA is an advanced PID controller, FBTUNE, which handles tuning of the feedback gains, has functions for pretuning and adaptation, and FFTUNE has functions for tuning of feedforward gains and gain scheduling.

Controller Structure Foxboro uses a controller structure where integral action is implemented with positive feedback around a lag as illustrated in Figure 3.3. This implementation gives a controller in series form, see (3.8). A controller with a special structure called PIDτ is also available in the system. This controller is a PID controller where the integral action is implemented with positive feedback around a lag with a time delay as shown in Figure 8.10. This arrangement gives a controller with more phase lead than an ordinary PID controller. Since phase lead is also associated with high gain it is necessary to provide good filtering if there is measurement noise. The controller can be interpreted as a controller where the future output is predicted with a combination of past controller inputs and controller outputs, see the discussion of the PPI controller in Section 8.4. The controller PIDτ can also be regarded as a special form of a Smith predictor. The controller PIDτ gives significant improvement of performance for lag-dominated processes but it requires careful tuning.

Pattern Recognition Adaptation based on pattern recognition can be viewed as an automation of the procedure used by an experienced process engineer when he tunes a controller. The following description follows the presentation in [Bristol and Kraus, 1984]. The control error after process perturbations are analyzed and the controller parameters are modified. If the controller parameters are reasonable, a transient error response of the type shown in Figure 9.12

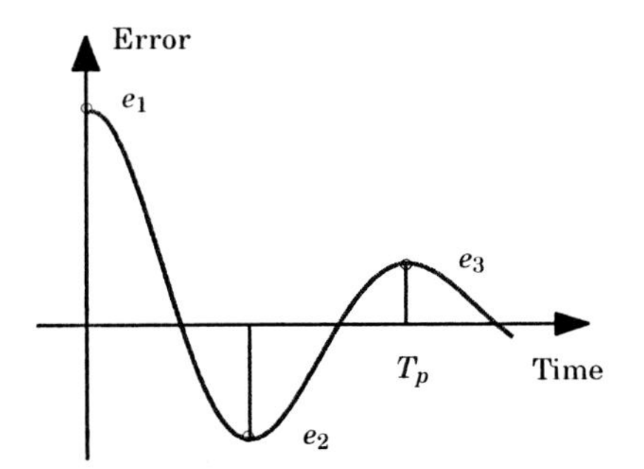

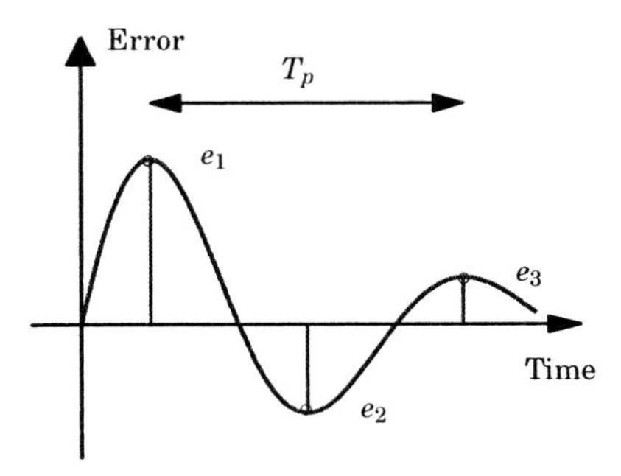

Figure 9.12 Response in control error to a step change of set point (left curve) and load (right curve).

is obtained. Heuristic rules are used to detect that a proper disturbance has occurred and to detect peaks e_1, e_2, e_3, and oscillation period T_p. Heuristics are also used to change the controller parameters if the response is overdamped. The transient is characterized quantitatively in terms of two parameters, overshoot (o) and damping (D), which are defined as

$$o = \left|\frac{e_2}{e_1}\right|, \qquad D = \frac{e_3 - e_2}{e_1 - e_2}, \tag{9.14}$$

where e_1, e_2, and e_3 are the peaks of the transients shown in Figure 9.12. Note that the definition of damping used is equal to the square root of the decay ratio (2.49). Quarter amplitude damping thus corresponds to $D = 0.5$.

The controller parameters are adjusted using heuristic rules to obtain desired damping and overshoot. Some rules are discussed in Section 6.3, and the effect of controller parameters on the transient are illustrated in the tuning map in Figure 6.7.

Pre-Tuning is Foxboro's notation for auto-tuning. The controller has a set of parameters that must be given either by the user from prior knowledge of the loop or estimated using the pre-tune function.

- Initial values of PB, T_i', and T_d'.
- Noise band *(NB)*. The controller starts adaptation whenever the error signal exceeds two times NB.
- Maximum wait time ($W_{\max}$). The controller waits for a time of $W_{\max}$ for the occurrence of the second peak.

If the user is unable to provide the required parameters, a pre-tune function that estimates these quantities can be activated. To activate the pre-tune function, the controller must first be put in manual. When the pre-tune function is activated, a step input is generated. The process parameters static gain K_p, dead time L, and time constant T are then obtained from a simple analysis of the process reaction curve. The controller parameters are calculated using a Ziegler-Nichols-like formula:

$$PB = 120K_pL/T, \quad T_i' = 1.5L, \quad T_d' = T_i/6. \tag{9.15}$$

Maximum wait time, $W_{\max}$, is also determined from the step response by $W_{\max} = 5L$.

The noise band is determined during the last phase of the pre-tune mode. The control signal is first returned to the level before the step change. With the controller still in manual and the control signal held constant, the output is passed through a high-pass filter. The noise band is calculated as an estimate of the peak-to-peak amplitude of the output from the high-pass filter. The estimated noise band is also used to adjust derivative action.

There are a number of optional parameters. If these are not supplied by the user then the default values will be used. The optional parameters are as follows (default values in parenthesis):

- Maximum allowed damping (0.3)
- Maximum allowed overshoot (0.5)
- Derivative factor (1). The derivative term is multiplied by the derivative factor. This allows the derivative influence to be adjusted by the user. Setting the derivative factor to zero results in PI control.
- Change Limit (10). This factor limits the controller parameters to a certain range. Thus, the controller will not set the PB, T_i' and T_d' values higher than ten times or lower than one tenth of their initial values if the default of 10 is used for the change limit.

The pretuning has been improved in the later Foxboro products. The process is excited with a doublet pulse, see Figure 2.33, instead of the step used in the original system. An FOTD or SOTD model is fitted to the data from the experiment, as described in [Shinskey, 1994]. The controller parameters are calculated from the model based on a novel robust analytic design method described in [Hansen, 2003]. Adaptation of feedback gain parameters are still done using the pattern recognition method.

The controllers in the Foxboro DCS system have lead-lag filters for feedforward from measured disturbances. The feedforward gains are tuned by fitting a low-order continuous-time model using the method of moments, see [Hansen, 2003].

ABB

Adaptation in the ABB's systems has its roots in a auto-tuners based on relay feedback first developed by the company NAF in the early 1980's. The first system was part of a small (about 50 loops) DCS system called SDM-20™ introduced in 1982 and a single-loop controller ECA-40™ introduced in 1986. These systems also used auto-tuning to build gain schedules. The company NAF went through a series of acquisitions, SattControl, Alfa Laval Automation, and is now part of ABB. The adaptive techniques were developed by adding continuous adaptation, adaptation of feedback and feedforward gains, and diagnostics. These features were all introduced in the ECA600™ which was announced in 1988. The technology is now an integral part of the ABB DCS system Industrial IT System 800xA™, which also has facilities for fuzzy control and for model predictive control. There are several types of PID controllers; the advanced versions give access to more parameters. There is also a PPI controller for systems with time delay.

Essential parts of the technology are described in [Åström and Hägglund, 1984c; Åström and Hägglund, 1984a; Åström and Hägglund, 1988; Åström and Hägglund, 1990; Åström and Hägglund, 1995a; Åström and Hägglund, 1995b; Hägglund, 1999; Hägglund and Åström, 2000; ABB, 2002].

Bi-directional Data Flow Distributed control systems are traditionally programmed graphically using a block oriented language. One drawback with traditional systems is that data-flow is unidirectional. This leads to unpredictable latency in the system, which is particularly noticeable in large systems and when back calculations to avoid windup are propagated through several loops. An interesting novel feature of the ABB System 800xA is a data structure called *control connection* that permits bi-directional data flow between the control modules. This feature makes it possible to implement windup protection in an elegant way which avoids latency even in complex systems with many cascades.

Relay Auto-Tuning The auto-tuning is performed using the relay method discussed in Section 2.7. The tuner is typically operated as follows. The process is brought to a desired operating point, either by the operator in manual mode or by a previously tuned controller in automatic mode. When the loop is stationary, the operator presses a tuning button. After a short period, when the noise level is measured automatically, a relay with hysteresis is introduced in the loop, and the PID controller is temporarily disconnected (see Figure 9.4). The width of the hysteresis is set automatically, based on measurement of the noise level in the process. The lower the noise level, the lower the amplitude required from the measured signal. The relay amplitude is controlled so that the oscillation is kept at a minimum level above the noise level. When an oscillation with constant amplitude and period is obtained, the relay experiment is interrupted and $P(i\omega_0)$, i.e., the value of the transfer function P at oscillation frequency ω_0, is calculated using describing function analysis.

Control Structures and Controller Design Several PID and PPI controllers are available in the ABB systems. The advanced versions give the user access to many parameters. The PID algorithm in the ECA600™ controller is of series form, the controllers in the DCS system use the parallel form.

The identification procedure provides a process model in terms of one point $P(i\omega_0)$ on the Nyquist curve. There is also a test to determine if process dynamics is lag dominated. The frequency ω_0 depends on the hysteresis in the relay. It is typically less than ω_{180}, which is advantageous; see Section 7.5. By introducing the PID controller $C(i\omega)$ in the control loop, it is possible to move the point corresponding to ω_0 on the Nyquist curve of the loop transfer to a desired location. In the normal case the desired point is

$$P(i\omega_0)C(i\omega_0) = 0.5e^{-i135\pi/180}. \tag{9.16}$$

Since there are three parameters, K, T_i, and T_d, and the design criterion (9.16) only specifies two parameters the additional constraint

$$T_i' = 4T_d'. \tag{9.17}$$

is introduced.

The normal procedure can give very high gains for lag-dominated systems. If this is detected a PI controller with the conservative tuning

$$K = 0.5/|P(i\omega_0)|, \quad T_i = 4/\omega_0 \tag{9.18}$$

is used. A different tuning can also be used for processes which are delay dominated. A PI controller with the tuning rule

$$K = 0.25/|P(i\omega_0)|, \quad T_i = 1.6/\omega_0 \tag{9.19}$$

is then used.

In the early versions of the controller (ECA-40^{TM} and ECA-600^{TM}) the user can influence tuning by selecting normal/PI/time-delay. Later versions of the controller obtain more information about the process by making a step change after the relay experiment. This gives the static gain of the process and gain ratio κ, and tuning can be improved without any user interaction. In the ABB 800xA^{TM} system this is accomplished by using the tuning rules in [Åström and Hägglund, 1995a]. Further improvements are possible by using the results in Chapter 7.

Gain Scheduling Gain scheduling was introduced in the early controllers SDM-20^{TM} and ECA-40^{TM}. It was very easy to build the schedules by using auto-tuning and the feature was well accepted by the users. Gain scheduling therefore became a standard feature of almost all controllers. The users can select the scheduling variable as the control signal, the measured process output, or an external signal. It is important that the scheduling variables do not change too quickly, filtering and hysteresis are used for signals like the process output that can change rapidly. Three sets of parameter values were available in the early systems but larger tables can be used in the later versions. The parameters are obtained by using the auto-tuner once at every operating condition. The ranges of the scheduling variable where different parameters are used can also be given by the user.

Adaptive Feedback Information from the relay feedback experiment is used to initialize the adaptive controller. Figure 9.7 shows the principle of the adaptive controller. The key idea is to track the point on the Nyquist curve obtained by the relay auto-tuner. It is performed in the following way. The control signal u and the measurement signal y are filtered through narrow band-pass filters centered at frequency ω_0. This frequency is obtained from the relay experiment. The signals are then analyzed in a least-squares estimator, which provides an estimate of the point $P(i\omega_0)$.

Adaptive Feedforward Feedforward from measured disturbances can frequently improve performance significantly. Adaptive feedforward has been a feature in all controllers starting with the ECA400^{TM}. Diagnostics for on-line assessment of the potential value of feedforward is an active research topic; see [Petersson *et al.*, 2001], [Petersson *et al.*, 2002], and [Petersson *et al.*, 2003].

The adaptive feedforward control is based on the simple model

$$y(t) = au(t - 4h) + bv(t - 4h), \tag{9.20}$$

where y is the measurement signal, u is the control signal and v is the disturbance signal that should be fed forward. The sampling interval h is determined from the relay experiment as $h = T_0/8$, where T_0 is the oscillation period. The parameters a and b are estimated recursively by a least-squares algorithm. The feedforward compensator has the simple structure

$$\Delta u_{ff}(t) = k_{ff}(t)\Delta v(t), \tag{9.21}$$

where the feedforward gain k_{ff} is calculated from the estimated process parameters

$$k_{ff}(t) = -0.8\,\frac{\hat{b}(t)}{\hat{a}(t)}. \tag{9.22}$$

The Man/Machine Interface The auto-tuners based on relay feedback can be implemented with very simple man-machine interfaces. In many cases it is sufficient to provide the controllers with just one button to initiate tuning. Gain scheduling can also be implemented in a very user-friendly fashion. Many of the problems normally associated with implementation of adaptive controllers can be avoided because the auto-tuner gives good initial values.

Industrial experience has also indicated that there is a significant advantage to combine adaptation with diagnostics and supervision. For example, it is meaningless to tune a controller if there is a bad actuator in the loop.

Emerson Process Management

The adaptive techniques used in Emerson's systems go back to the DCS systems Provox^{TM} and RS3^{TM}, where the Fisher-Rosemount Intelligent Tuner and Gain Scheduler were introduced. Use of adaptation has been expanded in the Delta V system. Fairly detailed information about the techniques used is available in the book [Blevins *et al.*, 2003], which also contains many references. The system has facilities for auto-tuning, gain scheduling, and adaptation. There is also software for fuzzy control and for model predictive control.

The automatic tuning is based on relay feedback. The range of the relay oscillation is typically a few percent of the full signal range. An estimate of the apparent time delay is obtained by analysing the initial portion of the first step. When an estimate of the time delay is available it is also possible to obtain an FOTD model. The parameters K_p, T, and L of the FOTD model can be displayed. Tuning is typically accomplished in a few periods of the oscillation. Since a FOTD model is available it is possible to use several tuning techniques. The available options include Ziegler-Nichols tuning, IMC tuning, and Lambda tuning to mention a few. The system is structured so that an inexperienced user has few choices, but an experienced user has many options. There is also a built-in simulator so that tuning can be tested against the process model before committing it to the process.

Adaptive control is based on data from the process during normal operation; excitation can also be provided. The system consists of a supervisor, an excitation generator, adaptors for gain, integration time, and derivative time, and a safety net. The goal of the adaptation is to obtain a well damped slightly oscillatory response. The approach is similar to that used in the Foxboro Exact.

Gain scheduling is done by estimating static process characteristics. Interpolation is done using fuzzy techniques.

Honeywell

Honeywell products using adaptive control started with the single-loop controller UDC 6000™ which had an adaptive function called Accutune. The adaptive techniques were developed further and they are essential components of Honeywells DCS system TDC 3000™.

UDC 6000™ Adaptation in the UDC 6000™ combines model-based procedures and rule-based procedures. Modeling is based on a step-response experiment. The user brings the process variable to a point some distance away from the desired set point in manual and waits for steady state. Switching to tuning mode initiates an open-loop step response experiment, where the size of the step is calculated to be so large that it is supposed to take the process variable to the set point.

During the experiment, the process variable and its derivative are continuously monitored. Dead time L is calculated as the time interval between the step change and the moment the process variable crosses a certain small limit.

If the derivative of the process variable continuously decreases from the start, it is concluded that the process is of first order and an FOTD model is determined from a few points on the step response. The calculations can be performed before the steady-state is reached, and it is claimed that the process is identified in a time less than one third of the time constant.

If the derivative of the process variable increases to a maximum and then decreases, the process is identified as a second-order process and an SOTD model is determined from the step response. The controller is then switched to automatic mode and controlled to the set point using preliminary controller parameters when the maximum slope of the process output has been reached, bu it is necessary to wait for steady state to obtain the complete model. More details about the modeling procedure are given in [Åström and Hägglund, 1995b].

When the model has been obtained the controller parameters are calculated from the model and the controller is switched to automatic control mode.

The controller used is on series form with the transfer function

$$C(s) = K\frac{(1+sT_i')(1+sT_d')}{sT_i'(1+0.125sT_d')}.$$

Notice that the filter time constant is 1/8th of the derivative time. Controller design is based on pole placement of the Dahlin-Higham type procedure where the process poles are cancelled. There are several different versions depending on system order and time delay. For systems with time delay the closed loop time constant is chosen as $T_{cl} = L + T/3$. The UDC 6000 controller also has continuous adaptation which is activated when the process variable changes more than 0.3 percent from the set point or if the set point changes more than a prescribed value.

Honeywell LOOPTUNE™ The DCS system TDC3000 has a wide range of controllers; Basic Controllers, Extended Controllers, Multifunction Controllers, Process Managers and Application Modules. LOOPTUNE is a software package in the system that tunes loops with PID controllers.

The tuning algorithm does not rely on any particular model of the system. Performance is evaluated using the quadratic loss function

$$J = \frac{1}{N} \sum_{t=1}^{N} ((1-\rho)(y(t) - y_{sp}(t))^2 + \rho(u(t) - u(t-1))^2), \tag{9.23}$$

where N is the evaluation horizon and ρ a weighting factor that balances control error against actuator changes. The controller parameters are changed one at a time and the loss function is evaluated over a given time horizon N. A large value of N is required to obtain a reliable estimate but the evaluation takes long time. Process knowledge can be used to improve the search for good controller parameters by biasing the search towards higher controller gain and lower integration time.

Yokogawa SLPC-181, 281

The Yokogawa SLPC-181 and 281 both use a process model as a first-order system with dead time for calculating the PID parameters. A nonlinear programming technique is used to obtain the model. The PID parameters are calculated from equations developed from extensive simulations. The exact equations are not published.

Two different controller structures are used.

$$\begin{aligned} 1:\quad u &= K\left(-y + \frac{1}{T_i}\int e\,dt - T_d\frac{dy_f}{dt}\right) \\ 2:\quad u &= K\left(e + \frac{1}{T_i}\int e\,dt - T_d\frac{dy_f}{dt}\right) \end{aligned} \tag{9.24}$$

where y_f is generated by filtering y with a first order filter having time constant T_d/N. The first structure is recommended if load disturbance rejection is most important, and structure 2 if set-point responses are most important. The set point can also be passed through two filters in series:

$$\text{Filter 1: } \frac{1+\alpha_i sT_i}{1+sT_i} \qquad \text{Filter 2: } \frac{1+\alpha_d sT_d}{1+sT_d} \tag{9.25}$$

where α_i and α_d are parameters set by the user, mainly to adjust the overshoot of the set-point response. The effects of these two filters are essentially

Table 9.2 Set-point response specifications used in the Yokogawa SLPC-181 and 281.

Type	Features	Criteria
1	no overshoot	no overshoot
2	5% overshoot	ITAE minimum
3	10% overshoot	IAE minimum
4	15% overshoot	ISE minimum

equivalent to set-point weighting. It can be shown that $\alpha_i = b$, where b is the set-point weighting factor.

The user specifies the type of set-point response performance according to Table 9.2. A high overshoot will, of course, yield a faster response. The controller has four adaptive modes:

Auto mode. The adaptive control is on. PID parameters are automatically updated.

Monitoring mode. In this mode, the computed model and the PID parameters are only displayed. This mode is useful for validating the adaptive function or checking the process dynamics variations during operation.

Auto startup mode. This is used to compute the initial PID parameters. An open loop step response is used to estimate the model.

On-demand mode. This mode is used to make a set-point change. When the on-demand tuning is requested, a step change is applied to the process input in closed loop. The controller estimates the process model using the subsequent closed-loop response.

The controller constantly monitors the performance of the system by computing the ratio of the variances of process output and model output. This ratio is expected to be about 1. If it is greater than 2 or less than 0.5, a warning message for retuning of the controller is given. Dead time and feedforward compensation are available for the constant gain controller, but they are not recommended by the manufacturer to be used in conjunction with adaptation.

Techmation Protuner

The Protuner is a process analyzer from Techmation Inc. It consists of a software package for personal computers and an interface module with cables to be connected to the process output and the control signal of the control loop to be analyzed. The Protuner monitors a step-response experiment, calculates the frequency response of the process based on the experimental data, and suggests controller parameters based on several methods for controller tuning.

Prior Information Before the process analysis is performed, the user must provide some information about the process and the controller. This is done using a couple of "Set-up menus." The following process information must be given:

- The ranges of the control and the measurement signals.

- It must be determined if the process is stable or if it has integral action.

To be able to set relevant controller parameters, the following data about the controller must be provided:

- P-type (gain or proportional band)
- I-type (seconds, seconds/repeat, minutes, or minutes/repeat)
- Controller structure (ideal, series, or parallel)
- Sampling rate
- Filter time constant (if there is a low-pass filter connected to the measurement signal).

Before the tuning experiment can be performed, the user must also specify a sample time. This is the time during which data will be collected during the experiment. It is important to choose the sample time long enough, so that the step response settles before the sample time has ended. In case of an open-loop experiment of an integrating process, the response must reach a constant rate of change when the experiment ends.

Determining the Process Model The tuning procedure is based on a step-response experiment. It can be performed either in open or closed loop. The open-loop experiment is recommended. When the user gives a start command, the process output and the control signal are displayed on the screen, with a time axis that is given by the sample time defined by the user. The user then makes a step change in the control signal. If the experiment is performed in closed loop a step is instead introduced in the set point.

There are several facilities for editing the data obtained from the step-response experiment. Outliers can be removed, and data can be filtered. These features are very useful because they make it possible to overcome problems that are often encountered when making experiments on industrial processes.

When the data has been edited the Protuner calculates the frequency response of the process. The result can be displayed in a Bode diagram, a Nyquist diagram, or a Nichols diagram. The static gain, the dominant time constant, and the apparent dead time are also displayed, as well as the ultimate gain and the ultimate period.

Design Calculations The controller parameters are calculated from the frequency response. A special technique is used. This is based on cancellation of process poles by controller zeros. The integral time and the derivative time are first determined to perform this cancellation. The gain is then determined to meet predetermined gain and phase margins.

The Protuner provides several design options. Controller parameters are given for the following closed-loop responses slow (critically damped), medium (slightly underdamped) and fast (decay ratio 0.38) responses. The different design options are obtained by specifying different values of the gain and the phase margins. The Protuner provides different controller parameters depending on whether set point or load disturbances are considered. Both P, PI, and

PID controller parameters are provided. The set-point weightings for proportional and derivative action and the high-frequency gain at the derivative part must be supplied by the user.

Evaluation It is possible to evaluate the performance of the closed-loop system in several ways. The combined frequency response, i.e., the frequency response of the loop transfer function $G_l(i\omega) = P(i\omega)C(i\omega)$, can be plotted in a Bode diagram, a Nyquist diagram, or a Nichols diagram. In this way, the phase and amplitude margins or the M_s value can be checked.

The Protuner also has a simulation facility. It is possible to simulate the closed-loop response of the process and the suggested controller. To do this, it is necessary to provide some additional controller parameters, namely, set-point weightings b and c, and derivative gain limitation factor N. Using the simulation facility, it is also possible to investigate the effects of noise and to design filters to reduce these effects.

Some Personal Reflections

Adaptive techniques have been used extensively in industry since the mid 1980s. The techniques are proven useful and the products continue to develop, but there is clearly a potential to improve current products.

Several lessons can be learned from the results of Chapter 7. One observation is that it is useful to characterize process dynamics with three parameters. Dynamics can then be classified as delay dominant, balanced, or lag dominant. Tight control can be obtained by using an FOTD model for systems with balanced and delay-dominated dynamics but control performance can be improved significantly for lag-dominated systems by using a better model. In Section 2.7 it was also shown that it is difficult to obtain an SOTD model from a step response experiment. Hence, it is not possible to design auto-tuners for tight control based on a step response or on knowledge of ultimate gain and ultimate frequency. An indication of this is that Foxboro switched to using a doublet instead of a step. The doublet can actually be regarded as a short version of an experiment with relay feedback.

It is highly desirable to accomplish tuning in a short time, as is illustrated by the Honeywell UDC 6000™. One advantage of the relay auto-tuner is that tuning often is accomplished in a time that is much shorter than the average settling time of the system. In particular, the time can be much shorter for systems with lag-dominated dynamics. An interesting question is therefore what information can be derived from an experiment with relay feedback. The Emerson experience indicates that at least an FOTD model can be determined from an experiment with relay feedback. To explore in detail the information that can be deduced from a relay experiment is therefore an interesting and useful research task. If improved models are obtained it is also possible to use the algorithms presented in Chapter 7 that give tight control. The potential gains are particularly large for lag-dominated process.

The model-free approaches to adaptive control have many attractive features but their main disadvantage is that tuning takes a long time. Tuning can be made more effective by using iterative feedback tuning, which also computes estimates of the gradient of the loss function; see Section 9.7. Adaptive controllers based on more elaborate models like the self-tuning controller discussed in Section 9.3 is an alternative to model based control. Such controllers work very well but so far they have required very knowledgeable users. There may be a possibility to make them simpler to use by exploiting the information obtained from automatic tuning. This may also be the road to introduce adaptation in the model predictive controllers.

Tools like Techmations Protuner, which permit simulation of a process with different controller settings, are very useful for the advanced user. Many components to build such a system are already available in current DCS systems. It is therefore natural to provide such tools as an integral part of the systems.

9.9 Summary

An essential feature of feedback is that it can be used to design systems that are insensitive to process variations. When there are large variations, performance may be improved by adjusting the controller parameters. Adaptive techniques are therefore increasingly being used in PID controllers to adapt the controller parameters to the changes in process dynamics or disturbances. In this chapter we have given a broad presentation of a variety of adaptive methods covering automatic tuning, gain scheduling, and continuous adaptation. The techniques are used in several ways.

In automatic tuning the controller parameters are adjusted on demand from the user. Gain scheduling can be used when there is a measured scheduling variable that correlate well with the process changes. The controller parameters are obtained from a table, which gives controller parameters as a function of the scheduling variable. Auto-tuning can be used to build the table. Adaptive control can be used when a scheduling variable is not available.

Model based and feature based methods are discussed, particular attention is given to use of relay feedback for auto-tuning, parameter estimation, and iterative feedback tuning.

The adaptive controller derives the knowledge required from the input and output of the process. Adaptive control is less robust than gain scheduling and it requires supervisory functions. Supervision of adaptive controllers is therefore discussed.

A short presentation of some industrial adaptive controller where adaptive methods have been used successfully are also discussed.

9.10 Notes and References

Controllers with automatic tuning grew out of research on adaptive control. Overviews of adaptive techniques are found in [Dumont, 1986; Åström, 1987a; Bristol, 1970; Åström, 1990]. More detailed treatments are found in the books [Harris and Billings, 1981; Hang *et al.*, 1993b; Åström and Wittenmark, 1995]. Overviews of different approaches and different products are found in [Isermann, 1982; Gawthrop, 1986; Kaya and Titus, 1988; Morris, 1987; Yamamoto, 1991; Åström *et al.*, 1993].

Many different approaches are used in the automatic tuners. The systems described in [Nishikawa *et al.*, 1984; Kraus and Myron, 1984; Takatsu *et al.*, 1991] are based on transient response techniques. The paper [Hang and Sin, 1991] is based on cross correlation. The use of orthonormal series representation of the step response of the system is proposed in [Zervos *et al.*, 1988; Huzmezan *et al.*, 2003]. Pattern recognition, which was the basis for Foxboros EXACT™ controller, is discussed in [Bristol, 1967; Bristol, 1970; Bristol *et al.*, 1970; Bristol, 1977; Bristol and Kraus, 1984; Bristol, 1986; Porter *et al.*, 1987; Anderson *et al.*, 1988; Klein *et al.*, 1991; Pagano, 1991; Swiniarski, 1991]. Auto-tuning based on relay feedback is treated in [Åström and Hägglund, 1984b; Åström and Hägglund, 1988; Hägglund and Åström, 1991; Schei, 1992; Hang *et al.*, 1993a; Leva, 1993; Schei, 1994; Voda and Landau, 1995]. Iterative feedback tuning is discussed in [Hjalmarsson *et al.*, 1998]. It is more effective than direct search because gradient information is used.

Traditional adaptive techniques based on system identification and control design have also been applied to PID control. Identification is often based on estimation of parameters in a transfer function model. Examples of this approach are given in [Hawk, 1983; Hoopes *et al.*, 1983; Yarber, 1984a; Yarber, 1984b; Cameron and Seborg, 1983]. There are also systems where the controller is updated directly as in [Radke and Isermann, 1987; Marsik and Strejc, 1989; Rad and Gawthrop, 1991]. Supervision of adaptive controllers is discussed in [Isermann and Lachmann, 1985; Sullivan, 1996; Clarke and Hinton, 1997; Liu, 1998; Hägglund and Åström, 2000].

An overview of several products that use adaptation is given by [Van Doren, 2003]. Several tuning aids are implemented in hand-held computers or as software in PCs where the user is entering the process information through a keyboard; see [Blickley, 1988; Tyreus, 1987; Yamamoto, 1991].

The papers [McMillan *et al.*, 1993b; McMillan *et al.*, 1993a] describe the Fisher Rosemount products for tuning and gain scheduling. The implementations in the Delta V™ DCS system is described in the book [Blevins *et al.*, 2003]. The Yokogawa systems are discussed in [Takatsu *et al.*, 1991] and [Yamamoto, 1991].

There have been comparisons of different auto-tuners and adaptive controllers, but few results from those studies have reached the public domain. Some papers that deal with the issue are [Nachtigal, 1986a; Nachtigal, 1986b; Dumont, 1986; Dumont *et al.*, 1989]. Some operational experience is described in [Higham, 1985; Callaghan *et al.*, 1986].

10

Loop and Performance Assessment

10.1 Introduction

The design, tuning, and implementation of control strategies and controllers is only the first phase in the solution of a control problem. The second phase includes operation, supervision, and maintenance. This phase has traditionally been handled manually, but the interest for automatic supervisory functions has increased significantly in recent years because of the reduction of personnel in the process industry.

This chapter treats methods for commissioning, supervision, and diagnosis of control loops. The adaptation methods presented in Chapter 9 were divided into two categories, tuning on demand and continuous adaptation. Procedures for supervision and diagnosis can be classified in the same way. We call them loop assessment and performance assessment. Loop assessment procedures are used to investigate properties of the control loop, e.g., signal levels, noise levels, nonlinearities, and equipment conditions. Performance assessment procedures are used to supervise the control loops during operation and ensure that they meet the specifications. Failure to meet the specifications may be caused by equipment problems, nonlinearities, or other variations in process dynamics or the surroundings.

The chapter begins with a presentation of problems occurring in valves. These problems are identified as one of the major reasons for bad control loop performance. Sections 10.3 and 10.4 treat loop assessment and performance assessment, respectively. Tuning and diagnosis have many aspects in common. These aspects are discussed in Section 10.5.

10.2 Valves

Control valves are subject to wear. After some time in operation, this wear results in friction and hysteresis that deteriorates the control performance. Furthermore, valves are often both nonlinear and over-sized. Therefore, valves

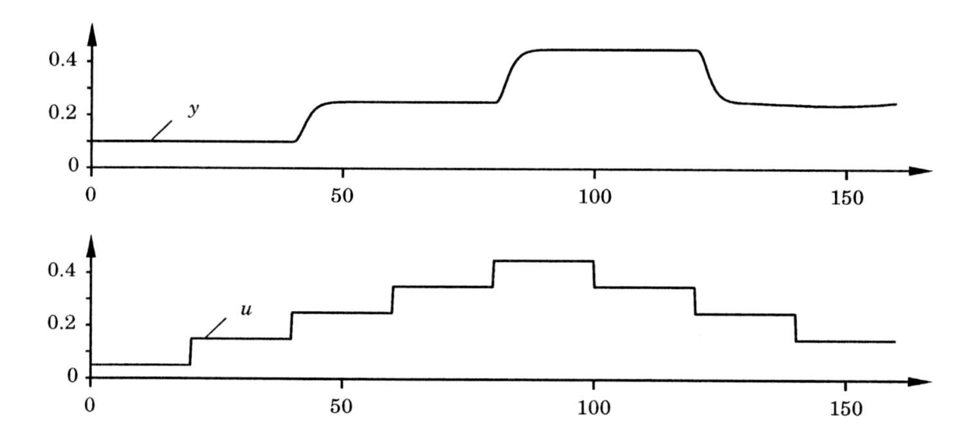

Figure 10.1 Procedure to check the amount of valve friction. The upper diagram shows process output y and the lower diagram shows control signal u.

have been identified as the major source of problems at the loop level in process control. We therefore devote a section to these problems.

Friction in the Valve

High friction in the valve is a common cause of problems. There is, of course, always static friction (stiction) in the valve, but if valve maintenance is insufficient, friction may be so large that the control performance degrades. The amount of friction can easily be measured by making small changes in the control signal and observing how the process outputs react. The procedure is shown in Figure 10.1. In the figure, the process output only responds to the control signal when the changes in the control signal are large enough to overcome the static friction.

Friction in the valve results in stick-slip motion. This phenomenon is illustrated in Figure 10.2. Suppose that the valve is stuck at a certain position due to friction. If there is a control error, the integral action of the controller will cause the controller output to increase until the pressure in the actuator is high enough to overcome the static friction. At this moment, the valve moves (slips) to a new position where it is stuck again. This valve position is normally such that the process output is moved to the other side of the set point, which means that the procedure is repeated. The process output will therefore oscillate around the set point. The pattern in Figure 10.2, where the measurement signal is close to a square wave and the control signal is close to a triangular wave, is typical for stick-slip motion.

Many operators detune the controller when they see oscillations like the one in Figure 10.2, since they believe that the oscillations are caused by a bad controller tuning. Unfortunately, most adaptive controllers do the same. What should be done when a control loop starts to oscillate is to first determine the cause of the oscillation. A good way to do this is presented in Figure 10.3.

The first problem to determine is whether the oscillations are generated inside or outside the control loop. This can be done by disconnecting the feed-

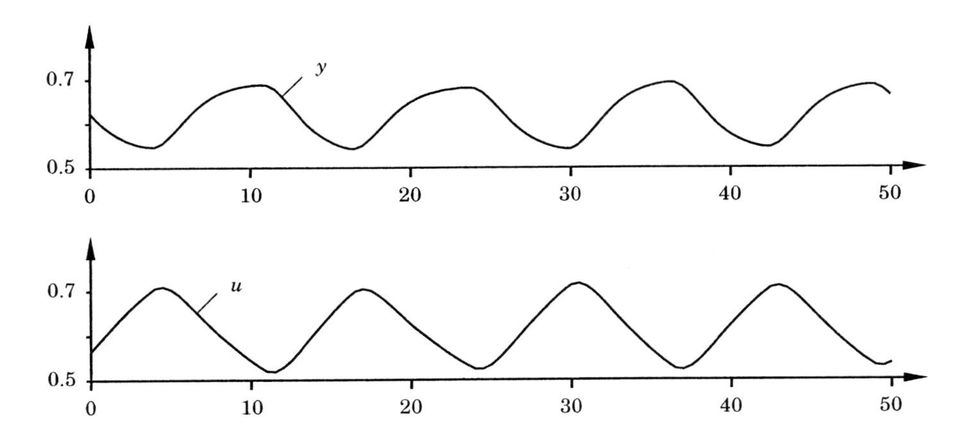

Figure 10.2 Stick-slip motion caused by valve friction and integral action. The upper diagram shows process output y, and the lower diagram shows control signal u.

back, e.g., by switching the controller to manual mode. If the oscillation is still present, the disturbances must be generated outside the loop, otherwise they were generated inside the loop. There might be a situation when the control loop oscillates because of valve friction even when the controller is in manual mode, namely, if the friction occurs in the pilot valve of the positioner instead of the valve itself.

If the disturbances are generated inside the loop, the cause can be either friction in the valve or a badly tuned controller. Whether friction is present or not can be determined by making small changes in the control signal and checking if the measurement signal follows, as shown in Figure 10.1. If friction is causing the oscillations, the solution to the problem is valve maintenance.

If the disturbances are generated outside the control loop, one should try, of course, to find the source of the disturbances and try to eliminate it. This is not always possible, even if the source is found. One can then try to feed the disturbances forward to the controller and in this way reduce their effect on the actual control loop. See Section 5.6.

Hysteresis in the Valve

Because of wear, there is often hysteresis (backlash) in the valve or actuator. The amount of hysteresis can be measured as shown in Figure 10.4. The experiment starts with two step changes in the control signal in the same direction. The hysteresis gap will close if the first step is sufficiently large. This means that the second step is performed without hysteresis. The third step is then made in the opposite direction. The control signal then has to pass the whole gap before the valve moves. If the last two steps are of the same size, the hysteresis is $\Delta y/K_p$, where Δy is the difference between the process outputs after the second and the third step (see Figure 10.4), and K_p is the static process gain (also easily obtained from Figure 10.4).

The hysteresis can also be determined from a continuous sweep over parts of the operating range. Figure 10.5 shows the process outputs from a process

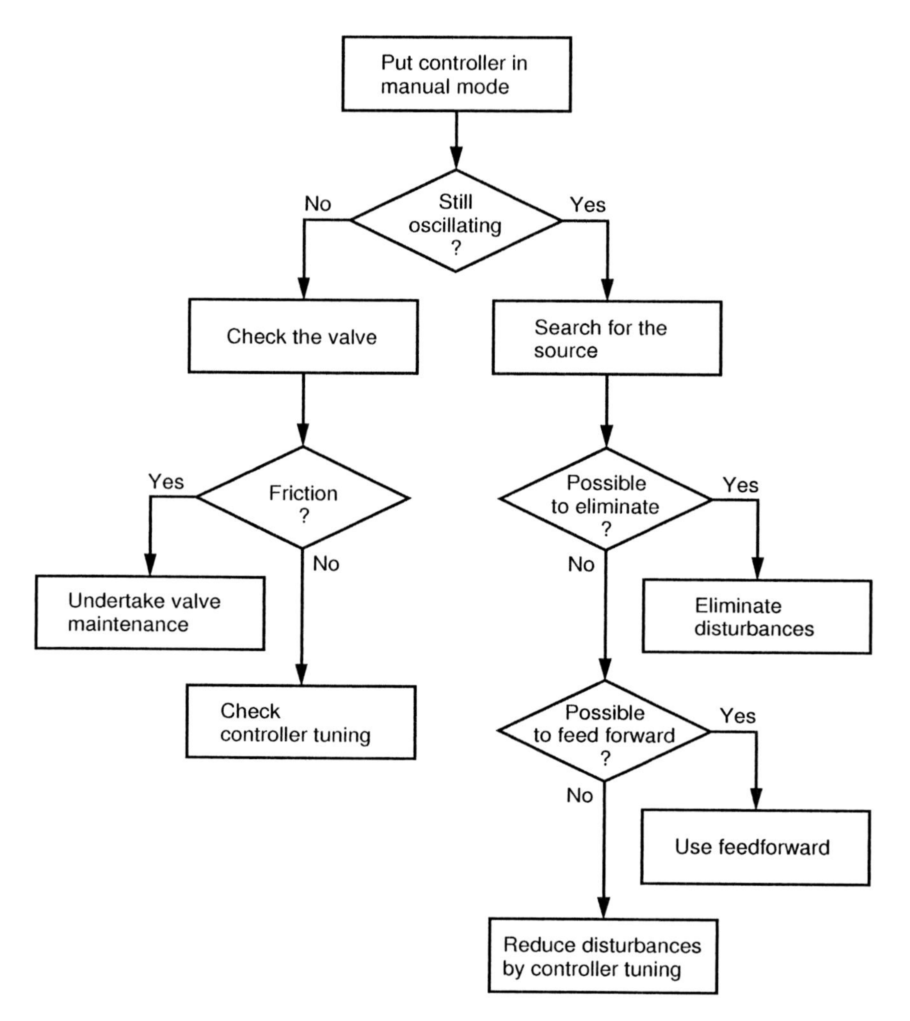

Figure 10.3 Diagnosis procedure to discover the cause of oscillations and recommended actions to eliminate them.

with friction and a process with hysteresis, respectively, when the process input is ramped from zero to one and then back to zero again. The corresponding phase plots are presented in Figure 10.6. One can easily measure the amount of hysteresis from the phase plot. Sweeps of this type are conveniently done during commissioning.

Figure 10.7 shows closed-loop control of a process with 10 percent hysteresis in the valve. The process is

$$P(s) = \frac{1}{(1+0.05s)^2}e^{-0.3s},$$

and the controller is a PI controller with parameters $K = 0.35$ and $Ti = 0.15$. The control signal has to travel through the gap in order to move the valve.

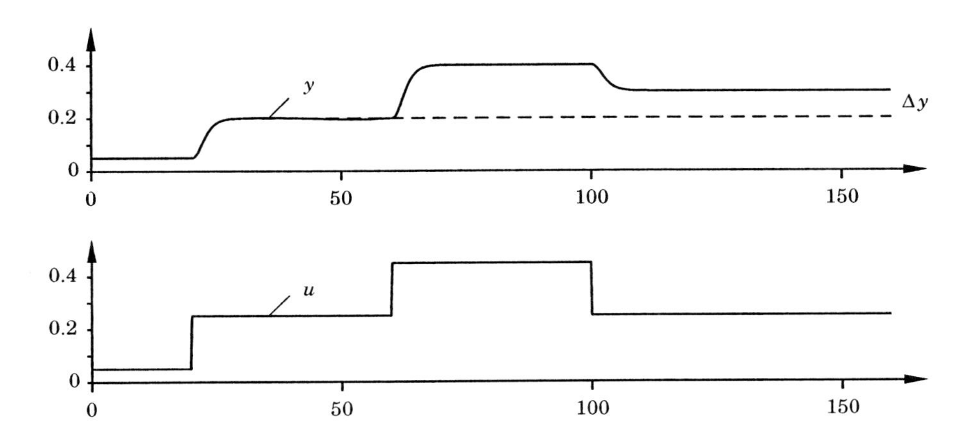

Figure 10.4 Procedure to check valve hysteresis. The upper diagram shows process output y, and the lower diagram shows control signal u.

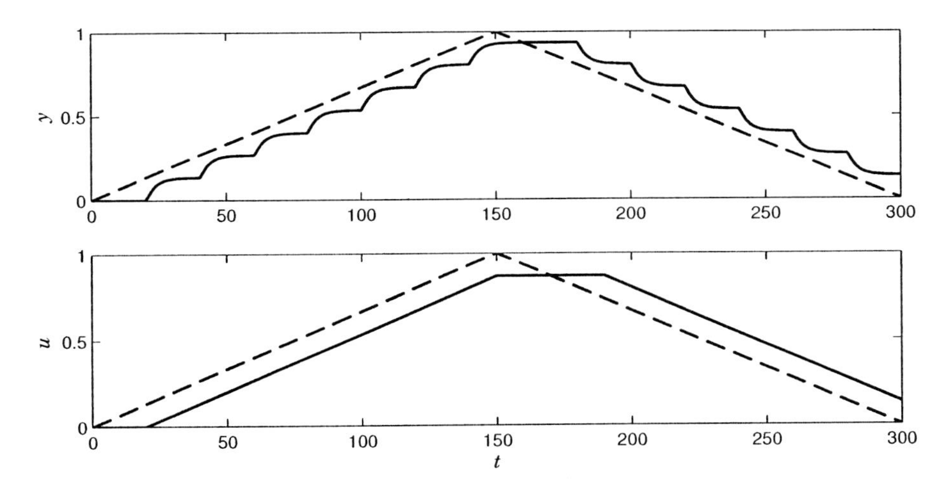

Figure 10.5 Process outputs (solid lines) and control signals (dashed lines) for process with friction (upper graph) and hysteresis (lower graph).

Therefore, we get the typical linear drifts in the control signal as shown in Figure 10.7.

If a relay auto-tuner is applied to a process with hysteresis, the estimated process gain will be smaller than the true value. This gives a too large controller gain. An auto-tuner based on a step-response experiment will work properly if the gap is closed before the step-response experiment is performed. (Compare with the second step in Figure 10.4).

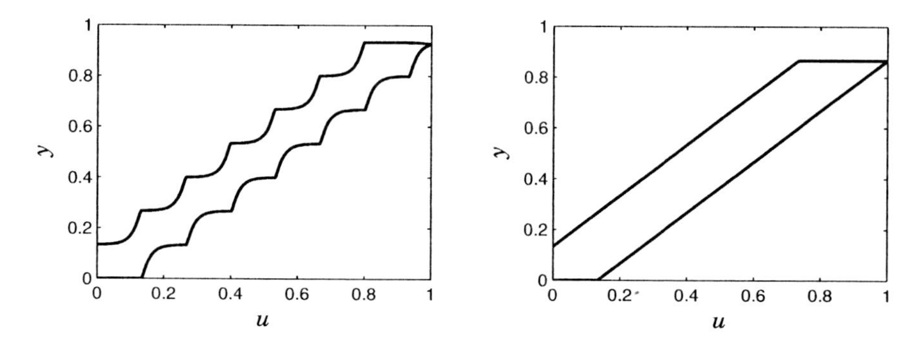

Figure 10.6 Phase plots of the signals in Figure 10.5 for the process with friction (left) and hysteresis (right).

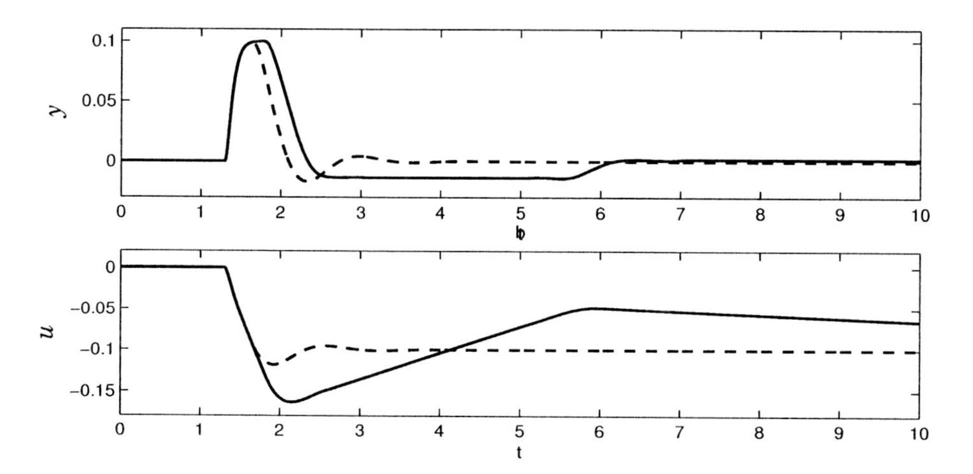

Figure 10.7 Closed-loop control with valve hysteresis. The upper diagram shows process output y, and the lower diagram shows control signal u. The dotted lines show control without hysteresis. The solid lines show control with a hysteresis of 10 percent (0.1).

10.3 Loop Assessment

This section suggests tests that are useful to perform on the control loop. These tests should be performed regularly, and especially in connection with controller tuning. The tests for friction and hysteresis, presented in Section 10.2, are two important loop assessment procedures. The experiments suggested in Section 2.7 to obtain the process dynamics are also loop assessment procedures for tuning the controllers. The checks and tests added in this section are basic, but often forgotten or neglected.

Signal Ranges

The signal range of the measurement signal is related to the resolution of the sensor. A large signal range means that the resolution becomes low. To obtain

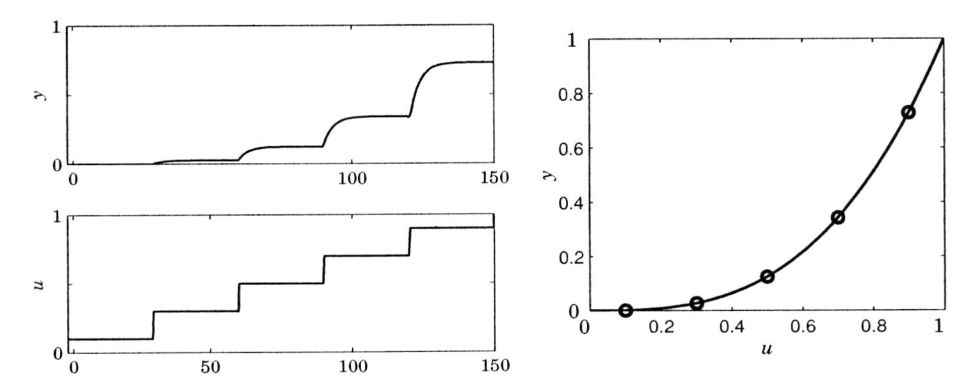

Figure 10.8 The left diagrams show a procedure to determine the static process characteristic. Control signal u is changed stepwise, and the corresponding changes in process output y are determined. The right diagram shows the static process characteristic, i.e., process output y as function of control signal u.

a high resolution, it is therefore important to restrict the signal range to those values that are relevant for the control.

If the final control element is a valve, the output range is determined by the size of the valve. Valves are normally over-sized. The main reasons are insecurity among engineers combined with a fear of installing a valve that is too small to deliver the maximum possible flows.

A large valve has not the same accuracy as a smaller one. The friction and backlash problems discussed in the previous section are more severe if the valve is over-sized.

If the signal ranges are properly chosen and if the process is linear, the ideal static process gain is $P(0) = 1$. If the static gain is one, the measurement signal reaches its maximum value when the control signal is at its maximum value. Because of over-sized valves, the static process gain is often larger than one in process control applications.

Static Input-Output Relations

From a control point of view, it is desirable to have a linear static input-output relation. This relation is, however, often nonlinear, mainly because of a nonlinear valve characteristic. Nonlinearities may also occur in sensors or in the process itself.

If the process is nonlinear, the control may be improved using gain scheduling or other forms of linearization. As pointed out in Section 9.3, it is important to understand the cause of the nonlinearity in order to determine a suitable gain-scheduling reference.

The static characteristic of the process can be obtained by determining the static relation between the control signal and the measured signal. This can be done by performing step changes in the control signal and measuring the corresponding changes in process output; see Figure 10.8.

The characteristic shown in Figure 10.8 is obviously nonlinear. It has a higher gain at larger control signals. If the stationary values of the measured signal are plotted against the control signal, we obtain the static process characteristic. See Figure 10.8. A plot like this reveals whether gain scheduling is suitable or not.

Disturbances

Another important issue to consider before tuning the controller is the disturbances acting on the control loop. We have pointed out that it is important to know if the major disturbances are set-point changes (the servo problem) or load disturbances (the regulator problem).

It is also important to investigate the level of the measurement noise and its frequency content. Compare with Section 2.6. If the noise level is high, it may be necessary to filter the measurement signal before it enters the control algorithm. This is an easy way to get rid of high-frequency noise. If there are disturbances with a large frequency content near the ultimate frequency, it is not possible to use low-pass filtering to remove them. Feedforward is one possibility, if the disturbances can be measured at their source. Notch filters can be used if the noise is concentrated in a narrow frequency range. See Section 2.6 where noise modeling and measurements were discussed.

10.4 Performance Assessment

The loop assessment, followed by appropriate actions like valve maintenance, selection of signal ranges, linearization of nonlinearities, and controller tuning, should leave the control loop in good shape.

After some time in operation, the performance may, however, deteriorate because of variations in the process and the operation. Therefore, it is important to supervise the control loops and detect these degradations. This supervision has traditionally been made by humans, but the reduction of personnel in the process industry combined with increasing quality demands have been a driving force behind developing procedures for automatic performance monitoring and assessment. This section provides some procedures for automatic supervision of control loop performance.

The Static Input-Output Relation

If a detector for stationarity is available, it is simple to keep a statistic for the fraction of time that the system is stationary. The static input-output relation can then be obtained simply by logging the process input and output during stationary conditions. To obtain good data the signals should be filtered with respect to the time scale of the closed loop. Graphs like the ones shown in Figure 10.9 are then obtained. From these curves it can be determined whether the major variations in the output are due to set-point changes or load disturbances, i.e., whether we are dealing with a servo problem or a regulation problem. We have a servo problem if the experimental data gives a well-defined curve and a regulation problem if there is no definite relation between inputs and outputs. A simple statistic of the fraction of the total time when there are

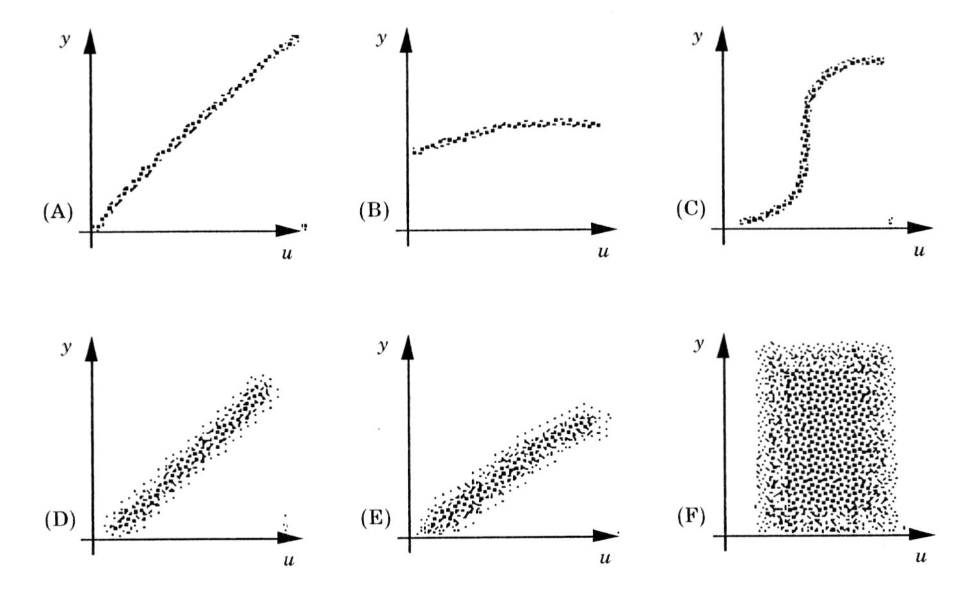

Figure 10.9 Examples of static input-output data logged during normal operation. The results shown in A, B, and C indicate a pure servo problem. The results in F indicate a pure regulation problem. Case D and E are mixed cases. Case B indicates poor resolution of the sensor, and case E indicates poor actuator sizing.

set-point changes or transients due to set-point changes is also a useful indicator. Of course, there are also systems that are mixtures of servo and regulation problems.

For a servo problem the variations in the static gain of a system can also be determined. This gives a valuable indication as to whether gain scheduling is required. The static gain curve can also be used for diagnostic purposes. Changes in the curve indicate changes in the process. By comparing the slope of the static gain curve with the incremental process gain measured during tuning or adaptation, we can also get indications of whether there is some hysteresis in the loop or not. It also indicates if actuators are properly sized.

Model-Based Diagnosis

Most automatic supervisory procedures are, in principle, based on the idea shown in Figure 10.10. If a model of the process is available, the control signal can be fed to the input of the process model. By comparing the output of the model with the true process output, one can detect when the process dynamics change. If the model is good, the difference between the model output and the process output (e) is small. If the process dynamics change, e will no longer be small, since the two responses to the control signal are different.

Harris Index

One of the most widely applied supervisory functions is based on the Harris index. The idea is to calculate the variance of the process output, either on line or off line, and then compare it with the minimum variance obtainable. The

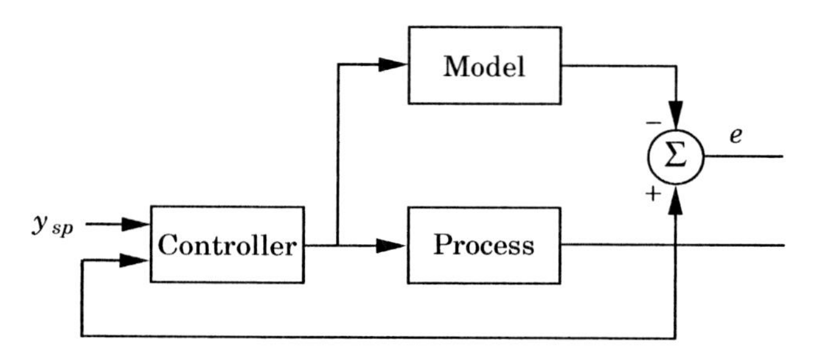

Figure 10.10 Model-based fault detection.

problem was discussed in Section 2.6. The Harris index is defined as

$$I_H = 1 - \frac{\sigma_{MV}^2}{\sigma_y^2},$$

where σ_{MV}^2 is the minimum variance of the process output, and σ_y^2 is the actual process output variance. The Harris index, I_H, takes values between zero and one. If the index is close to zero, the actual variance is close to the minimum variance, which means that the control loop behaves satisfactorily. If the actual variance is large, the Harris index is close to one.

The method requires that the minimum variance σ_{MV}^2 is known. A nice feature of the method is that the minimum variance can be determined from the deadtime only, which means that the modeling can be made relatively simple. A drawback is that the minimum variance normally cannot be achieved with a controller as simple as the PID controller, which means that it is difficult to determine reasonable values of the Harris index. Furthermore, even if it is possible to obtain minimum variance control, this control is often undesirable since it may be very aggressive.

For these reasons, many variations of the Harris index have been presented where the minimum variance σ_{MV}^2 is replaced with the variance obtained using other design objectives and where the limitations to the PID control structure are taken into account. The main drawback of these approaches is that they require a more accurate process model.

The performance monitoring tools based on the Harris index approach provides information about the loop performance compared to some ideal performance. There is no intention to detect any causes of possible bad performance. There are other performance monitoring tools that do not look at the overall performance, but instead try to detect certain types of problems. Some of these are discussed in the following subsections.

Oscillating Control Loops

The most serious problem at the loop level is that many control loops oscillate. There are several possible causes of these oscillations; see Section 10.2. One reason might be that an oscillating load is disturbing the loop. Low-frequency load disturbances are eliminated efficiently by the controller, since a controller with integral action gives a high loop gain at low frequencies. Since the process normally has a low-pass character, high-frequency load disturbances are

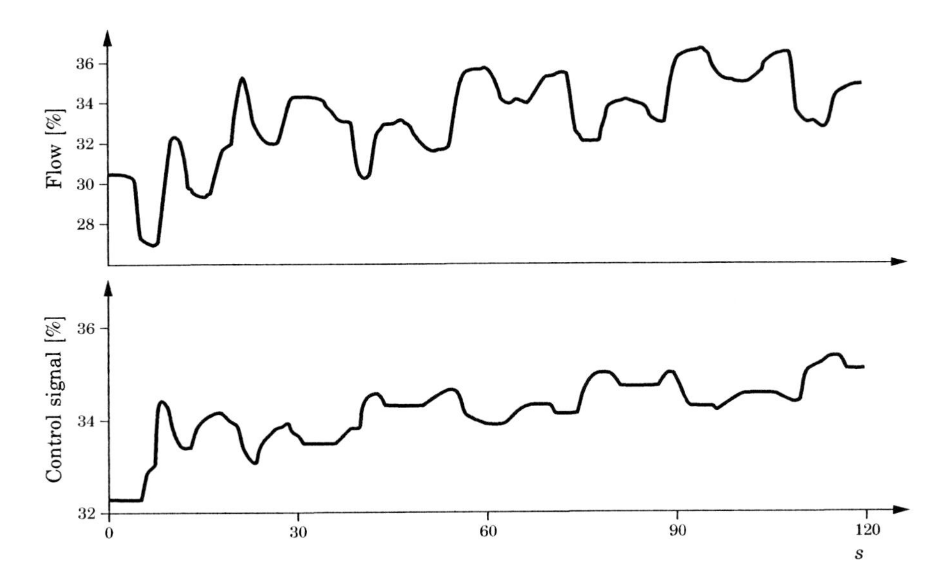

Figure 10.11 Stick-slip motion in a flow control loop.

filtered by the process. Therefore, high-frequency components in the measurement signal are normally not introduced in the process but in the sensor or on the connections between the sensor and the controller. Since they do not contain any valuable information about the status of the process, they should be filtered out by the controller. It is also important not to transfer these signals to the controller output, since they may cause wear on the actuating equipment.

Disturbances with much energy near the ultimate frequency ω_u are too fast to be treated efficiently by the controller, and they are too slow to be filtered out. These disturbances might even be amplified because of the feedback.

A badly tuned controller may be another reason for oscillations, in particular in nonlinear plants where a change in operating point might result in a too high loop gain. However, controllers in process control plants are often tuned conservatively, and bad controller tuning is not the most likely cause of oscillations.

The most common reason for oscillations in control loops is, however, friction in the valve, resulting in "stick-slip" motion as discussed in Section 10.2.

Detection Oscillations in control loops can be detected in several ways. One way is to make a spectral analysis of the measured signal and look for peaks in the spectrum. A difficulty is that the oscillations often are far from pure sine waves, which means that no distinct peaks appear in the spectrum.

Figure 10.11 shows a recording from a flow control loop in a paper mill with high valve stiction. The figure shows the result of a step change in the set point. The controller used was a PI controller with gain $K = 0.30$ and integral time $T_i = 34$ s. Notice that the oscillations are far from a pure sine wave. A retuning of the controller gave controller parameters $K = 0.19$ and $T_i = 2$ s. Notice that the integral time was decreased from 34 s to 2 s! A step

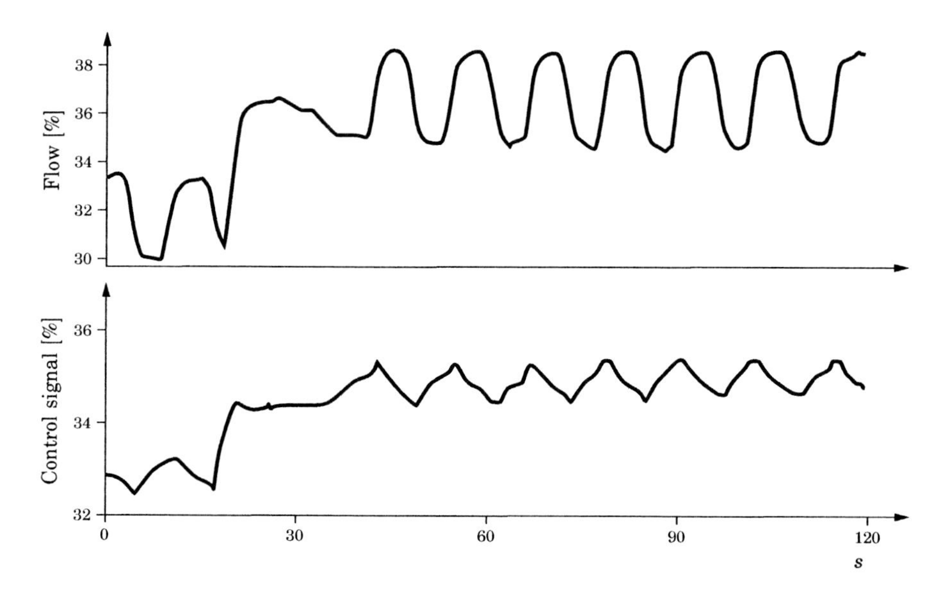

Figure 10.12 Stick-slip motion in a flow control loop – retuned controller.

response experiment using the new controller settings is shown in Figure 10.12. The settling time is significantly shorter than in Figure 10.11. It is also more obvious that the oscillations really are caused by friction, since the typical pattern of the measurement signal is close to a square wave and the control signal is close to a triangular wave.

Another approach to detect oscillations is to investigate the characteristics of the control error. The idea behind this detection procedure is to study the magnitude of the integrated absolute error (IAE) between successive zero crossings of the control error, i.e.,

$$IAE = \int_{t_{i-1}}^{t_i} |e(t)|dt, \tag{10.1}$$

where t_{i-1} and t_i are two consecutive instances of zero crossings. It is assumed that the controller has integral action, so that the average error is zero.

During periods of good control, the magnitude of the control error is small and the times between the zero crossings are relatively short. This means that the IAE values calculated from (10.1) are small when control is good.

When a load disturbance occurs, the magnitude of $e(t)$ increases, and there is a relatively long period without zero crossings. This means that the corresponding IAE value becomes large.

When the control loop starts to oscillate, there will be a high frequency of large IAE values. This observation is used to detect oscillations in the control loop.

EXAMPLE 10.1—PULP CONCENTRATION CONTROL

The following example is taken from a pulp concentration control section in a paper mill, where pulp is diluted with water to a desired concentration.

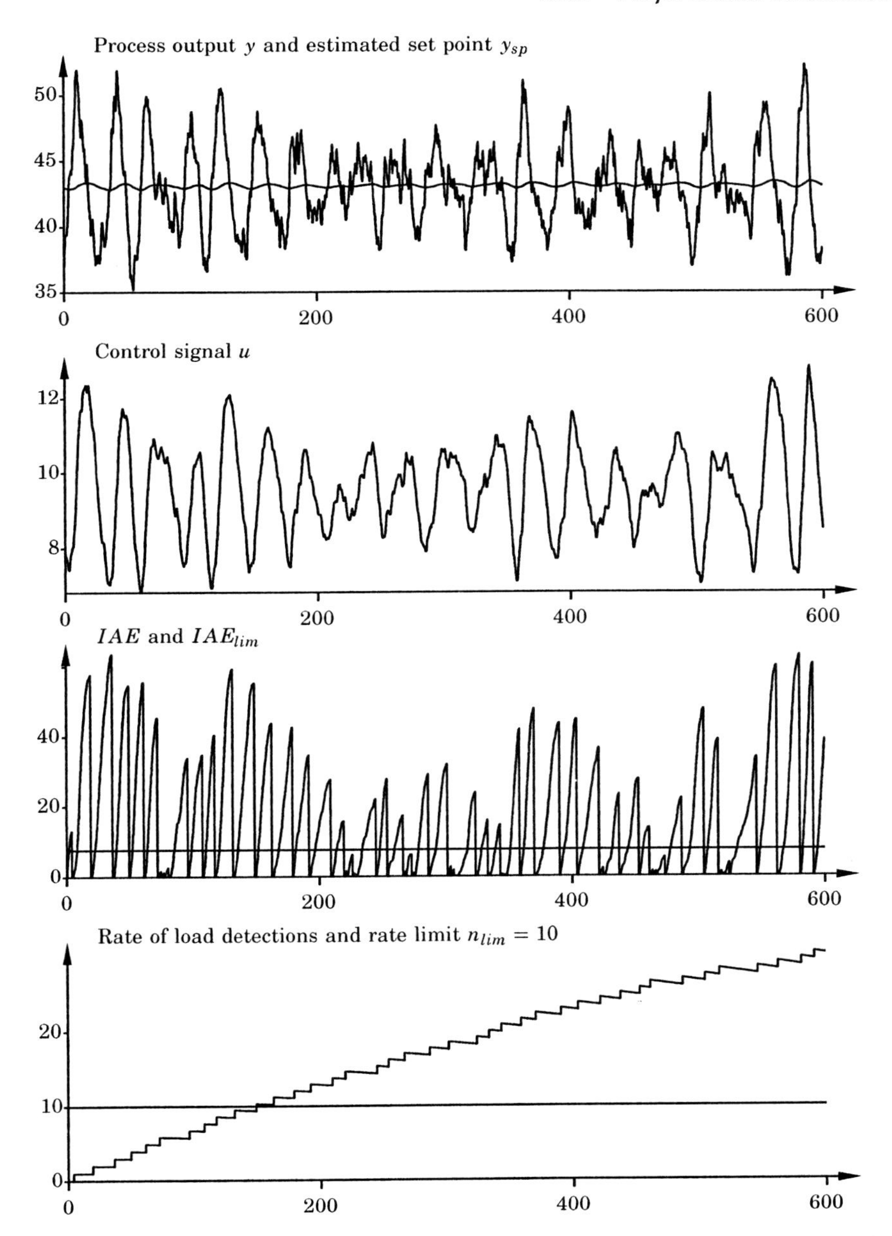

Figure 10.13 The oscillation detection procedure applied on a pulp concentration control loop.

The water valve had too high friction, and an oscillation detection procedure was connected to the controller. The controller was a PI controller with gain $K = 0.33$ and integral time $T_i = 24$ s.

Figure 10.13 shows 10 minutes of data from the concentration control loop. The first graph shows the process output, the pulp concentration in percent. Because of high friction in the water valve, the process is oscillating with an amplitude of a few percent. The first graph also shows an estimate of the set point, since this variable was not recorded. The estimate is simply obtained by a low-pass filtering of the process output.

The second graph shows the control signal in percent. It is obvious that the controller tries to eliminate the oscillation but without success.

The third graph shows the IAE calculated between successive zero crossings of the control error. The graph also shows IAE_{lim}, which is the limit of what is considered large values of IAE. In this implementation, the value of IAE_{lim} is determined automatically from the controller parameters in each loop. The IAE values are significantly larger than IAE_{lim}, indicating that the loop is oscillating.

The fourth graph finally shows the rate of load detections and the rate limit $n_{lim} = 10$. The rate exceeds the rate limit after about three minutes, and the detection procedure gives an alarm.

This example shows how the oscillation detection procedure manages to detect oscillations in control loops. The actual oscillations are easily noticed in Figure 10.13. However, process operators seldom have access to these kinds of graphs, but are often left with a bar graph with a low resolution. The present oscillation had been present for a long time without being discovered by the process operators. □

Diagnosis Since a control loop may oscillate for various reasons, it is important not only to detect the oscillation, but also to find the reason for oscillations. This can be done manually as described in Section 10.2.

Attempts have also been made to develop procedures for automatic diagnosis. Here, the difference in the spectrum can be used. When a control loop oscillates because of too high loop gain, the control error is often close to a sine wave, resulting in one single peak in the spectrum. The same holds in most cases when the loop is oscillating because of external disturbances. However, when the control loop is oscillating because of valve stiction, several peaks in the spectrum can be found.

Sluggish Control Loops

Oscillations in control loops are common, but the opposite situation is also common, namely, that the control loops are sluggish because of conservative tuning. This causes unnecessarily large and long deviations from the set point at load disturbances.

The main reason for the controllers being conservatively tuned is lack of time. The engineers tune the controllers until they are considered "good enough." They do not have the time to optimize the control. Many controllers are tuned once they are installed, and then never again. To retain stability when operating conditions change, the controllers are tuned for the "worst case." A better solution would, of course, be to use gain scheduling and perhaps adaptation. When a controller is retuned, it is mostly because the process conditions cause oscillatory control. In other words, when the controllers are retuned, they are detuned. When the process conditions change to sluggish control, the controller is normally not retuned again.

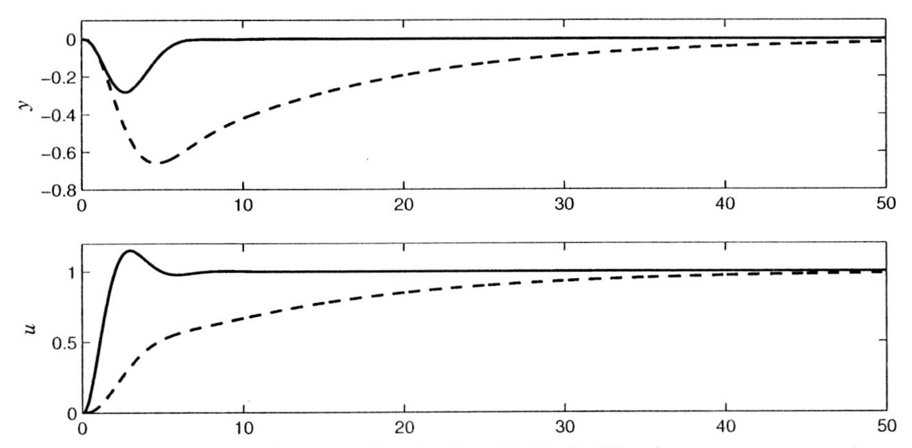

Figure 10.14 A good (solid lines) and a sluggish (dashed lines) response to a step change in load at the process input.

Detection Figure 10.14 shows two responses to load disturbances in the form of step changes at the process input. One response is good, with a quick recovery without any overshoot. The second response is very sluggish. One feature that characterizes the second response is that there is a long period where both process output y and control signal u drift slowly in the same direction. This feature is used for detection.

Both responses have an initial phase where the two signals go in opposite directions, i.e., $\Delta u \Delta y < 0$, where Δu and Δy are the increments of the two signals. What characterizes the sluggish response is that after this initial phase there is a very long time period where the correlation between the two signal increments is positive. This observation forms the base for the Idle index, which expresses the relation between the times of positive and negative correlation between the signal increments.

To form the Idle index, the time periods when the correlations between the signal increments are positive and negative, respectively, are first calculated. The following procedures are updated every sampling instant

$$t_{\text{pos}} = \begin{cases} t_{\text{pos}} + h & \text{if } \Delta u \Delta y > 0 \\ t_{\text{pos}} & \text{if } \Delta u \Delta y \leq 0 \end{cases}$$

$$t_{\text{neg}} = \begin{cases} t_{\text{neg}} + h & \text{if } \Delta u \Delta y < 0 \\ t_{\text{neg}} & \text{if } \Delta u \Delta y \geq 0, \end{cases}$$

where h is the sampling period. The Idle index I_I is then defined by

$$I_I = \frac{t_{\text{pos}} - t_{\text{neg}}}{t_{\text{pos}} + t_{\text{neg}}}. \qquad (10.2)$$

Note that I_I is bounded to the interval $[-1, 1]$. A positive value of I_I close to 1 means that the control is sluggish. The Idle index for the sluggish response in

Figure 10.14 is $I_I = 0.82$. A negative value of I_I close to -1 may be obtained in a well-tuned control loop. The Idle index for the good response in Figure 10.14 is $I_I = -0.63$. However, negative Idle indices close to -1 are also obtained in oscillatory control loops. Therefore, it is desirable to combine the Idle index calculation with an oscillation detection procedure like the one described above.

Calculation of the Idle index can be made both off line and on line using a recursive version. Since the method is based on the characteristics of signal increments, it is sensitive to noise. Therefore, it is important to filter the signals properly before they are differentiated.

EXAMPLE 10.2—CONTROL OF A HEAT EXCHANGER
This example is taken from an industrial heat exchanger. The control objective is to control the water temperature on the secondary side by controlling the water steam flow on the primary side.

The upper graphs in Figure 10.15 show load responses obtained with a conservatively tuned PI controller. The controller parameters were $K = 0.01$ and $T_i = 30s$. The signals are relatively noisy because of the low resolution, 1 percent, of the controller output. The control is sluggish. This is also well reflected by the Idle index, which was calculated to $I_I = 0.8$.

The controller structure was changed to a PID controller and tuned properly, resulting in the controller parameters $K = 0.025$, $T_i = 8s$, and $T_d = 2s$. The improved control behavior is illustrated in the lower graphs in Figure 10.15. The recovery after load disturbances is significantly faster, still without any noticeable overshoot. The integral gain k_i is increased by almost a factor of ten. The improvements are also demonstrated by the Idle, which index that was reduced to $I_I = 0.3$. □

10.5 Integrated Tuning and Diagnosis

The diagnosis procedures are related to the adaptive techniques in several ways. We have pointed out the importance of checking valves before applying an automatic tuning procedure. If not done, the automatic tuning procedure will not provide the appropriate controller parameters. For this reason, it would be desirable to have these checks incorporated in the automatic tuning procedures. Such devices are not yet available, and the appropriate checks, therefore, must be made by the operator.

The on-line detection methods are related to the continuous adaptive controller. The adaptive controller monitors the control loop performance and changes the controller parameters, if the process dynamics change. The performance assessment procedures also monitor the control-loop performance. They give an alarm instead of changing the controller parameters if the process dynamics change. As an example, in Figure 10.3 we have seen that it is important to determine *why* the performance has changed before actions are taken. Most adaptive controllers applied to a process with stiction will detune the controller, since they interpret the oscillations as caused by a badly tuned controller. Consequently, it is desirable to supply the adaptive controllers with

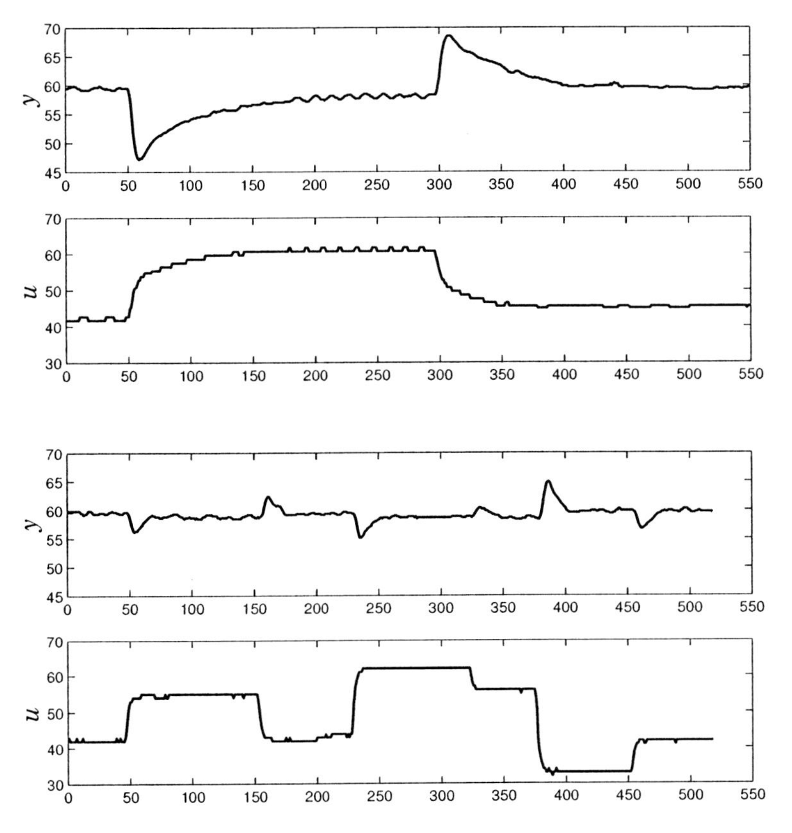

Figure 10.15 Control of a heat exchanger. The graphs show responses to load disturbances for a sluggish control loop with Idle index $I_I = 0.8$ (upper), and a properly tuned loop with Idle index $I_I = 0.3$ (lower).

on-line detection methods, so that reasons for bad control-loop performance, other than poor controller tuning, are detected. The lack of these kinds of detection procedures in adaptive controllers are perhaps the major reason for the relatively few applications of continuous adaptive control available today.

10.6 Summary

It is important to make an assessment of the control loop before tuning the controller. This assessment includes checks of equipment such as sensors and valves, signal ranges, nonlinearities, and disturbances.

When the loop assessment and the controller tuning is performed, the control loop should behave well. Due to changes in the process and its operation, the control loop may degrade after some time in operation. It is therefore important to supervise the control loops. This is traditionally done by humans, but methods for automatic supervision are becoming more and more used in process control.

In this section, some examples of loop and performance monitoring tools have been presented. The section has only provided a short overview of the area. It has focused on methods for the single loops only. In recent years, many attempts have been made to derive methods for the performance monitoring of process sections including several control loops. However, these procedures are seldom general, but often developed for specific plants.

10.7 Notes and References

Early work on fault-detection was done by [Himmelblau, 1978]. Problems associated with the control valves were brought to a broader audience in the early nineties; see [Ender, 1993; Bialkowski, 1994]. At that time there was also an awareness that it was beneficial to assess the performance of the control loops; see [Shinskey, 1990; Shinskey, 1991a; Åström, 1991]. The Harris index [DeWries and Wu, 1978], [Harris, 1989] is based on comparison with performance obtained by minimum variance control [Åström, 1970]. The concept has been extended and applied in various process control applications; see e.g. [Desborough and Harris, 1992; Stanfelj *et al.*, 1993; Harris *et al.*, 1996; Kozub and Garcia, 1993; Kozub and Garcia, 1996; Harris *et al.*, 1996; Owen *et al.*, 1996; Lynch and Dumont, 1996; Harris *et al.*, 1999; Thornhill *et al.*, 1999]. The oscillation detection procedure is described in [Hägglund, 1995] and [Thornhill and Hägglund, 1997], and the Idle index is presented in [Hägglund, 1999]. Good surveys of the area are presented in [Qin, 1998; Huang and Shah, 1999; Horch, 2000]. A method for reducing the effect of friction in valves was developed by [Hägglund, 2002].

11

Interaction

11.1 Introduction

So far we have focused on control of simple loops with one sensor, one actuator, and one controller. In practical applications, a control system can have many loops, sometimes thousands. In spite of this, a large control system can often be dealt with loop by loop since the interaction between the loops is negligible. There are, however, situations when there may be considerable interaction between different control loops. A typical case is when several streams are blended to obtain a desired mixture. In such a case it is clear that the loops interact. Other cases are control of boilers, paper machines, distillation towers, chemical reactors, heat exchangers, steam distribution networks, drive systems, and systems for air-conditioning. Processes that have many control variables and many measured variables are called multi-input multi-output (MIMO) systems. Because of the interactions it may be difficult to control such systems loop by loop.

A reasonably complete treatment of multivariable systems is far outside the scope of this book. In this chapter we will briefly discuss some issues in interacting loops that are of particular relevance for PID control. Section 11.2 gives simple examples that illustrate what may happen in interacting loops. In particular it is shown that controller parameters in one loop may have significant influence on dynamics of other loops. Bristol's relative gain array, which is a simple way to characterize the interactions, is also introduced. The problem of pairing inputs and outputs is discussed, and it is shown that the interactions may generate zeros of a multivariable system. In Section 11.3 we present a design method based on decoupling, which is a natural extension of the tuning methods for single-input single-output systems. Section 11.4 presents problems that occur in drive systems with parallel motors. The chapter ends with a summary and references.

11.2 Interaction of Simple Loops

In this section we will illustrate some effects of interaction in the simplest case

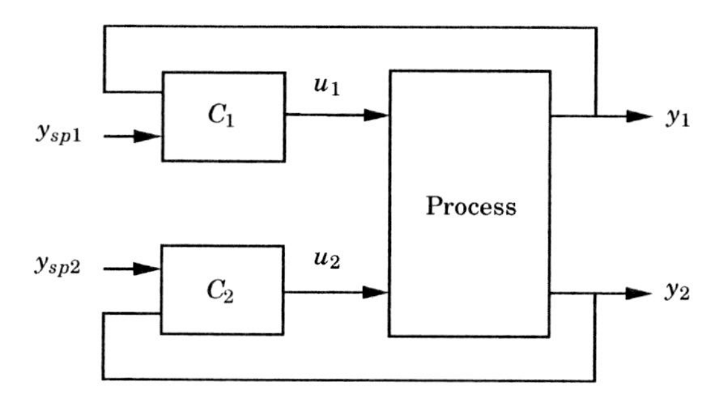

Figure 11.1 Block diagram of a system with two inputs and two outputs (TITO).

of a system with two inputs and two outputs. Such a system is called a TITO system. The system can be represented by the equations

$$\begin{aligned} Y_1(s) &= p_{11}(s)U_1(s) + p_{12}U_2(s) \\ Y_2(s) &= p_{21}(s)U_1(s) + p_{22}U_2(s), \end{aligned} \tag{11.1}$$

where $p_{ij}(s)$ is the transfer function from the j:th input to the i:th output. The transfer functions p_{11}, $p_{12}(s)$, $p_{21}(s)$, and p_{22} can be combined into the matrix

$$P(s) = \begin{pmatrix} p_{11}(s) & p_{12}(s) \\ p_{21}(s) & p_{22}(s) \end{pmatrix}, \tag{11.2}$$

which is called the transfer function or the matrix transfer function of the system. Some effects of interaction will be illustrated by an example.

Example 11.1—Effects of Interaction

Consider the system described by the block diagram in Figure 11.1. The system has two inputs and two outputs. There are two controllers, the controller C_1 controls the output y_1 by the input u_1 and C_2 controls the output y_2 by the input u_2. One effect of interaction is that the tuning of one loop can influence the other loop. This is illustrated in Figure 11.2, which shows a simulation of the first loop when C_1 is a PI controller and $C_2 = k_2$ is a proportional controller.

The example shows that the gain of the second loop has a significant influence on the behavior of the first loop. The response of the first loop is good when the second loop is disconnected, $k_2 = 0$, but the system becomes more sluggish when the gain of the second loop is increased. The system is unstable for $k_2 = 0.8$.

Simple analysis gives insight into what happens. In the particular case the system is described by

$$\begin{aligned} Y_1(s) &= \frac{1}{(s+1)^2}U_1(s) + \frac{2}{(s+1)^2}U_2(s) \\ Y_2(s) &= \frac{1}{(s+1)^2}U_1(s) + \frac{1}{(s+1)^2}U_2(s). \end{aligned}$$

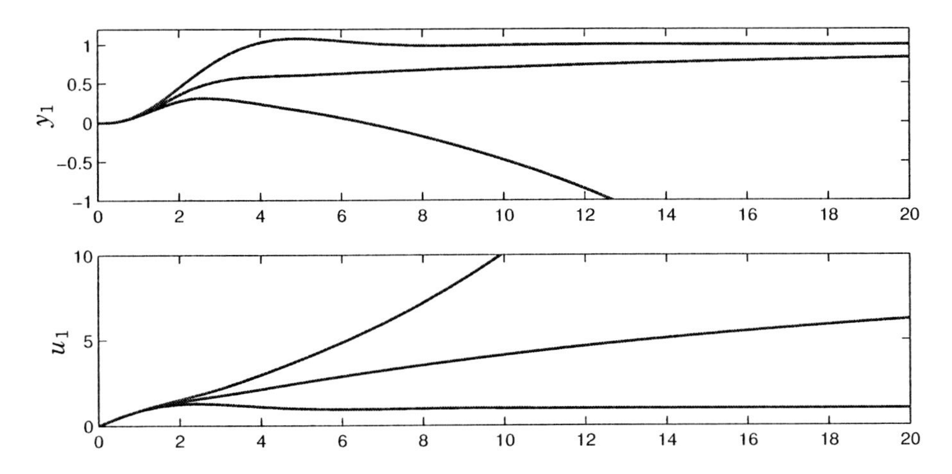

Figure 11.2 Simulation of responses to steps in set points for loop 1 of the system in Figure 11.1. Controller C_1 is a PI controller with gains $k_1 = 1$, $k_i = 1$, and the C_2 is a proportional controller with gains $k_2 = 0$, 0.8, and 1.6.

The feedback in the second loop is $U_2(s) = -k_2 Y_2(s)$. Introducing this in the second equation gives

$$U_2(s) = -\frac{k_2}{s^2 + 2s + k_2 + 1} U_1(s),$$

and insertion of this expression for $U_2(s)$ in the first equation gives

$$Y_1(s) = g_{11}^{cl}(s) U_1(s) = \frac{s^2 + 2s + 1 - k_2}{(s+1)^2(s^2 + 2s + 1 + k_2)} U_1(s).$$

This equation shows clearly that the gain k_2 in the second loop has a significant effect on the dynamics relating u_1 and y_1. The static gain is

$$g_{11}^{cl}(0) = \frac{1 - k_2}{1 + k_2}.$$

Notice that the gain decreases as k_2 increases and that the gain becomes negative for $k_2 > 1$. □

The example indicates that there is a need to have some way to determine if interactions may cause difficulties. A simple measure of interaction will now be discussed.

Bristol's Relative Gain Array

A simple way to investigate the effect of the interaction is to investigate how the static process gain of one loop is influenced by the gains in the other loops. Consider first the system with two inputs and two outputs shown in Figure 11.1. We will investigate how the static gain in the first loop is influenced by the controller in the second loop. To avoid making specific assumptions about the controller, Bristol assumed that the second loop was in perfect control, meaning that the output of the second loop is zero. It then follows from (11.1) that

$$\begin{aligned} Y_1(s) &= p_{11}(s)U_1(s) + p_{12}U_2(s) \\ 0 &= p_{21}(s)U_1(s) + p_{22}U_2(s). \end{aligned}$$

Eliminating $U_2(s)$ from the first equation gives

$$Y_1(s) = \frac{p_{11}(s)p_{22}(s) - p_{12}(s)p_{21}(s)}{p_{22}(s)} U_1(s).$$

The ratio of the static gains of loop 1 when the second loop is open and when the second loop is closed is thus

$$\lambda = \frac{p_{11}(0)p_{22}(0)}{p_{11}(0)p_{22}(0) - p_{12}(0)p_{21}(0)}. \tag{11.3}$$

Parameter λ is called *Bristol's interaction index* for TITO systems. Notice that the index refers to static conditions. In practice this can also be interpreted as interaction for low-frequency signals. There is no interaction if $p_{12}(0)p_{21}(0) = 0$, which implies that $\lambda = 1$. Small or negative values of λ indicate that there are interactions. Consider, for example, the system in Example 11.1 where the interaction index is $\lambda = -1$, which indicates that interactions pose severe difficulties.

The interaction index can be generalized to systems with many inputs and many outputs. The idea is to compare the static gains for one output when all other loops are open with the gains when all other outputs are zero. The result can be summarized in *Bristol's relative gain array* (RGA) which is defined as

$$R = P(0). * P^{-T}(0), \tag{11.4}$$

where $P(0)$ is the static gain of the system, $P^{-T}(0)$ the transpose of the inverse of $P(0)$, and $.*$ denotes component-wise multiplication of matrices. The element r_{ij} is the ratio between the open-loop and closed-loop static gains from the input signal u_j to the output y_i. It can be shown that the matrix R is symmetric and that all rows and columns sum to one. Notice that Bristol's relative gain array only captures the behavior of the process at low frequencies.

For the system (11.1) the relative gain array becomes

$$R = \begin{pmatrix} \lambda & 1-\lambda \\ 1-\lambda & \lambda \end{pmatrix}, \tag{11.5}$$

where λ is the interaction index (11.3). There is no interaction if $\lambda = 1$. This means that the second loop has no impact on the first loop and vice versa. If λ is between 0 and 1 the closed loop has higher gain than the open loop. The effect is most severe for $\lambda = 0.5$. If λ is larger than 1 the closed loop has lower gain than the open loop. When λ is negative the gain of the first loop changes sign when the second loop is closed. The effect of the interactions is thus severe.

Pairing

To control a system loop by loop we must first decide how the controllers should be connected, i.e., if y_1 in Figure 11.1 should be controlled by u_1 or u_2. This is called the *pairing problem*.

The relative gain array can be used as a guide for pairing. There is no interaction if $\lambda = 1$. If $\lambda = 0$ there is also no interaction, but the loops should be interchanged. The loops should be interchanged when $\lambda < 0.5$. If $0 < \lambda < 1$ the gain of the first loop increases when the second loop is closed, and if $\lambda > 1$ the closed-loop gain is less than the open-loop gain. Bristol recommended that pairing should be made so that the corresponding relative gains are positive and as close to one as possible. Pairing of signals with negative relative gains should be avoided. If the gains are outside the interval $0.67 < \lambda < 1.5$, decoupling can improve the control significantly. We illustrate pairing with an example.

EXAMPLE 11.2—PAIRING OF SIGNALS

Consider the system in Example 11.1. The static gain matrix is

$$P(0) = \begin{pmatrix} 1 & 2 \\ 1 & 1 \end{pmatrix}.$$

Its inverse is

$$P^{-1}(0) = \begin{pmatrix} -1 & 2 \\ 1 & -1 \end{pmatrix},$$

and the relative gain array becomes

$$R = P(0).*P^{-T}(0) = \begin{pmatrix} 1 & 2 \\ 1 & 1 \end{pmatrix} .* \begin{pmatrix} -1 & 1 \\ 2 & -1 \end{pmatrix} = \begin{pmatrix} -1 & 2 \\ 2 & -1 \end{pmatrix},$$

which means that $\lambda = -1$. The pairing rule says that y_1 should be paired with u_2.

When $u_1 = -k_2 y_2$ the relation between u_2 and y_1 becomes

$$Y_1(s) = g_{12}^{cl}(s)U_2(s) = \frac{2s^2 + 4s + 2 + k_2}{(s+1)^2(s^2 + 2s + 1 + k_2)} U_2(s),$$

and the static gain is

$$g_{12}^{cl}(0) = \frac{2 + k_2}{1 + k_2}.$$

The gain decreases with increasing k_2, but it is never negative for $k_2 > 0$. There is interaction but not as severe as for the pairing of y_1 with u_1. The properties of the closed-loop system are illustrated in Figure 11.3. A comparison with Figure 11.2 shows that there is a drastic reduction in the interaction when the inputs are switched. □

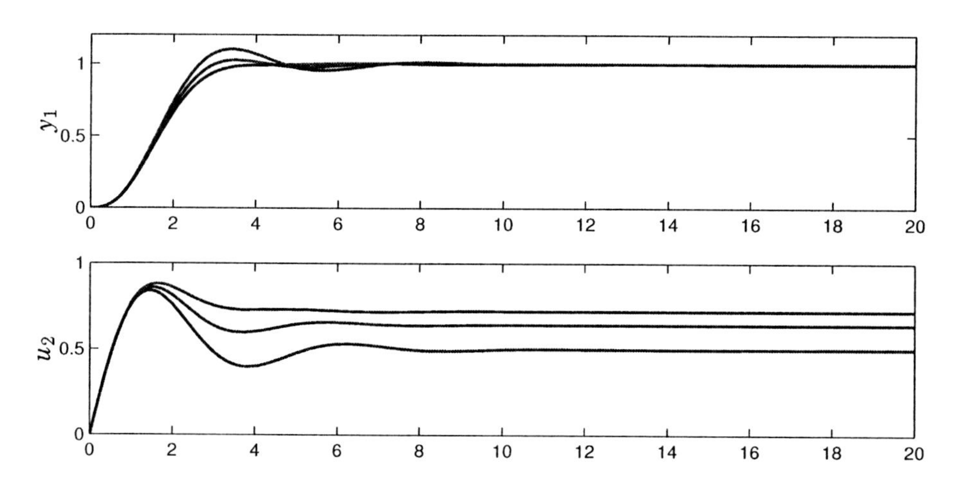

Figure 11.3 Simulation of responses to a step in the set point for y_1 of the system in Figure 11.1 when the loops are switched so that the controller for y_1 is $U_2 = C_1(s)(Y_{sp1} - Y_1)$ and the controller for y_2 is $u_1 = -k_2 y_2$ with $k_2 = 0$, 0.8, and 1.6. The controller C_1 is a PI controller with gains $k_1 = 1$, $k_i = 1$.

Multivariable Zeros

In Section 4.3 we found that right-half plane zeros imposed severe restriction on the achievable performance. For single-input single-output systems the zeros can be found by inspection. For multivariable systems zeros can, however, also be created by interaction. One definition of zeros that also works for multivariable systems is that the zeros are the poles of the inverse system. The zeros of the system (11.1) are given by

$$\det P(s) = p_{11}(s)p_{22}(s) - p_{12}(s)p_{21}(s) = 0. \tag{11.6}$$

Zeros in the right half plane are of particular interest because they impose limitations on the achievable performance. We illustrate this with an example.

EXAMPLE 11.3—ROSENBROCK'S SYSTEM
Consider a system with the transfer function

$$P(s) = \begin{pmatrix} p_{11}(s) & p_{12}(s) \\ p_{21}(s) & p_{22}(s) \end{pmatrix} = \begin{pmatrix} \dfrac{1}{s+1} & \dfrac{2}{s+3} \\ \dfrac{1}{s+1} & \dfrac{1}{s+1} \end{pmatrix}. \tag{11.7}$$

The dynamics of the subsystems are very benign. There are no dynamics limitations in control of any individual loop. The relative gain array is

$$R = \begin{pmatrix} 1 & 2/3 \\ 1 & 1 \end{pmatrix} .* \begin{pmatrix} 3 & -3 \\ -2 & 3 \end{pmatrix} = \begin{pmatrix} 3 & -2 \\ -2 & 3 \end{pmatrix},$$

which shows that there are significant interactions. Using the rules for pairing we find that it is reasonable to pair u_1 with y_1 and u_2 with y_2. Since $\lambda > 1.5$

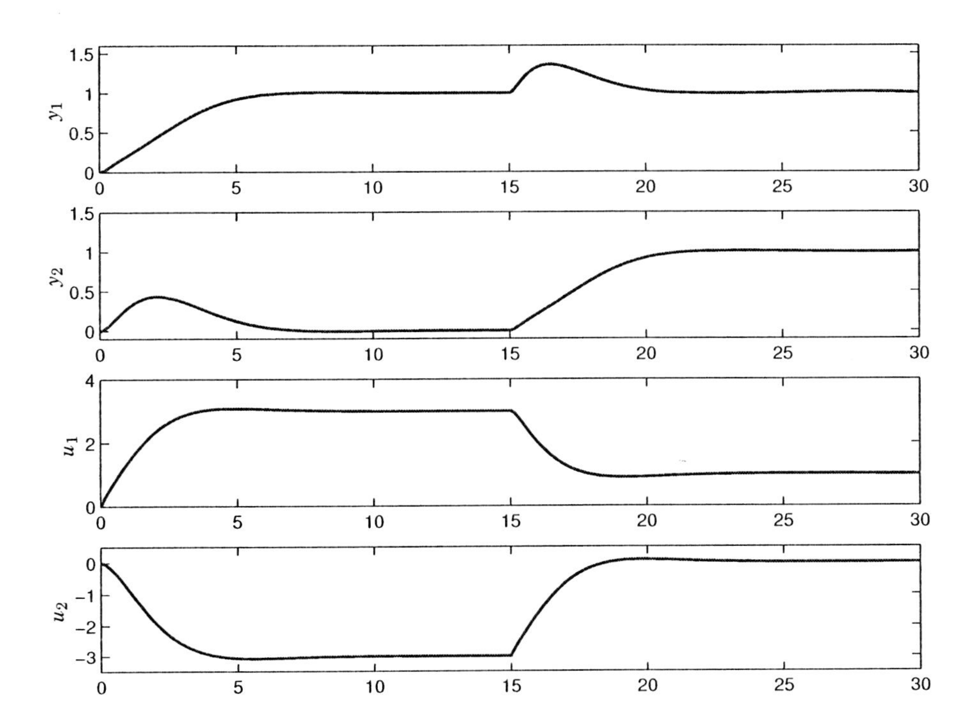

Figure 11.4 Step responses of the process (11.7) with PI control of both loops. Both PI controllers have gains $k = 2$ and $k_i = 2$. A step in y_{sp1} is first applied at time 0, and a step in y_{sp2} is then applied at time 15.

we can expect difficulties because of the interaction. It follows from (11.6) that the zeros of the system are given by

$$\det P(s) = \frac{1}{s+1}\Big(\frac{1}{s+1} - \frac{2}{s+3}\Big) = \frac{1-s}{(s+1)^2(s+3)} = 0.$$

There is a zero at $s = 1$ in the right half plane, and we can therefore expect difficulties when control loops are designed to have bandwidth larger than $\omega_0 = 1$.

Consider, for example, the problem of controlling the variable y_1. If the second loop is open we can achieve very fast response with a PI controller. When the second loop is closed there will, however, be severe performance limitations due to the interactions, and the control loop has to be detuned. Figure 11.4 shows responses obtained with controllers having gains $k = 2$ and $k_i = 2$ in both loops. In the figure we have first made a unit step in the set point of the first controller and then a set-point change in the second controller. The figure shows that there are considerable interactions. The system becomes unstable if the gain is increased by a factor of 3. □

Example 11.3 illustrates that an innocent-looking multivariable system may have zeros in the right half plane. The opposite is also possible, as is illustrated by the next example.

EXAMPLE 11.4—BENEFICIAL INTERACTION
Consider the system

$$P(s) = \begin{pmatrix} p_{11}(s) & p_{12}(s) \\ p_{21}(s) & p_{22}(s) \end{pmatrix} = \begin{pmatrix} \dfrac{s-1}{(s+1)(s+2)} & \dfrac{s}{(s+1)(s+2)} \\ \dfrac{-6}{(s+1)(s+2)} & \dfrac{s-2}{(s+1)(s+2)} \end{pmatrix}. \tag{11.8}$$

The system has the relative gain array

$$R = \begin{pmatrix} 1 & 0 \\ 0 & 1 \end{pmatrix},$$

which indicates that y_1 should be paired with u_1 and that y_2 should be paired with u_2. The multivariable system has no zeros. We thus have the interesting situation that there are severe limitations to control either the first or the second loop individually because of the right-half plane zeros in the elements p_{11} and p_{22}. Since the multivariable system does not have any right-half plane zeros it is possible to control the multivariable system with high bandwidth. This is illustrated in Figure 11.4, where both loops are controlled with PI controllers having gains $k = 100$ and $k_i = 2000$. Notice the fast response of the system. One difficulty is, however, that the system becomes unstable if one of the loops is broken. □

11.3 Decoupling

Decoupling is a simple way to deal with the difficulties created by interactions between loops. The idea is to design a controller that reduces the effects of the interaction. Ideally, changes in one set point should only affect the corresponding process output. This can be accomplished by a precompensator that mixes the signals sent from the controller to the process inputs. The details will be given for systems with two inputs and two outputs, but the method can be applied to signals with many inputs and many outputs.

Assume that the process has the transfer function (11.2) and that $P(0)$ is nonsingular. We first introduce a static decoupler $u = D\bar{u}$, where D is a constant matrix

$$D = \begin{pmatrix} d_{11} & d_{12} \\ d_{21} & d_{22} \end{pmatrix}.$$

The transfer function from $\bar{u}$ to y is then given by $P(s)D$. The choice

$$D = P^{-1}(0) = \frac{1}{\det P(0)} \begin{pmatrix} p_{22}(0) & -p_{12}(0) \\ -p_{21}(0) & p_{11}(0) \end{pmatrix} \tag{11.9}$$

makes $P(0)D$ the identity matrix. The system $P(s)D$ is thus statically decoupled, and the coupling is small for low frequencies. The coupling remains small if the system is controlled by decoupled controllers, provided that the bandwidths of the control loops are sufficiently small.

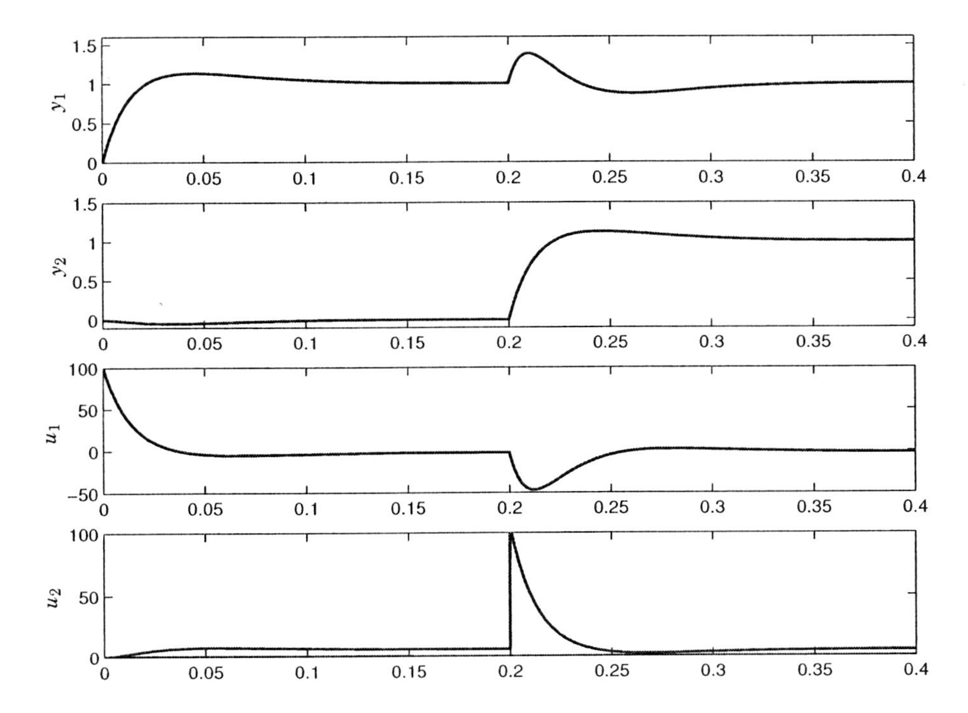

Figure 11.5 Step responses of PI control of the process (11.8) when both loops are closed. PI controllers with gains $k = 100$ and $k_i = 2000$ are used in both loops.

Assuming that the controllers are PID controllers we find that the statically decoupled controller is described by

$$\begin{pmatrix} U_1(s) \\ U_2(s) \end{pmatrix} = \begin{pmatrix} d_{11} & d_{12} \\ d_{21} & d_{22} \end{pmatrix} \begin{pmatrix} \bar{c}_1(s)Y_{sp1}(s) - c_1(s)Y_1(s) \\ \bar{c}_2(s)Y_{sp2}(s) - c_2(s)Y_2(s) \end{pmatrix},$$

where U is the control signal, Y the process output, and Y_{sp} the set point. The controllers are PID controllers with set-point weighting, hence,

$$c_i = k_{Pi} + \frac{k_{Ii}}{s} + k_{Di}s, \quad \bar{c}_i = b_i k_{Pi} + \frac{k_{Ii}}{s},$$

where b_i is the set-point weight. The set-point weights influence the interaction between the loops. Choosing $b_i = 0$ gives the smallest interaction.

The Decoupled System

The transfer function of the decoupled system is $Q(s) = P(s)D$, where

$$q_{11}(s) = \frac{p_{11}(s)p_{22}(0) - p_{12}(s)p_{21}(0)}{\det P(0)}$$
$$q_{12}(s) = \frac{p_{12}(s)p_{11}(0) - p_{12}(0)p_{11}(s)}{\det P(0)}$$
$$q_{21}(s) = \frac{p_{21}(s)p_{22}(0) - p_{21}(0)p_{22}(s)}{\det P(0)}$$
$$q_{22}(s) = \frac{p_{22}(s)p_{11}(0) - p_{21}(s)p_{12}(0)}{\det P(0)}.$$

It follows from the construction that $Q(0)$ is the identity matrix. A Taylor series expansion of the transfer function $Q(s)$ for small $|s|$ gives

$$Q(s) \approx \begin{pmatrix} 1 & \kappa_{12}s \\ \kappa_{21}s & 1 \end{pmatrix}$$

for some constants κ_{12} and κ_{21}. For low frequencies ω, the diagonal elements of $Q(s)$ are equal to one, and the off-diagonal elements are proportional to s. If the bandwidth of the decentralized PID controller is sufficiently low, the off-diagonal terms will thus be small, and the system will be approximately decoupled. The closed-loop system can be described by

$$\begin{pmatrix} 1 + q_{11}c_1 & q_{12}c_2 \\ q_{21}c_1 & 1 + q_{22}c_2 \end{pmatrix} Y = \begin{pmatrix} q_{11}\bar{c}_1 & q_{12}\bar{c}_2 \\ q_{21}\bar{c}_1 & q_{22}\bar{c}_2 \end{pmatrix} Y_{sp},$$

where the dependency on s is suppressed to simplify the notation. This equation can be written as

$$Y = \bar{H}Y_{sp},$$

where

$$\bar{h}_{11} = \frac{q_{11}\bar{c}_1(1 + q_{22}c_2) - q_{12}q_{21}\bar{c}_1c_2}{(1 + q_{11}c_1)(1 + q_{22}c_2) - q_{12}q_{21}c_1c_2}$$
$$\bar{h}_{12} = \frac{q_{12}\bar{c}_2(1 + q_{22}c_2) - q_{12}q_{22}\bar{c}_2c_2}{(1 + q_{11}c_1)(1 + q_{22}c_2) - q_{12}q_{21}c_1c_2}$$
$$\bar{h}_{21} = \frac{q_{21}\bar{c}_1(1 + c_1q_{11}) - q_{11}q_{21}c_1\bar{c}_1}{(1 + q_{11}c_1)(1 + q_{22}c_2) - q_{12}q_{21}c_1c_2}$$
$$\bar{h}_{22} = \frac{q_{22}\bar{c}_2(1 + q_{11}c_1) - q_{12}q_{21}c_1\bar{c}_2}{(1 + q_{11}c_1)(1 + q_{22}c_2) - q_{12}q_{21}c_1c_2}.$$

Since we designed the controllers so that the interactions are small, the term $q_{12}q_{21}$ is smaller than $q_{11}q_{22}$. The matrix $\bar{H}$ can then be approximated by

$$\bar{H} \approx H = \begin{pmatrix} \dfrac{q_{11}\bar{c}_1}{1 + q_{11}c_1} & \dfrac{q_{12}\bar{c}_2}{1 + q_{11}c_1} \\ \dfrac{q_{21}\bar{c}_1}{1 + q_{22}c_2} & \dfrac{q_{22}\bar{c}_2}{1 + q_{22}c_2} \end{pmatrix}.$$

The diagonal elements of H are the same as for SISO control design. The standard methods for design of PI controllers presented in Chapters 6 and 7 can be used to find the controllers c_1 and c_2. By analysing the off-diagonal elements we can estimate how severe the interactions are. The controllers may have to be detuned to make sure that the interactions are tolerable. The interaction can be reduced arbitrarily by making the control loops sufficiently slow. The interaction analysis also gives the performance loss due to the interaction. If much performance is lost it is advisable to consider other design methods.

Estimating Effects of Interaction

A simple way to estimate the effects of the interactions will now be developed. The off-diagonal elements of H are given by

$$h_{12} = \frac{q_{12}\bar{c}_2}{1 + q_{11}c_1}$$
$$h_{21} = \frac{q_{21}\bar{c}_1}{1 + q_{22}c_2}.$$

Notice that $q_{11}(0) = q_{22}(0) = 1$ and that $q_{12} \approx \kappa_{12}s$ and $q_{21}(s) \approx \kappa_{21}s$ for small s. Since the controllers have integral action, we have for small s

$$h_{12}(s) \approx \frac{\kappa_{12}k_{I2}s}{k_{I1}}, \qquad h_{21}(s) \approx \frac{\kappa_{21}k_{I1}s}{k_{I2}}.$$

The interaction is thus very small at low frequencies, and we can thus guarantee that the interaction is arbitrarily small by having sufficiently slow controllers. To estimate the maximum of the interaction, we observe that

$$h_{12} = q_{12}\bar{c}_2 S_1, \quad h_{21} = q_{21}\bar{c}_1 S_2,$$

where $S_1 = (1+q_{11}c_1)^{-1}$ and $S_2 = (1+q_{22}c_2)^{-1}$ are the sensitivity functions for the loops when the interaction is neglected. A crude estimate of the interaction terms is thus

$$\max_\omega |h_{12}(i\omega)| \approx |\kappa_{12}| k_{I2} M_{s1}$$
$$\max_\omega |h_{21}(i\omega)| \approx |\kappa_{21}| k_{I1} M_{s2},$$

where M_{s1} and M_{s2} are the maximum sensitivities of the individual loops and where we have also used the estimate

$$q_{12}(s) \approx \kappa_{12}s, \quad q_{21}(s) \approx \kappa_{21}s$$

and

$$\bar{c}_1 \approx k_{I1}/s, \quad \bar{c}_2 \approx k_{I2}/s.$$

The interaction can thus be captured by the interaction indices

$$\kappa_1 = |\kappa_{12}k_{I2}| M_{s1}, \quad \kappa_2 = |\kappa_{21}k_{I1}| M_{s2}. \tag{11.10}$$

The index κ_1 describes how the second loop influences the first loop, and κ_2 describes how the first loop influences the second loop. Note that the term κ_{12} depends on the system and the integral gain k_{I2} in the second loop. Interaction can thus be reduced by making the integral gains lower. The estimates are not precise because of the approximations made. They are not reliable when there is a significant difference in the bandwidths of the loops.

Examples

The design method will be illustrated by two examples. We will start by investigating Rosenbrock's system.

EXAMPLE 11.5—ROSENBROCK'S SYSTEM
Consider the system in Example 11.3 where the process has the transfer function (11.7). We have

$$D = P^{-1}(0) = \begin{pmatrix} 1 & 2/3 \\ 1 & 1 \end{pmatrix}^{-1} = \begin{pmatrix} 3 & -2 \\ -3 & 3 \end{pmatrix}.$$

If we introduce static decoupling, the compensated transfer function becomes

$$Q(s) = \begin{pmatrix} \dfrac{3(1-s)}{(s+1)(s+3)} & \dfrac{4s}{(s+1)(s+3)} \\ 0 & \dfrac{1}{s+1} \end{pmatrix} \approx \begin{pmatrix} 1 - 7s/3 & 4s/3 \\ 0 & 1-s \end{pmatrix}.$$

The interaction is given by $\kappa_{12} = 4/3$ and $\kappa_{21} = 0$. Since $\kappa_{21} = 0$, interaction gives no performance limitations for the second loop. There are, however, limitations because of the right half-plane zero at $s = 1$. Designing a PI controller that maximizes integral gain subject to the constraints that the maximum sensitivity M_{s1} and the maximum complementary sensitivity M_{p1} are less than 1.6, gives $k_{P1} = 0.2975$ and $k_{I1} = 0.3420$.

Since $\kappa_{12} = 4/3$ there are constraints on the design of the first loop because of the coupling. Requiring that the coupling κ_1 be less than 0.5 and the maximum sensitivity M_{s2} be less than 1.6, we find that the integral gain of the second loop k_{I2} must be less than $\kappa_1/(\kappa_{12}M_{s1}M_{s2}) = 0.23$. To design a PI controller, we use a placement procedure where the fast process pole $s = -1$ is canceled. The gain in the second loop is then $k_{P2} = 0.23$.

Figure 11.6 shows the frequency responses of h_{11}, h_{12}, and h_{22}. The largest magnitude of the term h_{12} is 0.26, which is half of the estimated value. The reason for the discrepancy is that the the simple estimate $q_{12} \approx \kappa_{12}s$ overestimates the term.

Figure 11.7 shows simulations of set point responses for the closed-loop system. The solid lines show the responses for controllers with set-point weighting $b_1 = 0$ and $b_2 = 0$. The dashed line shows the responses for controllers with error feedback. The plots show the proposed design with set-point weighting ($b_1 = b_2 = 0$). A unit step in the set point of the first controller is applied at time $t = 0$, and a step in the set point of the second controller is then applied at time $t = 20$. Figure 11.7 shows the step responses for a controller without set-point weighting. The figure clearly indicates the advantage of set-point weighting for multivariable systems. The reason why there is such a large difference is that the control signal is much smoother with set-point weighting.

The effect of set-point weighting is illustrated also in Figure 11.6, which shows the frequency response of the closed-loop system with (solid) and without (dashed) set-point weighting. The interaction increases considerably when no set-point weighting is applied. □

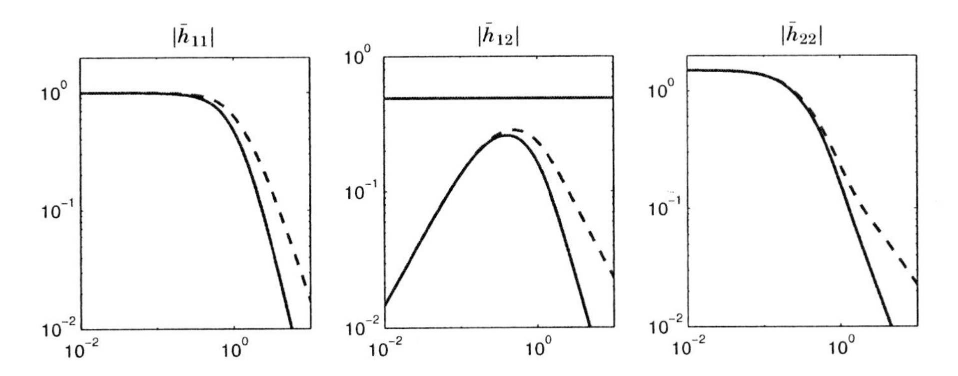

Figure 11.6 Frequency responses of the closed-loop system with set-point weighting (dashed) and without (solid). Note that without set-point weighting the interaction $|\bar{h}_{12}(i\omega)|$ is larger and extends to higher frequencies.

Distillation columns are typical industrial processes where interaction is significant. The next example deals with such a case.

EXAMPLE 11.6—THE WOOD–BERRY BINARY DISTILLATION COLUMN
The Wood–Berry binary distillation column is a multivariable system that has been studied extensively. A simple model of the system is given by the transfer function

$$P(s) = \begin{pmatrix} \dfrac{12.8e^{-s}}{16.7s+1} & \dfrac{-18.9e^{-3s}}{21.0s+1} \\ \dfrac{6.60e^{-7s}}{10.9s+1} & \dfrac{-19.4e^{-3s}}{14.4s+1} \end{pmatrix}.$$

Designing a static decoupler we find that

$$Q(s) = P(s)P^{-1}(0) \approx \begin{pmatrix} 1-11.7s & -12.31s \\ -0.5138s & 1-17.3s \end{pmatrix}.$$

Hence, $\kappa_{12} = -12.31$ and $\kappa_{21} = -0.5138$. Designing PI controller for the diagonal elements by maximizing integral gain subject to the robustness constraint $M_s = 1.6$ gives $k_1 = 2.3481$, $k_{i1} = 1.5378$, $k_2 = 0.5859$, and $k_{i2} = 0.2978$. The sensitivity frequencies are $\omega_{s1} = 0.30$ and $\omega_{s2} = 0.11$. Notice that the second loop is slower than the first loop. We have $\kappa_1 = 5.8$ and $\kappa_2 = 1.26$, which indicates that the interaction imposes constraints on the achievable performance and it is necessary to detune the controllers. This is illustrated by the dashed curves in the simulation shown in Figure 11.8. To reduce the interactions we will detune the controllers by decreasing the integral gains. As a first attempt we will reduce both integral gains by a factor of four. This implies that the integrated error for load disturbances is four times larger than for an uncoupled loop. Using the simple gain reduction rule developed in Section 7.9 we find that the proportional gains should then be reduced by a factor of two; see (7.27). The solid lines in Figure 11.8 show that the responses give a significant reduction of the interactions. The interaction can be reduced further at the price of lowered performance. □

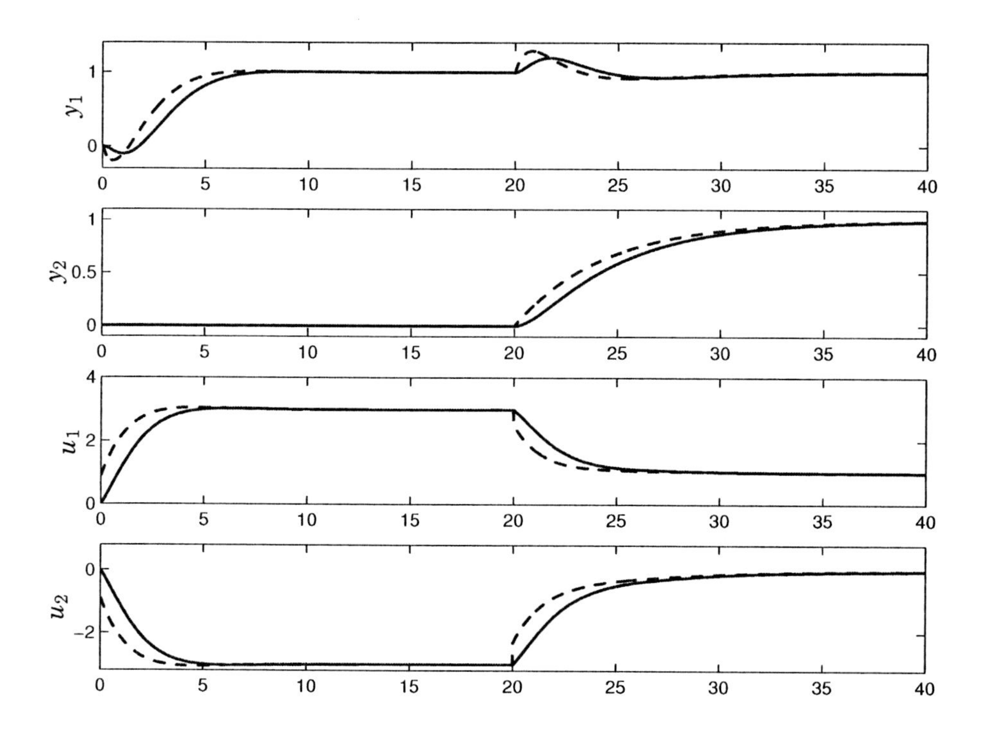

Figure 11.7 Simulation of the design method applied to Rosenbrock's system. The figure shows the response of the outputs to steps in the command signals. The PI controllers have gains $k_{P1} = 0.30$, $k_{I1} = 0.34$, $k_{P2} = 0.23$, $k_{I2} = 0.23$. The dashed lines show results with error feedback, and the solid lines show results with zero set-point weights.

11.4 Parallel Systems

Systems that are connected in parallel are quite common, particularly in drive systems. Typical examples are motors that are driving the same load, power systems, and networks for steam distribution. Control of such systems requires special consideration. To illustrate the difficulties that may arise we will consider the situation with two motors driving the same load. A schematic diagram of the system is shown in Figure 11.9.

Let ω be the angular velocity of the shaft, J the total moment of inertia, and D the damping coefficient. The system can then be described by the equation

$$J\frac{d\omega}{dt} + D\omega = M_1 + M_2 - M_L, \tag{11.11}$$

where M_1 and M_2 are the torques from the motors and M_L is the load torque.

Proportional Control

Assume each motor is provided with a proportional controller. The control

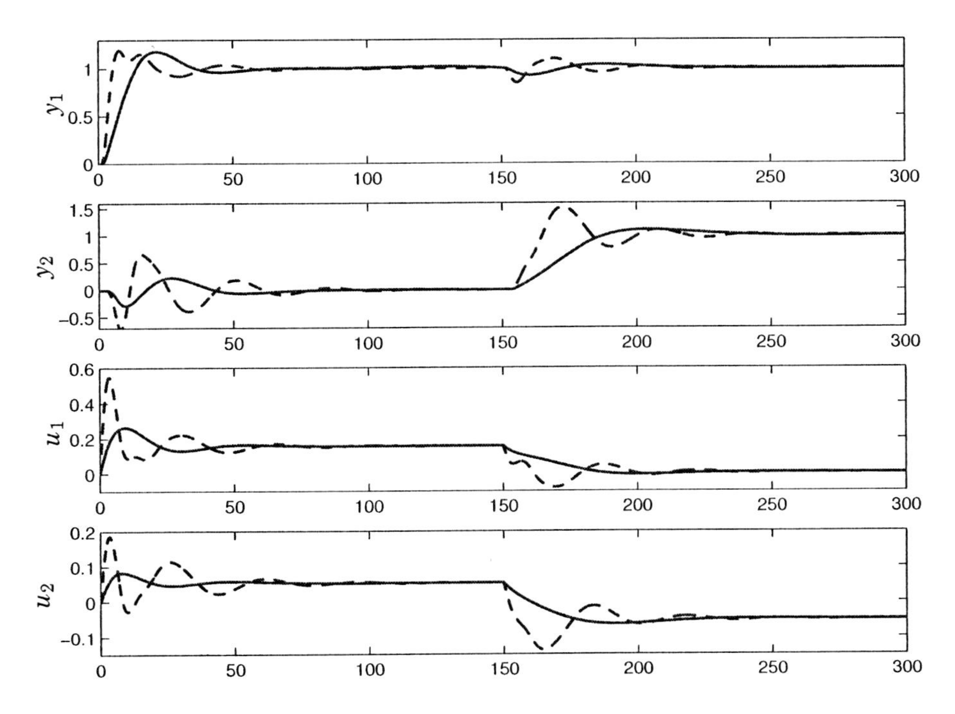

Figure 11.8 Simulation of decoupling control of Wood-Berry's distillation column. The figure shows the response of the outputs to steps in the command signals. The dashed curves show responses with PI controllers having gains $k_{P1} = 2.348$, $k_{I1} = 1.537$, $k_{P2} = 0.586$, and $k_{I2} = 0.298$. The solid lines show responses with detuned PI controllers. The gains are $k_{P1} = 1.119$, $k_{I1} = 0.384$, $k_{P2} = 0.293$, and $k_{I2} = 0.0745$. The set-point weights are zero in all cases.

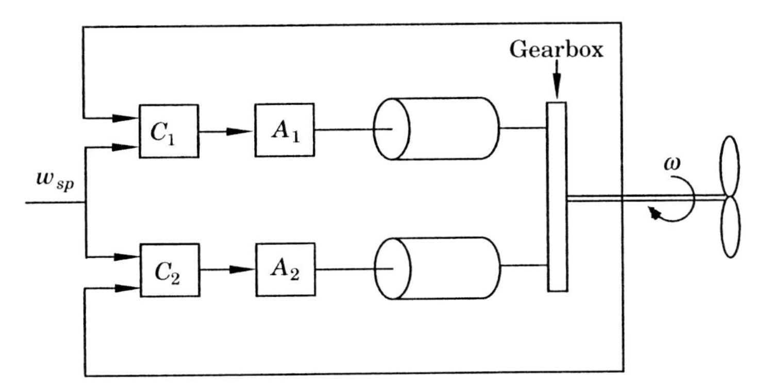

Figure 11.9 Schematic diagram of two motors that drive the same load.

strategies are then

$$\begin{aligned} M_1 &= M_{10} + K_1(\omega_{sp} - \omega) \\ M_2 &= M_{20} + K_2(\omega_{sp} - \omega). \end{aligned} \tag{11.12}$$

In these equations the parameters M_{10} and M_{20} give the torques provided by each motor when $\omega = \omega_{sp}$ and K_1 and K_2 are the controller gains. It follows from (11.11) and (11.12) that

$$J\frac{d\omega}{dt} + (D + K_1 + K_2)\omega = M_{10} + M_{20} - M_L + (K_1 + K_2)\omega_{sp}.$$

The closed-loop system is, thus, a dynamical system of first order. After perturbations, the angular velocity reaches its steady state with a time constant

$$T = \frac{J}{D + K_1 + K_2}.$$

The response speed is thus given by the sum of the damping and the controller gains. The stationary value of the angular velocity is given by

$$\omega = \omega_0 = \frac{K_1 + K_2}{D + K_1 + K_2}\omega_{sp} + \frac{M_{10} + M_{20} - M_L}{D + K_1 + K_2}.$$

This implies that there normally will be a steady-state error. Similarly, we find from (11.12) that

$$\frac{M_1 - M_{10}}{M_2 - M_{20}} = \frac{K_1}{K_2}.$$

The ratio of the controller gains will indicate how the load is shared between the motors.

Proportional and Integral Control

The standard way to eliminate a steady-state error is to introduce integral action. In Figure 11.10 we show a simulation of the system in which the motors have identical PI controllers. The set point is changed at time 0. A load disturbance in the form of a step in the load torque is introduced at time 10, and a pulse-like measurement disturbance in the second motor controller is introduced at time 20. When the measurement error occurs the balance of the torques is changed so that the first motor takes up much more of the load after the disturbance. In this particular case the second motor is actually breaking. This is highly undesirable, of course.

To understand the phenomenon we show the block diagram of the system in Figure 11.11. The figure shows that there are two parallel paths in the system that contain integration. This is a standard case where observability and controllability is lost. Expressed differently, it is not possible to change the signals M_1 and M_2 individually from the error. Since the uncontrollable state is an integrator, it does not go to zero after the disturbance. This means that the torques can take on arbitrary values after disturbance. For example, it may happen that one of the motors takes practically all the load, clearly an undesirable situation.

How to Avoid the Difficulties

Having understood the reason for the difficulty, it is easy to modify the controller as shown in Figure 11.12. In this case only one controller with integral

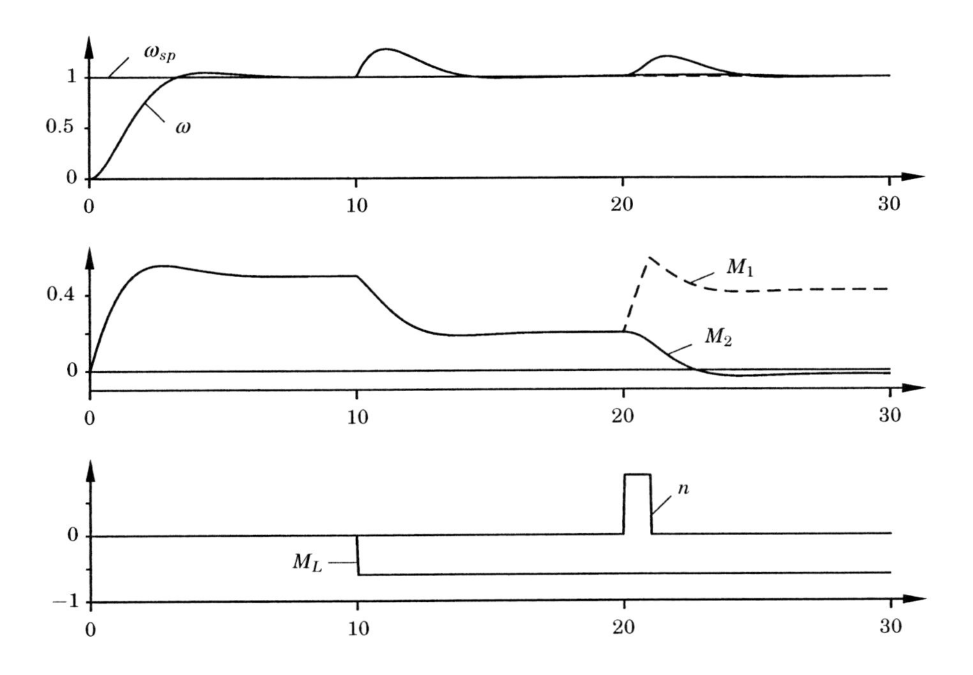

Figure 11.10 Simulation of a system with two motors with PI controllers that drive the same load. The figure shows set point ω_{sp}, process output ω, control signals M_1 and M_2, load disturbance M_L, and measurement disturbance n.

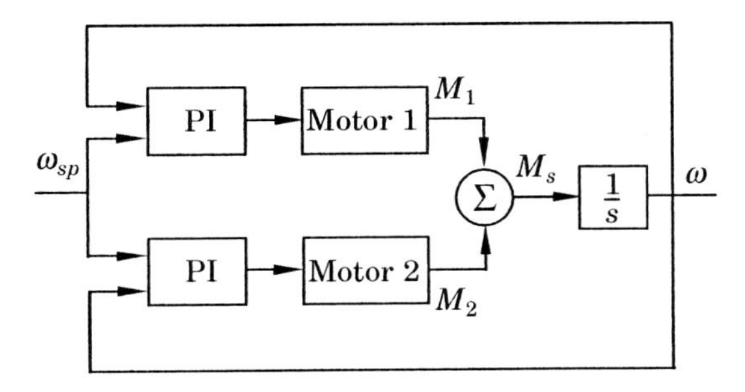

Figure 11.11 Block diagram for the system in Figure 11.10.

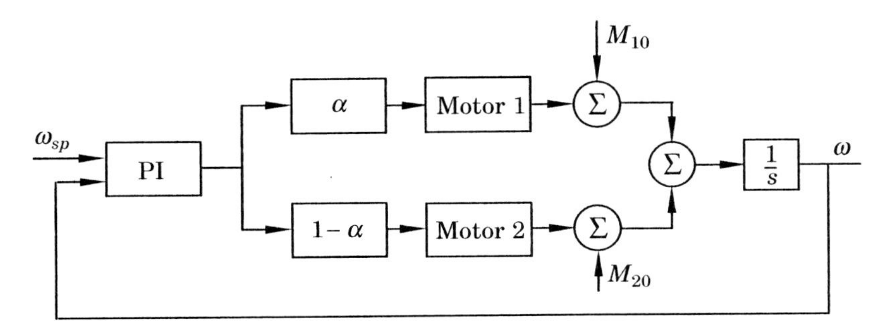

Figure 11.12 Block diagram of an improved control system.

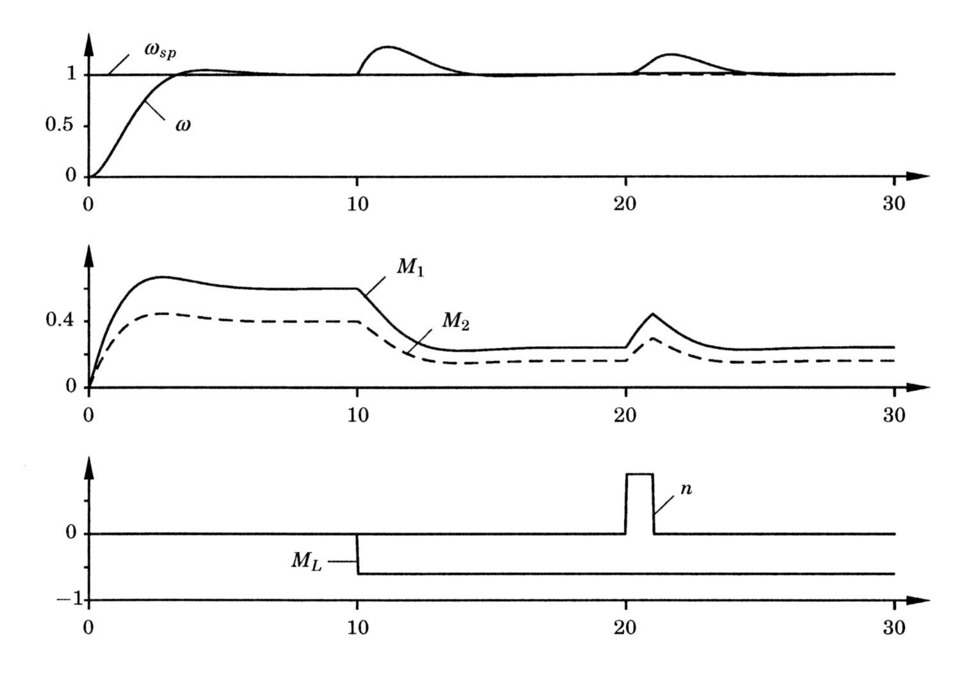

Figure 11.13 Simulation of the system with the modified controller. The figure shows set point ω_{sp}, process output ω, control signals M_1 and M_2, load disturbance M_L, and measurement disturbance n.

action is used. The output of this drives proportional controllers for each motor. A simulation of such a system is shown in Figure 11.13. The difficulties are clearly eliminated.

The difficulties shown in the examples with two motors driving the same load are accentuated even more if there are more motors. Good control in this case can be obtained by using one PI controller and distributing the outputs of this PI controller to the different motors, each of which has a proportional controller. An alternative is to provide one motor with a PI controller and let the other have proportional control. To summarize, we have found that there may be difficulties with parallel systems having integral action. The difficulties are caused by the parallel connection of integrators that produce unstable subsystems that are neither controllable nor observable. With disturbances these modes can change in an arbitrary manner. The remedy is to change the control strategies so there is only one integrator.

11.5 Summary

Even if a large control system may have many sensors and many actuators it can often be controlled by simple controllers of the PID type. This is particularly easy when there is little interaction in the system. In this chapter we have presented simple measures of interaction. They can be used to judge if the control problem can be solved using simple loops. Bristol's relative gain array can also be used to find pairs of inputs and outputs that are suitable for single-loop control. A simple design method that can be applied to systems with interaction has also been presented. This method combines static decoupling with the methods for design of single-loop controllers presented earlier in the book. Control of drive systems with parallel motors has also been discussed. For such systems there are particular problems with controllers having integral action.

11.6 Notes and References

Some fundamental issues related to interaction in systems are treated in [Rijnsdorp, 1965a; Rijnsdorp, 1965b; McAvoy, 1983]. The relative gain array was introduced in [Bristol, 1966]. It has been used widely and successfully in the process industries [Shinskey, 1981; McAvoy, 1983]. The most well-known results on the RGA are that a plant with large or negative elements in its RGA is difficult to control and that input and output variables should be paired such that the diagonal elements of the RGA are as close as possible to unity [Grosdidier *et al.*, 1985; Skogestad and Morari, 1987]. The RGA is based on the static gain of the process; an extension to account for dynamics is given in [McAvoy, 1983] . An alternative measure called the steady-state interaction indices was developed in [Chang and Davison, 1987] and it may provide a more accurate representation. Static and dynamic decoupling are treated in many textbooks in process control, e.g., [Seborg *et al.*, 2004]. Recent contributions to the design of decoupled PID control include the work by [Adusumilli *et al.*, 1998]. Detuning for multi-variable PID control, as discussed in the paper, was treated in a heuristic setting by [Niederlinski, 1971]. The particular method presented in Section 11.3 is based on [Åström *et al.*, 2002], other methods for design of non-interacting systems are given in [Yuzu *et al.*, 2002] and [Wang *et al.*, 2003]. Control of systems with strong interaction between many loops requires techniques that are very different from those discussed in this chapter; see [Cutler and Ramaker, 1980] and [Seborg *et al.*, 1986]. Multivariable systems are treated in standard textbooks on process control such as [Luyben, 1990; Marlin, 2000; Bequette, 2003; Seborg *et al.*, 2004]. There are also books that focus on multivariable systems: see [Shinskey, 1981; Skogestad and Postlethwaite, 1996].

12

Control Paradigms

12.1 Introduction

Process control systems are normally complex with many control variables and many measured signals. The bottom-up approach is one way to design such systems. In this procedure the system is built up from simple components. The systems can be implemented in many different ways. Originally, it was done by interconnection of separate boxes built of pneumatic or electronic components. Today, the systems are typically implemented in distributed control systems consisting of several hierarchically connected computers. The software for the distributed control system is typically constructed so that programming can be done by selecting and interconnecting the components. The key component, the PID controller, has already been discussed in detail. In this chapter, we present some of the components required to build complex automation systems. We also present some of the key paradigms that guide the construction of complex systems.

A collection of paradigms for control is used to build complex systems from simple components. The components are controllers of the PID type, linear filters, and static nonlinearities. Typical nonlinearities are amplitude and rate limiters and signal selectors. Feedback is an important paradigm. Simple feedback loops are used to keep process variables constant or to make them change in specified ways. Feedback has been discussed extensively in the previous chapters. Another important paradigm is feedforward. This was discussed in Chapter 5. The key problem is to determine the control variables that should be chosen to control given process variables. Another problem is that there may be interaction between different feedback loops. This was discussed in Chapter 11.

Section 12.2 gives an overview of the problem to design complex systems, and the two approaches top-down and bottom-up design are compared. This section also gives an overview and presents the outline of the chapter. The chapter ends with an example to illustrate how the different components and paradigms can be used. The process considered is a chemical reactor, and the design is given in Section 12.9. Some important observations made in the chapter are finally summarized in Section 12.10.

12.2 Bottom-Up and Top-Down Approaches

There are two general approaches for designing a complex system: bottom up and top down. In the *bottom-up* or Lego approach the system is designed by combining small subsystems. The *top-down* approach starts with a general overall design that is refined successively. In practice the approaches are often combined. In both approaches we need knowledge about the elementary building blocks or components of the system. The bottom-up approach requires principles for combining basic components, and the top-down approach requires principles for refining or decomposing a high-level objective so that it can be accomplished by the basic system components. Several components and control principles for composition and decomposition have been described earlier in the book. In this section we will give an overview of the approaches, and in later sections we will describe components and paradigms that have not been discussed previously.

The Bottom-Up Approach

Large control systems can be built from controllers, filters, and nonlinear elements. The components can either be separate pieces of hardware or function blocks implemented in software that can be combined graphically using cut and paste. Controllers and filters have been discussed in Chapters 3, 5, 9, and 11. The nonlinear elements will be discussed in Section 12.6.

Control principles like feedback, feedforward, and model following have been discussed extensively in Chapters 3 and 5. Other important control principles such as *repetitive control*, *cascade control*, *mid-range control*, *split-range control*, *ratio control*, and *selector control* will be discussed in Sections 12.3, 12.4, 12.5, and 12.6.

An advantage with the bottom-up approach is that the system can be commissioned and tuned loop by loop. There may be difficulties when the loops are interacting. The disadvantage is that it is not easy to judge if additional loops will bring benefits. The system can also be unwieldy when loops are added.

Top-Down Solutions

Top-down paradigms often start with a problem formulation in terms of an optimization problem. Paradigms that support a top-down approach are optimization, state feedback, observers, predictive control, and linearization. In the top-down approach it is natural to deal with many inputs and many outputs simultaneously. Since this is not the main topic of this book we will only give a brief discussion. The top-down approach often leads to the controller structure shown in Figure 12.1. In this system all measured process variables y together with the control variables u are sent to an observer, which uses the sensor information and a mathematical model to generate a vector $\hat{x}$ of good estimates of internal process variables and important disturbances. The estimated state $\hat{x}$ is then compared with the ideal state x_m produced by the feedforward generator, and the difference is fed back to the process. The feedforward generator also gives a feedforward signal u_{ff}, which is sent directly to the process inputs. The controller shown in Figure 12.1 is useful for process segments where there are several inputs and outputs that interact, but the

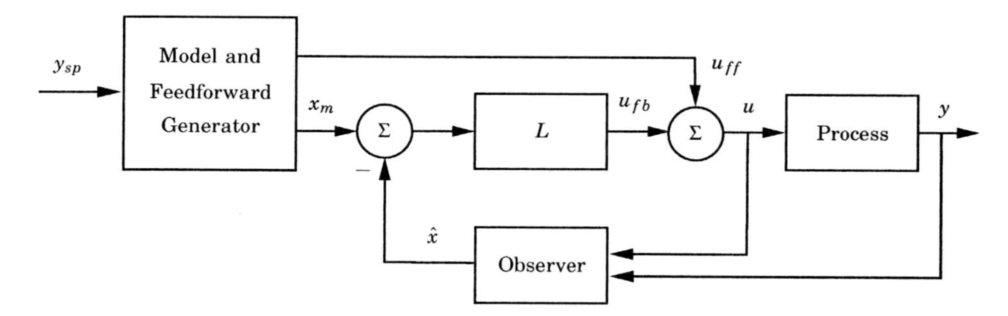

Figure 12.1 Block diagram of a controller based on model following, state feedback, and an observer.

system becomes very complicated when there is a large number of inputs and outputs. In such a case it may be better to decompose the system into several subsystems.

An advantage with the top-down approach is that the total behavior of the system is taken into account. A systematic approach based on mathematical modeling and simulation makes it easy to understand the fundamental limitations. Commissioning of the system is, however, difficult because many feedback loops have to be closed simultaneously. When using the top-down approach it is therefore good practice to first tune loops based on simulation, possibly also hardware in the loop simulation.

Soft Computing

Because of the widespread use of computers in control there has also been an influence on control from computer science. Two particular paradigms that originated from artificial intelligence are *neural networks* and *fuzzy control*, which both emerged from research in artificial intelligence. These paradigms are presented in Section 12.7 and Section 12.8. This branch of computer science is also called *soft computing*.

12.3 Repetitive Control

Attenuation of disturbances has been an essential theme in this book. For PID control we have focused on elimination of constant or slow disturbances. In this section we will show that similar ideas can be used to eliminate other types of disturbances, particularly periodic disturbances. Problems of this type are common when there are cyclic operations.

In Section 4.3 it was shown that attenuation of disturbances is captured by the transfer function from load disturbance to process output

$$G_{yd} = \frac{P}{1 + PC}, \tag{12.1}$$

where P is the process transfer function and C the controller transfer function, respectively. By designing a controller that has high gain at a particular fre-

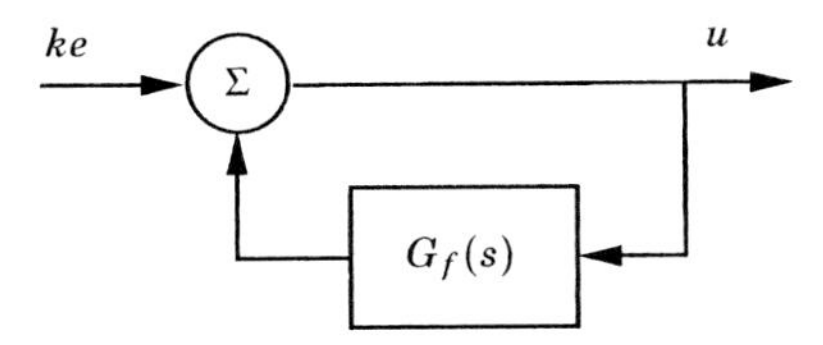

Figure 12.2 Block diagram of a controller with positive feedback of a filtered signal.

quency disturbances with that frequency are effectively reduced. The control error is zero in steady-state if the gain is infinite.

Consider the controller in Figure 12.2. Intuitively the system works as follows. The filter G_f filters out the signal component that we would like to eliminate, and the output of G_f is fed back to the input with positive feedback. The net effect is to create a high gain for the frequencies in the pass band of the filter G_f.

Constant and Sinusoidal Disturbances

To investigate the properties of the system analytically we observe that the controller has the transfer function

$$C(s) = \frac{k}{1 - G_f(s)}. \tag{12.2}$$

When $G_f(s)$ is a low-pass filter with transfer function

$$G_f(s) = \frac{1}{1 + sT},$$

we find that

$$C(s) = k\Big(1 + \frac{1}{sT}\Big),$$

which is the transfer function of a PI controller. Notice that the controller transfer function $C(s)$ has infinite gain at zero frequency, which implies that the steady-state error is zero for constant disturbances.

When $G_f(s)$ is the band-pass filter

$$G_f(s) = \frac{2\zeta\omega_0 s}{s^2 + 2\zeta\omega_0 s + \omega_0^2},$$

we find that

$$C(s) = k\frac{2\zeta\omega_0 s}{s^2 + \omega_0^2}.$$

Notice that this transfer function has infinite gain for $s = i\omega_0$, which implies that the steady-state error is zero for a sinusoidal disturbance of frequency ω_0.

Periodic Disturbances

Periodic disturbances can be reduced by choosing

$$G_f(s) = e^{-sL},$$

where L is the period of the disturbance. With this filter we find

$$C(s) = \frac{k}{1 - e^{-sL}}. \tag{12.3}$$

The relation between control error and control variable is

$$u(t) = ke(t) + u(t - L).$$

The control action at time t is thus a sum of the control error and the control signal at time $t - L$.

The controller has infinite gain for $s = 2n\pi i/L$, $n = 0, 1, \ldots$. A controller of this type is particularly useful when disturbances or set-point variations are periodic.

The transfer function from load disturbance to output (12.1) is

$$G_{yd}(s) = \frac{P(s)}{1 + P(s)C(s)} = \frac{P(s)(1 - e^{-sL})}{1 - e^{-sL} + kP(s)}.$$

The relation between load disturbance and output is then

$$(1 - e^{-sL} + kP(s))Y(s) = P(s)(1 - e^{-sL})D(s).$$

Notice that the time function corresponding to $(1 - e^{-sL})D(s)$ is

$$d(t) - d(t - L),$$

which vanishes if D is a periodic disturbance with period L. The steady-state error caused by a periodic disturbance is thus zero.

The effective disturbance rejection does, however, come at a price that is illustrated with the following example.

EXAMPLE 12.1—AN EXTREME CASE

Consider a process with the transfer function

$$P(s) = e^{-sL},$$

with the controller

$$C(s) = \frac{1}{1 - e^{-sL}}.$$

that attenuates periodic disturbances.

The loop transfer function

$$G_l(s) = \frac{e^{-sL}}{1 - e^{-sL}}$$

is periodic with period $2\pi/L$, and its gain is infinite for $\omega = 2n\pi/L$. The frequency response is

$$G_l(i\omega) = -\frac{1}{2} - i\frac{\sin \omega L}{2(1-\cos \omega L)} = -\frac{1}{2} - i\frac{1}{\tan(\omega L/2)}.$$

The Nyquist curve is a vertical line through the point $G_l = -0.5$ and a half circle to the right. This curve is transversed once for $0 \le \omega \le 2\pi/L$ and infinitely many times when ω increases towards infinity.

The system has the gain margin 2 and the phase margin is 60°. The sensitivity functions are

$$\begin{aligned} S(s) &= 1 - e^{-sL} \\ T(s) &= e^{-sL}, \end{aligned}$$

and we find that $M_s = 2$ and $M_t = 1$.

A superficial look at traditional robustness measures like gain margin $g_m = 2$, phase margin $\varphi_m = 60°$, and maximum sensitivities $M_s = 2$ and $M_t = 1$ may indicate that the system is robust to process perturbations.

The fact that $T(i\omega) = 1$ for all frequencies is, however, an indication that the system has unusual properties. Further insight is obtained by analysing the effect of parameter variations. The system has only one parameter, the time delay L, and we will investigate the effects of variations in the time delay. To use the robustness inequality (4.32) we will convert time delay variations to an additive process perturbation. Assume that the time delay changes from L to $L + \delta L$, then

$$e^{-s(L+\delta L)} = e^{-sL}e^{-s\delta L} = e^{-sL} + e^{-sL}(e^{-s\delta L} - 1).$$

A variation in the time delay can thus be represented by the additive perturbation

$$\Delta P(s) = e^{-sL}(e^{-s\delta L} - 1).$$

Hence $|\Delta P(i\omega)| = |e^{-i\omega\delta L} - 1|$.

Since $|P(i\omega)| = 1$, the robustness inequality (4.32) becomes

$$\frac{|\Delta P(i\omega)|}{|P(i\omega)|} = |e^{-i\omega\delta L} - 1| < \frac{1}{|T(i\omega)|} = 1.$$

This inequality is not satisfied for any $\delta L > 0$ because the left-hand side is 2 and the right hand side is 1, and we cannot guarantee stability for an arbitrary small perturbation in the time delay. □

The example shows that the effective attenuation of periodic disturbances comes at the cost of the system being extremely sensitive to parameter variations. A compromise between disturbance attenuation can be made by replacing $G_f(s)$ in Figure 12.2 by $\alpha G_f(s)$ with $\alpha < 1$. The controllers obtained for

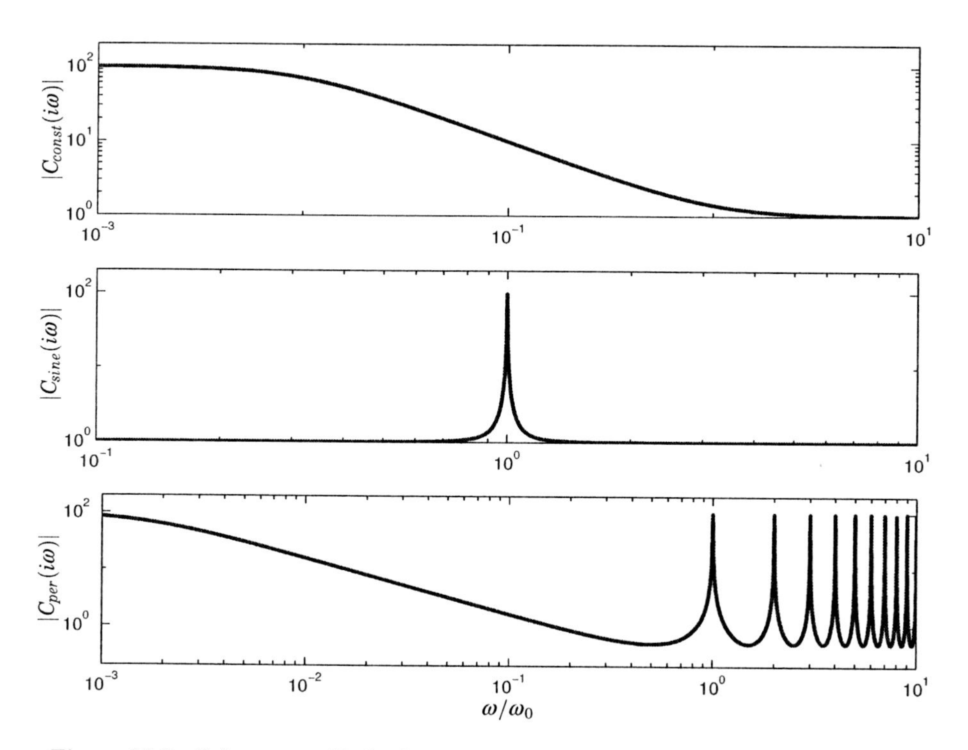

Figure 12.3 Gain curves of Bode plots for the controllers C_{const} (top), C_{sine} (middle), and C_{per} (bottom). The parameter α is 0.99 in all cases, which means that the largest gains of the controllers are 100. For the band-pass filter we have $\zeta = 0.1$, and for the repetitive controller we have $T = 2\pi/\omega_0$.

constant, sinusoidal, and periodic signals then become

$$
\begin{aligned}
C_{\text{const}}(s) &= \frac{1 + sT}{1 - \alpha + sT} \\
C_{\text{sine}}(s) &= \frac{s^2 + 2\zeta\omega_0 s + \omega_0^2}{s^2 + 2(1-\alpha)\zeta\omega_0 s + \omega_0^2} \\
C_{\text{per}}(s) &= \frac{1}{1 - \alpha e^{-sT}}.
\end{aligned}
$$

The largest gains of the transfer functions are $1/(1-\alpha)$ in all cases. Choosing $\alpha < 1$ diminishes disturbance attenuation but improves the robustness. The properties of the controllers C_{const}, C_{sine}, and C_{per} are illustrated in Figure 12.3 which shows the gain curves of the Bode plots for the controllers. The controller C_{const} has high gain for low frequencies, the controller C_{sine} has high gain for $\omega = \omega_0$, and the controller C_{per} has high gain for the frequencies ω_0, $2\omega_0$, $3\omega_0$, etc.

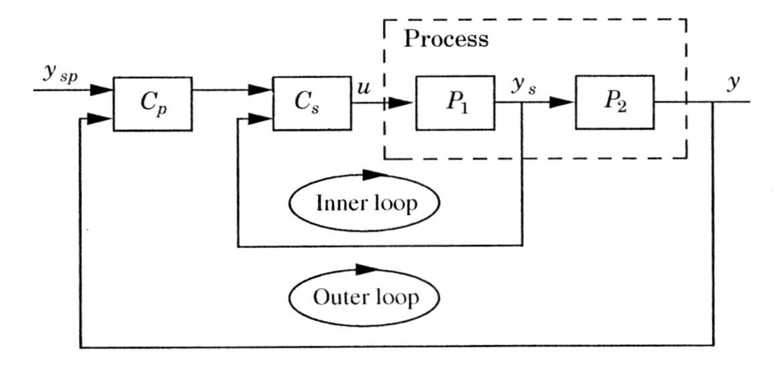

Figure 12.4 Block diagram of a system with cascade control.

12.4 Cascade Control

Cascade control can be used when there are several measurement signals and one control variable. It is particularly useful when there are significant dynamics, e.g., long dead times or long time constants, between the control variable and the process variable. Tighter control can then be achieved by using an intermediate measured signal that responds faster to the control signal. Cascade control is built up by nesting the control loops, as shown in the block diagram in Figure 12.4. The system in this figure has two loops. The inner loop is called *the secondary loop*; the outer loop is called *the primary loop*. The reason for this terminology is that the outer loop deals with the primary measured signal. It is also possible to have a cascade control with more nested loops. The performance of a system can be improved with a number of measured signals, up to a certain limit. If all state variables are measured, it is often not worthwhile to introduce other measured variables. In such a case the cascade control is the same as state feedback. We will illustrate the benefits of cascade control by an example.

EXAMPLE 12.2—IMPROVED LOAD DISTURBANCE REJECTION
Consider the system shown in Figure 12.4. Let the transfer functions be

$$P_1 = \frac{1}{s+1}$$

and

$$P_2 = \frac{1}{(s+1)^3}.$$

Assume that a load disturbance enters at the input of the process. There are significant dynamics from the control variable to the primary output. The secondary output does respond much faster than the primary output. Thus, cascade control can be expected to give improvements.

The dashed lines in Figure 12.5 show the response obtained with conventional feedback using a PI controller with the parameters $K = 0.37$ and $T_i = 2.2$. Since the response of the secondary measured variable to the control

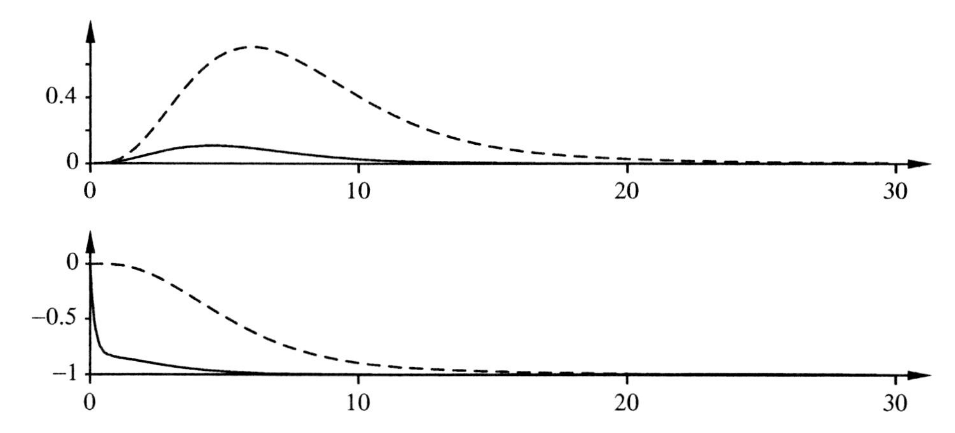

Figure 12.5 Responses to a load disturbance for a system with (solid line) and without (dashed line) cascade control. The upper diagram shows process output y, and the lower diagram shows control signal u.

signal is quite fast, it is possible to use high loop gains in the secondary loop. If the controller in the inner loop is proportional with gain K_s, the dynamics from the set point of C_s to process output becomes

$$G(s) = \frac{K_s}{(s+1+K_s)(s+1)^3}.$$

With $K_s = 5$ in the inner loop and PI control with $K = 0.55$ and $T_i = 1.9$ in the outer loop, the responses shown in solid lines Figure 12.5 are obtained. The figure shows that the disturbance response is improved substantially by using cascade control. Notice in particular that the control variable drops very much faster with cascade control. The main reason for this is the fast inner feedback loop, which detects the disturbance much faster than the outer loop.

The secondary controller is proportional, and the loop gain is 5. A large part of the disturbance is eliminated by the inner loop. The remaining error is eliminated at a slower rate through the action of the outer loop. In this case integral action in the inner loop will always give an overshoot in the disturbance response. □

Choice of Secondary Measured Variables

It is important to be able to judge whether cascade control can give improvement and to have a methodology for choosing the secondary measured variable. This is easy to do if we just remember that the key idea of cascade control is to arrange a tight feedback loop around a disturbance. In the ideal case the secondary loop can be so tight that the secondary loop is a perfect servo wherein the secondary measured variable responds very quickly to the control signal. The basic rules for selecting the secondary variable are:

- There should be a well-defined relation between the primary and secondary measured variables.

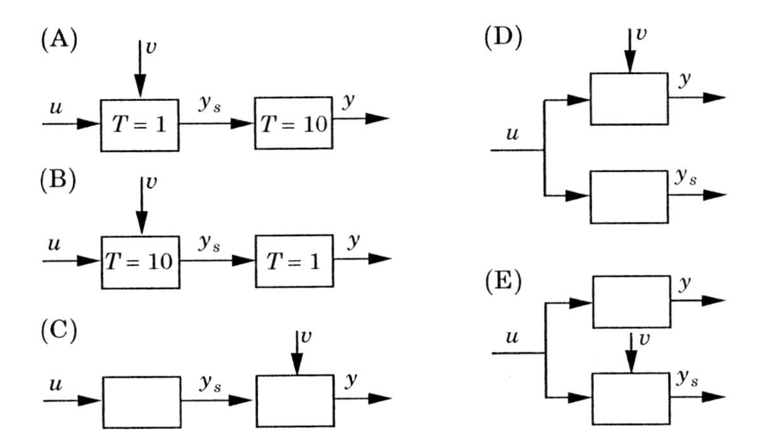

Figure 12.6 Examples of different process and measurement configurations.

- Essential disturbances should act in the inner loop.
- The inner loop should be faster than the outer loop. The typical rule of thumb is that the average residence times should have a ratio of at least five.
- It should be possible to have a high gain in the inner loop.

A common situation is that the inner loop is a feedback around an actuator. The reference variable in the inner loop can then represent a physical quantity, like flow, pressure, torque, velocity, etc., while the control variable of the inner loop could be valve pressure, control current, etc. This is also a typical example where feedback is used to make a system behave in a simple predictive way. It is also a very good way to linearize nonlinear characteristics.

A number of different control systems with one control variable and two measured signals are shown in Figure 12.6. In the figure the control variable is represented by u, the primary measured variable by y, the secondary measured variable by y_s, and the essential disturbance is v. With the rules given above it is only case A that is suitable for cascade control.

Choice of Control Modes

When the secondary measured signal is chosen it remains to choose the appropriate control modes for the primary and secondary controllers and to tune their parameters. The choice is based on the dynamics of the process and the nature of the disturbances. It is difficult to give general rules because the conditions can vary significantly. In critical cases it is necessary to analyze and simulate. It is, however, useful to have an intuitive feel for the problems.

Consider the system in Figure 12.4. To have a useful cascade control, it is necessary that the process P_2 be slower than P_1 and that the essential disturbances act on P_1. We assume that these conditions are satisfied. The secondary controller can often be chosen as a pure proportional controller or a PD controller. In some cases integral action can be useful to improve rejection of low-frequency disturbances. With controllers that lack integral action, there may be a static error in the secondary loop. This may not be a serious drawback. The secondary loop, as a rule, is used to eliminate fast disturbances. Slow disturbances can easily be eliminated by the primary loop, which will typically have integral action. There are also drawbacks to using integral control in the secondary loop. With such a system there will always be an overshoot in the response of the primary control loop. Integral action is needed if the process P_2 contains essential time delays and the process P_1 is such that the loop gain in the secondary loop must be limited.

The special case when the process P_2 is a pure integrator is quite common. In this case integral action in the inner loop corresponds to proportional control in the outer loop. If integral action is used in the inner loop, the proportional action in the outer loop must be reduced. This is a significant disadvantage for the performance of the system. A good remedy is to remove the integrator in the inner loop and to increase the gain in the outer loop.

Tuning and Commissioning

Cascade controllers must be tuned in a correct sequence. The outer loop should first be put in manual when the inner loop is tuned. The inner loop should then be put in automatic when tuning the outer loop. The inner loop is often tuned for critical or over-critical damping or equivalently for a small sensitivity (M_s). If this is not done there is little margin for using feedback in the outer loop.

Commissioning of cascade loops also requires some considerations. The following procedure can be used, starting with both controllers in manual mode.

1. Adjust the set point of the secondary controller to the value of the secondary process variable.
2. Set the secondary controller in automatic with internal set point selected.
3. Adjust the primary controller so that its set point is equal to the process variable and so that its control signal is equal to the set point of the secondary controller.
4. Switch the secondary controller to external set point.
5. Switch the primary controller to automatic mode.

The steps given above are automated to different degrees in different controllers. If the procedure is not done in the right way there will be switching transients.

Integral Windup

If integral action is used in both the secondary and primary control loops, it is necessary to have a scheme to avoid integral windup. The inner loop can be handled in the ordinary way, but it is not a trivial task to avoid windup in the outer loop. There are three situations that must be covered:

1. The control signal in the inner loop can saturate.
2. The secondary control loop may be switched to internal set point.
3. The secondary controller is switched from automatic to manual mode.

The feedback loop, as viewed from the primary controller, is broken in all these cases, and it is necessary to make sure that its integral mode is dealt with properly. This problem is solved automatically in a number of process controllers that have cascade control capabilities, but if we build up the cascade control using two independent controllers we have to solve the problem ourselves. This requires being able to inject a tracking signal into the primary controller.

If the output signal of the secondary controller is limited, the process variable of the secondary controller should be chosen as the tracking signal in the primary controller. This also requires a digital transfer from the secondary to the primary controller telling it when the tracking is to take place.

In the case where the secondary controller switches to work according to its local set point instead of the external one from the primary controller, the local set point should be sent back to the primary controller as a tracking signal. In this way one can avoid both integrator windup and jumps in the transition to cascade control.

When the secondary controller switches over to manual control, the process variable from the secondary controller should be sent back to the primary controller as a tracking signal.

Some Applications

Cascade control is a convenient way to use extra measurements to improve control performance. The following examples illustrate some applications.

EXAMPLE 12.3—VALVE POSITIONERS
Control loops with pneumatic valves are a very common application. In this case the inner loop is a feedback around the valve itself where the valve position is measured. The inner loop reduces the influences of pressure variations and various nonlinearities in the pneumatic system. □

EXAMPLE 12.4—MOTOR CONTROL
Figure 12.7 is a block diagram of a typical motor control system. This system has three cascaded loops. The innermost loop is a current loop where the current is measured. The next loop is the velocity loop, which is based on measurement of the velocity. The outer loop is a position loop. In this case integral action in the velocity loop is equivalent to proportional action in the position loop. Furthermore, it is clear that the derivative action in the position loop is equivalent to proportional action in the velocity loop. From this it follows directly that there is no reason to introduce integral action in the velocity controller or derivative action in the position controller. □

EXAMPLE 12.5—HEAT EXCHANGER
A schematic diagram of a heat exchanger is shown in Figure 12.8. The purpose of the control system is to control the outlet temperature on the secondary side by changing the valve on the primary side. The control system shown uses cascade control. The secondary loop is a flow control system around the valve. The control variable of the primary loop is the set point of the flow

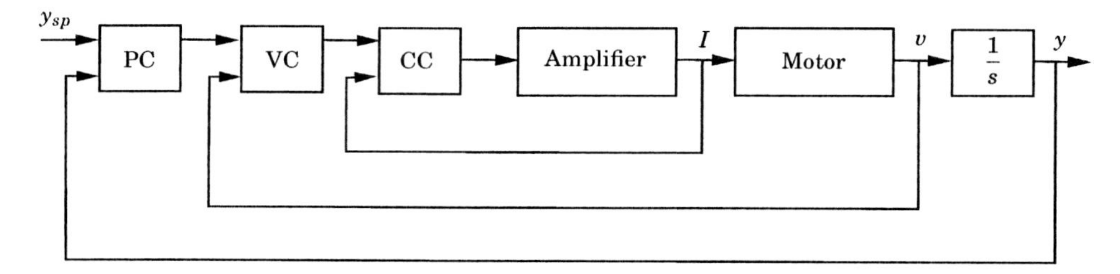

Figure 12.7 Block diagram of a system for position control. The system has three cascaded loops with a current controller (CC) with feedback from current (I), a velocity controller (VC) with feedback from velocity (v), and a position controller (PC) with feedback from position (y).

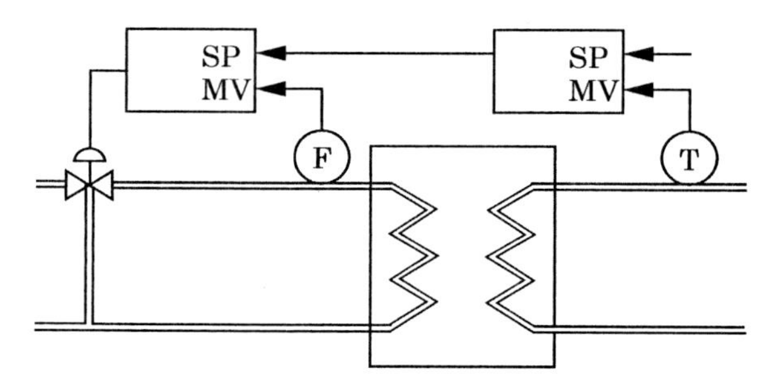

Figure 12.8 Schematic diagram of a heat exchanger with cascade control.

controller. The effect of nonlinearities in the valve, as well as flow and pressure disturbances, is thus reduced by the secondary controller. □

12.5 Mid-Range and Split-Range Control

Cascade control is a strategy where one control signal and two measurement signals are used to meet the control objective. The dual situation is when two control signals are used to control one measurement signal. The two control signals are sometimes used one at a time. This is the case in split-range control. In other situations it is necessary to use the two control signals simultaneously. A common situation is mid-range control or mid-ranging. Mid-range and split-range control are discussed in this section.

Mid-Range Control

The problem treated by mid-range control is illustrated in Figure 12.9. The figure illustrates an example where two valves are used to control a flow. One valve, v_1, is small but has a high resolution. The other valve, v_2, is large but has a low resolution.

Suppose that the small valve v_1 is in the middle of its operating range and that only small disturbances are acting on the system. In this case, one controller that manipulates valve v_1 is able to take care of the control problem.

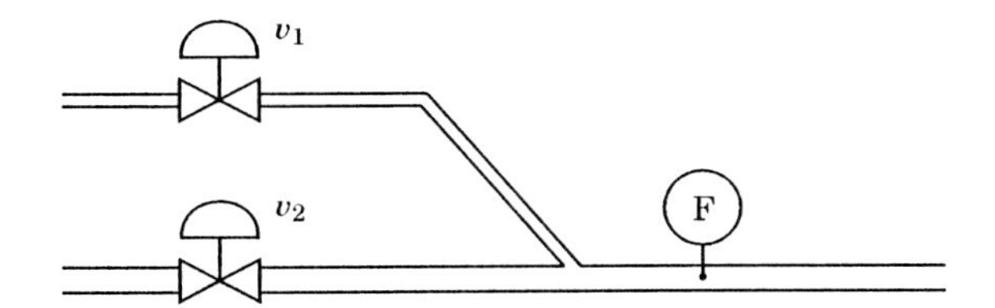

Figure 12.9 Two valves are used to control the flow.

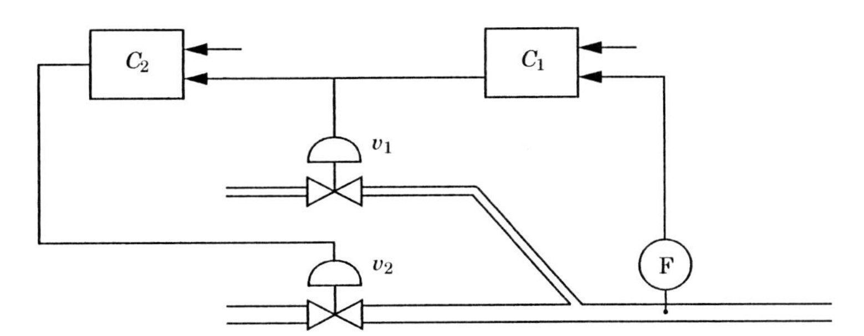

Figure 12.10 Mid-range control

However, when larger disturbances occur, valve v_1 will saturate. In this case, the larger valve v_2 must also be manipulated.

The mid-range control strategy is illustrated in Figure 12.10. Controller C_1 takes the set point y_{sp} and flow signal y as inputs and manipulates the small valve v_1. A second controller, C_2, takes the control signal from C_1 as input and tries to control it to a set point u_{sp} in the middle of its operating range by manipulating the large valve v_2. If both controllers have integral action, the flow will be at the set point y_{sp} and the valve v_1 will be at the set point u_{sp} in steady state.

A block diagram of the mid-range control strategy is given in Figure 12.11. Process P_1 and controller C_1 together form a fast feedback loop. The mid-ranging controller C_2 controls the valve position of controller C_1 via the process output y. This means that the output of controller C_1 is controlled by driving the process output y away from the set point. If this is done slowly, the deviation

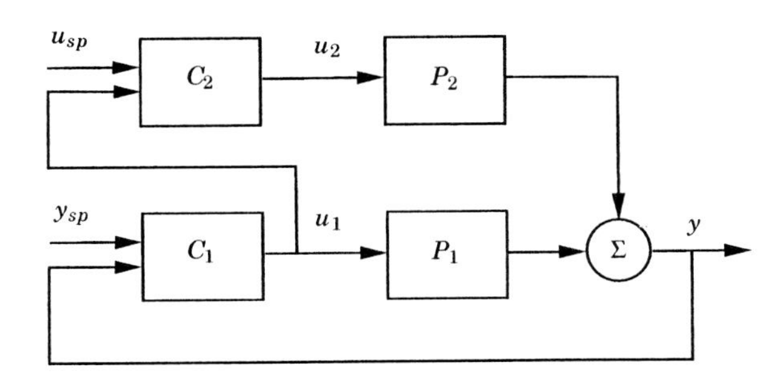

Figure 12.11 Block diagram of a system with mid-range control.

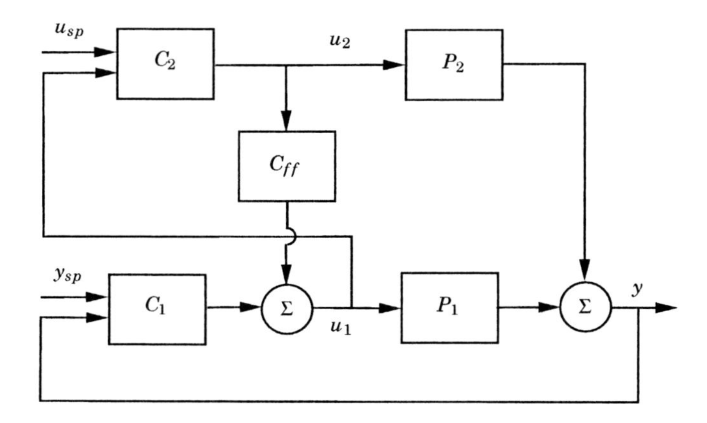

Figure 12.12 Block diagram of a system with mid-range control.

from the set point can be kept small. If not, it is recommended to use the structure given in Figure 12.12.

In Figure 12.12 a feedforward signal is added from control signal u_2 to controller C_1. If the feedforward compensator is

$$C_{ff}(s) = -\frac{P_2(s)}{P_1(s)},$$

controller C_2 will perform the mid-ranging control without any disturbance of the process output y.

It is likely that the small valve will saturate. In spite of this, it is not necessary that the controller C_1 has anti-windup. Since the control signal is controlled by the controller C_2, controller C_2 prevents controller C_1 from winding up.

Split-Range Control

In split-range control, the control is shared by two controllers that perform the control one at a time. Systems of this type are common, e.g., in connection with heating and cooling. One physical device is used for heating and another for cooling. The heating and cooling systems often have different static and dynamic characteristics. The principle of split-range control is illustrated in Figure 12.13, which shows the static relation between the measured variables and the control variables. When the temperature is too low, it is necessary to supply heat. The heater, therefore, has its maximum value when the measured variable is zero. It then decreases linearly until mid-range, where no heating is supplied. Similarly, there is no cooling when the measured variable is below mid-range. Cooling, however, is applied when the process variable is above mid-range, and it then increases.

There is a critical region when switching from heating to cooling. To avoid both heating and cooling at the same time, there is often a small dead zone where neither heating nor cooling is supplied. Switching between the different control modes may cause difficulties and oscillations.

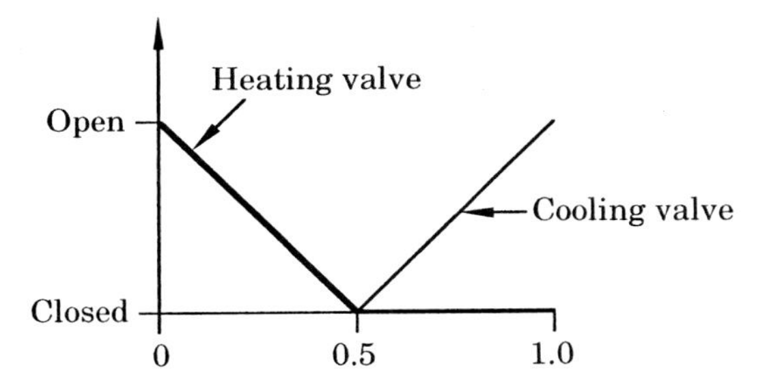

Figure 12.13 Illustration of the concept of split-range control.

Split-range control is commonly used in systems for heating and ventilation. It is also useful applications when the control variable ranges over a very large range. The flow is then separated into parallel paths, each controlled with a valve.

12.6 Nonlinear Elements

Nonlinear elements have been discussed before. In Section 3.5 we used a limiter to avoid integral windup in a controller with integral action. In Chapter 9 it was shown that controllers could be tuned by relay feedback and that performance could be improved by gain scheduling. In this section we describe more nonlinear elements and also present some control paradigms that guide the use of these elements.

Linearization

The nonlinearity in sensors and actuators can be compensated in a straightforward way. Consider, for example, an actuator that has the characteristics

$$v = f(u),$$

where v is the actual process input signal, and u is the control signal. To compensate for the nonlinearity we simply compute the control signal u_c as if the actuator was linear with unit gain. The control law

$$u = f^{-1}(u_c),$$

where f^{-1} is the inverse of the actuator nonlinearity, then gives

$$v = f(u) = f(f^{-1}(u_c)) = u_c.$$

The actuated process signal is then identical to u_c as was desired.

The same idea can be applied to sensors. Consider, for example, a sensor that has the nonlinearity $g(x)$. By designing a linear controller based on the assumption that the sensor is linear with unit gain and feeding the signal

$$y_c = g^{-1}(y)$$

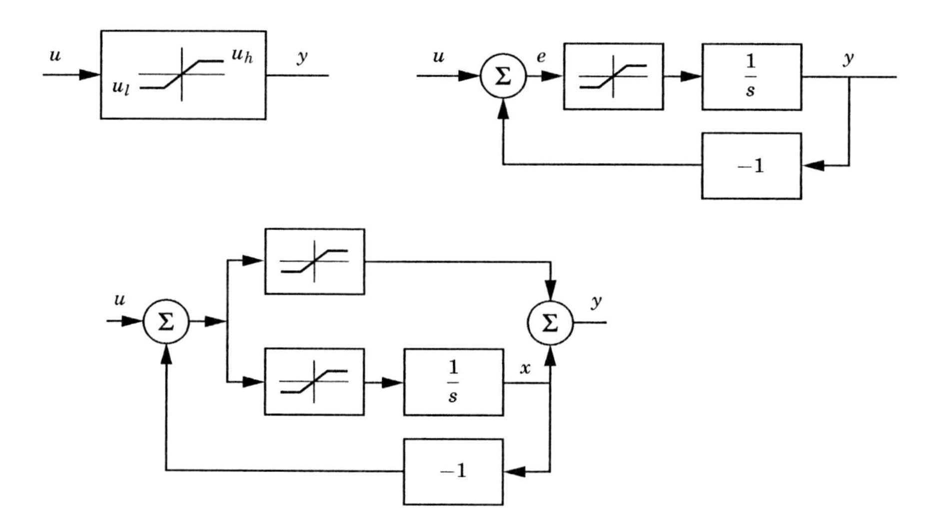

Figure 12.14 Block diagram of a simple amplitude limiter (upper left), a rate limiter (upper right), and a jump and rate limiter or a ramp unit (lower).

to the controller, the sensor nonlinearity is eliminated.

Similar ideas can be applied to process nonlinearities, but the compensation is not ideal because of dynamics. There is a technique for compensating for nonlinearities called feedback linearization, but this is outside the scope of this book. There are also situations when the nonlinearities are beneficial.

Limiters

Since all physical values are limited, it is useful to have limiting devices in control systems too. Limiters are used in many different ways. They can be used to limit the command signals so that we are not generating set points that are demanding larger or faster changes than a system can cope with.

A block diagram of a simple amplitude limiter is shown in upper left part of Figure 12.14. The limiter can mathematically be described as the static nonlinearity

$$y = \operatorname{sat}(u, u_l, u_h) = \begin{cases} u_l & \text{if } u \le u_l \\ u & \text{if } u_l < u < u_h \\ u_h & \text{if } u \ge u_h \end{cases}.$$

where u_l and u_h are the saturation limits.

It is also useful to limit the rate of change of signals. This can be done with the *rate limiter* or the *ramp unit* shown in the upper right part of Figure 12.14. The output follows the input signal if the rate of change of the input is smaller than the rate limit. In steady state the inputs and the outputs are identical because there is integral action in the system. Since the output is generated by an integrator with limited input signal, the rate of change of the output will be limited to the bounds given by the limiter. It is possible to use different limits for increasing or decreasing rates.

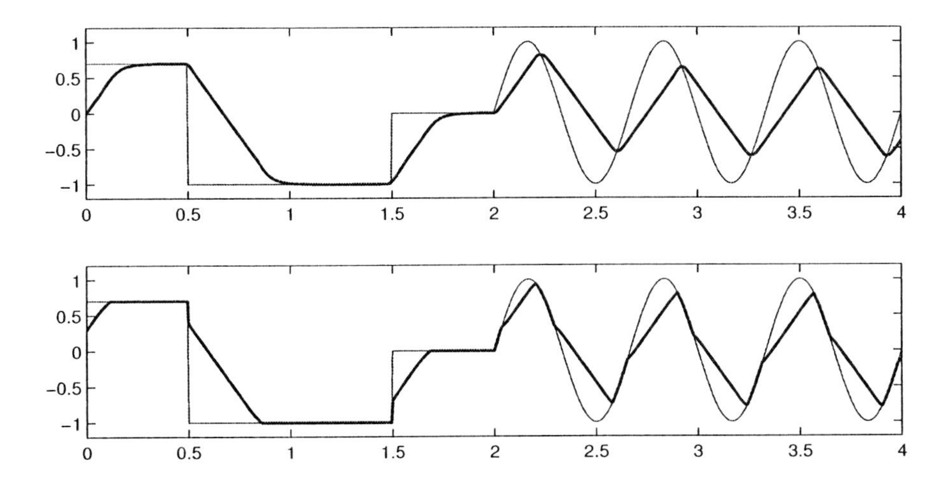

Figure 12.15 Simulation of a rate limiter (upper), and a jump and rate limiter (lower). The thin line shows the input to the limiter and the thick line shows the output of the limiter

A more sophisticated limiter called a *jump and rate limiter* is shown in the lower part of Figure 12.14. The output will follow the input for small changes in the input signal. At large changes, the output will follow the input with a limited rate. The jump and rate limiter can be described by

$$\begin{aligned}\frac{dx}{dt} &= \operatorname{sat}(u - x, -a, a)\\ y &= x + \operatorname{sat}(u - x, -a, a),\end{aligned}$$

If $|u - x| \leq a$ it follows from the equations describing the system that $y = u$, and if $u \geq x + a$ it follows that $dx/dt = a$. Thus, the output signal will approach the input signal at the rate a.

The properties of the different limiters are illustrated in the simulation shown in Figure 12.15. The input signal consists of a few steps and a sinusoid. The upper curve shows a rate limiter where the rate limit is 4. The figure shows that the rate of change of the output is limited. The response to a sinusoidal input shows clearly that the rate limiter gives a phase lag. The lower curve shows the response of a jump and rate limiter. Notice that the output follows rapid changes in the input as long as the difference between x and u are less than the jump limit, which is 0.5. The rate is limited to 4.

Surge Tank Control

The control problems that were discussed in Chapter 4 were all regulation problems where the task was to keep a process variable as close to a given set point as possible. There are many other control problems that also are important. Surge tank control is one example. The purpose of a surge tank is to act as a buffer between different production processes. Flow from one

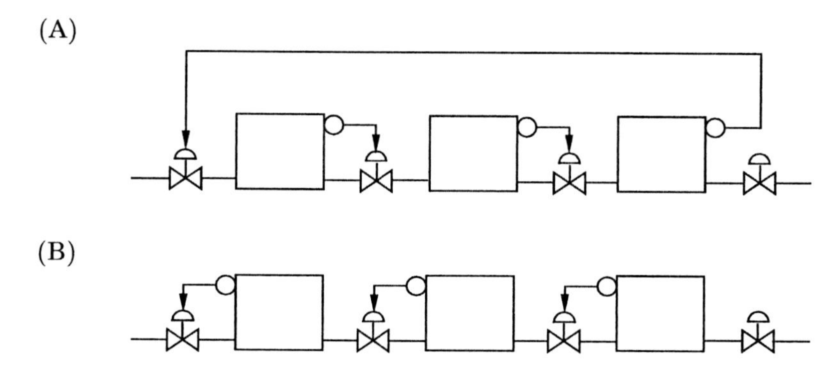

Figure 12.16 Different structures for surge tank control. The material flow is from the left to the right. The scheme in A is called control in the direction of the flow. The scheme in B is called control in the direction opposite to the flow.

process is fed to another via the surge tank. Variations in production rate can be accommodated by letting the level in the surge tank vary. Conventional level control, which attempts to keep the level constant, is clearly not appropriate in this case. To act as a buffer the level should indeed change. It is, however, important that the tank neither become empty nor overflow.

There are many approaches to surge tank control. A common, simple solution is to use a proportional controller with a low gain. Controllers with dead zones or nonlinear PI controllers are also used. Gain scheduling is a better method. The scheduling variable is chosen as the tank level. A controller with low gain is chosen when the level is between, e.g., 10 percent and 90 percent, and a controller with high gain is used outside the limits. There are also special schemes for surge tank control.

In many cases there are long sequences of surge tanks and production units, as illustrated in Figure 12.16. Two different control structures, control in the direction of the flow or opposite to the flow, are shown in the figure. Control in the direction opposite to the flow is superior because then all control loops are characterized by first-order dynamics. With control in the direction of the flow, it is easy to get oscillations or instabilities because of the feedback from the end of the chain to the beginning.

Ratio Control

Ratio control is applied when the control objective is to keep the ratio between two variables, often flows, at a certain ratio a. In combustion, for example, it is desired to control the fuel-to-air supply ratio, in order for the combustion to be as efficient as possible. Blending of chemicals is another example where it is desired to keep the ratio between different flows constant. In in-line blending systems, when there are no downstream mixing tanks, this is of special importance. If the composition is not maintained, quality problems may occur.

Ratio control is normally solved in the way shown in Figure 12.6. There are two control loops. The main loop consists of process P_1 and controller C_1. Output y_1 is the main flow, and the external set point r_1 is the desired main flow. In the second loop, consisting of process P_2 and controller C_2, it is attempted to

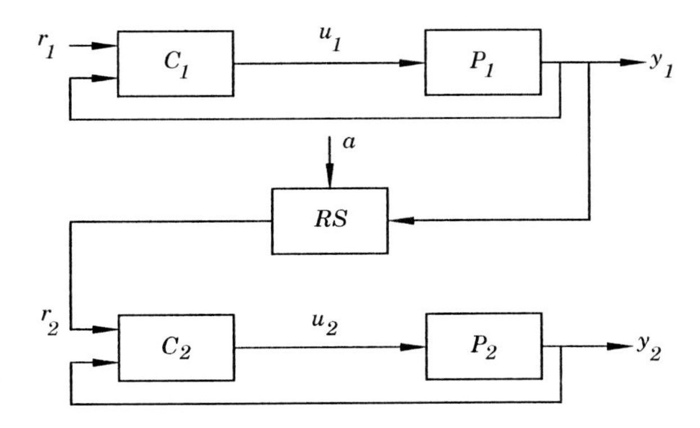

Figure 12.17 Ratio control using a Ratio station *(RS)* applied to main flow y_1.

control the flow y_2 so that the ratio y_2/y_1 is equal to ratio a. In Figure 12.6 this is obtained using a Ratio station where set point r_2 is determined by

$$r_2(t) = a y_1(t), \tag{12.4}$$

i.e., simply by multiplying the main flow y_1 with the desired ratio a.

In Equation 12.4, parameter a is assumed to be constant. This is not necessary. The desired ratio a is often time-varying. In combustion, for example, the ratio a is often adjusted based on O_2 measurements in the exhaust.

Provided the controllers have integral action, the solution given in Figure 12.6 will work in steady state, i.e., $y_1 = r_1$ and $y_2 = a y_1$. However, the simple Ratio station is not efficient during transients. The second flow y_2 will always be delayed compared to the desired flow $a y_1$. The length of this delay is determined by the dynamics of the second loop.

When set point r_1 is increasing, the delay causes an under-supply of the media corresponding to flow y_2, and conversely when r_1 is decreasing there is an excess of the media corresponding to flow y_2. There are cases when it is important never to get any under-supply of one of the two media. In the combustion case, one gets an under-supply of air during the transient part when the external set point increases, but an excess of air when the set point decreases. To prevent the fuel from not being fully burnt by an under-supply of air, the solution in Figure 12.6 has to be complemented with some logic using selectors. This is discussed in the next section.

The main drawback with the simple Ratio station approach shown in Figure 12.6 is that the secondary flow y_2 is delayed compared to the desired flow $a y_1$. This problem can be solved if not only y_1 is used to form the secondary set point, but also the main set point r_1. The structure, called the Blend station, is shown in Figure 12.18.

In the Blend station, the secondary set point is determined as

$$r_2(t) = a\left(\gamma r_1(t) + (1-\gamma) y_1(y)\right). \tag{12.5}$$

Gain γ is a weighting factor that determines the relation between set point r_1

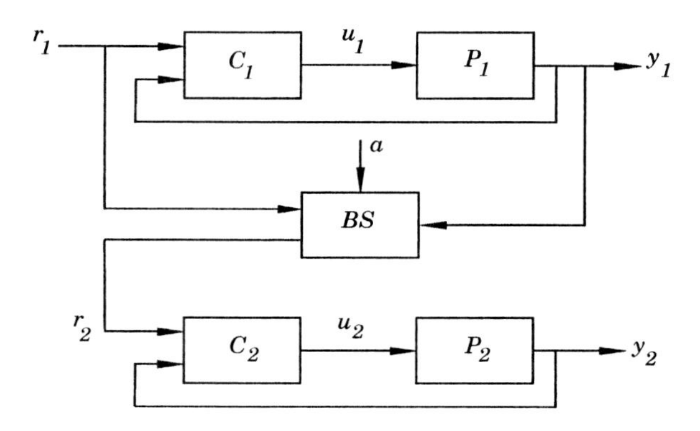

Figure 12.18 Ratio control using the Blend station *(BS)*.

and main flow y_1 when forming secondary set point r_2. When $\gamma = 0$, the Blend station is identical to the Ratio station.

EXAMPLE 12.6—PULP BLEACHING CONTROL

The Ratio station and the Blend station have been applied on a bleaching section in a paper mill. The pulp is bleached by adding Hydrosulphite to the pulp flow. The goal is to keep the ratio between the pulp flow and the Hydrosulphite flow constant.

The upper diagram in Figure 12.19 shows control using the Ratio station. The pulp flow controller, C_1, is a PI controller with setting $K_1 = 0.2$ and $T_{i1} = 4s$. The Hydrosulphite controller, C_2, is also a PI controller with setting $K_2 = 0.078$ and $T_{i2} = 1.07s$. The figure shows responses to two set-point changes in the pulp flow. The Hydrosulphite flow is scaled with the desired ratio and translated, so that the desired flow rates become identical. The figure shows that the Ratio station provides the correct ratio in steady state, but also that there is a deviation between the two flows during the transients. The Hydrosulphite flow is delayed compared to the pulp flow.

The lower diagram in Figure 12.19 shows the results obtained when using the Blend station with gain factor $\gamma = 0.75$. Here, the difference between the two flows is almost eliminated.

□

Selector Control

Selector control can be viewed as the inverse of split-range control. In split range there is one measured signal and several actuators. In selector control there are many measured signals and only one actuator. A selector is a static device with many inputs and one output. There are two types of selectors: *maximum* and *minimum*. For a maximum selector the output is the largest of the input signals.

There are situations where several controlled process variables must be taken into account. One variable is the primary controlled variable, but it is also required that other process variables remain within given ranges. Selector

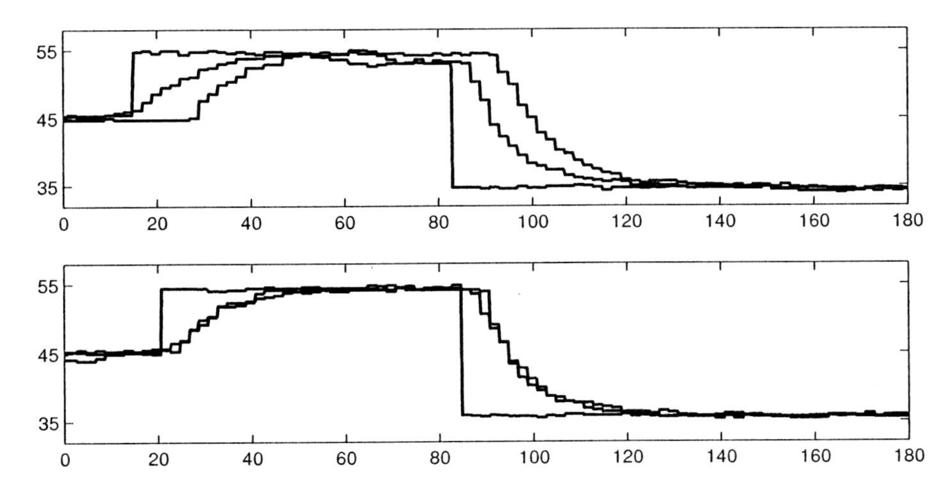

Figure 12.19 Ratio control of a pulp bleaching process using the original Ratio station (upper) and the Blend station with gain $\gamma = 0.75$ (lower). The figure shows two changes in the pulp set point, the pulp flow (fastest response) and the Hydrosulphite flow (slowest response).

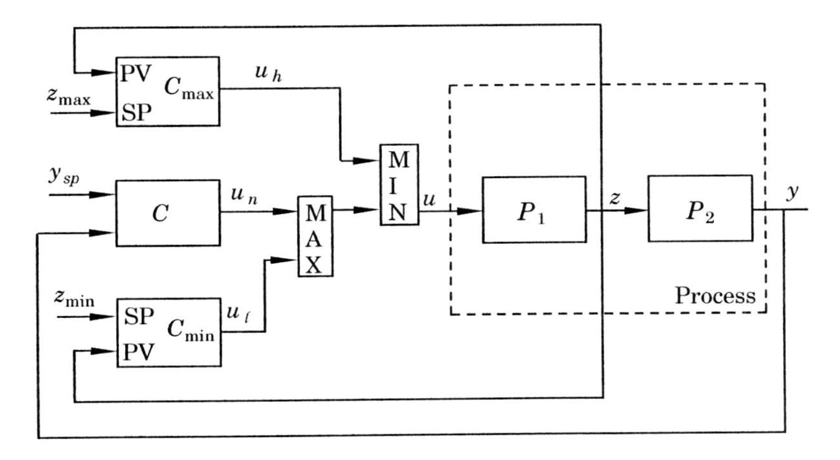

Figure 12.20 Selector control.

control can be used to achieve this. The idea is to use several controllers and to have a selector that chooses the controller that is most appropriate. One example of use is where the primary controlled variable is temperature and we must ensure that pressure does not exceed a certain range for safety reasons.

The principle of selector control is illustrated in Figure 12.20. The primary controlled variable is the process output y. There is an auxiliary measured variable z that should be kept within the limits $z_{\min}$ and $z_{\max}$. The primary controller C has process variable y, set point y_{sp}, and output u_n. There are also secondary controllers with measured process variables that are the auxiliary variable z and with set points that are bounds of the variable z. The outputs of these controllers are u_h and u_l. The controller C is an ordinary PI or PID controller that gives good control under normal circumstances. The output of the minimum selector is the smallest of the input signals; the output of the maximum selector is the largest of the inputs.

Under normal circumstances the auxiliary variable is larger than the minimum value $z_{\min}$ and smaller than the maximum value $z_{\max}$. This means that the output u_h is large and the output u_l is small. The maximum selector, therefore, selects u_n, and the minimum selector also selects u_n. The system acts as if the maximum and minimum controller were not present. If the variable z reaches its upper limit, the variable u_h becomes small and is selected by the minimum selector. This means that the control system now attempts to control the variable z and drive it towards its limit. A similar situation occurs if the variable z becomes smaller than $z_{\min}$.

In a system with selectors, only one control loop at a time is in operation. The controllers can be tuned in the same way as single-loop controllers. There may be some difficulties with conditions when the controller switches. With controllers having integral action, it is also necessary to track the integral states of those controllers that are not in operation. Selector control is very common in order to guarantee that variables remain within constraints. The technique is commonly used in the power industry for control in boilers, power systems, and nuclear reactors. The advantage is that it is built up of simple nonlinear components and PI and PID controllers. An alternative to selector control is to make a combination of ordinary controllers and logic. The following example illustrates the use of selector control.

EXAMPLE 12.7—AIR-FUEL CONTROL
In the previous section we discussed air-fuel control using ratio control. When the Ratio station is used, there may be lack of air because the set point of the air controller increases first when the fuel controller has increased the oil flow. One way to solve this problem is to use the Blend station. However, the system cannot compensate for perturbations in the air channel. This problem can be treated using selectors, such as is shown in Figure 12.21. The system uses one minimum and one maximum selector. There is one PI controller for fuel flow and one PI controller for the air flow. The set point for the air controller is the larger of the command signal and the fuel flow. This means that the air flow will increase as soon as more energy is demanded. Similarly, the set point to the fuel flow is the smaller of the demand signal and the air flow. This means that when demand is decreased, the set point to the dual flow controller will immediately be decreased, but the set point to the air controller will remain high until the oil flow has actually decreased. The system thus ensures that there will always be an excess of air. □

Median Selectors

A median selector is a device with many inputs and many outputs. Its output selects the input that represents the current median of the input signals. A special case is the two-out-of-three selector, commonly used for highly sensitive systems. To achieve high reliability it is possible to use redundant sensors and controllers. By inserting median selectors it is possible to have a system that will continue to function even if several components fail.

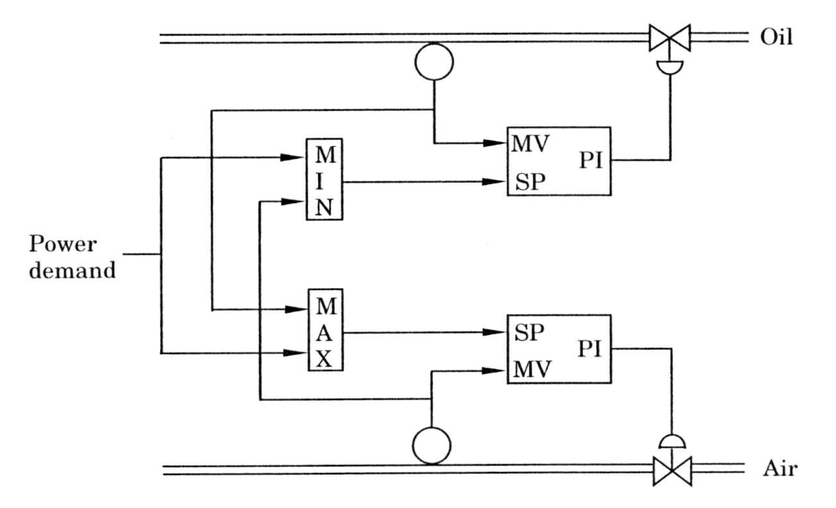

Figure 12.21 Air-fuel controller based on selectors.

12.7 Neural Network Control

In the previous section, we have seen that simple nonlinearities can be used very effectively in control systems. In this and the following section, we will discuss some techniques based on nonlinearities, where the key idea is to represent functions of several variables in a compact way. The ideas have been introduced under the names of *neural* and *fuzzy control*. At first sight, these methods may seem quite complicated, but once the colorful language is stripped off we find that the algorithms have natural representations as implementations of nonlinear functions. It is a nontrivial problem to find good representations of a nonlinear function. If we simply try to grid the variables and use an interpolation we find that the number of entries in the table for representing the function grows very rapidly with the number of variables. For example, if n variables are gridded in N points each we find that the number of entries are N^n. For a function of five variables with $N = 100$ we find that 10^{10} entries are required. Another useful property of neural networks is that there are methods to fit the parameters of the function to data.

Neural Networks

Neural networks originated in attempts to make simple models for neural activity in the brain and attempts to make devices that could recognize patterns and carry out simple learning tasks. A brief description that captures the essential idea follows.

A Simple Neuron A schematic diagram of a simple neuron is shown in Figure 12.22. The system has many inputs and one output. If the output is y

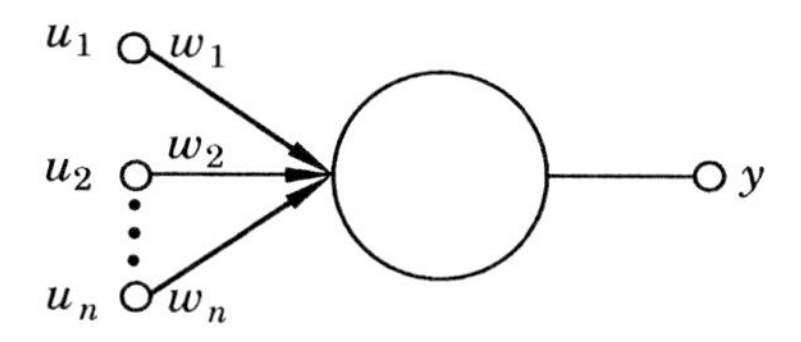

Figure 12.22 Schematic diagram of a simple neuron.

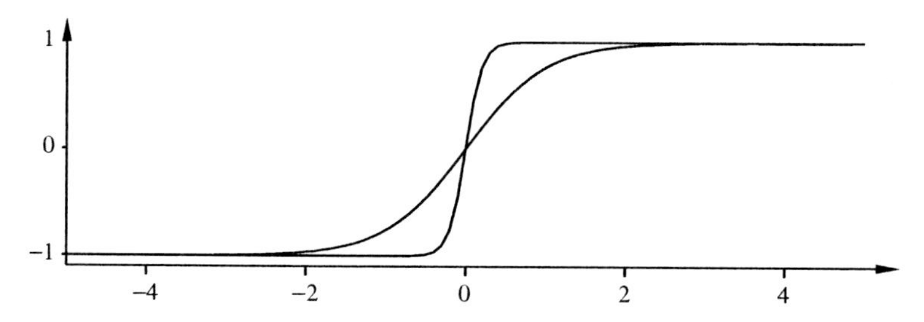

Figure 12.23 Sigmoid functions.

and the inputs are $u_1, u_2, \ldots, u_n$ the input-output relation is described by

$$y = f(w_1u_1 + w_2u_2 + \cdots + w_nu_n) = f\left(\sum_{k=1}^{n} w_iu_i\right), \tag{12.6}$$

where the numbers w_i are called weights. The function f is a so-called sigmoid function, illustrated in Figure 12.23. Such a function can be represented as

$$f(x) = \sinh \alpha x = \frac{e^{\alpha x} - e^{-\alpha x}}{e^{\alpha x} + e^{-\alpha x}} \tag{12.7}$$

where α is a parameter. This model of a neuron is thus simply a nonlinear function. Some special classes of functions can be approximated by (12.6).

Neural Networks More complicated models can be obtained by connecting neurons together as shown in Figure 12.24. This system is called a neural network or a neural net. The adjective feedforward is often added to indicate that the neurons are connected in a feedforward manner. There are also other types of neural networks. In the feedforward network, the input neurons are connected to a layer of neurons, the outputs of the neurons in the first layer are connected to the neurons in the second layer, and so on, until we have the outputs. The intermediate layers in the net are called hidden layers.

Each neuron is described by Equation (12.6). The input-output relation of a neural net is thus a nonlinear static function. Conversely, we can consider a neural net as one way to construct a nonlinear function of several variables. The neural network representation implies that a nonlinear function of several variables is constructed from two components: a single nonlinear function, the sigmoid function (12.7), which is a scalar function of one variable; and linear operations. It is thus a simple way to construct a nonlinearity from simple operations. A key reason why neural networks are interesting is that practically

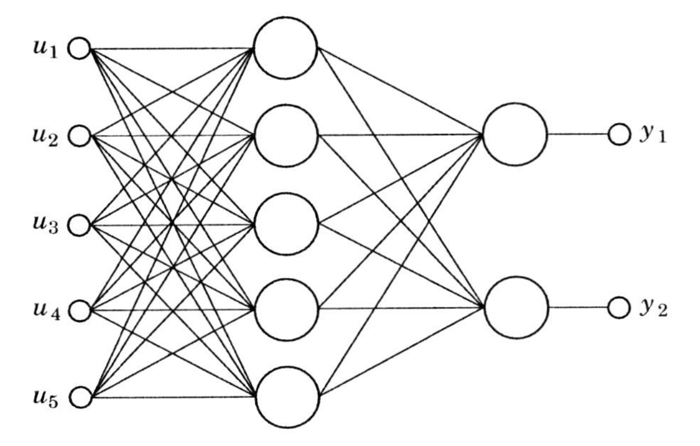

Figure 12.24 A feedforward neural network.

all continuous functions can be approximated by neural networks having one hidden layer. It has been found practical to use more hidden layers because then fewer weights can be used. Another practical feature of the sigmoidal functions is that the approximations are local.

Learning Notice that there are many parameters (weights) in a neural network. Assuming that there are n neurons in a layer, if all neurons are connected, n^2 parameters are then required to describe the connections between two layers.

Another interesting property of a neural network is that there are so-called learning procedures. This is an algorithm that makes it possible to find parameters (weights) so that the function matches given input-output values. The parameters are typically obtained recursively by giving an input value to the function and the desired output value. The weights are then adjusted so that the data is matched. A new input-output pair is then given, and the parameters are adjusted again. The procedure is repeated until a good fit has been obtained for a reasonable data set. This procedure is called training a network. A popular method for training a feedforward network is called back propagation. For this reason the feedforward net is sometimes called a back-propagation network. Fitting a neural network to experimental data is illustrated in Figure 12.25. A nice feature is that it is possible to find both the function and its inverse. The inverse function is useful when compensating for nonlinearities in sensors and actuators.

Control Applications A feedforward neural network can be viewed as a nonlinear function of several variables with a training procedure. The function has many parameters (weights) that can be adjusted by the training procedure so that the function will match given data. Even if this is an extremely simplistic model of a real neuron, it is a useful system component. In process control we can often make good use of nonlinear functions. Sensor calibration is one case. There are many situations where an instrument has many different sensors, the outputs of which must be combined nonlinearly to obtain the desired measured value. Nonlinear functions can also be used for pattern recognition.

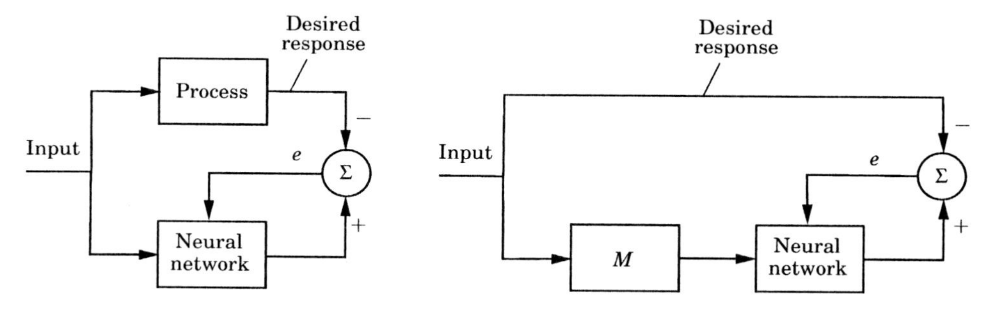

Figure 12.25 Illustration of training of a simple feedforward network. The block diagram on the left shows training of a function, and the figure on the right shows training of an inverse function.

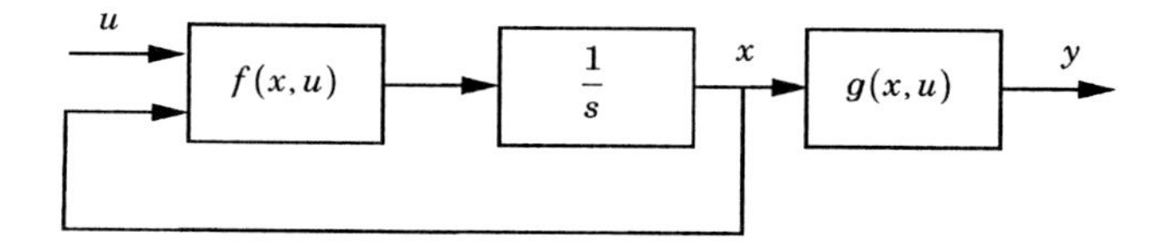

Figure 12.26 Implementation of a nonlinear dynamical system using integrators and neural networks.

It is also possible to model dynamic systems by combining the neural network with integrators as is illustrated in Figure 12.26. The system in the figure implements the nonlinear system

$$\frac{dx}{dt} = f(x, u)$$
$$y = g(x, u),$$

where the nonlinear functions are represented by neural networks.

12.8 Fuzzy Control

Fuzzy control is an old control paradigm that has received a lot of attention recently. In this section we will give a brief description of the key ideas. We will start with fuzzy logic, which has inspired the development.

Fuzzy Logic

Ordinary Boolean logic deals with quantities that are either true or false. Fuzzy logic is an attempt to develop a method for logic reasoning that is less sharp. This is achieved by introducing linguistic variables and associating them with *membership functions*, which take values between 0 and 1. In fuzzy control the logical operations *and*, *or*, and *not* are operations on linguistic variables. These operations can be expressed in terms of operations on the membership functions of the linguistic variables. Consider two linguistic variables with the

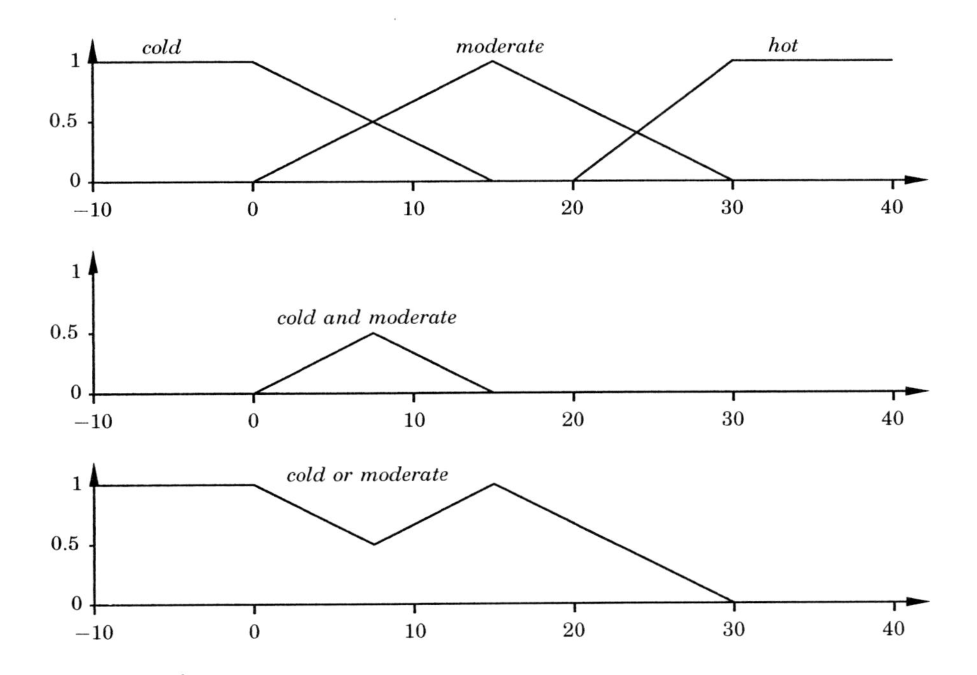

Figure 12.27 Illustration of fuzzy logic. The upper diagram shows the membership functions of *cold*, *moderate*, and *hot*. The middle diagram shows the membership functions for *cold and moderate* the lower diagram shows the membership functions for *cold or moderate*.

membership functions $f_A(x)$ and $f_B(x)$. The logical operations are defined by the following operations on the membership functions.

$$
\begin{aligned}
f_{A \text{ and } B} &= \min(f_A(x), f_B(x)) \\
f_{A \text{ or } B} &= \max(f_A(x), f_B(x)) \\
f_{\text{not } A} &= 1 - f_A(x).
\end{aligned}
$$

A linguistic variable, where the membership function is zero everywhere except for one particular value, is called a crisp variable.

Assume, for example, that we want to reason about temperature. For this purpose we introduce the linguistic variables *cold*, *moderate*, and *hot*, and we associate them with the membership functions shown in Figure 12.27. The membership function for the linguistic variables *cold and moderate* and *cold or moderate* are also shown in the figure.

A Fuzzy Controller

A block diagram of a fuzzy PD controller is shown in Figure 12.28. The control error, which is a continuous signal, is fed to a linear system that generates the derivative of the error. The error and its derivative are converted to so-called linguistic variables in a process called "fuzzification." This procedure converts continuous variables to a collection of linguistic variables. The number of linguistic variables is typically quite small, for example: negative large

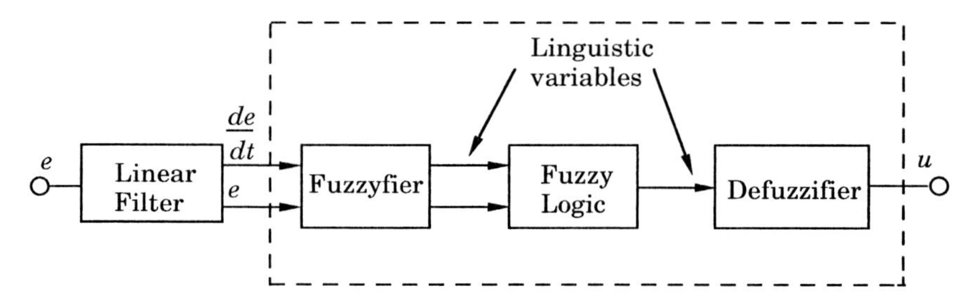

Figure 12.28 A fuzzy PD controller.

(*NL*), negative medium (*NM*), negative small (*NS*), zero (*Z*), positive small (*PS*), positive medium (*PM*), and positive large (*PL*). The control strategy is expressed in terms of a function that maps linguistic variables to linguistic variables. This function is defined in terms of a set of rules expressed in fuzzy logic. As an illustration we give the rules for a PD controller where the error and its derivative are each characterized by three linguistic variables (*N*, *Z*, *P*) and the control variable is characterized by five linguistic variables (*NL*, *NM*, *Z*, *PM*, and *PL*).

```
Rule 1: If e is N and de/dt is P then u is Z
Rule 2: If e is N and de/dt is Z then u is NM
Rule 3: If e is N and de/dt is N then u is NL
Rule 4: If e is Z and de/dt is P then u is PM
Rule 5: If e is Z and de/dt is Z then u is Z
Rule 6: If e is Z and de/dt is N then u is NM
Rule 7: If e is P and de/dt is P then u is PL
Rule 8: If e is P and de/dt is Z then u is PM
Rule 9: If e is P and de/dt is N then u is Z}
```

These rules can also be expressed in table form; see Table 12.1. The membership functions representing the linguistic variables normally overlap (see Figure 12.27). Due to this, several rules contribute to the control signal. The linguistic variable representing the control signal is calculated as a weighted sum of the linguistic variables of the control signal. The linguistic variable representing the control signal is then mapped into a real number by an operation called "defuzzification." More details are given in the following.

Fuzzy Inference Many different shapes of membership functions can be used. In fuzzy control it is common practice to use overlapping triangular shapes like the ones shown in Figure 12.27 for both inputs and control variables. Typically only a few membership functions are used for the measured variables.

Fuzzy logic is only used to a moderate extent in fuzzy control. A key issue is to interpret logic expressions of the type that appears in the description of the fuzzy controller. Some special methods are used in fuzzy control. To describe these we assume that f_A, f_B, and f_C are the membership functions associated with the linguistic variables A, B, and C. Furthermore let x and y represent

Table 12.1 Representation of the fuzzy PD controller as a table.

		$\frac{de}{dt}$ P	Z	N
	N	Z	NM	NL
e	Z	PM	Z	NM
	P	PL	PM	Z

measurements. If the values x_0 and y_0 are measured, they are considered as crisp values. The fuzzy statement

```
If x is A and y is B
```

is then interpreted as the crisp variable

$$z^0 = \min(f_A(x_0), f_B(y_0))$$

where *and* is equivalent to minimization of the membership functions. The linguistic variable u defined by

```
If x is A or y is B then u is C
```

is interpreted as a linguistic variable with the membership function

$$f_u(x) = z^0 f_C(x).$$

If there are several rules, as in the description of the PD controller, each rule is evaluated individually. The results obtained for each rule are combined using the or operator. This corresponds to taking the maximum of the membership functions obtained for each individual rule.

Figure 12.29 is a graphical illustration for the case of the first two rules of the PD controller. The figure shows how the linguistic variable corresponding to each rule is constructed and how the control signal is obtained by taking the maximum of the membership functions obtained from all rules.

The inference procedure described is called "product-max." This refers to the operations on the membership functions. Other inference procedures are also used in fuzzy control. The and operation is sometimes represented by taking the product of two membership functions and the or operator by taking a saturated sum. Combinations of the schemes are also used. In this way it is possible to obtain "product-max" and "min-sum" inferences.

Defuzzification Fuzzy inference results in a control variable expressed as a linguistic variable and defined by its membership function. To apply a control signal we must have a real variable. Thus, the linguistic variable defining the control signal must be converted to a real number through the operation of "defuzzification." This can be done in several different ways. Consider a

Rule 1: If e is N and de/dt is P then u is Z

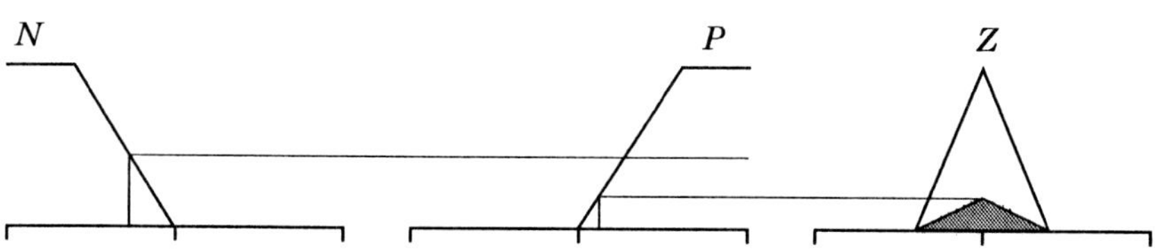

Rule 2: If e is N and de/dt is Z then u is NM

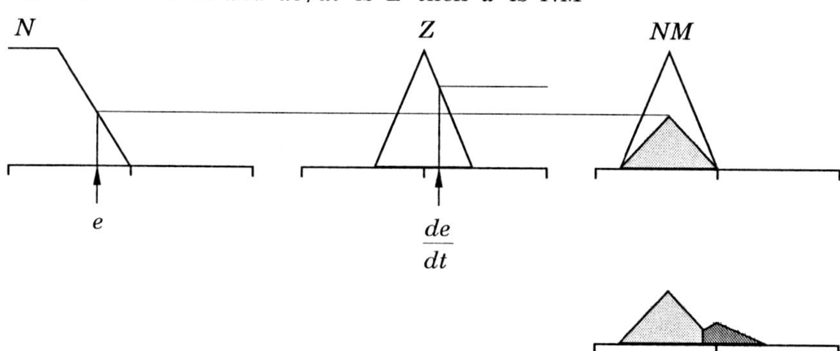

Figure 12.29 Illustration of fuzzy inference with two rules using the min-max rule.

linguistic variable A with the membership function $f_A(x)$. Defuzzification by mean values gives the value

$$x_0 = \frac{\int x f_A(x)dx}{\int f_A(x)dx}.$$

Defuzzification by the centroid gives a real variable x_0 that satisfies

$$\int_{-\infty}^{x_0} f_A(x)dx = \int_{x_0}^{\infty} f_A(x)dx.$$

Nonlinear Control

Having gone through the details, we return to the fuzzy PD controller in Figure 12.28. We first notice that the operations fuzzification, fuzzy logic, and defuzzification can be described in a very simple way. Stripping away the vocabulary and considering the final result, a fuzzy controller is nothing but a nonlinear controller. The system in Figure 12.28 can in fact be expressed as

$$u = F\left(e, \frac{de}{dt}\right),$$

where F is a nonlinear function of two variables. Thus, the fuzzy PD controller is a controller where the output is a nonlinear function of the error e and its derivative de/dt. In Figure 12.30 we give a graphic illustration of the

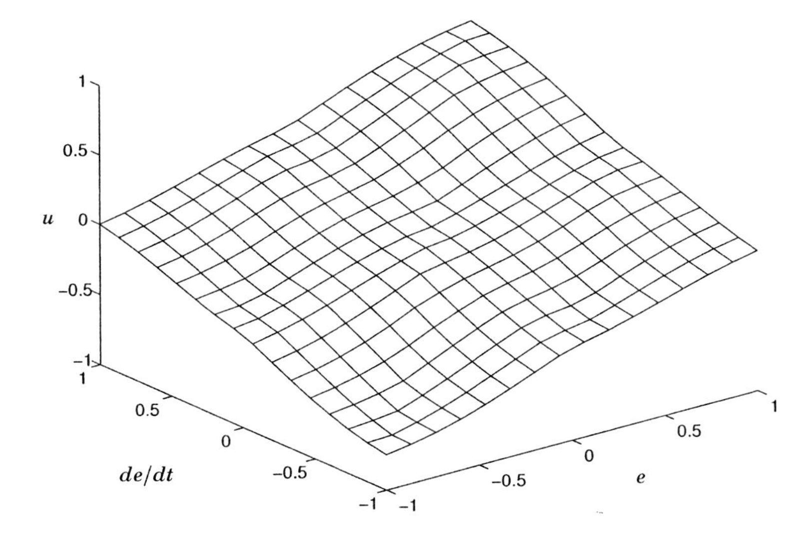

Figure 12.30 Graphic illustration of the nonlinearity of the fuzzy controller showing control signal u as function of control error e and its derivative.

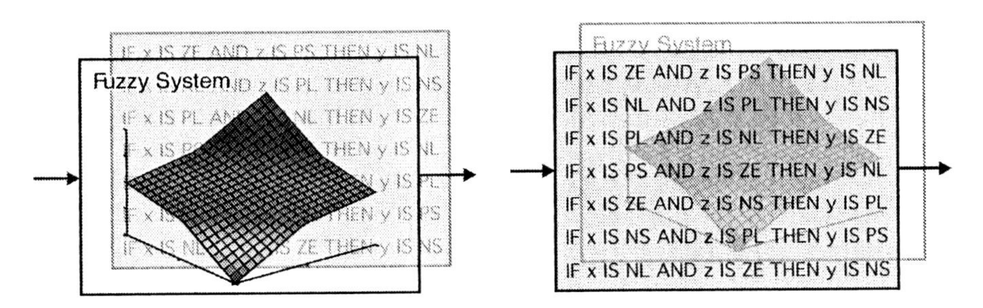

Figure 12.31 Two views of a fuzzy controller. The figure on the left shows that the fuzzy controller can be viewed as a nonlinear controller. The figure on the right instead emphasizes the rules.

nonlinearity defined by given rules for the PD controller with standard triangular membership functions and product fuzzification. The figure shows that the function is close to linear. In this particular case the fuzzy controller will behave similarly to an ordinary linear PD controller.

Fuzzy control may be considered as a way to represent a nonlinear function. This is illustrated in Figure 12.31. Notice that it is still necessary to deal with the generation of derivatives or integrals, integral windup, and all the other matters in the same way as for ordinary PID controllers. We may also inquire as to when it is useful to introduce the nonlinearities and what shape they should have.

Representation of a nonlinearity by fuzzification, fuzzy logic, and defuzzification is not very different from representation of a nonlinear function as a table with an interpolation procedure. Roughly speaking, the function values correspond to the rules; the membership functions and the fuzzification and defuzzification procedures correspond to the interpolation mechanism. To illustrate this we consider a function of two variables. Such a function can be visualized as a surface in two dimensions. A linear function is simply a tilted plane. This function can be described completely by three points on a plane, i.e., three rules. More complex surfaces or functions are obtained by using more function values. The smoothness of the surface is expressed by the interpolation procedures.

From the point of view of control, the key question is understanding when nonlinearities are useful and what shape they should have. These are matters where much research remains to be done. There are cases where the nonlinearities can be very beneficial but also cases where the nonlinearities cause problems. It is also a nontrivial task to explore what happens. A few simulations of the behavior is not enough because the response of a nonlinear system is strongly amplitude dependent.

Let us also point out that the properties of the controller in Figure 12.28 are strongly influenced by the linear filter used. It is thus necessary to limit the high-frequency gain of the approximation of the derivative. It is also useful to take derivatives of the process output instead of the error, as was discussed in Section 3.3. Other filters can also be used; by adding an integrator to the output of the system in Figure 12.28, we obtain a fuzzy PI controller.

Applications

The representation of the control law as a collection of rules for linguistic variables has a strong intuitive appeal. It is easy to explain heuristically how the control system works. This is useful in communicating control strategies to persons with little formal training. It is one reason why fuzzy control is a good tool for automation of tasks that are normally done by humans. In this approach it is attempted to model the behavior of an operator in terms of linguistic rules. Fuzzy control has been used in a number of simple control tasks for appliances. It has also been used in controllers for processes that are complicated and poorly known. Control of a cement kiln is one example of this type of application. Fuzzy control has also been used for controller tuning.

12.9 System Structuring

In this section we illustrate how complex control systems can be built from simple components by using the paradigms we have discussed. The problem is quite complex. It involves selection of measured variables and control variables, and it requires significant physical understanding of the process.

The Process

The process to consider is a chemical reactor. A schematic diagram is shown in Figure 12.32. Two substances A and B are mixed in the reactor. They react to form a product. The reaction is exothermic, which means that it will generate heat. The heat is dissipated through water that is circulating in cooling pipes

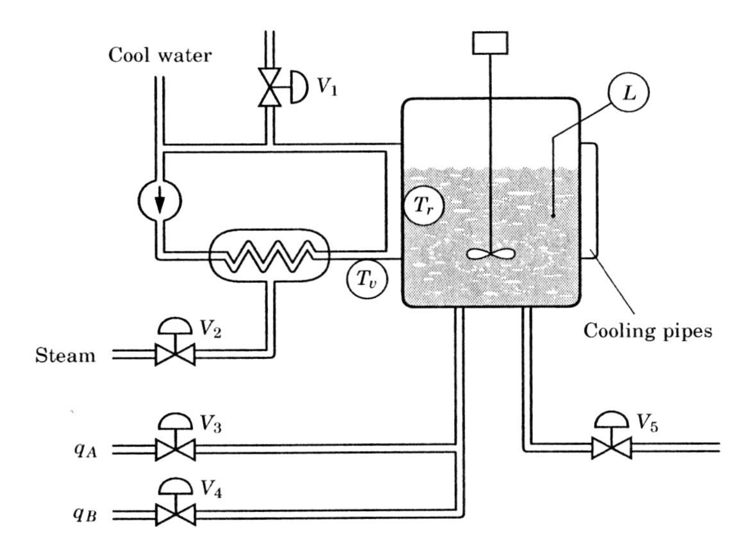

Figure 12.32 Schematic diagram of a chemical reactor.

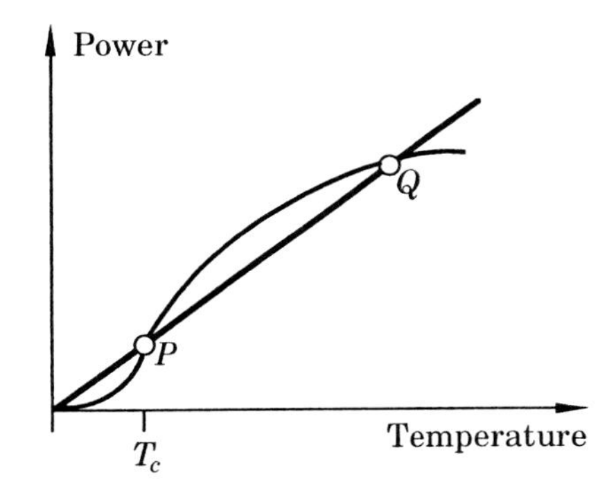

Figure 12.33 Static process model for the exothermic reactor.

in the reactor. The reaction is very fast; equilibrium is achieved after a time that is much shorter than the residence time of the reactor. The flow q_A of substance A is considerably larger than q_B. Efficiency of the reaction and the heat generation is essentially proportional to the flow q_B.

A static process model is useful in order to understand the control problem. Figure 12.33 shows the efficiency and the heat generation as a function of temperature. A model of this type was derived in Section 2.5. In the figure we have drawn a straight line that corresponds to the cooling power. There are equilibria where the power generated by the reaction is equal to the cooling power represented at points P and Q in the figure. The point P corresponds to an unstable equilibrium. It follows from Figure 12.33 that if the temperature is increased above P the power generated by the reaction is larger than the cooling power. Temperature will thus increase. The catalyst in the reactor may be damaged if the temperature becomes too high. Similarly, if the temperature decreases below point P it will continue to decrease and the reaction stops. This phenomenon is called "freezing." Freezing starts at the surface of the cooling tube and will spread rapidly through the reactor. If this happens the reactor must be switched off and restarted again.

Design Requirements

There are considerable risks in running an exothermic reactor. The reactor can explode if the temperature is too high. To reduce the risk of explosion, the reactors are placed in special buildings far away from the operator. Because of the risk of explosion, it is not feasible to experiment with controller tuning. Consequently, it is necessary to compute controller setting beforehand and verify that the settings are correct before starting the reactor. Safety is the overriding requirement of the control system. It is important to guarantee that the reaction temperature will not be too high. It is also important to make sure that process upsets do not lead to loss of coolant flow and that stirring does not lead to an explosion. It is also desirable to operate the reactor efficiently. This means that freezing must be avoided. Besides, it is desirable to keep the efficiency as high as possible. Because of the risks, it is also necessary to automate start and stop as well as normal operation. It is desirable to avoid having to run the reactor under manual control. In this particular case the operator can set two variables: the reactor temperature and the ratio between the flows q_A and q_B. The reaction efficiency and the product quality can be influenced by these two variables.

Controller Structure

The reactor has five valves. Two of them, V_1 and V_2, influence the coolant temperature. The flow of the reactor is controlled by V_3 and V_4, and the product flow is controlled by the valve V_5. In this particular application the valve V_5 is controlled by process steps downstream. (Compare this with the discussion of surge tanks in Section 12.6).

There are five measured signals: the reactor temperature T_r, the level in the reactor tank L, the cooling temperature T_v, and the flows q_A and q_B. The physical properties of the process give a natural structuring of the control system. A mass balance for the material in the reactor tank shows that the level is essentially influenced by the flow q_A and the demanded production. It follows from the stoichiometry of the reaction that the ratio of the flows q_A and q_B should be kept constant for an efficient reaction. The reactor temperature is strongly influenced by the water temperature, by the temperature of the coolant flow, and the flows q_A and q_B. Coolant temperature is influenced by the valve V_1 that controls the amount of flow and by the steam valve V_2.

This simple physical discussion leads to the diagram shown in Figure 12.34, which shows the causality of the variables in the process. The valve V_5 can be regarded as a disturbance because it is set by downstream process units. Figure 12.34 suggests that there are three natural control loops:

- Level control: Controlling the tank level with valve V_3.
- Temperature control: Control of the reactor temperature with valves V_1 and V_2.

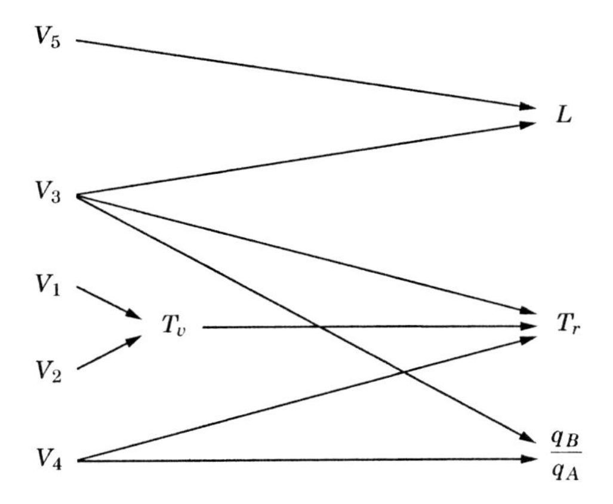

Figure 12.34 Causality diagram for the process variable.

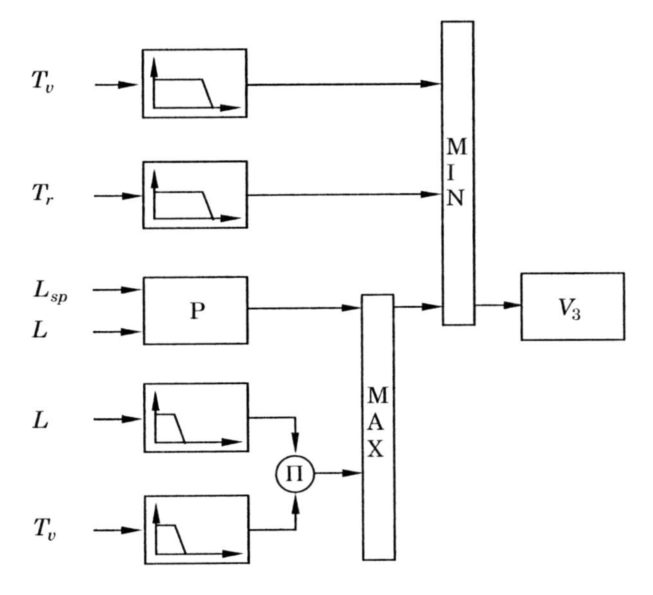

Figure 12.35 Block diagram for the level control through valve V_3.

- Flow ratio control: Control of ratio q_B/q_A with valve V_4.

These control loops are discussed in detail.

Level Control

The block diagram for the level control is shown in Figure 12.35. The primary function is a proportional feedback from the level to the flow q_A, which is controlled by the valve V_3. The reactor is also used as a surge tank to smooth out the difference between actual production and commanded production. The level in the tank will vary during normal operations. Reasonable limits are that the level should be between 50 percent and 100 percent. If the proportional band of the controller is chosen as 50 percent, the control variable will be

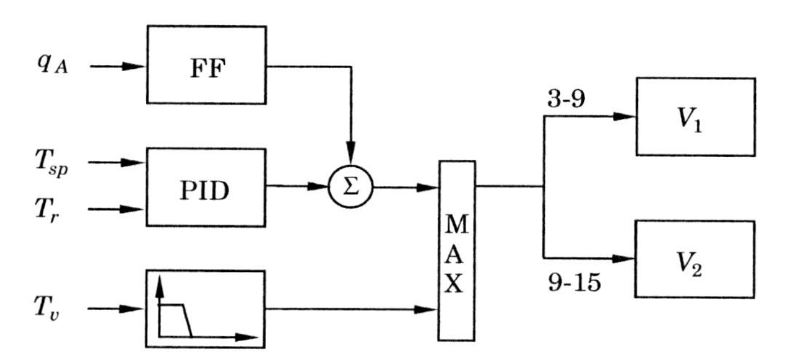

Figure 12.36 Block diagram showing temperature control through valves V_1 and V_2.

fully closed when the tank is full and half-open when the tank is half-full. It is important that the reactor temperature remains within given bounds. The flow q_A is constrained, therefore, by two selectors based on measurements of the temperature in the reactor tank (T_r) and the coolant temperature (T_v). When starting the reactor the level is kept at the lower limit until the coolant temperature becomes sufficiently high. This is achieved by a combination of limiters, multipliers, and selectors, as shown in Figure 12.35.

Temperature Control

Figure 12.36 gives a block diagram for controlling the reactor temperature. Since the chemical reaction is fast compared to temperature and flow dynamics, the reactor can be viewed as a heat exchanger from the control point of view. During normal conditions the temperature is controlled by adjusting the coolant flow through the valve V_1. The primary control function is a feedback from temperature to the valves V_1 and V_2. The set point in this control loop can be adjusted manually. The parameters of this control loop can be determined as follows. The transfer function from coolant flow to the reactor temperature is approximately given by

$$G(s) = \frac{K_p}{(1+sT_1)(1+sT_2)}, \tag{12.8}$$

where the time constant typically has values $T_1 = 300$ s and $T_2 = 50$ s. The following rough calculation gives approximate values of the controller parameter. A proportional controller with gain K gives the loop transfer function

$$G_0(s) = \frac{KK_p}{(1+sT_1)(1+sT_2)}. \tag{12.9}$$

The characteristic equation of the closed loop becomes

$$s^2 + s\left(\frac{1}{T_1} + \frac{1}{T_2}\right) + \frac{1+KK_p}{T_1T_2} = 0.$$

The closed system is thus of second order. The relative damping ζ and the undamped natural frequency ω are given by

$$2\zeta\omega = \frac{1}{T_1} + \frac{1}{T_2} \approx \frac{1}{T_2} \tag{12.10}$$

and

$$2\zeta\omega^2 = \frac{1+KK_p}{T_1T_2}. \tag{12.11}$$

The approximation in the first expression is motivated by $T_1 \gg T_2$. With a relative damping $\zeta = 0.5$ the Equation (12.10) then gives $\omega \approx 1/T_2$. Furthermore, it follows from Equation (12.11) that

$$1 + KK_p = \frac{T_1}{T_2} = \frac{300}{50} = 6.$$

The loop gain is thus essentially determined by the ratio of the time constants. The controller gain becomes

$$K = \frac{5}{K_p},$$

and the closed-loop system has the undamped natural frequency.

$$\omega = 1/T_2 = 0.02 \text{ rad/s}.$$

If PI control is chosen instead, it is reasonable to choose a value of the integration time

$$T_1 \approx 5T_2.$$

Control can be improved by using derivative action. The achievable improvement depends on the time constant of the temperature sensor. In typical cases this time constant is between 10 s and 40 s. If it is as low as 10 s it is indeed possible to obtain improved control by introducing a derivative action in the controller. The derivative time can be chosen to eliminate the time constant T_2. We then obtain a system with the time constants 300 s and 10 s. The gain can then be increased so that

$$1 + KK_p = \frac{300}{10} = 30$$

and the undamped natural frequency of the system then becomes $\omega \approx 0.1$ rad/s. If the time constant of the temperature sensor is around 40 s, the derivative action gives only marginal improvements.

The heat generated by the chemical reaction is proportional to the flow q_A. To make sure that variations in q_A are compensated rapidly we have also introduced a feedforward from the flow q_A. This feedforward will only operate when the tank level is larger than 50 percent in order to avoid freezing when the reactor is started.

To start the reaction the reactor must be heated so that the temperature in the reaction vessel is larger than T_c (compare with Figure 12.33). This is done by using the steam valve V_2. Split-range control is used for the steam and water valves (compare Section 12.6). The water valve is open for low signals (3–9 PSI), and the steam valve is open for large pressures (9–15 PSI).

To avoid having the reactor freeze, it is necessary to make sure that the reaction temperature is always larger than T_c. This is the reason for the extra feedback from water temperature to T_v through a maximum selector. This feedback makes sure that the steam valve opens if the temperature in the coolant flow becomes too low. Cascade control would be an alternative to this arrangement.

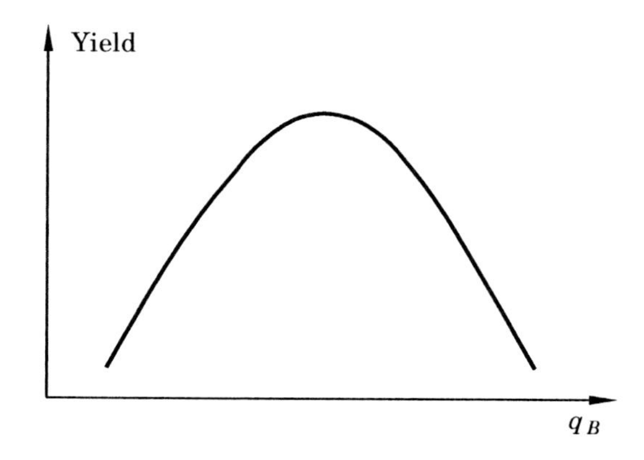

Figure 12.37 Reaction yield as a function of q_B at constant q_A.

Flow Ratio Control

The ratio of the flows q_A and q_B must be kept constant. Figure 12.37 shows how the efficiency of the reaction depends on q_B when q_A is kept constant. The flow q_B is controlled with a ratio control system (as shown in Figure 12.38), which is the primary control function. The reaction rate depends strongly on q_B. To diminish the risk of explosion, there is a nonlinearity in the feedback that increases the gain when q_B/q_A is large. The flow loop has several selectors. At startup it is desirable that substance B not be added until the water temperature has reached the critical value T_c and the reactor tank is half-full. To achieve this the feedback from water temperature and tank level has been introduced through limiters and a minimum selector. There are also limiters and a selector that closes valve V_4 if flow q_A is lost. There is also a direct feedback from q_A through limiters and selectors and a feedback from the reactor temperature that closes valve V_4, if the reactor temperature becomes too high.

Override Control of the Outlet Valve

The flow out of the reactor is determined by valve V_5. This valve is normally controlled by process steps downstream. The control of the reactor can be improved by introducing an override, which depends on the state of the reactor. When starting the reactor, it is desirable to have the outlet valve closed until the reactor tank is half-full and the reaction has started. This is achieved by introducing the tank level and the tank temperature to the set point of the valve controller via limiters and minimum selectors as is shown in Figure 12.39. The valve V_5 is normally controlled by q_{sp}. The minimum selector overrides the command q_{sp} when the level L or the temperature T_r are too low.

12.10 Summary

In this chapter we have illustrated how complex control systems can be built from simple components such as PID controllers, linear filters, gain schedules,

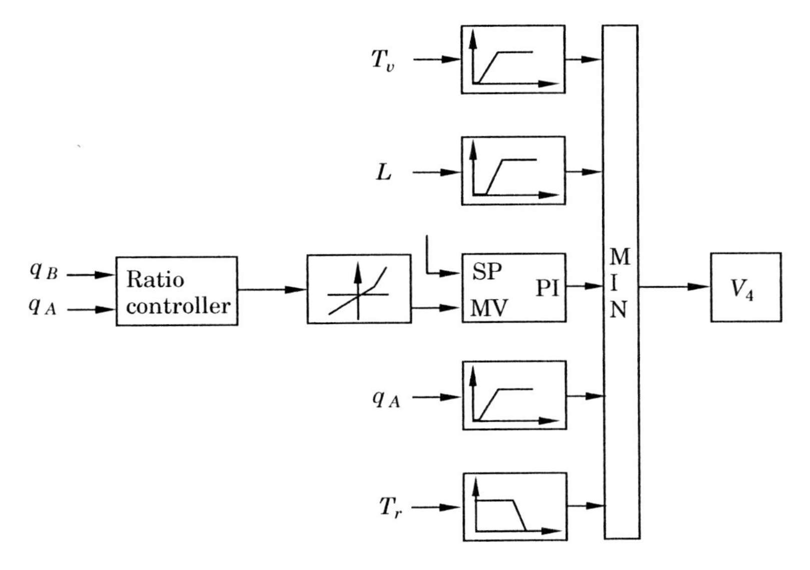

Figure 12.38 Block diagram for controlling the mixing ratio q_B/q_A through valve V_4.

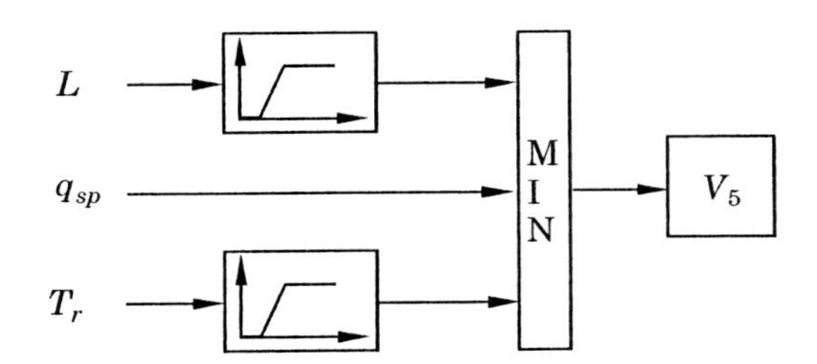

Figure 12.39 Block diagram for controlling the outflow of the reactor through valve V_5.

and simple nonlinear functions. A number of control paradigms have been introduced to guide system design.

The primary linear control paradigms are feedback by PID control and feedforward. Cascade control can be used to enhance control performance through the use of extra measurements. State feedback may be viewed as an extreme case of cascade control where all states of a system are measured. Observers can be used to infer values of variables that are not measured by combining mathematical models with available measurements. Mid-range and split-range control are paradigms for control when there are several control signals but only one measured signal. These paradigms are the dual of cascade control. Repetitive control is a technique that is efficient for cases where the disturbances are periodic. The idea is to create a high loop gain at the frequency of the disturbance.

We also discussed several nonlinear components and related paradigms including nonlinear functions, gain schedules, limiters, and selectors. Recall that it was shown in Section 3.5 how PID controllers could be enhanced by simple nonlinear functions to avoid windup. Ratio control is a nonlinear strategy that admits control of two process variables so that their ratio is constant. In Section 9.3 we showed how gain schedules could be used to cope with changes in process dynamics. Gain schedules and nonlinear functions are also useful for control of buffers, where the goal is not to keep constant levels in the buffers but to allow them to vary within given ranges. Selector control is another important paradigm that is used for constraint control where certain process variables have to be kept within given constraints. Neural and fuzzy techniques were also discussed briefly. It was shown that they could be interpreted both as rule-based control and as nonlinear control.

We also gave an example how the components and the paradigms could be used to develop a control system for a chemical process.

12.11 Notes and References

Many aspects of the material in this chapter are found in classical textbooks on process control such as [Buckley, 1964; Shinskey, 1988; Bequette, 2003; Seborg *et al.*, 2004] and in the books [Shinskey, 1981; Klefenz, 1986] which focus on energy systems. A more specialized presentation is given in [Hägglund, 1991].

The methods discussed in this chapter can all be characterized as bottom-up procedures in the sense that a complex system is built up by combining simple components. An interesting view of this is given in [Bristol, 1980]. A top-down approach is another possibility. A discussion of this, which is outside the scope of this book, is found in [Seborg *et al.*, 1986] and [Morari and Zafiriou, 1989].

Cascade and feedforward control are treated in the standard texts on control. A presentation with many practical aspects is found in [Tucker and Wills, 1960]. Selector control is widely used in practice. A general presentation is given in [Åström, 1987b]. It is difficult to analyse nonlinear systems. A stability analysis of a system with selectors is given in [Foss, 1981]. The Blend station is presented in [Hägglund, 2001].

Fuzzy control has been around for a long time; see [Mamdani, 1974; Mamdani and Assilian, 1974; King and Mamdani, 1977; Tong, 1977]. It has received a lot of attention particularly in Japan: see [Zadeh, 1988; Tong, 1984; Sugeno, 1985; Driankov *et al.*, 1993; Wang, 1994]. The technique has been used for automation of complicated processes that have previously been controlled manually. Control of cement kilns is a typical example; see [Holmblad and Østergaard, 1981]. There has been a similar development in neural networks; see, for example, [Hecht-Nielsen, 1990; Pao, 1990; Åström and McAvoy, 1992]. There was a lot of activity in neural networks during the late 1960s, which vanished rapidly. There was a rapid resurgence of interest in the 1980s. There are a lot of exaggerations both in fuzzy and neural techniques, and no balanced view of the relevance of the fields for control has yet emerged. The paper [Willis *et al.*, 1991] gives an overview of possible uses of neural networks for process control, and the paper [Pottman and Seborg, 1993] describes an application to control of pH. The papers [Lee, 1990; Huang, 1991; Swiniarski, 1991] describe applications to PID controllers and their tuning. There have also been attempts to merge fuzzy and neural control; see [Passino and Antsaklis, 1992] and [Brown and Harris, 1994]. Section 12.9 is based on [Buckley, 1970].

13

Implementation

13.1 Introduction

PID controllers were originally implemented using analog techniques. Early systems used pneumatic relays, bellows, and needle-valve constrictions. Electric motors with relays and feedback circuits and operational amplifiers were used later. Many of the features like anti-windup and derivation of process output instead of control error were incorporated as "tricks" in these implementations.

It is now common practice to implement PID controllers using microprocessors, and some of the old tricks have been rediscovered. Several digital PID controllers in use today have features that are inherited from old techniques when the controllers were implemented using pneumatic devices. This is a typical example of the fact that ideas sometimes change at a much slower rate than hardware. Several additional issues must be considered in connection with digital implementations. The most important ones have to do with sampling, discretization, and quantization.

This chapter presents some implementation issues related to PID control. Section 13.2 gives a short overview of the early analog pneumatic and electronic implementations. Section 13.3 treats computer implementation aspects such as sampling, prefiltering, and discretization of the PID algorithm. Velocity algorithms, or incremental algorithms, are needed in applications where the integration is performed outside the controller. The most common application is electrical motors. These algorithms, which are shown to be useful even when the integration is performed inside the controller, are presented in Section 13.4. Operational aspects, such as bumpless transfers at mode switches and parameter changes, are presented in Section 13.5. A controller may have different outputs depending on which actuating device is used. Controller outputs are discussed in Section 13.6. The chapter ends with a summary and references.

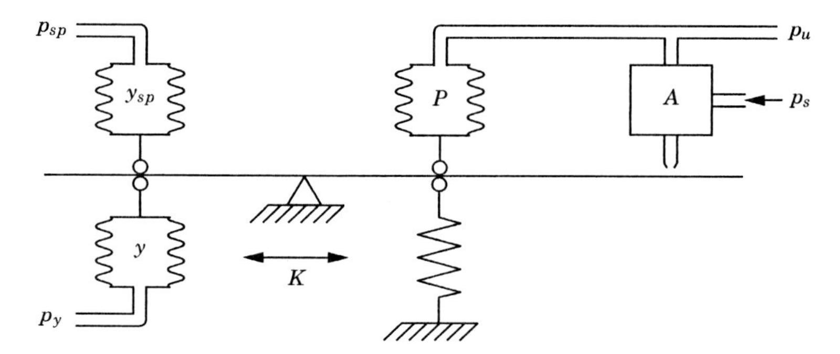

Figure 13.1 Schematic diagram of a pneumatic P controller based on the force balance principle.

13.2 Analog Implementations

The early implementations of PID controllers were all analog. This section presents the pneumatic controller implementation and the analog electronic implementation.

The Pneumatic Controller

This section presents the basic function of pneumatic controllers. To make this clear, lots of details have been removed from the presentation and the drawings. We refer to the references for details.

The Pneumatic P Controller A schematic diagram of a pneumatic P controller based on force balance is shown in Figure 13.1. The system consists of a beam that can rotate around a pivot point. The beam is provided with three bellows, a spring, a position sensor, and a pneumatic amplifier. The bellows can exert forces on the beam proportional to the pressure in the bellows. The position sensor is a flapper valve, which gives a pressure signal that is approximately inversely proportional to the distance between the nozzle and the beam. The pneumatic amplifier A can amplify pneumatic signals.

To understand the operation of the system it is assumed that the forces of all bellows are proportional to the air pressure in the bellows. The two left bellows receive pressures p_{sp} and p_y proportional to the set point and the measured variable, respectively. The pressure amplifier A receives supply pressure p_s and provides output pressure p_u, which is the controller output. The right bellow labeled P is the feedback bellow or the proportional bellow. In the P controller the pressure in this bellow, p_p, is equal to the output pressure p_u.

A torque balance gives the following relation between the pressures:

$$p_u - bias = K(p_{sp} - p_y). \tag{13.1}$$

The *bias* term is the force given by the spring. The gain K is determined by the position of the balance point, and can therefore be chosen by adjusting this point. Equation 13.1 is obviously the equation for a P controller.

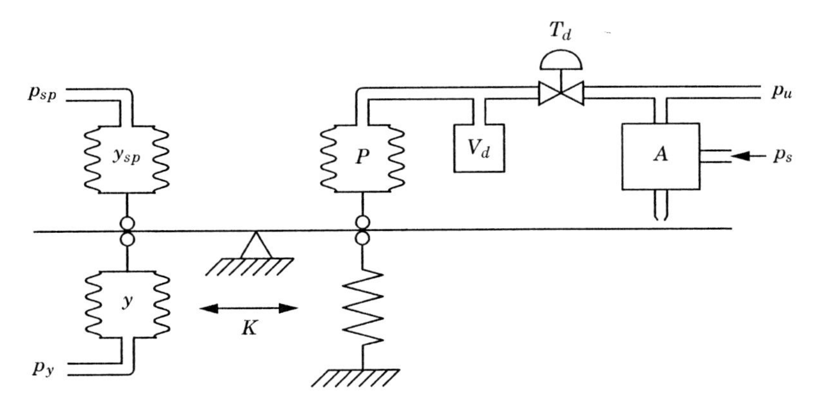

Figure 13.2 Schematic diagram of a pneumatic PD controller based on the force balance principle.

Suppose, for example, that the set-point pressure p_{sp} increases. The beam will then rotate in positive direction, leading to a decrease in the outflow from the nozzle valve. This will cause an increase of the output pressure p_u.

The Pneumatic PD Controller A pneumatic PD controller is shown in Figure 13.2. In this controller, a valve and a volume V_d is introduced between the amplifier A and the feedback bellow P. Because of this valve, it is no longer true that $p_p = p_u$, but the following dynamic relation between the two pressures holds:

$$P_p(s) = \frac{1}{1 + sT_d} P_u(s). \tag{13.2}$$

The value of the time constant T_d can be adjusted by the valve position.

Since a counteraction caused by the feedback bellow P is delayed compared with the P controller, a change in p_y or p_{sp} will initially result in a larger reaction in the output pressure p_u.

A torque balance gives the following relations between the pressures:

$$p_p - bias = K(p_{sp} - p_y).$$

From (13.2) this gives the following output pressure;

$$P_u(s) = bias + K(1 + sT_d)(P_{sp} - P_y),$$

which is the equation of a PD controller with derivative time T_d.

The Pneumatic PID Controller A pneumatic PID controller is shown in Figure 13.3. In this controller, the spring is replaced by a bellow labeled I. This bellow is connected to the pressure p_p through a volume V_i and a valve labeled T_i. The pressure in the bellow I is

$$P_i(s) = \frac{1}{1 + sT_i} P_p(s), \tag{13.3}$$

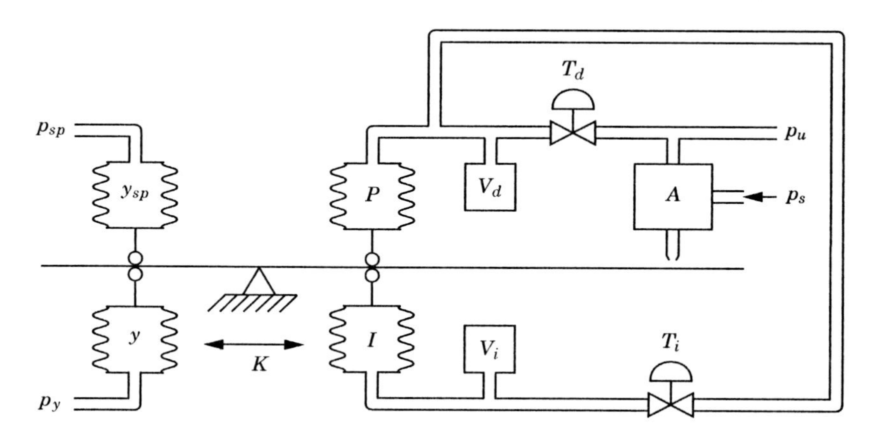

Figure 13.3 Schematic diagram of a pneumatic PID controller based on the force balance principle.

where the time constant T_i may be adjusted by the valve labeled T_i.

A torque balance gives the following relations between the pressures:

$$p_p - p_i = K(p_{sp} - p_y).$$

From (13.2) and (13.3) this gives the following output pressure:

$$P_u(s) = K\frac{(1 + sT_i)(1 + sT_d)}{sT_i}(P_{sp}(s) - P_y(s)). \tag{13.4}$$

This equation shows that the system is a PID controller on interacting form (see Section 3.2) with gain K, integral time T_i, and derivative time T_d.

The idea of using feedback in the controller was a major invention. Both the flapper valve and the pneumatic amplifier are strongly nonlinear. The arrangement with the feedback loop implies that the input-output relation of the controller does not change much even if the component changes, provided that the gain is sufficiently large. This idea, which is called force feedback, gave drastic improvements in the performance of the controllers. A typical example of the impact of feedback.

The Analog Electronic Controller

A PID controller may be implemented by analog electronic components in many ways. This section presents some basic implementations based on operational amplifiers. Lots of details have been left out for the sake of simplicity. As for the pneumatic controllers, we refer to the references for details.

The Electronic PI Controller An electronic PI controller is shown in Figure 13.4.

An approximate relation between the input voltage e and the output voltage u is obtained by

$$u = -\frac{Z_1}{Z_0}e,$$

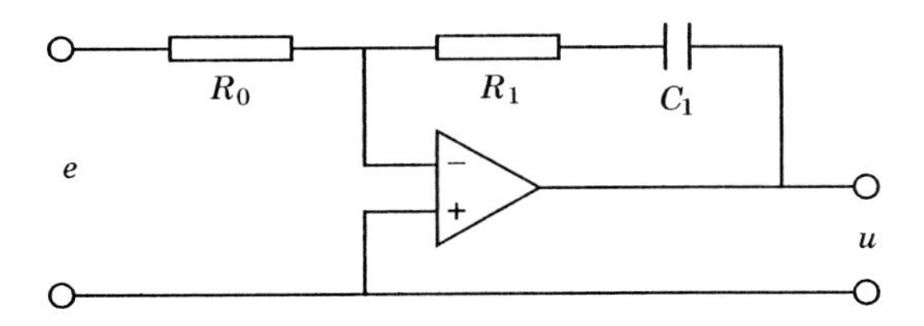

Figure 13.4 Schematic diagram of an electronic PI controller based on feedback around an operational amplifier.

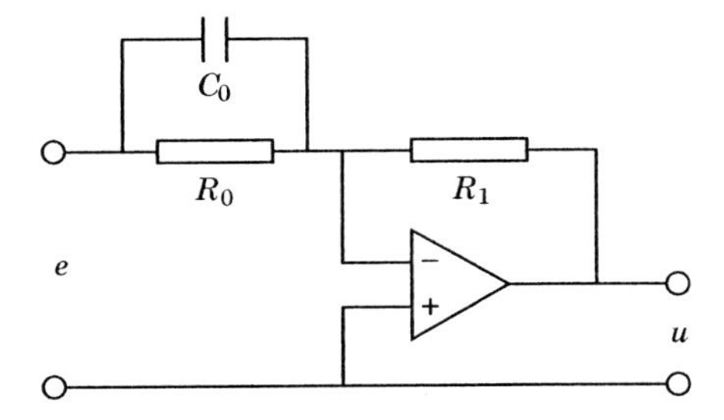

Figure 13.5 Schematic diagram of an electronic PD controller based on feedback around an operational amplifier.

where Z_0 is the impedance between the negative input of the amplifier and the input voltage e, and Z_1 is the impedance between the zero input of the amplifier and the output voltage u. These impedances are

$$Z_0 = R_0$$
$$Z_1 = R_1 + \frac{1}{C_1 p},$$

where p is the differential operator. This gives the following relation between the input voltage e and the output voltage u:

$$u = -\frac{Z_1}{Z_0} e = -\frac{R_1}{R_0} \left(1 + \frac{1}{R_1 C_1 p}\right) e.$$

This is a PI controller with parameters

$$K = \frac{R_1}{R_0} \qquad T_i = R_1 C_1.$$

A P controller is obtained by removing the capacitor.

The Electronic PD Controller An electronic PD controller is shown in Figure 13.5.

The impedances between the negative input of the amplifier and the input and output voltages, respectively, become

$$Z_0 = \frac{R_0}{1 + R_0 C_0 p}$$
$$Z_1 = R_1.$$

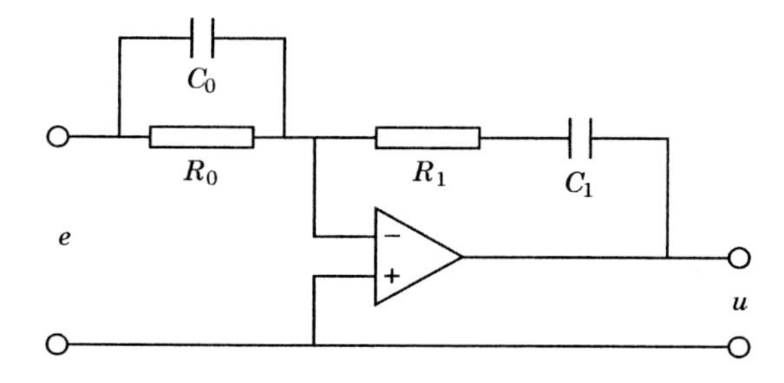

Figure 13.6 Schematic diagram of an electronic PID controller based on feedback around an operational amplifier.

This gives the following relation between the input voltage e and the output voltage u:

$$u = -\frac{Z_1}{Z_0} e = -\frac{R_1}{R_0} (1 + R_0 C_0 p)\, e.$$

This is a PD controller with parameters

$$K = \frac{R_1}{R_0} \qquad T_d = R_0 C_0.$$

A P controller is obtained by removing the capacitor.

The Electronic PID Controller An electronic PID controller may be obtained by combining the two previous schemes. This is shown in Figure 13.6.

The impedances between the negative input of the amplifier and the input and output voltages, respectively, become

$$Z_0 = \frac{R_0}{1 + R_0 C_0 p}$$
$$Z_1 = R_1 + \frac{1}{C_1 p}.$$

This gives the following relation between the input voltage e and the output voltage u:

$$u = -\frac{Z_1}{Z_0} e = -\frac{R_1}{R_0} \frac{(1 + R_0 C_0 p)(1 + R_1 C_1 p)}{R_1 C_1 p} e.$$

This is a PID controller on interacting form with parameters

$$K = \frac{R_1}{R_0} \qquad T_i = R_1 C_1 \qquad T_d = R_0 C_0.$$

13.3 Computer Implementations

Most controllers are implemented nowadays in computers. There are some topics that have to be considered due to the fact that the signals are sampled at discrete time instances. These topics are treated in this section.

Sampling

When the controller is implemented in a computer, the analog inputs are read, and the outputs are set with a certain sampling period. This is a drawback compared to the analog implementations, since the sampling introduces dead time in the control loop.

When a digital computer is used to implement a control law, the ideal sequence of operation is the following.

1. Wait for clock interrupt
2. Read analog input
3. Compute control signal
4. Set analog output
5. Update controller variables
6. Go to 1

With this implementation, the delay is minimized. If the analog input is read with a sampling period h, the average delay of the measurement signal is $h/2$. The computation time is often short compared to the sampling period. This means that the total delay is about $h/2$. However, most controllers and instrument systems do not organize the calculation in this way. Therefore, the delays introduced because of the sampling are often several sampling periods.

Aliasing

The sampling mechanism introduces some unexpected phenomena, which must be taken into account in a good digital implementation of a PID controller. To explain these, consider the signals

$$s(t) = \cos(n\omega_s t \pm \omega t)$$

and

$$s_a(t) = \cos(\omega t),$$

where $\omega_s = 2\pi/h$ [rad/s] is the sampling frequency. Well-known formulas for the cosine function imply that the values of the signals at the sampling instants $[kh, k = 0, 1, 2, ...]$ have the property

$$s(kh) = \cos(nkh\omega_s \pm \omega kh) = \cos(\omega kh) = s_a(\omega kh).$$

The signals s and s_a thus have the same values at the sampling instants. This means that there is no way to separate the signals if only their values at the sampling instants are known. Signal s_a is, therefore, called an *alias* of signal s. This is illustrated in Figure 13.7. A consequence of the aliasing effect is that a high-frequency disturbance after sampling may appear as a low-frequency signal. In Figure 13.7 the sampling period is 1 s, and the sinusoidal disturbance has a period of 6/5 s. After sampling, the disturbance appears as a sinusoid with the frequency

$$f_a = 1 - \frac{5}{6} = 1/6 \text{ Hz}.$$

This low-frequency signal with time period 6 s is seen in the figure.

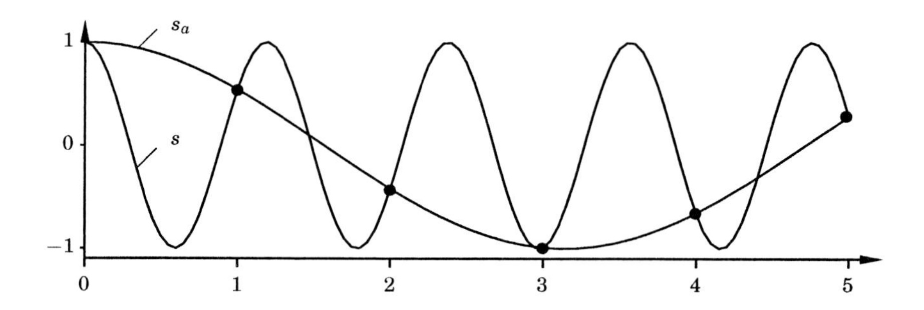

Figure 13.7 Illustration of the aliasing effect. The diagram shows signal s and its alias s_a.

Prefiltering

The aliasing effect can create significant difficulties if proper precautions are not taken. High frequencies, which in analog controllers normally are effectively eliminated by low-pass filtering, may, because of aliasing, appear as low-frequency signals in the bandwidth of the sampled control system. To avoid these difficulties, an analog prefilter (which effectively eliminates all signal components with frequencies above half the sampling frequency) should be introduced. Such a filter is called an antialiasing filter. A second-order Butterworth filter is a common antialiasing filter. Higher-order filters are also used in critical applications. An implementation of such a filter using operational amplifiers is shown in Figure 13.8. The selection of the filter bandwidth is illustrated by the following example.

EXAMPLE 13.1—SELECTION OF PREFILTER BANDWIDTH

Assume it is desired that the prefilter attenuate signals by a factor of 16 at half the sampling frequency. If the filter bandwidth is ω_b and the sampling frequency is ω_s, we get

$$(\omega_s/2\omega_b)^2 = 16.$$

Hence,

$$\omega_b = \frac{1}{8}\,\omega_s.$$

□

Notice that the dynamics of the prefilter will be combined with the process dynamics.

Discretization

To implement a continuous-time control law, such as a PID controller in a digital computer, it is necessary to approximate the derivatives and the integral that appear in the control law. A few different ways to do this are presented below.

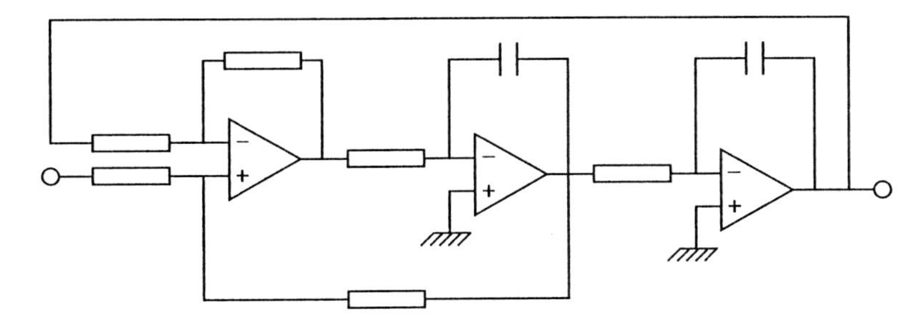

Figure 13.8 Circuit diagram of a second-order Butterworth filter.

Proportional Action The proportional term is

$$P = K(by_{sp} - y).$$

This term is implemented simply by replacing the continuous variables with their sampled versions. Hence,

$$P(t_k) = K\left(by_{sp}(t_k) - y(t_k)\right), \tag{13.5}$$

where $\{t_k\}$ denotes the sampling instants, i.e., the times when the computer reads the analog input.

Integral Action The integral term is given by

$$I(t) = \frac{K}{T_i}\int_0^t e(s)ds.$$

It follows that

$$\frac{dI}{dt} = \frac{K}{T_i}e. \tag{13.6}$$

There are several ways of approximating this equation. Approximating the derivative by a forward difference gives

$$\frac{I(t_{k+1}) - I(t_k)}{h} = \frac{K}{T_i}e(t_k).$$

This leads to the following recursive equation for the integral term

$$I(t_{k+1}) = I(t_k) + \frac{Kh}{T_i}e(t_k). \tag{13.7}$$

If the derivative in Equation 13.6 is approximated instead by a backward difference, the following is obtained:

$$\frac{I(t_k) - I(t_k - 1)}{h} = \frac{K}{T_i}e(t_k).$$

This leads to the following recursive equation for the integral term:

$$I(t_{k+1}) = I(t_k) + \frac{Kh}{T_i}\, e(t_{k+1}). \tag{13.8}$$

Another simple approximation method is due to Tustin. This approximation is

$$I(t_{k+1}) = I(t_k) + \frac{Kh}{T_i}\,\frac{e(t_{k+1}) + e(t_k)}{2}. \tag{13.9}$$

Yet another method is called ramp equivalence. This method gives exact outputs at the sampling instants if the input signal is continuous and piecewise linear between the sampling instants. The ramp equivalence method gives the same approximation of the integral term as the Tustin approximation, i.e., Equation 13.9.

Notice that all approximations have the same form, i.e.,

$$I(t_{k+1}) = I(t_k) + b_{i1}e(t_{k+1}) + b_{i2}e(t_k), \tag{13.10}$$

but with different values of parameters b_{i1} and b_{i2}.

Derivative Action The derivative term with the classical first-order filter is given by Equation 3.14, i.e.,

$$\frac{T_d}{N}\frac{dD}{dt} + D = -KT_d\,\frac{dy}{dt}. \tag{13.11}$$

This equation can be approximated in the same way as the integral term.

Approximating the derivative by a forward difference gives

$$\frac{T_d}{N}\frac{D(t_{k+1}) - D(t_k)}{h} + D(t_k) = -KT_d\,\frac{y(t_{k+1}) - y(t_k)}{h}.$$

This can be rewritten as

$$D(t_{k+1}) = \left(1 - \frac{Nh}{T_d}\right) D(t_k) - KN\left(y(t_{k+1}) - y(t_k)\right). \tag{13.12}$$

If the derivative in Equation 13.11 is approximated by a backward difference, the following equation is obtained:

$$\frac{T_d}{N}\frac{D(t_k) - D(t_{k-1})}{h} + D(t_k) = -KT_d\,\frac{y(t_k) - y(t_{k-1})}{h}.$$

This can be rewritten as

$$D(t_k) = \frac{T_d}{T_d + Nh}\, D(t_{k-1}) - \frac{KT_dN}{T_d + Nh}\left(y(t_k) - y(t_{k-1})\right). \tag{13.13}$$

Using the Tustin approximation to approximate the derivative term gives

$$D(t_k) = \frac{2T_d - Nh}{2T_d + Nh}\, D(t_{k-1}) - \frac{2KT_dN}{2T_d + Nh}\left(y(t_k) - y(t_{k-1})\right). \tag{13.14}$$

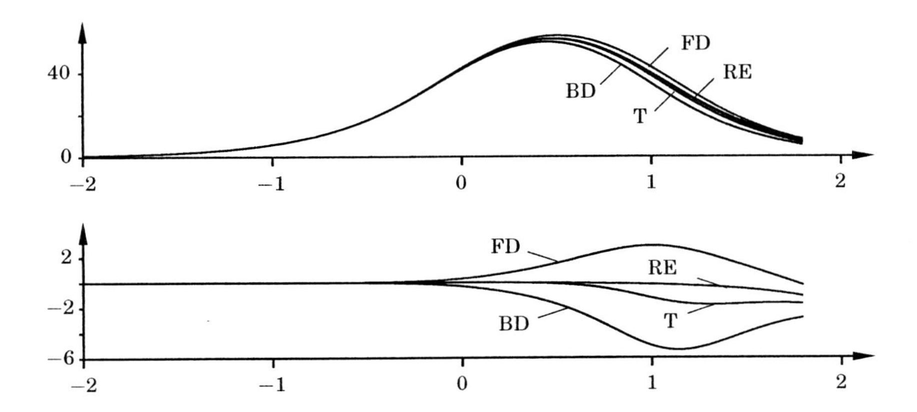

Figure 13.9 Phase curves for PD controllers obtained by different discretizations of the derivative term $sT_d/(1+sT_d/N)$ with $T_d = 1, N = 10$ and a sampling period 0.02. The discretizations are forward differences (FD), backward differences (BD), Tustin's approximation (T), and ramp equivalence (RE). The lower diagram shows the differences between the approximations and the true phase curve.

Finally, the ramp equivalence approximation is

$$D(t_k) = e^{-Nh/T_d} D(t_{k-1}) - \frac{KT_d(1-e^{-Nh/T_d})}{h}\,(y(t_k) - y(t_{k-1})). \tag{13.15}$$

All approximations have the same form,

$$D(t_k) = a_d D(t_{k-1}) - b_d\,(y(t_k) - y(t_{k-1})), \tag{13.16}$$

but with different values of parameters a_d and b_d.

The approximations of the derivative term are stable only when $|a_d| < 1$. When using the forward difference approximation stability requires that $T_d > Nh/2$. The approximation becomes unstable for small values of T_d. The other approximations are stable for all values of T_d. Notice, however, that Tustin's approximation and the forward difference approximation give negative values of a_d if T_d is small. This is undesirable because the approximation will then exhibit ringing. The backward difference approximation give good results for all values of T_d, including $T_d = 0$.

For reasonable fast sampling there are only small differences between the approximations as long as they are stable. There are, however, practical differences. In a general-purpose controller it is desirable that derivative action can be switched off. A natural way to do this is to set $T_d = 0$. This can easily be accomplished when the derivative is approximated by a backward difference. All other methods will either give instability or overflow for $T_d = 0$. The backward difference is therefore a reasonable choice for approximating the derivative.

Figure 13.9 shows the phase curves for the different discrete time approximations. Tustin's approximation and the ramp equivalence approximation give the best agreement with the continuous time case, the backward approximation gives less phase advance, and the forward approximation gives more phase advance. The forward approximation is seldom used because of the problems with instability for small values of derivative time T_d. Tustin's algorithm is used quite frequently because of its simplicity and its close agreement with the continuous time transfer function. The backward difference is used when an algorithm that is well behaved for small T_d is needed.

All approximations of the PID controller can be represented as

$$R(q)u(kh) = T(q)y_{sp}(kh) - S(q)y(kh), \tag{13.17}$$

where q is the forward shift operator, and the polynomials R, S, and T are of second order. The polynomials R, S, and T have the forms

$$\begin{aligned} R(q) &= (q-1)(q-a_d) \\ S(q) &= s_0q^2 + s_1q + s_2 \\ T(q) &= t_0q^2 + t_1q + t_2, \end{aligned} \tag{13.18}$$

which means that Equation 13.17 can be written as

$$\begin{aligned} u(kh) = t_0y_{sp}(kh) &+ t_1y_{sp}(kh-h) + t_2y_{sp}(kh-2h) \\ &- s_0y(kh) - s_1y(kh-h) - s_2y(kj-2h) \\ &+ (1+a_d)u(kh-h) - a_du(kh-h). \end{aligned}$$

The coefficients in the S and T polynomials are

$$\begin{aligned} s_0 &= K + b_{i1} + b_d \\ s_1 &= -K(1+a_d) - b_{i1}a_d + b_{i2} - 2b_d \\ s_2 &= Ka_d - b_{i2}a_d + b_d \\ t_0 &= Kb + b_{i1} \\ t_1 &= -Kb(1+a_d) - b_{i1}a_d + b_{i2} \\ t_2 &= Kba_d - b_{i2}a_d. \end{aligned} \tag{13.19}$$

The coefficients in the polynomials for different approximation methods are given in Table 13.1.

Controller with Second Order Filter

A nice implementation of a PID controller is to combine a second order filtering of the measured signal with an ideal PID controller; see Section 3.3. We will now discuss how such controllers can be implemented. Let y be the measured signal and y_f the filtered signal. We have

$$Y_f(s) = G_f(s)Y(s) = \frac{1}{1 + sT_f + (sT_f)^2/2}Y(s). \tag{13.20}$$

Introducing the state variables $x_1 = y_f$ and $x_2 = T_f dy_f/dt$ the filter can be represented as

$$\begin{aligned} T_f\frac{dx_1}{dt} &= x_2 \\ T_f\frac{dx_2}{dt} &= 2(-x_1 - x_2 + y). \end{aligned} \tag{13.21}$$

Table 13.1 Coefficients in different approximations of the continuous time PID controller.

	Forward	Backward	Tustin	Ramp equivalence
b_{i1}	0	$\frac{Kh}{T_i}$	$\frac{Kh}{2T_i}$	$\frac{Kh}{2T_i}$
b_{i2}	$\frac{Kh}{T_i}$	0	$\frac{Kh}{2T_i}$	$\frac{Kh}{2T_i}$
a_d	$1-\frac{Nh}{T_d}$	$\frac{T_d}{T_d+Nh}$	$\frac{2T_d-Nh}{2T_d+Nh}$	e^{-Nh/T_d}
b_d	KN	$\frac{KT_dN}{T_d+Nh}$	$\frac{2KT_dN}{2T_d+Nh}$	$\frac{KT_d(1-e^{-Nh/T_d})}{h}$

The filtered derivative $dy_f/dt = x_2/T_f$ can be extracted from the filter and the controller is then given by

$$u = k(by_{sp} - y_f) + k_i \int_0^t (y_{sp}(\tau) - y_f(\tau))d\tau + k_d \frac{dy_f}{dt}. \tag{13.22}$$

If the PID controller (13.22) is implemented digitally, both $x_1 = y_f$ and $x_2 = T_f dy_f/dt$ have to be converted to digital form. This implementation is suitable for special-purpose systems. For general-purpose systems the filter can be implemented digitally. Assume that the sampling has period h and let the sampling instants be t_k. Approximating the derivative in (13.21) with a backward difference we find

$$\begin{aligned}
x_1(t) &= x_1(t-h) + \frac{hT_f}{T_f^2 + 2hT_f + 2h^2} x_2(t-h) \\
&\quad + \frac{2h^2}{T_f^2 + 2hT_f + 2h^2}(y(t) - x_1(t-h)) \\
x_2(t) &= \frac{T_f^2}{T_f^2 + 2hT_f + 2h^2} x_2(t-h) + \frac{2hT_f}{T_f^2 + 2hT_f + 2h^2}(y(t) - x_1(t-h)).
\end{aligned}$$

To obtain an algorithm which permits the parameter T_f to be zero we introduce the state variables

$$\begin{aligned}
y_1 &= x_1 \\
y_2 &= \frac{h}{T_f} x_2.
\end{aligned}$$

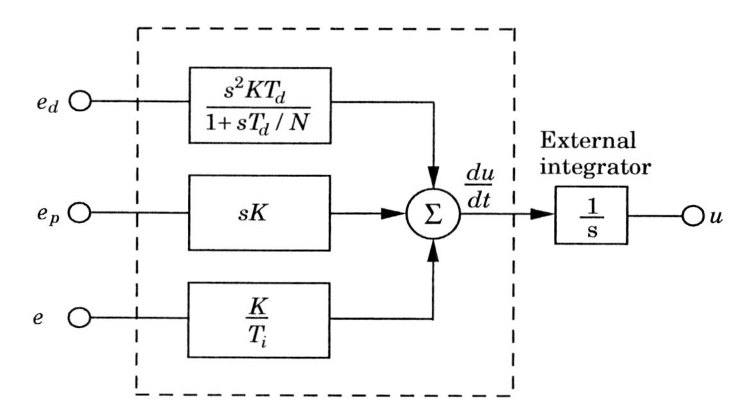

Figure 13.10 Block diagram of a PID algorithm in velocity form.

The equation for the controller can then be written as

$$\begin{aligned}
y_1(t) &= y_1(t-h) + p_1 y_2(t-h) + p_2\big(y(t) - y_1(t-h)\big) = y_1(t-h) + y_2(t) \\
y_2(t) &= p_1 y_2(t-h) + p_2\big(y(t) - y_1(t-h)\big) \\
v(t) &= K(by_{sp} - y_1) - p_4 y_2(t) + I(t-h) \\
&= K(by_{sp} - y_1) - p_2(K + p_4)y(t) \\
&\quad + \big(p_2(K+p_4) - K\big)y_1(t-h) + \big(K - p_1(k+p_4)\big)y_2(t-h) + I(t-h) \\
u(t) &= sat(v) \\
I(t) &= I(t-h) + p_3(y_{sp}(t) - y_1(t)) + p_5(u(t) - v(t)).
\end{aligned} \tag{13.23}$$

where the integral term has been approximated by a forward difference and protection for windup has been introduced. The parameters of the controller are given by

$$\begin{aligned}
p_1 &= \frac{T_f^2}{T_f^2 + 2hT_f + 2h^2} \quad p_2 = \frac{2h^2}{T_f^2 + 2hT_f + 2h^2} \\
p_3 &= \frac{Kh}{T_i} \quad p_4 = \frac{KT_d}{h} \quad p_5 = \frac{h}{T_t}
\end{aligned} \tag{13.24}$$

13.4 Velocity Algorithms

The algorithms described so far are called positional algorithms because the output of the algorithms is the control variable. In certain cases the control system is arranged in such a way that the control signal is driven directly by an integrator, e.g., a motor. It is then natural to arrange the algorithm in such a way that it gives the velocity of the control variable. The control variable is then obtained by integrating its velocity. An algorithm of this type is called a velocity algorithm. A block diagram of a velocity algorithm for a PID controller is shown in Figure 13.10.

Velocity algorithms were commonly used in many early controllers that were built around motors. In several cases, the structure was retained by the manufacturers when technology was changed in order to maintain functional compatibility with older equipment. Another reason is that many practical issues, like wind-up protection and bumpless parameter changes, are easy to implement using the velocity algorithm. This is discussed further in Sections 3.5 and 13.5. In digital implementations velocity algorithms are also called incremental algorithms.

Incremental Algorithm

The incremental form of the PID algorithm is obtained by computing the time differences of the controller output and adding the increments

$$\Delta u(t_k) = u(t_k) - u(t_{k-1}) = \Delta P(t_k) + \Delta I(t_k) + \Delta D(t_k).$$

In some cases integration is performed externally. This is natural when a stepper motor is used. The output of the controller should then represent the increments of the control signal, and the motor implements the integrator. The increments of the proportional part, the integral part, and the derivative part are easily calculated from Equations 13.5, 13.10, and 13.16:

$$\begin{aligned}
\Delta P(t_k) &= P(t_k) - P(t_{k-1}) = K\left(by_{sp}(t_k) - y(t_k) - by_{sp}(t_{k-1}) + y(t_{k-1})\right) \\
\Delta I(t_k) &= I(t_k) - I(t_{k-1}) = b_{i1}\, e(t_k) + b_{i2}\, e(t_{k-1}) \\
\Delta D(t_k) &= D(t_k) - D(t_{k-1}) = a_d \Delta D(t_{k-1}) - b_d\left(y(t_k) - 2y(t_{k-1}) + y(t_{k-2})\right).
\end{aligned}$$

One advantage with the incremental algorithm is that most of the computations are done using increments only. Short word-length calculations can often be used. It is only in the final stage where the increments are added that precision is needed.

Velocity Algorithms for Controllers without Integral Action

A velocity algorithm cannot be used directly for a controller without integral action because such a controller cannot keep the stationary value. This can be understood from the block diagram in Figure 13.11A, which shows a proportional controller in velocity form. Stationarity can be obtained for any value of the control error e, since the output from the derivation block is zero for any constant input. The problem can be avoided with the modification shown in Figure 13.11B. Here, stationarity is only obtained when $u = Ke + u_b$, where u_b is the bias term.

If a sampled PID controller is used, a simple version of the method illustrated in figure 13.11B is obtained by implementing the P controller as

$$\Delta u(t) = u(t) - u(t-h) = Ke(t) + u_b - u(t-h),$$

where h is the sampling period.

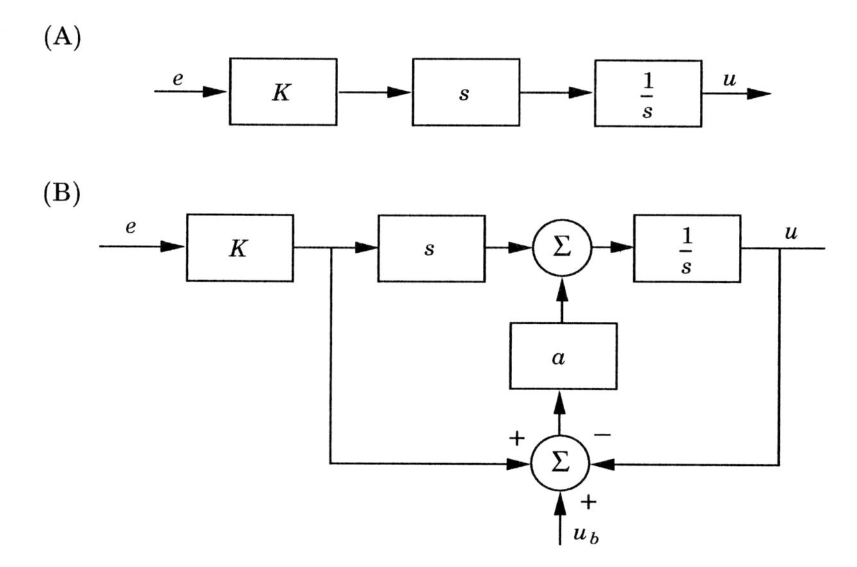

Figure 13.11 Illustrates the difficulty with a proportional controller in velocity form (A) and a way to avoid it (B).

Feedforward Control

Feedforward control was discussed in Chapter 5. In feedforward control, the control signal is composed of two terms,

$$u = u_{fb} + u_{ff}.$$

Here u_{fb} is the feedback component and u_{ff} is the feedforward component, either from a measurable disturbance or from the set point.

To avoid integrator windup, it is important that the antiwindup mechanism acts on the final control signal u, and not only on the feedback component u_{fb}. Unfortunately, many of the block-oriented instrument systems available today have the antiwindup mechanisms inside the feedback controller blocks, without any possibility to add feedforward signals to these blocks. Hence, the feedforward signals must be added after the controller blocks. This may lead to windup. Because of this, several tricks, like feeding the feedforward signal through high-pass filters, are used to reduce the windup problem. These strategies do, however, lead to a less effective feedforward.

Incremental algorithms are efficient for feedforward implementation. By first adding *the increments* of the feedback and feedforward components,

$$\Delta u = \Delta u_{fb} + \Delta u_{ff}$$

and then forming the control signal as

$$u(t) = u(t-h) + \Delta u(t),$$

windup is avoided. This requires that the feedback control blocks have inputs for feedforward signals.

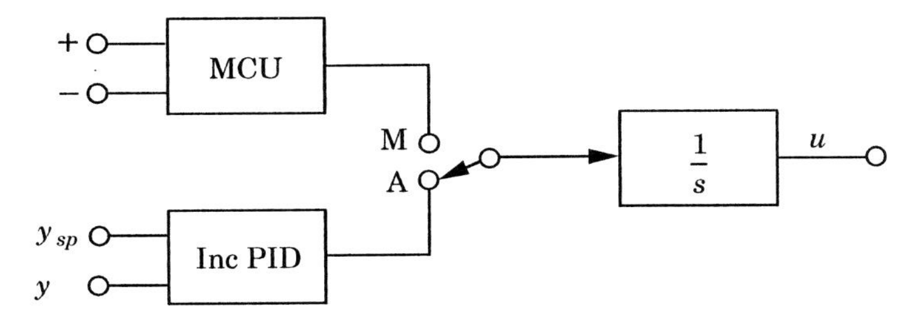

Figure 13.12 Bumpless transfer in a controller with incremental output. MCU stands for manual control unit.

13.5 Operational Aspects

Practically all controllers can be run in two modes: manual or automatic. In manual mode the controller output is manipulated directly by the operator, typically by pushing buttons that increase or decrease the controller output. A controller may also operate in combination with other controllers, such as in a cascade or ratio connection, or with nonlinear elements, such as multipliers and selectors. This gives rise to more operational modes. The controllers also have parameters that can be adjusted in operation. When there are changes of modes and parameters, it is essential to avoid switching transients. The way the mode switchings and the parameter changes are made depends on the structure chosen for the controller.

Bumpless Transfer Between Manual and Automatic

Since the controller is a dynamic system, it is necessary to make sure that the state of the system is correct when switching the controller between manual and automatic mode. When the system is in manual mode, the control algorithm produces a control signal that may be different from the manually generated control signal. It is necessary to make sure that the two outputs coincide at the time of switching. This is called *bumpless transfer*.

Bumpless transfer is easy to obtain for a controller in incremental form. This is shown in Figure 13.12. The integrator is provided with a switch so that the signals are either chosen from the manual or the automatic increments. Since the switching only influences the increments there will not be any large transients.

A similar mechanism can be used in the series, or interacting, implementation of a PID controller shown in Figure 3.3, see Figure 13.13. In this case there will be a switching transient if the output of the PD part is not zero at the switching instant.

For controllers with parallel implementation, the integrator of the PID controller can be used to add up the changes in manual mode. The controller shown in Figure 13.14 is such a system. This system gives a smooth transition between manual and automatic mode provided that the switch is made when the output of the PD block is zero. If this is not the case, there will be a switching transient.

It is also possible to use a separate integrator to add the incremental

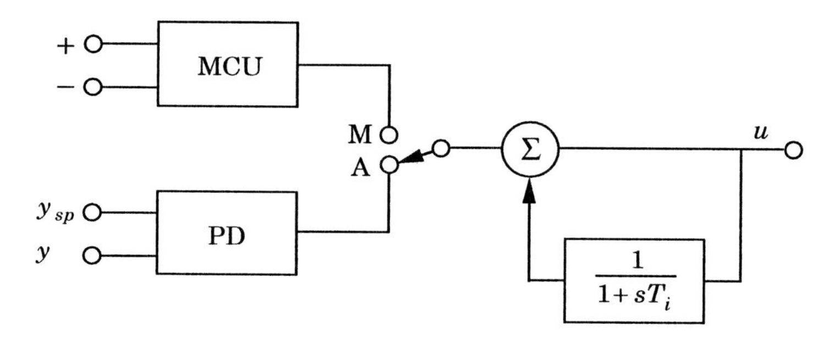

Figure 13.13 Bumpless transfer in a PID controller with a special series implementation.

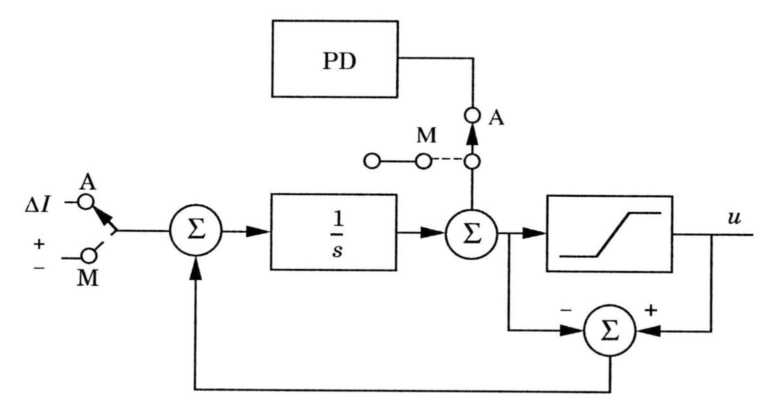

Figure 13.14 A PID controller where one integrator is used both to obtain integral action in automatic mode and to sum the incremental commands in manual mode.

changes from the manual control device. To avoid switching transients in such a system, it is necessary to make sure that the integrator in the PID controller is reset to a proper value when the controller is in manual mode. Similarly, the integrator associated with manual control must be reset to a proper value when the controller is in automatic mode. This can be realized with the circuit shown in Figure 13.15. With this system the switch between manual and automatic is smooth even if the control error or its derivative is different from zero at the switching instant. When the controller operates in manual mode, as is shown in Figure 13.15, the feedback from the output v of the PID controller tracks the output u. With efficient tracking the signal v will thus be close to u at all times. There is a similar tracking mechanism that ensures that the integrator in the manual control circuit tracks the controller output.

Bumpless Parameter Changes

A controller is a dynamical system. A change of the parameters of a dynamical system will naturally result in changes of its output. Changes in the output can be avoided, in some cases, by a simultaneous change of the state of the system. The changes in the output will also depend on the chosen realization. With a PID controller it is natural to require that there be no drastic changes in the

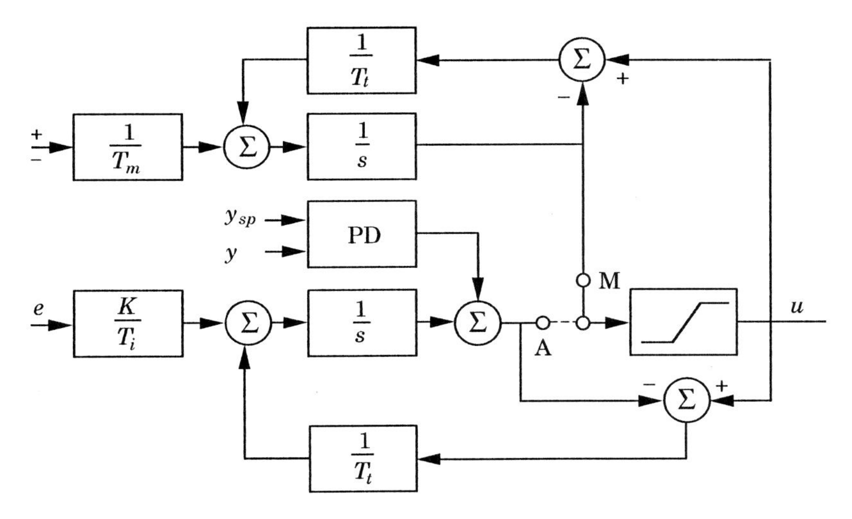

Figure 13.15 PID controller with parallel implementation that switches smoothly between manual and automatic control.

output if the parameters are changed when the error is zero. This will hold for all incremental algorithms because the output of an incremental algorithm is zero when the input is zero, irrespective of the parameter values. For a position algorithm it depends, however, on the implementation.

Assume that the state is chosen as

$$x_I = \int^t e(\tau)d\tau$$

when implementing the algorithm. The integral term is then

$$I = \frac{K}{T_i} x_I.$$

Any change of K or T_i will then result in a change of I. To avoid bumps when the parameters are changed, it is essential that the state be chosen as

$$x_I = \int^t \frac{K(\tau)}{T_i(\tau)} e(\tau)d\tau$$

when implementing the integral term.

With sensible precautions, it is easy to ensure bumpless parameter changes if parameters are changed when the error is zero. There is, however, one case where special precautions have to be taken, namely, if set-point weighting is used. To have bumpless parameter changes in such a case it is necessary that the quantity $P + I$ be invariant to parameter changes. This means that when parameters are changed, the state I should be changed as follows:

$$I_{\text{new}} = I_{\text{old}} + K_{\text{old}}(b_{\text{old}}\, y_{sp} - y) - K_{\text{new}}(b_{\text{new}}\, y_{sp} - y). \tag{13.25}$$

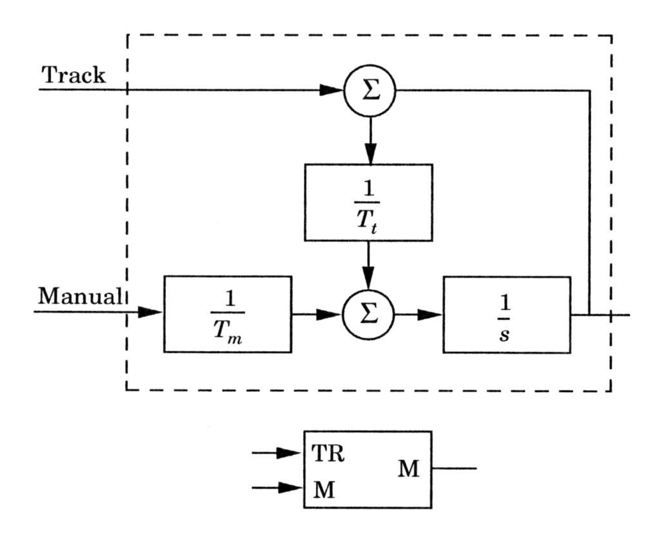

Figure 13.16 Manual control module.

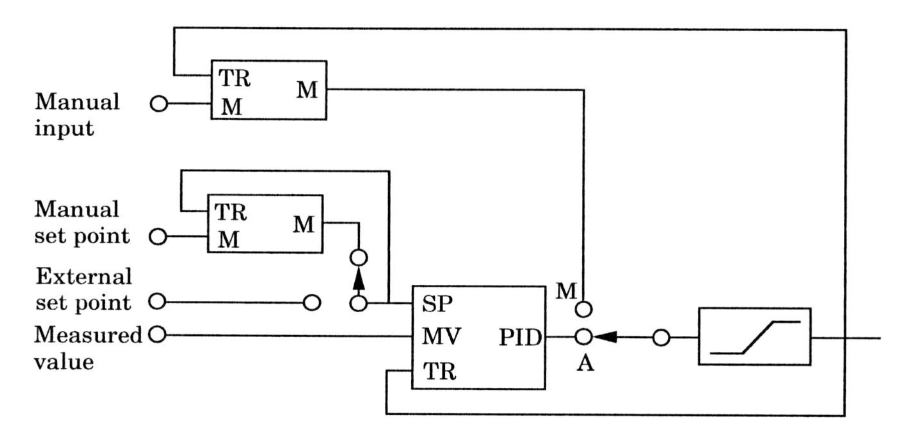

Figure 13.17 A reasonable complete PID controller with antiwindup, automatic-manual mode, and manual and external set point.

To build automation systems it is useful to have suitable modules. Figure 13.16 shows the block diagram for a manual control module. It has two inputs: a tracking input and an input for the manual control commands. The system has two parameters: the time constant T_m for the manual control input and the reset time constant T_t. In digital implementations it is convenient to add a feature so that the command signal accelerates as long as one of the increase-decrease buttons is pushed. Using the module for PID control and the manual control module in Figure 13.16, it is straightforward to construct a complete controller. Figure 13.17 shows a PID controller with internal or external set points via increase-decrease buttons and manual automatic mode. Notice that the system only has two switches.

Computer Code

As an illustration we will give computer codes for two PID controllers. A PID controller with first order filtering of the derivative term where the derivative term is approximated by backward differences is described by Equations 13.5, 13.7,13.9, and 13.13. Anti-windup is provided using the scheme described in Section 3.5. A skeleton code for the controller is given in Figure 13.18. The main loop has two states, the integral term `I`, and `x` which is used to implement derivative action. The parameters $p_1, \ldots, p_6$ are precomputed to save computing time in the main loop. These parameters have to be computed only when parameters are changed. The integral term is also reset as described by (13.25) to avoid transients when parameters are changed. The main loop in the control algorithm requires eight additions and six multiplications. Notice that the calculations are structured so that there are only three additions and two multiplications between reading the analog inputs are setting the digital output. The states are updated after setting the digital output.

A PID controller with second order filtering of the process variable is described by Equation 13.23, where the filter is implemented using backward differences and the integral term is approximated using forward differences. Anti-windup is obtained by the scheme shown in Figure 3.13. The algorithm has three states `y1`, `y2`, and `I`, which represent the states of the measurement filter and the integral term. The main loop in the control algorithm requires ten additions and seven multiplications. Using a second order filter only requires a marginal increase of computing time. The time between reading the analog inputs and setting the digital output can be reduced by changing the coordinates of the representation of the filter.

13.6 Controller Outputs

Analog Outputs

The inputs and outputs of a controller are normally analog signals, typically 0–20 mA or 4–20 mA. The main reason for using 4 mA instead of 0 mA as the lower limit is that many transmitters are designed for two-wire connection. This means that the same wire is used for both driving the sensor and transmitting the information from the sensor. It would not be possible to drive the sensor with a current of 0 mA. The main reason for using current instead of voltage is to avoid the influence of voltage drops along the wire due to resistance in the (perhaps long) wire. In pneumatic controllers, the standard range is 3–15 psi.

Thyristors and Triacs

In temperature controllers it is common practice to integrate the power amplifier with the controller. The power amplifier could be a thyristor or a triac. With a thyristor, an AC voltage is switched to the load at a given angle of the AC voltage. Since the relation between angle and power is nonlinear, it is crucial to use a transformation to maintain a linear relationship. A triac is

```
"Compute controller coefficients
    p1=K*b                                "set-point gain
    p2=K+K*Td/(Tf+h)                      "PD gain
    p3=Tf/(Tf+h)                          "filter constant
    p4=K*Td*h/((Tf+h)*(Tf+h))             "derivative gain
    p5=K*h/Ti                             "integral gain
    p6=h/Tt                               "anti-windup gain

"Bumpless parameter changes
    I=I+Kold*(bold*ysp-y)-Knew*(bnew*ysp-y)

"Control algorithm
    adin(ysp)                             "read set point
    adin(y)                               "read process variable
    v=p1*ysp-p2*y+x+I                     "compute nominal output
    u=sat(v,ulow,uhigh)                   "saturate output
    daout(u)                              "set analog output
    x=p3*x+p4*y                           "update derivative
     I=I+p5*(ysp-y)+p6*(u-v)              "update integral
```

Figure 13.18 Skeleton code for implementing a PID controller with first order filtering of the derivative term.

```
"Compute controller coefficients
    den=Tf*Tf+2*h*Tf+2*h*h                "denominator
    p1=Tf*Tf/den                          "filter constant
    p2=2*h*h/den                          "filter constant
    p3=K*h/Ti                             "integral gain
    p4=K*Td/h                             "derivative gain
    p5=h/Tt                               "anti-windup gain

"Bumpless parameter changes
    I=I+Kold*(bold*ysp-y1)-Knew*(bnew*ysp-y1)

"Control algorithm
    r=adin(ysp)                           "read set point
    y=adin(y)                             "read process variable
    x2=p1*y2+p2*(y-y1)                    "update filter state x2
    y1=y1+y2                              "update filter state x1
    v=K*(b*ysp-y1)-p4*y2+I                "compute nominal output
    u=sat(v,ulow,uhigh)                   "saturate output
    daout(u)                              "set analog output
    I=I+p3*(ysp-y1)+p5*(u-v)              "update integral
```

Figure 13.19 Skeleton code for implementing a PID controller with second order filtering of the measured signal.

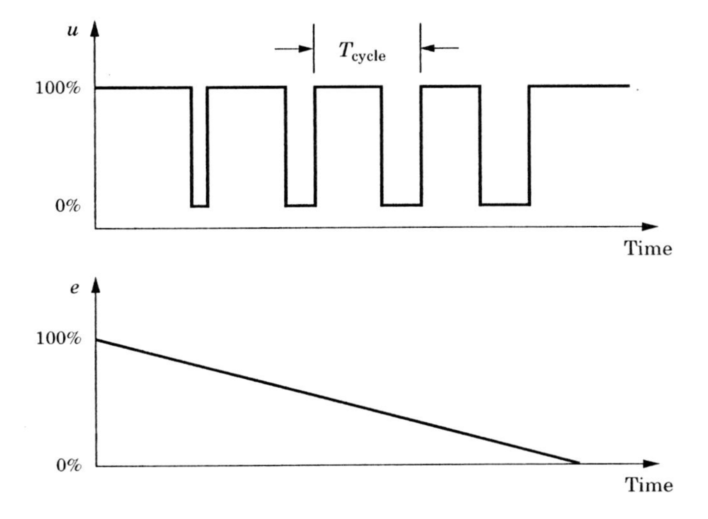

Figure 13.20 Illustration of controller output based on pulse width modulation.

also a device that implements switching of an AC signal, but only at the zero crossing. Such a device is similar to a pulse output.

Pulse Width Modulation

In some cases, such as with the triac, there is an extreme quantization in the sense that the actuator only accepts two values, on or off. In such a case, a cycle time T_{cycle} is specified, and the controller gives a pulse with width

$$T_{pulse}(t) = \frac{u(t) - u_{min}}{u_{max} - u_{min}} T_{cycle}. \tag{13.26}$$

A similar, but slightly different, situation occurs when the actuator has three levels: max, min, and zero. A typical example is a motor-driven valve where the motor can stand still, go forward, or go backward.

Figure 13.20 illustrates the pulse width modulation. The figure shows the output from a P controller with pulse width modulation for different values of the control error.

Three-Position Pulse Output

If a valve is driven by a constant-speed electrical motor, the valve can be in three states: "increase," "stop," and "decrease." Control of valves with electrical actuators is performed with a controller output that can be in three states. Three-position pulse output is performed using two digital outputs from the controller. When the first output is conducting, the valve position will increase. When the second output is conducting, the valve position will decrease. If none of the outputs are conducting, the valve position is constant. The two outputs must never be conducting at the same time.

There is normally both a dead zone and a dead time in the controller to ensure that the change of direction of the motor is not too frequent and not too

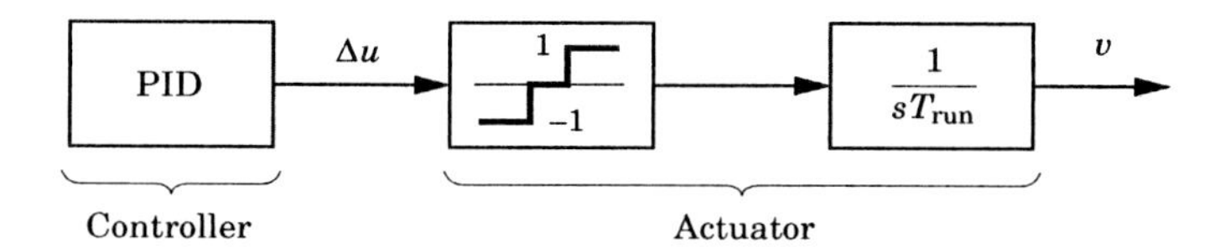

Figure 13.21 A PID controller with three-position pulse output combined with an electrical actuator.

fast. It means that the controller output is constant as long as the magnitude of the control error is within the dead zone and that the output is stopped for a few seconds before it is allowed to change direction.

A servo-motor is characterized by its running time T_{run}, which is the time it takes for the motor to go from one end position to the other. Since the servo-motor has a constant speed, it introduces an integrator in the control loop, where the integration time is determined by T_{run}. A block diagram describing a PID controller with three-position pulse output combined with an electrical actuator is shown in Figure 13.21. Suppose that we have a steady-state situation, where the output from the PID controller u is equal to the position v of the servo-motor. Suppose further that we suddenly want to increase the controller output by an amount Δu. As long as the increase-output is conducting, the output v from the servo-motor will increase according to

$$\Delta v = \frac{1}{T_{\text{run}}} \int_0^t 1\, dt = \frac{t}{T_{\text{run}}}.$$

To have Δv equal to Δu, the integration must be stopped after time

$$t = \Delta u T_{\text{run}}.$$

In a digital controller, this means that the digital output corresponding to an increasing valve position is to be conducting for n sampling periods, where n is given by

$$n = \frac{\Delta u T_{\text{run}}}{h},$$

where h is the sampling period of the controller.

To be able to perform a correct three-position pulse output, two buffers (`Buff_increase` and `Buff_decrease`) must be used to hold the number of pulses that should be sent out. A computer code for three-position pulse output is given in Figure 13.22. For the sake of simplicity, details such as dead zone and dead time are omitted in the code.

According to Figure 13.21, the controller output is Δu instead of u in the case of three-position pulse output. The integral part of the control algorithm is outside the controller, in the actuator. This solution causes no problems if the control algorithm really contains an integral part. P and PD control can not be obtained without information of the valve position, see Figure 13.11.

```
if delta_u > 0 then
    if valve_is_increasing then
        Buff_increase = Buff_increase + n;
    else
    Buff_decrease = Buff_decrease - n;
    if Buff_decrease < 0 then
        Buff_increase = - Buff_decrease;
        Buff_decrease = 0;
        valve_is_decreasing = false;
        valve_is_increasing = true;
        end;
    end;
else if delta_u < 0 then
    if valve_is_increasing then
        Buff_decrease = Buff_decrease + n;
    else
    Buff_increase = Buff_increase - n;
    if Buff_increase < 0 then
        Buff_decrease = - Buff_increase;
        Buff_increase = 0;
        valve_is_increasing = false;
        valve_is_decreasing = true;
        end;
    end;
end;
if Buff_increase > 0 then
    Increaseoutput = 1;
    Decreaseoutput = 0;
    Buff_increase = Buff_increase - 1;
else if Buff_decrease > 0 then
    Increaseoutput = 0;
    Decreaseoutput = 1;
    Buff_decrease = Buff_decrease - 1;
end;
```

Figure 13.22 Skeleton code for three-position pulse output.

13.7 Summary

In this chapter we have described implementation of PID controllers. We have followed the historical development starting with pneumatic and electronic implementation of analog controllers. Computer implementation are presented in detail including skeleton code. The reason for doing this is that many features of modern implementation have inherited several features of the old analog computers; the preference for the series form is one example.

It is interesting to consider the development of the controllers. During each phase of the development the technology has matured and improved, but knowledge has often been lost in the technology shifts. For example, it took quite a while before the importance of measurement filtering and anti-windup were appreciated in the computer implementations. One reason for it is that many details were not well documented and thus easily forgotten when technology changed. Another was that some good features were obtained automatically because of particular features of the technology. The important issues of operational aspects and human-machine interfaces have also discussed in this chapter.

13.8 Notes and References

The book [Holzbock, 1958] presents many early implementations of PID controllers using pneumatic, hydraulic, and electric technologies. Implementation of pneumatic controllers are discussed in [Lloyd and Anderson, 1971; Pavlik and Machei, 1960]. Electronic implementations are discussed by [Anderson, 1972]. It in interesting that all books mentioned above are written by equipment vendors. The paper by [Goff, 1966b] describes early efforts in digital implementation of PID controllers. Digital implementations are treated in detail in [Clarke, 1984; Hanselmann, 1987; Åström and Wittenmark, 1997]. The paper [Turnbull, 1988] gives a broad description of the development of Eurotherm's temperature controller spanning a period of more than two decades and technologies from electronic to digital. The book [Dote, 1972] describes implementation of controllers for motion control. Code for implementation on signal processors that admits very fast sampling is found in [Åström and Steingrímsson, 1991].

Bibliography

ABB (2002): *Control^{IT}: Analog Process Control — Objects and Design*, version 3.2 edition. 3BSE 028 809 R101 Rev B.

Adusumilli, S., D. E. Rivera, S. Dash, and K. Tsakalis (1998): "Integrated MIMO identification and robust PID controller design through loop shaping." In *Proc. American Control Conference (ACC'98)*, vol. 2, pp. 1230–1234. Philadelphia, PA.

Akkermans, S. and S. Stan (2002): "Digital servo IC for optical disc drives." *Control Engineering Practice*, **9:11**, pp. 1245–1253.

Allgower, F. and H. Zheng (2000): *Nonlinear model predictive control, progress in systems and control theory*. Birkhauser, Basel.

Anderson, K. L., G. L. Blankenship, and L. G. Lebow (1988): "A rule-based PID controller." In *Proc. IEEE Conference on Decision and Control*. Austin, Texas.

Anderson, N. A. (1972): *Instrumentation for Process Measurement and Control*. Chilton Book Company, Radnor, PA.

Anderssen, A. S. and E. T. White (1970): "Parameter estimation by the transfer function method." *Chemical Engineering Science*, **25**, pp. 1015–1021.

Anderssen, A. S. and E. T. White (1971): "Parameter estimation by the weighted moments method." *Chemical Engineering Science*, **26**, pp. 1203–1221.

Anonymous (1997): "Control loop optimization." Technical Report SSG-5253. Pulp and Paper Industries' Engineering Co, Sundsvall, Sweden.

Araki, M. (1984): "PID control systems with reference feedforward (PID-FF control systems)." In *SICE Annual Conference*, p. 31.

Araki, M. and H. Taguchi (1998): "Two-degree-of-freedom PID controller." *Systems, Control and Information*, **42**.

Åström, K. J. (1967): "Computer control of a paper machine — An application of linear stochastic control theory." *IBM Journal of Research and Development*, **11**.

Åström, K. J. (1970): *Introduction to Stochastic Control Theory*. Academic Press, New York.

Åström, K. J. (1977): "Frequency domain properties of Otto Smith regulators." *International Journal of Control*, **26**, pp. 307–314.

Åström, K. J. (1987a): "Adaptive feedback control." *Proc. IEEE*, **75**, February, pp. 185–217. Invited paper.

Åström, K. J. (1987b): "Advanced control methods—Survey and assessment of possibilities." In Morris *et al.*, Eds., *Advanced Control in Computer Integrated Manufacturing*, Proceedings 13th Annual Advanced Control Conference. Purdue University, West Lafayette, Indiana.

Åström, K. J. (1990): "Automatic tuning and adaptive control—Past accomplishments and future directions." In Prett and Garcia, Eds., *The Second Shell Process Control Workshop*, pp. 1–24. Butterworths, Stoneham, Massachusetts.

Åström, K. J. (1991): "Assessment of achievable performance of simple feedback loops." *International Journal of Adaptive Control and Signal Processing*, **5**, pp. 3–19.

Åström, K. J. (2000): "Limitations on control system performance." *European Journal on Control*, **6:1**, pp. 2–20.

Åström, K. J. (2001): "Control problems in paper making: Revisited." *Pulp & Paper Canada*, **102:6**, pp. 39–44.

Åström, K. J. and T. Bohlin (1965): "Numerical identification of linear dynamic systems from normal operating records." In *Proc. IFAC Conference on Self-Adaptive Control Systems*. Teddington, UK.

Åström, K. J., H. Elmqvist, and S. E. Mattsson (1998): "Evolution of continuous-time modeling and simulation." In Zobel and Moeller, Eds., *Proceedings of the 12th European Simulation Multiconference, ESM'98*, pp. 9–18. Society for Computer Simulation International, Manchester, UK.

Åström, K. J. and T. Hägglund (1984a): "Automatic tuning of simple regulators." In *Preprints 9th IFAC World Congress*, pp. 267–272. Budapest, Hungary.

Åström, K. J. and T. Hägglund (1984b): "Automatic tuning of simple regulators with specifications on phase and amplitude margins." *Automatica*, **20**, pp. 645–651.

Åström, K. J. and T. Hägglund (1984c): "Automatic tuning of simple regulators with specifications on phase and amplitude margins." *Automatica*, **20:5**, pp. 645–651.

Åström, K. J. and T. Hägglund (1988): *Automatic Tuning of PID Controllers*. ISA, Research Triangle Park, North Carolina.

Åström, K. J. and T. Hägglund (1990): "Practical experiences of adaptive techniques." In *American Control Conference (ACC '90)*. San Diego, California.

Åström, K. J. and T. Hägglund (1995a): "New tuning methods for PID controllers." In *European Control Conference*, pp. 2456–2462. Rome, Italy.

Åström, K. J. and T. Hägglund (1995b): *PID Controllers: Theory, Design, and Tuning.* ISA, Research Triangle Park, North Carolina.

Åström, K. J. and T. Hägglund (2001): "The future of PID control." *Control Engineering Practice*, **9**, pp. 1163–1175.

Åström, K. J., T. Hägglund, C. C. Hang, and W. K. Ho (1993): "Automatic tuning and adaptation for PID controllers—A survey." *Control Engineering Practice*, **1:4**, pp. 699–714.

Åström, K. J., C. C. Hang, and B. C. Lim (1994): "A new Smith predictor for controlling a process with an integrator and long dead-time." *IEEE Transactions on Automatic Control*, **39:2**.

Åström, K. J., K. H. Johansson, and Q. Wang (2002): "Design of decoupled PI controller for two-by-two systems." *IEE Proceedings on Control Theory and Applications*, **149**, January, pp. 74–81.

Åström, K. J. and T. J. McAvoy (1992): "Intelligent control." *Journal of Process Control*, **2:2**, pp. 1–13.

Åström, K. J., H. Panagopoulos, and T. Hägglund (1998): "Design of PI controllers based on non-convex optimization." *Automatica*, **34:5**, pp. 585–601.

Åström, K. J. and L. Rundqwist (1989): "Integrator windup and how to avoid it." In *Proceedings of the American Control Conference (ACC '89)*, pp. 1693–1698. Pittsburgh, Pennsylvania.

Åström, K. J. and H. Steingrímsson (1991): "Implementation of a PID controller on a DSP." In Ahmed, Ed., *Digital Control Applications with the TMS 320 Family*, Selected Application Notes, pp. 205–238. Texas Instruments.

Åström, K. J. and B. Wittenmark (1973): "On self-tuning regulators." *Automatica*, **9**, pp. 185–199.

Åström, K. J. and B. Wittenmark (1995): *Adaptive Control*, second edition. Addison-Wesley, Reading, Massachusetts.

Åström, K. J. and B. Wittenmark (1997): *Computer-Controlled Systems*, third edition. Prentice Hall.

Athans, M. and P. L. Falb (1966): *Optimal Control.* McGraw-Hill, New York.

Atherton, D. (1999): "PID controller tuning." *Computing & Control Engineering Journal*, April, pp. 44–50.

Atherton, D. P. (1975): *Nonlinear Control Engineering—Describing Function Analysis and Design.* Van Nostrand Reinhold Co., London, UK.

Bellman, R. (1957): *Dynamic Programming.* Princeton University Press, New Jersey.

Bennett, S. (1979): *A History of Control Engineering 1800-1930.* Peter Peregrinus, London.

Bennett, S. (1993): *A History of Control Engineering 1930-1955.* Peter Peregrinus, London.

Bennett, S. (2000): "Past of PID controllers." In *Proc. IFAC Workshop on Digital Control: Past, Present and Future of PID Control.* Terrassa, Spain.

Bequette, W. (2003): *Process Control: Modeling, Design and Simulation.* Prentice Hall.

Bialkowski, B. (1995): *Process Control Fundamentals for the Pulp & Paper Industry*, chapter Process control sample problems. Tappi Press.

Bialkowski, W. L. (1994): "Dream vs. reality - a view from both sides of the gap." *Pulp & Paper Canada*, **11**, pp. 19–27.

Black, H. S. (1977): "Inventing the negative feedback amplifier." *IEEE spectrum*, December, pp. 55–60.

Blevins, T. L., G. K. McMillan, W. K. Wojsznis, and B. M. W. (2003): *Advanced Control Unleashed.* ISA, Research Triangle Park, NC.

Blickley, G. (1990): "Modern control started with Ziegler-Nichols tuning." *Control Engineering*, October, pp. 11–17.

Blickley, G. J. (1988): "PID tuning made easy with hand-held computer." *Control Engineering*, November, p. 99.

Bode, H. W. (1945): *Network Analysis and Feedback Amplifier Design.* Van Nostrand, New York.

Bohlin, T. (1991): *Interactive system identification: Prospects and Pitfalls.* Springer, Berlin.

Boyd, S. P. and C. H. Barratt (1991): *Linear Controller Design – Limits of Performance.* Prentice Hall Inc., Englewood Cliffs, New Jersey.

Bristol, E. (1966): "On a new measure of interaction for multivariable process control." *IEEE Transactions on Automatic Control*, **11**, p. 133.

Bristol, E. H. (1967): "A simple adaptive system for industrial control." *Instrumentation Technology*, June.

Bristol, E. H. (1970): "Adaptive control odyssey." In *ISA Silver Jubilee Conference*, Paper 561–570. Philadelphia.

Bristol, E. H. (1977): "Pattern recognition: An alternative to parameter identification in adaptive control." *Automatica*, **13**, pp. 197–202.

Bristol, E. H. (1980): "After DDC: Idiomatic (structured) control." In *Proceedings American Institute of Chemical Engineering (AIChE).* Philadelphia.

Bristol, E. H. (1986): "The EXACT pattern recognition adaptive controller, a user-oriented commercial success." In Narendra, Ed., *Adaptive and Learning Systems*, pp. 149–163. Plenum Press, New York.

Bristol, E. H., G. R. Inaloglu, and J. F. Steadman (1970): "Adaptive process control by pattern recognition." *Instrum. Control Systems*, pp. 101–105.

Bristol, E. H. and T. W. Kraus (1984): "Life with pattern adaptation." In *Proc. 1984 American Control Conference.* San Diego, California.

Brown, G. S. and D. P. Campbell (1948): *Principles of Servomechanisms.* Wiley & Sons, New York.

Brown, M. and C. Harris (1994): *Neurofuzzy Adaptive Modelling and Control.* Prentice Hall.

Bryson, A. E. and Y. C. Ho (1969): *Applied Optimal Control Optimization, Estimation and Control.* Blaisdell Publishing Company.

Buckley, P. S. (1964): *Techniques of Process Control.* John Wiley & Sons, Inc.

Buckley, P. S. (1970): "Protective controls for a chemical reactor." *Chemical Engineering*, April, pp. 145–150.

Callaghan, P. J., P. L. Lee, and R. B. Newell (1986): "Evaluation of Foxboro controller." *Process Control Engineering*, **May**, pp. 38–40.

Callender, A., D. R. Hartree, and A. Porter (1936): "Time lag in a control system." *Philos. Trans. A.*, **235**, pp. 415–444.

Cameron, F. and D. E. Seborg (1983): "A self-tuning controller with a PID structure." *Int. J. Control*, **38:2**, pp. 401–417.

Cannon, R. H. (1967): *Dynamics of Physical Systems.* McGraw-Hill, New York.

Chang, T. N. and E. J. Davison (1987): "Steady-state interaction indices for decentralized unknown systems." In *Preprints 16th IEEE Conf. on Decision and Control*, vol. 2, pp. 881–887. Athens, Greece.

Chen, B.-S. and S.-S. Wang (1988): "The stability of feedback control with nonlinear saturating actuator: Time domain approach." *IEEE Transactions on Automatic Control*, **33**, pp. 483–487.

Chen, C.-L. (1989): "A simple method for on-line identification and controller tuning." *AIChE Journal*, **35:12**, pp. 2037–2039.

Chestnut, H. and R. W. Mayer (1959): *Servomechanisms and Regulating System Design.* Wiley, New York.

Chien, I. L. (1988): "IMC-PID controller design—an extension." In *IFAC Symposium, Adaptive Control of Chemical Processes*, pp. 155–160. Copenhagen, Denmark.

Chien, I.-L. and P. S. Fruehauf (1990): "Consider IMC tuning to improve controller performance." *Chemical Engineering Progress*, October, pp. 33–41.

Chien, K. L., J. A. Hrones, and J. B. Reswick (1952): "On the automatic control of generalized passive systems." *Trans. ASME*, **74**, pp. 175–185.

Choi, Y. and W. K. Chung (2004): *PID Trajectory Tracking Control for Mechanical Systems.* Springer, Berlin.

Clarke, D. W. (1984): "PID algorithms and their computer implementation." *Trans. Inst. Measurement and Control,* **6:6**, pp. 305–316.

Clarke, D. W. and C. E. Hinton (1997): "Adaptive control of materials-testing machines." *Automatica,* **33:6**, pp. 1119–1131.

Close, C. M. and D. K. Frederick (1993): *Modeling and Analysis of Dynamic Systems.* Houghton Mifflin.

Cohen, G. H. and G. A. Coon (1953): "Theoretical consideration of retarded control." *Trans. ASME,* **75**, pp. 827–834.

Cominos, P. and N. Munro (2002): "PID controllers: recent tuning methods and design to specifications." *IEE Proceedings; Control theory and applications,* **149:1**, pp. 46–53.

Coon, G. A. (1956a): "How to find controller settings from process characteristics." *Control Engineering,* **3**, pp. 66–76.

Coon, G. A. (1956b): "How to set three-term controller." *Control Engineering,* **3**, pp. 71–76.

Corripio, A. B. (1990): *Tuning of Industrial Control Systems.* ISA.

Cutler, C. R. and B. C. Ramaker (1980): "Dynamic matrix control—A computer control algorithm." In *Proceedings Joint Automatic Control Conference,* Paper WP5-B. San Francisco, California.

Dahlin, E. B. (1968): "Designing and tuning digital controllers." *Instruments and Control Systems,* **42**, June, pp. 77–83.

Desborough, L. and T. Harris (1992): "Performance assessment measures for univariate feedback control." *Canadian Journal of Chemical Engineeering,* **70**, pp. 1186–1197.

Desbourough, L. and R. Miller (2002): "Increasing customer value of industrial control performance monitoring - Honeywell's experience." In *Sixth International Conference on Chemical Process Control.* AIChE Symposium Series Number 326 (Volume 98).

Deshpande, P. B. and R. H. Ash (1981): *Elements of Computer Process Control with Advanced Control Applications.* ISA, Research Triangle Park, North Carolina.

DeWries, W. and S. Wu (1978): "Evaluation of process control effectiveness and diagnosis of variation in paper basis weight via multivariate time-series analysis." *IEEE Trans. on Automatic Control,* **23:4**, pp. 702–708.

Dote, Y. (1972): *Servo motor and motion control using digital signal processors.* Chilton Book Company, Radnor, PA.

Downs, J. J. (2001): "Linking control strategy design and model predictive control." pp. 411–422. Chemical Process Control - 6, Assessment and new directions for research (CPC-VI), Tucson, AZ.

Doyle, J. C., B. A. Francis, and A. R. Tannenbaum (1992): *Feedback Control Theory.* Macmillan, New York.

Dreinhofer, L. H. (1988): "Controller tuning for a slow nonlinear process." *IEEE Control Systems Magazine,* **8:2**, pp. 56–60.

Driankov, D., H. Hellendoorn, and M. Reinfrank (1993): *An Introduction to Fuzzy Control.* Springer-Verlag.

Dumont, G. A. (1986): "On the use of adaptive control in the process industries." In Morari and McAvoy, Eds., *Proceedings Third International Conference on Chemical Process Control–CPCIII.* Elsevier, Amsterdam.

Dumont, G. A., J. M. Martin-Sánchez, and C. C. Zervos (1989): "Comparison of an auto-tuned PID regulator and an adaptive predictive control system on an industrial bleach plant." *Automatica,* **25**, pp. 33–40.

Eckman, D. P. (1945): *Principles of industrial process control.* Wiley, New York.

Elgerd, O. I. and W. C. Stephens (1959): "Effect of closed-loop transfer function pole and zero locations on the transient response of linear control systems." *Applications and Industry,* **42**, pp. 121–127.

Elmqvist, H., S. E. Mattsson, and M. Otter (1998): "Modelica—The new object-oriented modeling language." In Zobel and Moeller, Eds., *Proceedings of the 12th European Simulation Multiconference, ESM'98,* pp. 127–131. Society for Computer Simulation International, Manchester, UK.

Ender, D. B. (1993): "Process control performance: Not as good as you think." *Control Engineering,* **40:10**, pp. 180–190.

Fertik, H. A. (1975): "Tuning controllers for noisy processes." *ISA Transactions,* **14**, pp. 292–304.

Fertik, H. A. and C. W. Ross (1967): "Direct digital control algorithms with anti-windup feature." *ISA Trans.,* **6:4**, pp. 317–328.

Foss, A. M. (1981): "Criterion to assess stability of a 'lowest wins' control strategy." *IEEE Proc. Pt. D,* **128:1**, pp. 1–8.

Foxboro, Inc. (1979): *Controller Tuning Guide,* PUB 342A.

Fröhr, F. (1967): "Optimierung von Regelkreisen nach dem Betragsoptimum und dem symmetrischen Optimum." *Automatik,* **12**, January, pp. 9–14.

Fröhr, F. and F. Orttenburger (1982): *Introduction to Electronic Control Engineering.* Siemens Aktiengesellschaft, Heyden & Son Ltd., London.

Gallun, S. E., C. W. Matthews, C. P. Senyard, and B. Slater (1985): "Windup protection and initialization for advanced digital control." *Hydrocarbon Processing,* June, pp. 63–68.

Garcia, C. E. and A. M. Morshedi (1986): "Quadratic programming solution of dynamic matrix control (QDMC)." *Chemical Engineering Communications,* **46**, pp. 73–87.

Gawthrop, P. J. (1986): "Self-tuning PID controllers: Algorithms and implementation." *IEEE Transactions on Automatic Control*, **31**, pp. 201–209.

Gelb, A. and W. E. V. Velde (1968): *Multiple-Input Describing Functions and Nonlinear System Design*. McGraw-Hill, New York.

Gerry, J. (1999): "Tuning process controllers start in manual." *InTech*, **May**, pp. 125–126.

Gerry, J. P. (1987): "A comparison of PID controller algorithms." *Control Engineering*, March, pp. 102–105.

Gibilaro, L. and F. Lees (1969): "The reduction of complex transfer function models to simple models using the method of moments." *Chem Eng Sci*, **24**, January, pp. 85–93.

Gille, J. C., M. J. Pelegrin, and P. Decaulne (1959): *Feedback Control Systems*. McGraw-Hill, New York.

Glattfelder, A. H., L. Guzzella, and W. Schaufelberger (1988): "Bumpless transfer, anti-reset-windup, saturating and override controls: A status report on self-tuning regulators." In *Proceedings of IMACS-88, Part 2*, pp. 66–72. Paris, France.

Glattfelder, A. H. and Schaufelberger (1983): "Stability analysis of single loop systems with saturation and antireset-windup circuits." *IEEE Transactions on Automatic Control*, **28**, pp. 1074–1081.

Glattfelder, A. H. and W. Schaufelberger (1986): "Start-up performance of different proportional-integral-anti-wind-up regulators." *International Journal of Control*, **44**, pp. 493–505.

Glattfelder, A. H. and W. Schaufelberger (2003): *Control systems with input and output constraints*. Springer.

Glover, K. (1990): "A tutorial on model reduction." In Willems, Ed., *From data to model*. Springer, Berlin.

Goff, K. W. (1966a): "Dynamics in direct digital control I—Estimating characteristics and effects of noisy signals." *ISA Journal*, **13**, November, pp. 45–49.

Goff, K. W. (1966b): "Dynamics in direct digital control II—A systematic approach to DDC design." *ISA Journal*, **13**, December, pp. 44–54.

Graham, D. and R. C. Lathrop (1953): "The synthesis of 'optimum' transient response: Criteria and standard forms." *Transactions of the AIEE*, **72**, November, pp. 273–288.

Grebe, J. J., R. H. Boundy, and R. W. Cermak (1933): "The control of chemical processes." *Trans. of American Institute of Chemical Engineers*, **29**, pp. 211–255.

Green, M. and D. J. N. Limebeer (1995): *Linear Robust Control*. Prentice Hall, Englewood Cliffs, N.J.

van der Grinten, P. M. E. M. (1963): "Finding optimum controller settings." *Control Engineering*, December, pp. 51–56.

Grosdidier, P., M. Morari, and P. Holt (1985): "Closed-loop properties from steady-state gain information." *Ind. Eng. Chem. Fundamentals*, **24**, pp. 221–235.

Haalman, A. (1965): "Adjusting controllers for a deadtime process." *Control Engineering*, **July 65**, pp. 71–73.

Habel, F. (1980): "Ein Verfahren zur Bestimmung der Parametern von PI-, PD- und PID-Reglern." *Regelungstechnik*, **28:6**, pp. 199–205.

Hägglund, T. (1991): *Process Control in Practice*. Chartwell-Bratt Ltd, Bromley, UK.

Hägglund, T. (1995): "A control-loop performance monitor." *Control Engineering Practice*, **3**, pp. 1543–1551.

Hägglund, T. (1996): "An industrial dead-time compensating PI controller." *Control Engineering Practice*, **4**, pp. 749–756.

Hägglund, T. (1999): "Automatic detection of sluggish control loops." *Control Engineering Practice*, **7**, pp. 1505–1511.

Hägglund, T. (2001): "The Blend station - a new ratio control structure." *Control Engineering Practice*, **9**, pp. 1215–1220.

Hägglund, T. (2002): "A friction compensator for pneumatic control valves." *J. of Process Control*, **12**, pp. 897–904.

Hägglund, T. and K. J. Åström (1991): "Industrial adaptive controllers based on frequency response techniques." *Automatica*, **27**, pp. 599–609.

Hägglund, T. and K. J. Åström (2000): "Supervision of adaptive control algorithms." *Automatica*, **36**, pp. 1171–1180.

Hägglund, T. and K. J. Åström (2002): "Revisiting the Ziegler-Nichols tuning rules for PI control." *Asian Journal of Control*, **4:4**, pp. 364–380.

Hägglund, T. and K. J. Åström (2004a): "Revisiting the Ziegler-Nichols step response method for PID control." *Journal of Process Control*, **14:6**, pp. 635–650.

Hägglund, T. and K. J. Åström (2004b): "Revisiting the Ziegler-Nichols tuning rules for PI control - part II, the frequency response method." *Asian Journal of Control*, **6:4**, pp. 469–482.

Hang, C. and K. J. Åström (2002): "Relay feedback auto-tuning of process controllers – a tutorial review." *Journal of Process Control*, **12**, pp. 143–162.

Hang, C. C., K. J. Åström, and W. K. Ho (1991): "Refinements of the Ziegler-Nichols tuning formula." *IEE Proceedings, Part D*, **138:2**, pp. 111–118.

Hang, C. C., K. J. Åström, and W. K. Ho (1993a): "Relay auto-tuning in the presence of static load disturbance." *Automatica*, **29:2**, pp. 563–564.

Hang, C. C., T. H. Lee, and W. K. Ho (1993b): *Adaptive Control.* ISA, Research Triangle Park, North Carolina.

Hang, C. C. and K. K. Sin (1991): "An on-line auto-tuning method based on cross-correlation." *IEEE Transactions on Industrial Electronics,* **38:6**, pp. 428–437.

Hanselmann, H. (1987): "Implementation of digital controllers—A survey." *Automatica,* **23:1**, pp. 7–32. Survey paper.

Hansen, Peter, D. (2000): "Robust adaptive PID controller tuning for unmeasured load rejection." In *IFAC Workshop on Digital Control – Past, present, and future of PID Control.* Terrassa, Spain.

Hansen, Peter, D. (2003): *Adaptive Tuning Methos of the Foxboro I/A System,* chapter Techniques for Adaptive Control, pp. 23–54. Elsevier, New York.

Hanus, R. (1988): "Antiwindup and bumpless transfer: a survey." In *Proceedings of IMACS-88, Part 2,* pp. 59–65. Paris, France.

Hanus, R., M. Kinnaert, and J.-L. Henrotte (1987): "Conditioning technique, a general anti-windup and bumpless transfer method." *Automatica,* **23**, pp. 729–739.

Harriott, P. (1964): *Process Control.* McGraw-Hill, New York, NY.

Harris, C. J. and S. A. Billings, Eds. (1981): *Self-tuning and Adaptive Control: Theory and Applications.* Peter Peregrinus, London.

Harris, T., C. Seppala, and L. Desborough (1999): "A review of performance monitoring and assessment techniques for univariate and multivariate control systems." *J. of Process Control,* **9**, pp. 1–17.

Harris, T. J. (1989): "Assessment of control loop performance." *Canadian Journal of Chemical Engineeering,* **67**, pp. 856–861.

Harris, T. J., F. Boudreau, and J. F. MacGregor (1996): "Performance assessment of multivariable feedback controllers." *Automatica,* **32:11**, pp. 1505–1518.

Hartree, D. R., A. Porter, A. Callender, and A. B. Stevenson (1937): "Time-lag in control systems—II." *Proceedings of the Royal Society of London,* **161**, pp. 460–476.

Hawk, Jr., W. M. (1983): "A self-tuning, self-contained PID controller." In *Proc. 1983 American Control Conference,* pp. 838–842. San Francisco, California.

Hazebroek, P. and B. L. van der Waerden (1950): "Theoretical considerations on the optimum adjustment of regulators." *Trans. ASME,* **72**, pp. 309–322.

Hazen, H. L. (1934): "Theory of servomechanisms." *JFI,* **218**, pp. 283–331.

Hecht-Nielsen, R. (1990): *Neurocomputing.* Addison-Wesley.

Higham, E. H. (1985): "A self-tuning controller based on expert systems and artificial intelligence." In *Proceedings of Control 85,* pp. 110–115. England.

Higham, J. D. (1968): "'Single-term' control of first- and second-order processes with dead time." *Control*, February, pp. 2–6.

Himmelblau, D. M. (1978): *Fault Detection and Diagnosis in Chemical and Petrochemical Processes.* Elsevier Scientific, Amsterdam.

Hjalmarsson, H., M. Gevers, and O. Lequin (1998): "Iterative feedback tuning: theory and applications." *IEEE Control Systems Magazine*, **18:4**, pp. 26–41.

Holmblad, L. P. and J. Østergaard (1981): "Control of a cement kiln by fuzzy logic." *F.L. Smidth Review*, **67**, pp. 3–11. Copenhagen, Denmark.

Holzbock, Werner, G. (1958): *Automatic Control Theory and Practice.* Reinhold Publishing Company, New York.

Hoopes, H. S., W. M. Hawk, Jr., and R. C. Lewis (1983): "A self-tuning controller." *ISA Transactions*, **22:3**, pp. 49–58.

Horch, A. (2000): *Condition monitoring of control loops.* PhD thesis, Royal Institute of Technology, Stochholm, Sweden.

Horowitz, I. (1993): *Quantitative feedback theory (QFT).* QFT Publications, Boulder, Colorado.

Horowitz, I. M. (1963): *Synthesis of Feedback Systems.* Academic Press, New York.

Howes, G. (1986): "Control of overshoot in plastics-extruder barrel zones." In *EI Technology*, No. 3, pp. 16–17. Eurotherm International, Brighton, UK.

Huang, B. and S. Shah (1999): *Performance assessment of control loops.* Springer-Verlag, London.

Huang, Z. (1991): "Auto-tuning of PID controllers using neural networks." In *Preprints IFAC International Symposium on Intelligent Tuning and Adaptive Control (ITAC 91).* Singapore.

Huzmezan, M., W. A. Gough, and G. Dumont (2003): *Adaptive predictive regulatory control with brainwave*, chapter Techniques for Adaptive Control, pp. 99–143. Elsevier, New York.

Hwang, S.-H. and H.-C. Chang (1987): "A theoretical examination of closed-loop properties and tuning methods of single-loop PI controllers." *Chemical Engineering Science*, **42**, pp. 2395–2415.

Ingimundarson, A. and T. Hägglund (2002): "Performance comparison between PID and dead-time compensating controllers." *Journal of Process Control*, **12**, pp. 887–895.

Isermann, R. (1980): "Practical aspects of process identification." *Automatica*, **16**, pp. 575–587.

Isermann, R. (1982): "Parameter adaptive control algorithms—A tutorial." *Automatica*, **18**, pp. 513–528.

Isermann, R. and K. Lachmann (1985): "Parameter adaptive control with configuration aids and supervision functions." *Automatica*, **21**, pp. 623–638.

Ivanoff, A. (1934): "Theoretical foundations of the automatic regulation of temperature." *J. Institute of Fuel*, **7**, pp. 117–138.

James, H. M., N. B. Nichols, and R. S. Phillips (1947): *Theory of Servomechanisms.* Mc-Graw-Hill, New York.

Johansson, R. (1993): *System Modeling and Identification.* Prentice Hall, Englewood Cliffs, New Jersey.

Kalman, R. E. (1960): "Contributions to the theory of optimal control." *Boletin de la Sociedad Matématica Mexicana*, **5**, pp. 102–119.

Kalman, R. E. (1961): "New methods and results in linear prediction and filtering theory." Technical Report 61-1. RIAS. 135 pp.

Kalman, R. E. and R. S. Bucy (1961): "New results in linear filtering and prediction theory." *Trans ASME (J. Basic Engineering)*, **83 D**, pp. 95–108.

Kapasouris, P. and M. Athans (1985): "Multivariable control systems with saturating actuators antireset windup strategies." In *Proc. Automatic Control Conference*, pp. 1579–1584. Boston, Massachusetts.

Kaya, A. and S. Titus (1988): "A critical performance evaluation of four single loop self-tuning control products." In *Proceedings of the 1988 American Control Conference.* Atlanta, Georgia.

Kaya, I. and D. P. Atherton (1999): "A new PI-PD Smith predictor for control of processes with long time delays." In *Preprints. 14th World Congress of IFAC*, pp. 283–288. Beijing, China.

Kessler, C. (1958a): "Das symmetrische Optimum, Teil I." *Regelungstechnik*, **6:11**, pp. 395–400.

Kessler, C. (1958b): "Das symmetrische Optimum, Teil II." *Regelungstechnik*, **6:12**, pp. 432–436.

King, P. J. and E. H. Mamdani (1977): "The application of fuzzy control systems to industrial processes." *Automatica*, **13**, pp. 235–242.

Klefenz, G. (1986): *Automatic Control of Steam Power Plants*, third edition. Bibliographisches Institut.

Klein, M., T. Marczinkowsky, and M. Pandit (1991): "An elementary pattern recognition self-tuning PI-controller." In *Preprints IFAC International Symposium on Intelligent Tuning and Adaptive Control (ITAC 91)*, vol. 1. Singapore.

Kouvaritakis, B. and M. Cannon (2001): *Nonlinear predictive control, theory and practice.* IEE, London.

Kozub, D. and C. Garcia (1993): "Monitoring and diagnosis of automated controllers in the chemical process industries." In *AIChE Meeting.* St Louis, MO.

Kozub, D. and C. Garcia (1996): "Controller performance monitoring and diagnosis: Experiences and challenges." In *Chemical Process Control V*, pp. 83–96. Tahoe City, CA.

Kramer, L. C. and K. W. Jenkins (1971): "A new technique for preventing direct digital control windup." In *Proc. Joint Automatic Control Conference*, pp. 571–577. St Louis, Missouri.

Kraus, T. W. and T. J. Myron (1984): "Self-tuning PID controller uses pattern recognition approach." *Control Engineering*, June, pp. 106–111.

Krikelis, N. J. (1984): "Design of tracking systems subject to actuators and saturation and integrator windup." *International Journal of Control*, **39:4**, pp. 667–682.

Kristensson, B. (2003): *PID Controllers Design and Evaluation*. PhD thesis, Chalmers, Gothenburg.

Kristiansson, B. and B. Lennartson (1999): "Optimal PID controllers including roll off and Smith predictor structure." In *Preprints. 14th World Congress of IFAC*, pp. 297–302. Beijing, China.

Kristiansson, B. and B. Lennartsson (2002): "Robust and optimal tuning of PI and PID controllers." *IEE Proceedings– Control theory and applications*, **149:1**, pp. 17–25.

Kulhavy, R., J. Lu, and T. Samad (2001): "Emerging technologies for enterprise optimization in the process industries." pp. 411–422. Chemical Process Control - 6, Assessment and new directions for research (CPC-VI), Tucson, AZ.

Küpfmüller, K. (1928): "Über die Dynamik der selbststätigen Verstärkungsregler." *ENT*, **5**, pp. 459–467.

Lee, C. C. (1990): "A self-learning rule-based controller with approximate reasoning and neural nets." In *Preprints 11th IFAC World Congress*. Tallinn, Estonia.

Leva, A. (1993): "PID autotuning algorithm based on relay feedback." *IEE Proceedings D*, **140:5**, pp. 328–338.

Liu, Z. (1998): "A frequency response based adaptive control for center-driven web winders." In *1998 American Control Conference*. Philadelphia, PA.

Ljung, L. (1998): *System Identification—Theory for the User*. Prentice Hall, Englewood Cliffs, New Jersey. Second Edition.

Ljung, L. and T. Söderström (1983): *Theory and Practice of Recursive Identification*. MIT Press, Cambridge, Massachusetts.

Lloyd, S. G. and G. D. Anderson (1971): *Industrial Process Control*. Fisher Controls Co., Marshalltown, Iowa.

L&N (1968): *Leeds & Northrup Technical Journal*. Spring Issue, Number 3.

Lopez, A. M., J. A. Miller, C. L. Smith, and P. W. Murrill (1967): "Tuning controllers with error-integral criteria." *Instrumentation Technology*, November, pp. 57–62.

Lopez, A. M., P. W. Murrill, and C. L. Smith (1969): "Tuning PI and PID digital controllers." *Instruments and Control Systems,* **42**, February, pp. 89–95.

Lu, J. (2004): "An efficient single-loop MPC algorithm for replacing PID." In *AIChE Annual Conference.* Austin, TX.

Lukas, M. P. (1986): *Distributed Process Control Systems—Their Evaluation and Design.* Van Nostrand Reinhold, New York.

Luyben, W. L. (1990): *Process Modeling, Simulation and Control for Chemical Engineers,* second edition. McGraw-Hill.

Lynch, C. B. and G. A. Dumont (1996): "Control loop performance monitoring." *IEEE Trans. Control Syst. Technol.,* **4**, pp. 185–192.

Maciejowski, J. M. (1989): *Multivariable Feedback Design.* Addison-Wesley, Reading, Massachusetts.

Maciejowski, J. M. (2002): *Predictive control with constraints.* Prentice Hall, Englewood Cliffs, NJ.

Mamdani, E. H. (1974): "Application of fuzzy algorithm for control of simple dynamic plant." *Proc. IEE,* **121**, pp. 1585–1588.

Mamdani, E. H. and S. Assilian (1974): "A case study on the application of fuzzy set theory to automatic control." In *Proceedings IFAC Stochastic Control Symposium.* Budapest, Hungary.

Mantz, R. J. and E. J. Tacconi (1989): "Complementary rules to Ziegler and Nichols' rules for a regulating and tracking controller." *International Journal of Control,* **49**, pp. 1465–1471.

Marlin, T. E. (2000): *Process Control.* McGraw-Hill.

Marsik, J. and V. Strejc (1989): "Application of identification-free algorithms for adaptive control." *Automatica,* **25**, pp. 273–277.

Marsili-Libelli, S. (1981): "Optimal design of PID regulators." *International Journal of Control,* **33:4**, pp. 601–616.

Mason, C. E. and G. A. Philbrick (1940): "Automatic control in the presence of process lags." *Transactions of the ASME,* **62**, pp. 295–308.

Matausek, M. and A. Micic (1996): "A modified Smith predictor for controlling a process with an integrator and long dead-time." *IEEE Transaction on Automatic Control,* **41**, pp. 1199–1203.

Matausek, M. and A. Micic (1999): "On the modified Smith predictor for controlling a process with an integrator and long dead-time." *IEEE Transaction on Automatic Control,* **44**, pp. 1603–1606.

Maxwell, J. C. (1868): "On governors." *Proceedings of the Royal Society of London,* **16**, pp. 270–283. Also published in "Mathematical Trends in Control Theory" edited by R. Bellman and R. Kalaba, Dover Publications, New York 1964, pp. 3–17.

McAvoy, T. J. (1983): *Interaction Analysis: Principles and Applications.* ISA, Research Triangle Park, North Carolina.

McMillan, G. K. (1983): *Tuning and Control Loop Performance,* second edition. ISA, Research Triangle Park, North Carolina.

McMillan, G. K. (1986): "Advanced control algorithms: Beware of false prophecies." *InTech,* January, pp. 55–57.

McMillan, G. K., W. K. Wojsznis, and G. T. Borders, Jr (1993a): "Flexible gain scheduler." In *Advances in Instrumentation and Control,* vol. 48 of *ISA Conference,* pp. 811–818.

McMillan, G. K., W. K. Wojsznis, and K. Meyer (1993b): "Easy tuner for DCS." In *Advances in Instrumentation and Control,* vol. 48 of *ISA Conference,* pp. 703–710.

Meyer, C., D. E. Seborg, and R. K. Wood (1976): "A comparison of the Smith predictor and conventional feedback control." *Chemical Engineering Science,* **31**, pp. 775–778.

Michael, A. J. and M. H. Moradi (2005): *PID Control: New Identification and Design Methods.* Springer, Berlin.

Miller, J. A., A. M. Lopez, C. L. Smith, and P. W. Murrill (1967): "A comparison of controller tuning techniques." *Control Engineering,* December, pp. 72–75.

Minorsky, N. (1922): "Directional stability of automatically steered bodies." *J. Amer. Soc. of Naval Engineers,* **34:2**, pp. 280–309.

Moore, C. F., C. L. Smith, and P. W. Murrill (1970): "Improved algorithm for direct digital control." *Instruments & Control Systems,* **43**, January, pp. 70–74.

Morari, M. and J. H. Lee (1991): "Model predictive control: The good, the bad, and the ugly." In *Chemical Process Control, CPCIV,* pp. 419–442. Padre Island, TX.

Morari, M. and E. Zafiriou (1989): *Robust Process Control.* Prentice-Hall, Englewood Cliffs, New Jersey.

Morris, H. M. (1987): "How adaptive are adaptive process controllers?" *Control Engineering,* **34-3**, pp. 96–100.

Nachtigal, C. L. (1986a): "Adaptive controller performance evaluation: Foxboro EXACT and ASEA Novatune." In *Proceedings ACC-86,* pp. 1428–1433.

Nachtigal, C. L. (1986b): "Adaptive controller simulated process results: Foxboro EXACT and ASEA Novatune." In *Proceedings ACC-86,* pp. 1434–1439.

Newton, Jr., G. C., L. A. Gould, and J. F. Kaiser (1957): *Analytical Design of Linear Feedback Controls.* John Wiley & Sons.

Nicholson, H., Ed. (1980): *Modelling of Dynamical Systems, Vol. 1.* Peter Peregrinus.

Nicholson, H., Ed. (1981): *Modelling of Dynamical Systems, Vol. 2.* Peter Peregrinus.

Niederlinski, A. (1971): "A heuristic approach to the design of linear multivaribale interacting control systems." *Automatica*, **7**, pp. 691–701.

Nishikawa, Y., N. Sannomiya, T. Ohta, and H. Tanaka (1984): "A method for auto-tuning of PID control parameters." *Automatica*, **20**, pp. 321–332.

Nyquist, H. (1932): "Regeneration theory." *Bell System Technical Journal*, **11**, pp. 126–147. Also published in "Mathematical Trends in Control Theory," edited by R. Bellman and R. Kalaba, Dover Publications, New York 1964, pp. 83–105.

O'Dwyer, A. (2003): *Handbook of PI and PID Controller Tuning Rules.* Imperial College Press, London.

Oldenburg, R. (1956): *Frequency Response.* MacMillan, New York.

Oldenburg, R. C. and H. Sartorius (1954): "A uniform approach to the optimum adjustment of control loops." *Transactions of the ASME*, **76**, November, pp. 1265–1279.

Oppelt, W. (1964): *Kleines Handbuch technischer Regelvorgänge.* Verlag Chemie, Weinheim.

Oquinnaike, B. A. and W. H. Ray (1994): *Process Dynamics, Modeling and Control (Topics in Chemical Engineering.* Oxford University Press, Oxford.

Owen, J. G., D. Read, H. Blekkenhorst, and A. A. Roche (1996): "A mill prototype for automatic monitoring of control loop performance." In *Proc. Control Syst. '96*, pp. 171–178. Halifax, Nova Scotia, Canada.

Pagano, D. (1991): "Intelligent tuning of PID controllers based on production rules system." In *Preprints IFAC International Symposium on Intelligent Tuning and Adaptive Control (ITAC 91).* Singapore.

Palmor, Z. J. and R. Shinnar (1979): "Design of sampled data controllers." *Ind. Eng. Chem. Process Design and Development*, **18:1**, pp. 8–30.

Panagopoulos, H. (2000): *PID Control Design, Extension, Application.* PhD thesis, Department of Automatic Control, Lund Institute of Technology, Lund, Sweden.

Panagopoulos, H. and K. J. Åström (2000): "PID control design and H_∞ loop shaping." *Int. J. Robust Nonlinear Control*, **10**, pp. 1249–1261.

Panagopoulos, H., K. J. Åström, and T. Hägglund (1997): "Design of PI controllers." In *Proc. 1997 IEEE International Conference on Control Applications*, pp. 417–422. Hartford, Connecticut.

Panagopoulos, H., K. J. Åström, and T. Hägglund (2002): "Design of PID controllers based on constrained optimisation." *IEE Proc. Control Theory Appl.*, **149:1**, pp. 32–40.

Pao, H. H. (1990): "Use of neural-net technology in control: A survey and a perspective." In *Preprints 11th IFAC World Congress.* Tallinn, Estonia.

Passino, K. M. and P. J. Antsaklis, Eds. (1992): *An Introduction to Intelligent and Autonomous Control.* Kluwer Academic Publishers.

Patwardhan, A. A., M. N. Karim, and R. Shah (1987): "Controller tuning by a least squares method." *AIChE Journal,* **33**, October, pp. 1735–1737.

Pavlik, E. and B. Machei (1960): *Ein kombiniertes Regelsystem fur die Verfahrensindustrie.* Oldenburg, Munchen.

Pemberton, T. J. (1972a): "PID: The logical control algorithm." *Control Engineering,* May, pp. 66–67.

Pemberton, T. J. (1972b): "PID: The logical control algorithm–II." *Control Engineering,* July, pp. 61–63.

Persson, P. (1992): *Towards Autonomous PID Control.* PhD thesis ISRN LUTFD2/TFRT--1037--SE, Department of Automatic Control, Lund Institute of Technology, Sweden.

Persson, P. and K. J. Åström (1992): "Dominant pole design—A unified view of PID controller tuning." In *Preprints 4th IFAC Symposium on Adaptive Systems in Control and Signal Processing,* pp. 127–132. Grenoble, France.

Persson, P. and K. J. Åström (1993): "PID control revisited." In *Preprints IFAC 12th World Congress.* Sydney, Australia.

Pessen, B. W. (1954): "How to "tune in" a three mode controller." *Instrumentation,* Second Quarter, pp. 29–32.

Petersson, M., K.-E. Årzén, and T. Hägglund (2001): "Assessing measurements for feedforward control." In de Carvalho, Ed., *European Control Conference - ECC'01,* pp. 432–437. Porto, Portugal.

Petersson, M., K.-E. Årzén, and T. Hägglund (2003): "A comparison of two feedforward control structure assessment methods." *International Journal of Adaptive Control and Signal Processing,* **17:7–9**, pp. 609–624.

Petersson, M., K.-E. Årzén, H. Sandberg, and L. de Maré (2002): "Implementation of a tool for control structure assessment." In *Proceedings of the 15th IFAC World Congress.* Barcelona, Spain.

Polonoyi, M. J. G. (1989): "PID controller tuning using standard form optimization." *Control Engineering,* March, pp. 102–106.

Pontryagin, L. S., V. G. Boltyanskii, R. V. Gamkrelidze, and E. F. Mischenko (1962): *The Mathematical Theory of Optimal Processes.* John Wiley, New York.

Porter, B., A. H. Jones, and C. B. McKeown (1987): "Real-time expert tuners for PI controllers." *IEE Proceedings Part D,* **134:4**, pp. 260–263.

Pottman, M. and D. E. Seborg (1993): "A radial basis function control strategy and its application to a pH neutralization process." In *Proceedings 2nd European Control Conference, ECC '93.* Groningen, The Netherlands.

Qin, S. (1998): "Control performance monitoring – a review and assessment." *Computers and Chemical Engineering*, **23**, pp. 173–186.

Qin, S. J. and T. A. Badgwell (2003): "A survey of industrial model predictive control technology." *Control Engineering Practice*, **11**, pp. 733–764.

Quevedo, J. and T. Escobet (2000): *Digital Control 2000 – Past, present, and future of PID Control.* Pergamon, Oxford.

Rad, A. B. and P. J. Gawthrop (1991): "Explicit PID self-tuning control for systems with unknown time delay." In *IFAC International Symposium ITAC 91 Preprint*, vol. 5. Singapore.

Radke, F. and R. Isermann (1987): "A parameter-adaptive PID controller with stepwise parameter optimization." *Automatica*, **23**, pp. 449–457.

Rake, H. (1980): "Step response and frequency response methods." *Automatica*, **16**, pp. 519–526.

Rawlings, J. B. (2000): "Tutorial overview of model predictive control." *IEEE Control Systems Magazine*, **20**.

Rawlings, J. B. and J. G. Ekerdt (2002): *Chemical Reactor Analysis and Design Fundamentals.* Nob Hill Publishing, WI.

Richalet, J., A. Rault, J. L. Testud, and J. Papon (1976): "Model predictive heuristic control: Applications to industrial processes." *Automatica*, **14**, pp. 413–428.

Rijnsdorp, J. (1965a): "Interaction in two-variable control systems for distillation columns – I." *Automatica*, **1**, p. 15.

Rijnsdorp, J. (1965b): "Interaction in two-variable control systems for distillation columns – II." *Automatica*, **1**, pp. 29–51.

Rivera, D. E., M. Morari, and S. Skogestad (1986): "Internal model control—4. PID controller design." *Ind. Eng. Chem. Proc. Des. Dev.*, **25**, pp. 252–265.

Ross, C. W. (1977): "Evaluation of controllers for deadtime processes." *ISA Transactions*, **16:3**, pp. 25–34.

Rovira, A. A., P. W. Murrill, and C. L. Smith (1969): "Tuning controllers for setpoint changes." *Instruments and Control Systems*, December, pp. 67–69.

Rundqwist, L. (1990): "Anti-reset windup for PID controllers." In *Preprints 11th IFAC World Congress.* Tallinn, Estonia.

Schei, T. S. (1992): "A method for closed loop automatic tuning of PID controllers." *Automatica*, **28:3**, pp. 587–591.

Schei, T. S. (1994): "Automatic tuning of PID controllers based on transfer function estimation." *Automatica*, **30:12**, pp. 1983–1989.

Seborg, D. E., T. F. Edgar, and D. A. Mellichamp (2004): *Process Dynamics and Control*, second edition. Wiley, New York, NY.

Seborg, D. E., T. F. Edgar, and S. L. Shah (1986): "Adaptive control strategies for process control: A survey." *AIChE Journal*, **32**, pp. 881–913.

Sell, N. J. (1995): *Process control fundamentals for the pulp & paper industry.* TAPPI PRESS.

Shearer, J. L. and B. T. Kulakowski (1990): *Dynamic Modeling and Control of Engineering Systems.* Macmillan, New York.

Shigemasa, T., Y. Iino, and M. Kanda (1987): "Two degrees of freedom PID auto-tuning controller." In *Proceedings of ISA Annual Conference*, pp. 703–711.

Shinskey, F. G. (1963): "Feedforward control applied." *ISA J.*, March, pp. 79–83.

Shinskey, F. G. (1981): *Controlling Multivariable Processes.* ISA, Research Triangle Park, North Carolina.

Shinskey, F. G. (1988): *Process-Control Systems. Application, Design, and Tuning*, third edition. McGraw-Hill, New York.

Shinskey, F. G. (1990): "How good are our controllers in absolute performance and robustness?" *Measurement and Control*, **23**, May, pp. 114–121.

Shinskey, F. G. (1991a): "Evaluating feedback controllers challenges users and vendors." *Control Engineering*, September, pp. 75–78.

Shinskey, F. G. (1991b): "Model predictors: The first smart controllers." *Instruments and Control Systems*, September, pp. 49–52.

Shinskey, F. G. (1994): *Feedback Controllers for the Process Industries.* McGraw-Hill, New York.

Shinskey, F. G. (1996): *Process-Control Systems. Application, Design, and Tuning*, 4th edition. McGraw-Hill, New York.

Shinskey, F. G. (2002): "PID-deadtime control of distributed processes." *Control Engineering Practice*, **9:11**, pp. 1177–1183.

Skogestad, S. (2003): "Simple analytic rules for model reduction and PID controller tuning." *Journal of Process Control*, **13:4**, pp. 291–309.

Skogestad, S. and M. Morari (1987): "Implications of large RGA-elements on control performance." *Ind. & Eng. Chem. Research*, **26:11**, pp. 2323–2330. Also see correction to Eq. 13 in Ind. & Eng. Chem. Research (27:5), 898 (1988).

Skogestad, S. and I. Postlethwaite (1996): *Multivariable feedback control: analysis and design.* Wiley, Chichester, UK.

Smith, C. L. (1972): *Digital Computer Process Control.* Intext Educational Publishers, Scranton, Pennsylvania.

Smith, C. L., A. B. Corripio, and J. J. Martin (1975): "Controller tuning from simple process models." *Instrumentation Technology*, December, pp. 39–44.

Smith, C. L. and P. W. Murrill (1966): "A more precise method for tuning controllers." *ISA Journal*, May, pp. 50–58.

Smith, O. J. M. (1957): "Closed control of loops with dead time." *Chemical Engineering Progress*, **53**, May, pp. 217–219.

Smith, O. J. M. (1958): *Feedback Control Systems.* McGraw-Hill, New York.

Söderström, T. and P. Stoica (1989): *System Identification.* Prentice-Hall, London, UK.

Stanfelj, N., T. Marlin, and J. MacGregor (1993): "Monitoring and diagnosing process control performance: The single-loop case." *Ind. Eng. Chem. Res.*, **32**, pp. 301–314.

Stephanopoulos (1984): *Chemical Process Control. An Introduction to Theory and Practice.* Prentice-Hall.

Stock, J. T. (1988): "Pneumatic process controllers: The ancestry of the proportional-integral-derivative controller." *Trans. of the Newcomen Society*, **59**, pp. 15–29.

Strejc, V. (1959): "Näherungsverfahren für Aperiodische Übertragscharacteristiken." *Regelungstechnik*, **7:7**, pp. 124–128.

Suda, N. *et al.* (1992): *PID Control.* Asakura Shoten Co., Ltd., Japan.

Sugeno, M., Ed. (1985): *Industrial Applications of Fuzzy Control.* Elsevier Science Publishers BV, The Netherlands.

Sullivan, G. A. (1996): "Adaptive control with expert system based supervisory funcitons." *Journal of Systems Science*, **27:9**, pp. 839–850.

Swiniarski, R. W. (1991): "Neuromorphic self-tuning PID controller uses pattern recognition approach." In *Preprints IFAC International Symposium on Intelligent Tuning and Adaptive Control (ITAC 91).* Singapore.

Taguchi, H. and M. Araki (2000): "Two-degree-of-freedom PID controllers–their functions and optimal tuning." In *IFAC Workshop on Digital Control – Past, present, and future of PID Control.* Terrassa, Spain.

Takahashi, Y., M. J. Rabins, and D. M. Auslander (1972): *Control and Dynamic Systems.* Addison-Wesley, Reading, MA.

Takatsu, H., T. Kawano, and K. Kitano (1991): "Intelligent self-tuning PID controller." In *Preprints IFAC International Symposium on Intelligent Tuning and Adaptive Control (ITAC 91).* Singapore.

Tan, L.-Y. and T. W. Weber (1985): "Controller tuning of a third-order process under proportional-integral control." *Industrial & Engineering Chemistry Process Design and Development*, **24**, pp. 1155–1160.

Thornhill, N. F. and T. Hägglund (1997): "Detection and diagnosis of oscillation in control loops." *Control Engineering Practice*, **5**, pp. 1343–1354.

Thornhill, N. F., M. Oettinger, and P. Fedenczuk (1999): "Refinery-wide control loop performance assessment." *Journal of Process Control*, **9**, pp. 109–124.

Tiller, M. M. (2001): *Introduction to Physical Modeling with Modelica.* Kluwer.

Tong, R. M. (1977): "A control engineering review of fuzzy system." *Automatica,* **13**, pp. 559–569.

Tong, R. M. (1984): "A retrospective view of fuzzy control systems." *Fuzzy Sets and Systems,* **14**, pp. 199–210.

Truxal, J. (1955): *Automatic Feedback Control System Synthesis.* McGraw-Hill, New York.

Tucker, G. K. and D. M. Wills (1960): *A Simplified Technique for Control System Engineering.* Minneapolis-Honeywell Regulator Company. Brown Instruments Division, Philadelphia, PA.

Turnbull, G. (1988): "Three-term control in EI." **6**, pp. 3–7. Published by Eurotherm International.

Tyreus, B. (1987): "TUNEX – an expert system for controller tuning." Technical Report. du Pont.

van der Grinten, P. M. E. M. (1963): "Determining plant controllability." *Control Engineering,* October, pp. 87–89.

Van Doren, V. J. (2003): *Techniques for Adaptive Control.* Elsevier, New York.

Vinnicombe, G. (2000): *Uncertainty and Feedback: $\mathcal{H}_\infty$ loop-shaping and the μ-gap metric.* Imperial College Press, London.

Voda, A. and I. D. Landau (1995): "A method for the auto-calibration of PID controllers." *Automatica,* **31:2**.

Vyshnegradskii, J. (1876): "Sur la théorie générale des régulateurs." *Compt. Rend. Acid. Sci. Paris,* **83**, pp. 318–321.

Walgama, K. S. and J. Sternby (1990): "Inherent observer property in a class of anti-windup compensators." *International Journal of Control,* **52:3**, pp. 705–724.

Wallén, A. (2000): *Tools for Autonomous Process Control.* PhD thesis ISRN LUTFD2/TFRT--1058--SE, Department of Automatic Control, Lund Institute of Technology, Sweden.

Wallén, A., K. J. Åström, and T. Hägglund (2002): "Loop-shaping design of PID controllers with constant Ti/Td ratio." *Asian Journal of Control,* **4:4**, pp. 403–409.

Wang, L. and W. R. Cluett (2000): *From Plant Data to Process Control: Ideas for Process Identification and PID Design.* Taylor & Francis, London.

Wang, L.-X. (1994): *Adaptive Fuzzy Systems and Control: Design and Stability Analysis.* Prentice Hall.

Wang, Q.-G., C. C. Hang, T. J. Hägglund, and K. K. Tan (2000): *Advances in PID Control.* Springer, Berlin.

Wang, Q.-G., A. Yu, and C. Min-Sen (2003): "Non-interacting control design for multivariable industrial processes." *Journal of Process Control*, **13**, pp. 253–265.

Webb, J. C. (1967): "Representative DDC systems." *Instruments & Control Systems*, **40**, October, pp. 78–83.

Wellstead, P. E. (1979): *Introduction to Physical System Modelling*. Academic Press.

Willis, M. J., C. Di Massimo, G. A. Montague, M. T. Tham, and A. J. Morris (1991): "Artificial neural networks in process engineering." *IEE Proceedings D*, **138:3**, pp. 256–266.

Wills, D. M. (1962a): "A guide to controller tuning." *Control Engineering*, August, pp. 93–95.

Wills, D. M. (1962b): "Tuning maps for three-mode controllers." *Control Engineering*, April, pp. 104–108.

Wolfe, W. A. (1951): "Controller settings for optimum control." *Transactions of the ASME*, **64**, pp. 413–418.

Wong, S. K. P. and D. E. Seborg (1988): "Control strategy for single-input single-output non-linear systems with time delays." *International Journal of Control*, **48:6**, pp. 2303–2327.

Yamamoto, S. (1991): "Industrial developments in intelligent and adaptive control." In *Preprints IFAC International Symposium on Intelligent Tuning and Adaptive Control (ITAC 91)*. Singapore.

Yamamoto, S. and I. Hashimoto (1991): "Present status and future needs: The view from Japanese industry." In *Chemical Process Control CPCIV*, pp. 1–28. Padre Island, TX.

Yarber, W. H. (1984a): "Electromax V plus, A logical progression." In *Proceedings, Control Expo 84.*

Yarber, W. H. (1984b): "Single loop, self-tuning algorithm applied." In *Preprints AIChE Anaheim Symposium.*

Young, R. E., R. B. Bartusiak, and R. B. Fontaine (2001): "Evolutions of an industrial nonlinear model predictive controller." pp. 399–410. Chemical Process Control - 6, Assessment and new directions for research (CPC-VI), Tucson, AZ.

Yuwana, M. and D. E. Seborg (1982): "A new method for on-line controller tuning." *AIChE Journal*, **28:3**, pp. 434–440.

Yuzu, Z., Q. G. Wang, and K. J. Åström (2002): "Dominant pole placement for multi-loop control systems." *Automatica*, **38**, pp. 1213–1220.

Zadeh, L. A. (1988): "Fuzzy logic." *IEEE Computer*, April, pp. 83–93.

Zadeh, L. A. and C. A. Desoer (1963): *Series in Systems Science*. McGraw-Hill, Inc., USA.

Zervos, C., P. R. Bélanger, and G. A. Dumont (1988): "On PID controller tuning using orthonormal series identification." *Automatica*, **24:2**, pp. 165–175.

Zhang, C. and R. J. Evans (1988): "Rate constrained adaptive control." *International Journal of Control*, **48:6**, pp. 2179–2187.

Zhou, J., J. Doyle, and K. Glover (1996): *Robust and optimal control.* Prentice Hall.

Zhuang, M. and D. P. Atherton (1991): "Optimal PID controller setting using integral performance criteria." *Proc. ACC*, pp. 3042–3043.

Ziegler, J. G. and N. B. Nichols (1942): "Optimum settings for automatic controllers." *Trans. ASME*, **64**, pp. 759–768.

Ziegler, J. G. and N. B. Nichols (1943): "Process lags in automatic-control circuits." *Transactions of the ASME*, **65:5**, pp. 433–443.

Index